MW01245340

Fisheries Subsidies, Sustainable Development and the WTO

Fisheries Subsidies, Sustainable Development and the WTO

Edited by
Anja von Moltke

London • Washington, DC

First published in 2011 by Earthscan

Earthscan Ltd, Dunstan House, 14a St Cross Street, London EC1N 8XA, UK
Earthscan LLC,1616 P Street, NW, Washington, DC 20036, USA

Earthscan publishes in association with the International Institute for Environment and Development

For more information on Earthscan publications, see www.earthscan.co.uk or write to earthinfo@earthscan.co.uk

ISBN: 978-1-84971-135-7

Typeset by MapSet Ltd, Gateshead, UK
Cover design by Clifford Hayes

Cover images:
Main photograph: Attente des pirogues à Nouakchott, Mauritanie, © Matthieu BERNARDON, IUCN
Bottom left: 'Ship' © Rob Bouwman/istockphoto.com
Bottom centre: Marché aux poissons, © Matthieu BERNARDON, IUCN
Bottom right: Photo: © WTO/Jay Louvion

A catalogue record for this book is available from the British Library

Library of Congress Cataloging-in-Publication Data

Fisheries Subsidies, Sustainable Development and the WTO / edited by Anja von Moltke.
p. cm.
Includes bibliographical references and index.
ISBN 978-1-84971-135-7 (hardback)
1. Fisheries subsidies. 2. Fishery management, International. 3. Fishery policy.
4. Sustainable fisheries. 5. World Trade Organization 6. Trade and Environment
7. Anja von Moltke.
SH334.F576 2010
338.3'727–dc22

2010004447

At Earthscan we strive to minimize our environmental impacts and carbon footprint through reducing waste, recycling and offsetting our CO_2 emissions, including those created through publication of this book. For more details of our environmental policy, see www.earthscan.co.uk.

Printed and bound in the UK by TJ International,
an ISO 14001 accredited company.
The paper used is FSC certified.

The United Nations Environment Programme

The United Nations Environment Programme (UNEP) is the overall coordinating environmental organization of the United Nations system. Its mission is to provide leadership and encourage partnerships in caring for the environment by inspiring, informing and enabling nations and people to improve their quality of life without compromising that of future generations. In accordance with its mandate, UNEP works to observe, monitor and assess the state of the global environment, improve the scientific understanding of how environmental change occurs and, in turn, how such change can be managed by action-oriented national policies and international agreements. UNEP's capacity building work thus centres on helping countries strengthen environmental management in diverse areas that include freshwater and land resource management, the conservation and sustainable use of biodiversity, marine and coastal ecosystem management and cleaner industrial production and eco-efficiency, among many others.

UNEP was founded in 1972 and is headquartered in Nairobi, Kenya. In partnership with a global array of collaborating organizations, UNEP has achieved major advances in the development of international environmental policy and law, environmental monitoring and assessment and the understanding of the science of global change. This work also supports the successful development and implementation of the world's major environmental conventions. In parallel, UNEP administers several multilateral environmental agreements (MEAs) including the Vienna Convention's Montreal Protocol on Substances that Deplete the Ozone Layer, the Convention on International Trade in Endangered Species of Wild Fauna and Flora (CITES), the Basel Convention on the Control of Transboundary Movements of Hazardous Wastes and their Disposal (SBC), the Convention on Prior Informed Consent Procedure for Certain Hazardous Chemicals and Pesticides in International Trade (Rotterdam Convention, PIC) and the Cartagena Protocol on Biosafety to the Convention on Biological Diversity as well as the Stockholm Convention on Persistent Organic Pollutants (POPs).

Division of Technology, Industry and Economics

The mission of the Division of Technology, Industry and Economics (DTIE) is to encourage decision-makers in government, local authorities and industry to develop and adopt policies, strategies and practices that are cleaner and safer, make efficient use of natural resources, ensure environmentally sound management of chemicals and reduce pollution and risks for humans and the environment. In addition, it seeks to enable implementation of conventions and international agreements and encourage the internalization of environmental costs. UNEP DTIE's strategy in carrying out these objectives is to influence decision-making through partnerships with other international organizations, governmental authorities, business and industry and non-governmental organizations; facilitate knowledge management through networks; support implementation of conventions and work closely with UNEP regional offices. The Division, with its Director and Division Office in Paris, consists of one centre and five branches located in Paris, Geneva and Osaka.

Economics and Trade Branch

The Economics and Trade Branch (ETB) is one of the five branches of DTIE. ETB seeks to support a transition to a green economy by enhancing the capacity of governments, businesses and civil society to integrate environmental considerations in economic, trade and financial policies and practices. In so doing, ETB focuses its activities on:

1 stimulating investment in green economic sectors;
2 promoting integrated policy assessment and design;
3 strengthening environmental management through subsidy reform;
4 promoting mutually supportive trade and environment policies; and
5 enhancing the role of the financial sector in sustainable development.

During the past decade, ETB has worked intensively on the issue of fisheries to promote integrated and well-informed responses to the need for fisheries policies reform. Through a series of workshops, analytic papers and country projects, ETB particularly seeks to improve the understanding of the impact of fisheries subsidies and to present policy options to address harmful impacts.

Contents

Foreword by Achim Steiner		*xi*
List of Figures, Tables and Boxes		*xiii*
Acknowledgements		*xvii*
List of Abbreviations		*xix*
Introduction		**xxiii**
1	**Fisheries Governance, Subsidies and Sustainable Development**	**1**
	The Need for a 'Green Economy' of Fishing	1
	The Scale and Dynamics of Trade in Fisheries Products	3
	Governance of Global Fisheries	5
	Subsidies in the Fisheries Sector	10
	Challenges for Reform	15
2	**Fisheries Subsidies and their Resource Impacts**	**19**
	Introduction	19
	Categories of Fisheries Subsidies	20
	Analytical Framework	26
	Analysis of Impacts by Category	32
	Summary	58
3	**National Experiences with Subsidies, their Impacts and Reform Processes**	**65**
	Introduction	65
	Fisheries Subsidies: The Senegalese Experience	67
	The Impact of Fisheries Subsidies on Tuna Sustainability and Trade in Ecuador	95
	Fisheries Subsidy Reform in Norway	113
	Common lessons from Senegal, Ecuador and Norway Cases	128

4 Emergence of an International Issue: History of Fisheries Subsidies in the WTO 131
Introduction 131
Phase I: Early Analysis and Preliminary International Action 132
Phase II: Globalization and the Shift of Focus to the WTO 136
Phase III: The WTO Negotiations Take Shape 143
Phase IV: Towards a High-Ambition Draft 155
Phase V: A Text on the Table Amidst Rising Politics and Delay 166
Conclusion 174

5 Four Key Issues at the WTO 183
Introduction 183
Special and Differential Treatment 187
The Special Case of Artisanal Fisheries 213
The Special Case of Access Agreements 235
Sustainability Criteria for Fisheries Subsidies 265

6 Conclusion and Way Forward 307
Challenge 1: Improving Fisheries Management and Implementing Sustainability Criteria 307
Challenge 2: Pursuing Equity and Sustainable Growth for Developing Countries 309
Challenge 3: Strengthening Capacities for National and Regional Action 310
Challenge 4: Improving Policy Integration and Coherence 312
Challenge 5: Increasing Inter-institutional Innovation Internationally 313
Challenge 6: Improve Stakeholder Involvement and Coordination in Domestic Processes 314
Challenge 7: Improving Transparency and Accountability in Subsidies Programmes 314
The Bottom Line: The Oceans Can't Wait 315

Annexes

Annex 1: Timeline – The Emergence of an International Process on Fisheries Subsidies 317
Annex 2: Summary of International Fisheries Related Agreements 321
Annex 3: Regional Fisheries Bodies 323
Annex 4: Regional Seas Programmes and Conventions 325
Annex 5: 2001 Doha Ministerial Declaration 327
Annex 6: 2002 Plan of Implementation of the World Summit on Sustainable Development (excerpts) 341
Annex 7: 2005 WTO Hong Kong Ministerial Declaration (excerpts) 343

Annex 8: WTO Chair's text on Fisheries Subsidies (TN/RL/W/213, 30 November 2007, Annex VIII, pp.87–93) 347
Annex 9: WTO Fisheries Roadmap (TN/RL/W/236, 19 December 2008, Annex VIII, pp.85–94) 353
Annex 10: WTO Agreement on Subsidies and Countervailing Measures (ASCM) (excerpts) 365
Annex 11: Developing Country Categories and S&D Benefits under the ASCM 389
Annex 12: Artisanal Fishing: A few Examples of Existing Usages and Definitions 391
Annex 13: Summary of Proposed Sustainability Criteria 395
Annex 14: Benchmarks for Rapid Overall Evaluations 399
Annex 15: Model WTO Language on Sustainability Criteria 401
Annex 16: Resource Material: Contents of the CD-ROM 405

Bibliography *413*
Index *439*

Foreword by Achim Steiner

Since 1997, when UNEP co-sponsored the first international conference on the problem of fisheries subsidies, a great deal has changed – but also very little. At the time the link between subsidies and overfishing was barely recognized and poorly understood. Governments were content to pay billions in support to their fishing industries without regard for anything but expanded production and the promise of short-term economic and political benefits. Meanwhile, within the newly created World Trade Organization the use of international trade rules to help address serious environmental problems was far from an accepted concept.

Today, as UNEP publishes this review from more than a decade of thinking and debate, the need to eliminate harmful fisheries subsidies is recognized as a global priority. Governments are increasingly focused on processes of domestic reform, while WTO negotiations are well advanced towards the first multilateral trade rules ever aimed at improving the environmental sustainability of international commerce.

These are welcome signs of progress that cannot come too soon. We face a serious challenge in terms of marine ecosystems, and particularly to the once-abundant fisheries that for millennia have provided food and livelihoods to human societies. Barely half a century after the onset of industrial scale fishing, 80 per cent of the world's fisheries have been fished up to, and often beyond, their biological carrying capacity. With even more powerful fishing technologies, major stocks are crashing, triggering economic and environmental collapse. Hundreds of millions of people – the great majority in the developing world – are already suffering the consequences, with signs that unless we change course there will be much worse to come.

In the midst of this gathering crisis, governments have been funnelling public funds into subsidies that encourage ever more fleet capacity, ever more fishing effort and inevitably ever fewer fish. The stark irrationality of these subsidies – equivalent in value to between US$15–34 billion per year, or approximately a quarter of industry revenue – has compelled international attention. Meanwhile, the WTO is under increasing scrutiny to deliver genuine 'win-win-win' results for trade, development and the environment. Fisheries subsidies are the first real test case.

UNEP has played a leading role in advancing the international dialogue on fisheries subsidies. In collaboration with partners among NGOs, academics and other intergovernmental organizations, including the WTO itself, UNEP has helped bring this important issue to light and promoted solutions through a substantial body of case studies, theoretical analyses and detailed policy proposals. Through multiple workshops, briefings and symposia, UNEP has helped convene a vital and interdisciplinary 'parallel process' alongside the formal WTO negotiations.

Activities such as these form an important cornerstone of UNEP's work towards a global Green Economy, motivating governments and businesses to significantly increase investment in the environment as an engine for economic recovery and sustainable growth, decent job creation and poverty reduction in the 21st century.

However, success is not yet at hand. International rules to limit fisheries subsidies must still be agreed and early efforts at national reform are just starting to produce concrete results. The valuable work reflected in this publication helps illuminate a clear path towards solutions to the fisheries subsidies challenge. It is time for the international community to harvest the lessons learnt and take decisive action to save not only the resource, but also marine ecosystems upon which millions of people depend for nutrition and employment.

Achim Steiner
Executive Director, UNEP

List of Figures, Tables and Boxes

Figures

1.1 Impact of Fisheries Subsidies on Environment and Trade 12
1.2 Government Financial Transfers to Marine Capture Fisheries in OECD Countries 2003 (US$m) 15
3.1 Impact of Fisheries Subsidies on Fishery Stocks (Japan, China, Taiwan, Thailand, Indonesia, Philippines, Spain) 101
3.2 Processed Tuna with Subsidized Raw Material versus Range of the Price of Canned Tuna in the US–American market (2001–2006) 105
3.3 Market Participation of the Processed Tuna from Thailand and Ecuador in the US–American market (2000–2008) 105
3.4 Elements of an Effective Fishery Management System 106
3.5 Government support to the fishing fleet 1980–2008 117
3.6 Average operating margin and total operating revenues for vessels 8 metres and above 1980–2007 124
3.7 Vessel numbers 1980–2008 125
4.1 Rapidly Rising Trade in Oceanic Fish Products 134
4.2 The breakthrough in Hong Kong was hailed in newspapers around the world 156
4.3 Newspaper headlines after crash of Geneva Ministerial, July 2008 171
5.1 Net Exports of Selected Agricultural Commodities in Developing Countries 190
5.2 Option 1 – S&D eligibility criteria without specific S&D subsidies list 209
5.3 Option 2 – S&D eligibility criteria plus specific S&D subsidies list 210

Tables

2.1 Matrix Template 32
2.2 Expected Impact of Subsidies to Fisheries Infrastructure on Fish Stocks 34
2.3 Expected Impact of Provision of Management Services on Fish Stocks 36

2.4 Expected Impact of Subsidies for Access to Foreign Countries' Waters on Fish Stocks 38
2.5 11 Decommissioning Schemes: Policy Conditions and Results 40
2.6 Expected Impact of Subsidies to Decommissioning of Vessels and License Retirement on Fish Stocks 46
2.7 Expected Impact of Subsidies to Capital Costs on Fish Stocks 50
2.8 Expected Impact of Subsidies to Variable Costs 52
2.9 Expected Impact of Subsidies to Income 56
2.10 Expected Impact of Price Support Subsidies on Fish Stocks 58
2.11 Impact of Eight Categories of Fisheries Subsidies on Fish Stocks 62
3.1 Catches versus MSY for Five Major Species 71
3.2 Export Trends by Product 1990–2006 (tons) 71
3.3 Fish Exports in 2006 According to the Nature of the Products 72
3.4 Evolution of the Consumption of Fuel by the Small-scale Fishery and the Level of Subsidies 77
3.5 Evolution of Annual Subsidies to the Industrial Fishery 78
3.6 Proposed Capacity Reduction Programme – Projected Costs 89
3.7 Subsidies for Fishery Activity in Ecuador 96
3.8 Tuna Catching Countries in the EPO 97
3.9 Subsidies in Important Fishing Regions 98
3.10 Effects of Subsidies on the Economy and Trade of the Fishing Industry 102
3.11 Determination of the Volume of Subsidized Processed Tuna in the Thai Market 104
3.12 Comparison of the Evaluation Criteria of UNEP-WWF and Measures Adopted by the Ecuadorian Fisheries Legislation System 107
3.13 Wild capture fisheries production, 2008 114
3.14 Timeline for Norwegian fisheries management changes 122
5.1 Top 20 exporters of fish commodities in 2002 compared to 2001 (US$ 1000) 192
5.2 Elements of a Definition of 'Artisanal Fishing' 225
5.3 Possible Risks of Distortion to Production or Trade 231
5.4 Summary of Legal Analysis: The Onward Transfer of Rights in the Context of Fisheries Access Agreements Falls Under the ASCM Definition of Subsidies of Art. 1 ASCM 251
5.5 Key Elements of Country Submissions 253
5.6 Various Approaches to Stock Assessment 284

Boxes

1.1 International Calls for 'Greening' the Economy 2
1.2 The Economic Need for 'Greening' Fisheries 4
1.3 Objectives of the FAO Code of Conduct for Responsible Fisheries (Art. 2) 8

1.4 FAO International Plan of Action for the Management of Fishing Capacity 9
2.1 Catch and Effort Controls 27
2.2 Property Rights in Fisheries 28
2.3 Territorial Use Rights and Community-based Management 28
2.4 Defining and Measuring Overcapacity 29
5.1 FAO Terminology for Describing Stock Conditions 270

Acknowledgements

This book is produced by the Economics and Trade Branch (ETB) of the Division of Technology, Industry and Economics (DTIE) of the United Nations Environment Programme (UNEP). It is a reflection of UNEP's work on fisheries subsidies during the past ten years.

Within UNEP, Anja von Moltke has been responsible for managing the programme on environmentally harmful subsidies. She initiated and implemented the UNEP work on fisheries subsidies, including analytical papers, country studies, publications, stakeholder consultations, workshops and symposia and was responsible for the production and editing of this book.

Hussein Abaza was Chief of the Economics and Trade Branch during the implementation of this project and supported it throughout. The research, compilation, drafting and editing of this book would not have been possible without the dedicated support within UNEP-ETB of Karin Bieri, Magdalene Swiderski, Sophie Kuppler, Katharina Peschen and Louise Gallagher. Benjamin Simmons provided legal advice on WTO-related issues. Administrative support was provided by Rahila Mughal and Desiree Leon. Others involved within UNEP include Moustapha Kamal Gueye, Charles Arden-Clarke, Fanny Demassieux, Ellik Adler, Ole Vestergaard and Jacqueline Alder. They all deserve particular thanks.

The country case studies in Chapter 3 have been prepared by the following authors: the Senegalese case study, commissioned and managed by UNEP, was written by Moustapha Deme, Pierre Failler and Thomas Binet, and based on earlier UNEP material and review by Karim Dahoo, Para Gora Ndeye and Moustapha Kamal Gueye. The Ecuadorian case study chapter is based on a UNEP-CPPS-Ecuadorian government study, 'Impacto de los Subsidios pesqueros en la sustentabilidad y el comercio del atun en el Ecuador'. The author of the study is Ivan Prieto. The project was managed by Alfonso Jalil, CPPS in collaboration with UNEP. Important contributions to the project report were made by Jimmy Anastacio, Martín Velasco, Rafael Trujillo, David Schorr, Pablo Guerrero and Ramón Montaño. The sub-chapter on Norwegian fisheries subsidies reform was prepared by Anthony Cox from the OECD based on work undertaken while he was in the Fisheries Policies Division, Trade and Agriculture Directorate, OECD. It is also reported in a forthcoming OECD study. The views expressed in the chapter do not necessarily reflect those of the OECD or its member countries.

Chapter 2 and Chapter 5 consist of selected papers which UNEP commissioned and brought forward in the context of its informal consultations feeding into the negotiating process at the WTO between 2002 and 2009. The extracts have been slightly adapted for the purposes of this book. Chapter 2 was originally written by Gareth Porter. Chapter 5's sub-chapter on 'Special and Differential Treatment' was authored by Vice Yu and Darlan Fonseca-Marti from the South Centre; 'The Special Case of Artisanal Fisheries' by David Schorr; 'The Special Case of Access Agreements' by Marcos Orellana and 'Sustainability Criteria for Fisheries Subsidies' by David Schorr and John Caddy. The original reports all underwent a detailed expert and government review process before their publication and have been discussed at various UNEP hosted workshops. The remaining chapters were authored by Anja von Moltke with support from the UNEP team.

Many people have contributed to the workshops and reports that UNEP has brought forward during the last decade and that formed the basis for this book. It would be impossible to name them all, but UNEP would particularly like to thank David Schorr for his partnership during the past years and for the contribution he has made to this book.

UNEP's work has also benefitted greatly from the close collaboration of: H.E. Guillermo Valles Galmés, Ambassador and Permanent Representative of Uruguay to the World Trade Organization (Chair of the WTO Negotiating Group on Rules); Clarisse Morgan and Gabrielle Marceau (WTO Secretariat); William Emerson, Audun Lem, Kevern Cochrane, Rebecca Metzner and Angel Gumy of the FAO, Anthony Cox and Ron Steenblik (OECD); Aimee Gonzales (WWF); Sebastian Mathew (ICSF); Markus Lehmann (CBD); Stephen Hall (World Fish Centre); Moustapha Kamal Gueye (ICTSD) (now UNEP); Courtney Sakai (OCEANA); Sadeq Bigdeli (WTI); Renaud Bailleux (Sub-Regional Fisheries Commission, Dakar); Nathalie Bernasconi (CIEL); Rashid Sumaila (UBC); Milton Houghton (Caribbean Regional Fisheries Mechanism); Stephen Mbithi Mwikya (Fresh Produce Exporters Association of Kenya); Elizabeth Havice, Will Martin and many others, especially representatives from the WTO missions in Geneva. Particular thanks go to Tony Lynch, Robert Kasper, Felipe Hees, Matthew Wilson and David Schorr for the detailed comments and contributions to the section on the history of the WTO negotiations in Chapter 4.

Thanks also go to the governments of Germany, Norway and New Zealand, which have funded particular pieces of the work reflected in this publication.

The views, comments and options expressed within this book are, however, those of the authors, and do not in any way reflect the views of UNEP or its member states.

List of Abbreviations

ACP	African, Caribbean and Pacific Group of States
ADB	Asian Development Bank
AIDCP	Agreement on the International Dolphin Conservation Program
APA	At-sea Processors Association
APEC	Asia-Pacific Economic Cooperation
ASCM	Agreement on Subsidies and Countervailing Measures (WTO)
CCAMLR	Commission for the Conservation of Antarctic Marine Living Resources
CFFA	Coalition for Fair Fisheries Agreements
CFP	Common Fisheries Policy (EU)
COFI	Committee on Fisheries (FAO)
CPPS	Permanent Commission for the South Pacific
CPUE	Catch per Unit Effort
CTE	Committee on Trade and Environment (WTO)
DC	Developed Countries
DGP	Dirrección General de Pesca (Ecuador)
DIGMER	General Direction of Merchant Navy
DIRNEA	National Directorate of Aquatic Spaces
DMD	Doha Ministerial Declaration (WTO)
DOPM	Direction de l'Océanographie et des Pêches Maritimes (Senegal)
DSU	Dispute Settlement Understanding (WTO)
DWF	Distant Water Fleet
DWFN	Distant Water Fishing Nation
DWP	Doha Work Programme (WTO)
EEZ	Exclusive Economic Zone
EFF	European Fisheries Fund
EPA	Economic Partnership Agreements
EPO	Eastern Pacific Ocean
EVSL	Early Voluntary Sectoral Liberalization (APEC)
FAD	Fish Aggregating Device
FAO	Food and Agriculture Organization (UN)
FAO SOFIA	FAO State of World Fisheries and Aquaculture
FERU	Fisheries Economics Research Unit (UBC)

FIFG	Financial Instrument for Fisheries Guidance (EU)
FIRMS	Fishery Resources Monitoring System (FAO)
FMS	Fishery Management System
FOB	Free on Board
FY	Financial Year
G8	Group of Eight Leading Industrialized Countries
GAO	General Accounting Office (US)
GATT	General Agreement on Tariffs and Trade
GFTs	Government Financial Transfers
GRULAC	Group of Latin American and Caribbean Countries
HELCOM	Baltic Marine Environment Protection Commission (Helsinki Commission)
HSVAR	High Seas Vessels Authorization Record
IATTC	Inter American Tropical Tuna Commission
ICTSD	International Centre for Trade and Sustainable Development
IFZ	Industrial Free Zone
IGO	Intergovernmental Organization
INP	National Fishing Institute (Ecuador)
IO	Indian Ocean
IPOA	International Plan of Action (FAO)
IPOA Capacity	International Plan of Action for the Management of Fishing Capacity (FAO)
IPR	Intellectual Property Rights
ITQ	Individual Transferable Quota system
IUCN	International Union for Conservation of Nature
IUU	Illegal, Unreported and Unregulated
IVQ	Individual Vessel Quota
LDC	Least Developed Country
LIFDC	Low Income Food Deficit Country
LRP	Limit Reference Point
MAFF	Ministry of Agriculture, Forestry and Fisheries (Japan)
MCS	Monitoring, Control and Surveillance
MEA	Multilateral Environmental Agreement
MEM	Ministry of Maritime Economy (Senegal)
MEMTM	Ministry of Maritime Economy and Maritime Transport (Senegal)
MEY	Maximum Economic Yield
MSC	Marine Stewardship Council
MSC	Mutual of Saving and Credit
MSY	Maximum Sustainable Yield
MTI	Marine Trophic Index
NAFTA	North American Free Trade Agreement
NCFA	Norwegian Coastal Fishers's Association
NFA	Norwegian Fishermen's Association
NGMA	Negotiating Group on Market Access (WTO)

NGO	Non-governmental Organization
NGR	Negotiating Group on Rules (WTO)
NTB	Non-tariff Barriers (WTO)
ODA	Overseas Development Assistance
OECD	Organization for Economic Co-operation and Development
OLADE	Latin American Energy Organization
R&D	Research and Development
RBM	Rights-based Fisheries Management
RFB	Regional Fisheries Body
RFMO	Regional Fishery Management Organization
RSC	Regional Seas Convention
SAP	Strategic action plan
SCM	Subsidies and Countervailing Measures (WTO)
S&DT	Special and Differential Treatment (WTO)
SFLP	Sustainable Fisheries Livelihood Programme
SIDS	Small Island Developing States
SPS	Sanitary and Phytosanitary (Measures)
SQS	Structural Quota System
SRP	Subsecretaria Regional de Pesca (Ecuador)
SVEs	Small, Vulnerable Economies
TAC	Total Allowable Catch
TBT	Technical Barriers to Trade (WTO)
TED	Turtle excluder device
ThRP	Threshold Reference Point
TRIPS	Trade-Related Aspects of Intellectual Property Rights (WTO)
TRP	Target Reference Point
UBC	University of British Columbia
UI	Unemployment Insurance
UNCED	United Nations Conference on Environment and Development
UNCLOS	United Nations Convention on the Law of the Sea
UNCSD	United Nations Commission for Sustainable Development
UNCTAD	United Nations Conference on Trade and Development
UNDP	United Nations Development Programme
UNEP	United Nations Environment Programme
UNEP DTIE	UNEP Division of Technology, Industry and Economics
UNEP-ETB	UNEP Economics and Trade Branch
UNFSA	United Nations Fish Stocks Agreement
UQS	Unit Quota System
WCPFC	Western and Central Pacific Fishery Commission
WCPO	Western and Central Pacific Ocean
WSSD	World Summit on Sustainable Development (Johannesburg 2001)
WTO	World Trade Organization
WTPO	World Tuna Purse Seine Organization
WWF	Worldwide Fund for Nature / World Wildlife Fund

Introduction

About the Issue

Fisheries subsidies reform lies at the intersection of globalization and environmental policy. Fishing is both a major form of human interaction with the marine environment and the source of significant international trade. Challenges to healthy fisheries thus threaten both marine ecosystem integrity and economic welfare. Scores of millions of people depend on fish and the fishing industry for their livelihoods, while oceanic fisheries are a principal source of animal protein in the diet of an estimated 1 billion people, including many living in developing countries. Yet despite this, of the three ecosystems that supply our food – croplands, rangelands and fisheries – it is fisheries that have experienced the greatest level of excessive exploitation.

Our vested economic interest in pursuing healthy fisheries is big. The annual loss of potential economic benefits due to overcapacity and overfishing is estimated to be US$50 billion (World Bank/FAO 2008). This enormous waste of natural capital runs directly counter to the 'green economy' which political leaders worldwide now acknowledge is crucial for a sustainable economic development. Indeed, promoting a green economy in the fisheries sector is an especially urgent challenge. Re-building fish stocks, adjusting industry capacity to the limits of marine resources and shifting investment away from fishing pressure and towards value-addition – all of these hold significant potential for promoting healthy marine ecosystems while reducing poverty and stimulating economic growth.

Despite the importance of the fisheries sector for food security and development, global fisheries collapse is now a realistic possibility. According to the FAO, 80 per cent of commercially exploited fish stocks worldwide have been fished up to or beyond their biological limits. This is largely a result of inadequate fisheries management, compounded by cross-cutting pressures from marine pollution, global population increase, climate change and even deforestation which can erode upriver spawning grounds, coastal wetlands, mangrove forests and even damage coral reefs. However, it is overfishing – driven by massive fleet overcapacity and an ever increasing demand for fish products – that is the root cause of a crisis of depletion that has now achieved global proportions.

Inappropriate subsidies, in turn, are among the main causes of overfishing. They are one of the key factors driving the overcapacity of fishing fleets by radically altering income structures and decoupling industry revenues from the diminishing economic returns that otherwise accompany fisheries depletion. Subsidies also distort competitive relationships, giving subsidized fleets both the ability to catch more fish and to enjoy a price advantage in international markets. In the dispassionate terms of economic science, these effects lead to 'inefficient' resource allocation decisions that ignore real environmental costs. Put more bluntly, inappropriate subsidies are helping finance a dangerous and inequitable race for the last fish.

The dedicated work of UNEP and many other institutions and organizations, including FAO, OECD, WWF and others, has succeeded in bringing the need for fisheries subsidies reform to global consciousness. As fisheries closures and other evidence of accelerating overfishing came to light in the early 1990s, fisheries diplomacy concentrated on producing a sequence of international instruments aimed at establishing standards and norms for the world's fisheries, culminating in the 1995 FAO Code of Conduct for Responsible Fisheries. In the same period, evidence was mounting that large scale subsidization to the fishing sector was a significant factor in fisheries depletion. By 2002, when world leaders gathered in Johannesburg at the World Summit on Sustainable Development (WSSD), the importance of the fisheries subsidies issue had gained wide recognition. The Johannesburg Plan of Implementation explicitly identified the elimination of harmful subsidies as one of the top eight priorities for achieving sustainable fisheries.

It is clear that the responsibility for subsidies reform ultimately lies with national governments. However, competitive realities have made it difficult for them to 'disarm unilaterally' in the absence of effective international cooperation. Moreover, there has been an undeniable need for an international mechanism to help generate momentum for restructuring fisheries subsidies. In the late 1990s, the World Trade Organization (WTO) was recognized as having a major role to play in fulfilling the pledges of the Earth Summit's Agenda 21 to 'make international trade and environmental policies mutually supportive in favour of sustainable development' and to 'remove or reduce those subsidies that do not conform with sustainable development objectives'. In 2001, trade ministers meeting in Doha, Qatar, launched negotiations that included the aim to 'clarify and improve WTO disciplines on fisheries subsidies, taking into account the importance of this sector to developing countries' (Para. 28). Moreover, Paragraph 31 of the Doha Declaration also recalls the aim of 'mutual supportiveness of trade and environment' and agrees on further negotiations on trade and environment. In 2005, WTO ministers meeting in Hong Kong issued a strengthened negotiating mandate that called for an enforceable ban on fisheries subsidies that contribute to overcapacity and overfishing.

The WTO negotiations have been characterized by uncertainty and slow movement – but real progress has been achieved nonetheless. Seven years on from the WSSD, the topic of fisheries subsidies has penetrated more deeply into the policy-making process at the WTO than any other environmental issue. Agreement has already been reached in principle that new WTO rules should include a ban on subsidies that contribute to overcapacity and overfishing, and negotiations have reached a critical phase, with the focus now on the details of proposed new disciplines. But much remains to be decided. WTO negotiators now face a clear challenge: to agree precisely which types of subsidies will be banned and to craft meaningful limits and conditions on subsidies that remain permitted. The ability of negotiators to meet this challenge will determine the ultimate strength of new WTO fisheries subsidies rules.

Given the achievements to date, the reform of fisheries subsidies has the potential to become one of the most successful international efforts to achieve environmental, economic and development policy coherence at the global level. Fundamentally, however, the fisheries subsidies problem presents an even greater challenge to national governments. With or without WTO rules, both developed and developing countries must give increased attention to reforming their own fisheries subsidies practices. As governments and industry compete to expand their share of the potential economic returns from fisheries, providing jobs and food for countless people, they face a bleak reality: the resource supporting this development is no longer assured its long-term sustainability.

About UNEP and Fisheries Subsidies Reform

UNEP attaches high priority to work on fisheries subsidies reform. Since 1997, UNEP has played a lead role in generating policy-relevant analysis and in facilitating effective dialogue between the trade and fisheries policy-making communities. It has contributed to the international discussion on subsidy reform through a series of workshops, analytical papers and country projects. Working in close collaboration with governments, inter-governmental organizations (IGOs), and non-governmental organizations (NGOs), UNEP has provided a forum for policy-makers, experts and various stakeholders at the interface of fishery and trade policy-making.

UNEP's work on fisheries subsidies is based on a UNEP Governing Council mandate to enhance the capacities of countries, especially developing countries and countries with economies in transition, to integrate environmental considerations into development planning and macro-economic policies, including trade policies. The fisheries subsidies work programme includes the provision of technical assistance to governments in the development of policy reform packages and measures for sustainable fisheries management. UNEP has carried out a number of country studies that highlight the risk of resource depletion from fisheries subsidies. The impacts of fisheries subsidies on overfishing have also been explored through several international workshops. UNEP feeds the outcomes of these discussions into the on-going WTO negotiations on new disciplines for fisheries subsidies to ensure that these will be based on concrete sustainability criteria.

About this Publication

Almost all publications on fisheries subsidies reform to date have either been very technical – aimed at the officials and experts directly involved in the fisheries subsidies debate – or highly simplified for consumption by the wider public. With the debate now at a significant degree of maturity and global visibility, UNEP has perceived a need for a comprehensive reference on fisheries subsidies for use by policy-makers in the fields of trade, fisheries and international environmental governance, as well as academics and interested laypeople who desire a synoptic overview of the fisheries subsidies issue and an introduction to its technical components.

The book includes UNEP's key publications related to the fisheries subsidies debate at the WTO produced between 2004 and 2009 and puts them into the broader context of globalization, sustainable development and efficient resources management. It also explains the historical development of a subject entering WTO negotiations and highlights the novelty of the approach taken in comparison to previous subsidy negotiations. In short, this book explains why and how the reform of fisheries subsidies has become one of the most concrete and potentially successful international efforts to achieve global environmental, economic and developmental policy coherence and highlight the challenges that still need to be addressed.

It is organized in five parts: Chapter 1 locates the fisheries subsidies discussion in the wider globalization, sustainable development and fisheries debate. It focuses on the importance of fisheries internationally, the emergence of significant instruments of global governance for fisheries management and the growth of international fish trade. It then sets the context for the debate on fisheries subsidies, by explaining different definitions of fisheries subsidies and illustrating their current scale and size. Chapter 2 provides a categorization of subsidies and discusses the resource impacts of different subsidies types under different management and biological conditions. Using a matrix approach as a systematic framework for analysis, this chapter concludes that almost all subsidies have the potential to be harmful under the prevalent conditions found in the vast majority of fisheries. Chapter 3 presents three national perspectives and experiences on fisheries policy and subsidization. Based on two UNEP-commissioned studies and one study carried out by the OECD, this chapter presents (i) the latest developments in the Senegalese fisheries sector focusing on subsidies, their impacts and measures taken to counteract fisheries depletion, (ii) the resource and trade impacts of fisheries subsidies to the tuna fisheries in Ecuador and (iii) the history and development of subsidy reform and the introduction of other market-based instruments in Norway. Chapter 4 documents the roots of the international debate on fisheries subsidies and illustrates its development into one of the most important issues related to Trade and the Environment. It shows how the international community responded to the problem and explains why and how the issue was pursued within the WTO. Chapter 5 then showcases UNEP's contribution to the technical discussions at the WTO over the past number of years. It discusses four issues that have been critical in the WTO negotiations, particularly for developing countries: Special and Differential Treatment for Developing Countries, Artisanal Fisheries, Access Agreements and Sustainability Criteria. It elaborates on specific technical aspects that have marked the process of developing workable disciplines on fisheries subsidies and outlines options for pursuing them further. Finally, Chapter 6 identifies seven overarching challenges in need of urgent attention by governments and stakeholders along the path towards sustainability and fisheries subsidies reform. Accompanying the book is a CD-ROM containing full-text versions of major contributions to the fisheries subsidies debates from UNEP as well as FAO, OECD, World Bank, WTO, WWF, ICTSD, UBC and many others.

Chapter 1

Fisheries Governance, Subsidies and Sustainable Development

On a global scale, the limits of the Earth's ecosystems have never before been tested as severely as they are being tested today. In the near term, it may be the impending collapse of marine fisheries that most quickly teaches us the costs of failure to respect the limits of our planet's natural resources. The world's fisheries are in deep trouble. Outsized fleets are hauling in catches that exceed the biological limits of fish stocks in every ocean. As marine ecosystems are disrupted by widespread unsustainable fishing practices, livelihoods and food security are being jeopardized, often in the most vulnerable communities of developing countries.

A proper solution to the problem of overfishing will have to be global, durable and equitable. Achieving it will require nothing less than constructing a new 'green economy' in the fisheries sector. This will mean altering the system of economic incentives that currently lead fishers and merchants to perpetuate an unsustainable race for dwindling stocks. Among the reforms needed will be a fundamental change in the way governments currently employ subsidies to the fisheries sector. The various facets of that reform and the technical and political challenges to accomplishing it are the subject of this book.

This chapter sets the fisheries subsidies issue in its broader context, looking first at the economic values at stake in the fisheries sector and then at the basic elements of international fisheries governance through which the international community has begun to confront the overfishing crisis. Against this background, the chapter then concludes with an overview of the fisheries subsidies problem itself.

The Need for a 'Green Economy' of Fishing

The dimensions of the global fisheries crisis are dramatic. Eighty per cent of commercially valuable fish stocks are now overexploited, fully exploited, significantly depleted or slowly recovering from depletion (FAO SOFIA, 2008). Of the top 10 commercially used species – roughly 30 per cent of world marine catches – most are overexploited or fully exploited. The biomass of many top predator species such as bluefin tuna and swordfish has been reduced to small fractions of historic levels. Species that were once the staples of regional economies, such as cod in the northwest Atlantic, are all but commercially extinct. Even along some of the most biologically bountiful coasts, such as off the coast of Senegal in West Africa, unsustainable fishing has already taken a toll on local food supplies. And the trend lines remain negative, with the percentage of the world's fisheries that are not overfished continuing to shrink each year.

All of these grim facts result from a rapid expansion of fishing capacity that has seen global fish landings increase more than 5 times over the past 50 years (FAO SOFIA, 2006). Today, the fishing power of fleets worldwide may be as much as 250 per cent higher than can be sustainably utilized. As is commonly said, there are simply 'too many boats chasing too few fish'. But even this apparently simple reality comes with historic complexities: while it is generally accepted that aggregate fishing capacity must be reduced, there remain more than a few developing countries whose fishing industries are underdeveloped and for whom some degree of continued expansion is an equitable necessity.

Political leaders worldwide have though come to acknowledge that resource-based economies cannot exploit the very basis of their existence and that investing in a 'green economy' is crucial for a sustainable economic development (see Box 1.1) This can be achieved through reshaping and refocusing policies, investment and spending on sustainability in the fisheries sector. A 'greening' of the economy is necessary in order to have better returns on natural, human and capital investments in the fisheries sector, while at the same time reducing greenhouse gas emissions, social disparities and extracting and using less natural resources. Positive resource and environmental effects of more sustainability focused spending in the sector include a re-building of fish stocks, keeping the global fisheries industry within the limits of the marine resources and a strengthening of value-addition instead of uncontrolled expansion of fishing effort.

Box 1.1: International Calls for 'Greening' the Economy

23. Stable and sustained long-term growth will require a smooth unwinding of the existing imbalances in current accounts...New sources of growth will have to be supported by investments in infrastructure, innovation and education to facilitate productivity growth, while ensuring sustainable use of resources in a greener economy, within a context of open markets...
60. ...A shift towards green growth will provide an important contribution to the economic and financial crisis recovery...

(G8 Declaration, 8 July 2009)

The need for a green fisheries economy is compelling in human terms. To begin with, fisheries are an important contributor to food-security, employment, and economic development. For the dozens of 'Low Income Food Deficit Countries' (which face the most serious food security concerns), fish provides an average of about 20 per cent of people's protein needs. In some Small Island Developing States (SIDS), as well as in some developing coastal states, this figure rises to above 50 per cent. And even where fish is consumed in smaller amounts, it provides essential micro-nutrients. In short, for hundreds of millions of people, fish are a critical source of nutrition. And demand for fish products is continuing to rise due to population growth, income growth and urbanization (Delgado et al 2003). Even the advent of aquaculture – heralded by some as the solution to the problem of limited fish supplies – has increased the pressure on wild capture fisheries as demand has risen rapidly for fishmeal to feed farmed fish. Since it can take up to four kilos of small fish to make the meal needed to produce 1 kg of a farmed product, this has raised new questions about how best to utilize wild fish to meet human protein needs.

Beyond supplying food directly, the fisheries sector also provides employment to scores of millions of people. Worldwide, the fisheries sector directly employs around 43.5 million people, with the great majority in developing countries (UNEP, 2005a). Over the last three decades this number has grown faster than the world's population (due in part to the rapid growth of aquaculture in China and elsewhere). Taking into account the people employed in processing and marketing fish products, estimates of employment in the sector rise up to 170 million. In developing countries, local processing and handling industries are often dominated by women, tying the fisheries economy directly to household welfare and social development in traditional communities. Meanwhile, in both developed and developing countries, the fishing industry is the foundation of local livelihoods in coastal communities. Almost 90 per cent of all fishers worldwide are artisanal and small-scale harvesting accounts for approximately 25 per cent of global catches (UNEP, 2005a).

Given the socio-economic stakes, the costs of unsustainable fishing are severe. The World Bank and FAO estimate that overfishing and overcapacity already cost approximately US$50 billion per year in lost economic benefits (see Box 1.2).

The Scale and Dynamics of Trade in Fisheries Products

The fisheries economy has already gone global. As of 2006 trade in fish and fishery products is a US$86 billion-per-year industry. Fisheries exports increased by 32.1 per cent from 2000 to 2006 in real terms (in other words, adjusted for inflation), with exports in 2006 accounting for 37 per cent of total fisheries and aquaculture production (FAO COFI, 2008). Fish is thus one of the most highly traded food and feed commodities in the world – the percentage of meat traded

Box 1.2: The Economic Need for 'Greening' Fisheries

Marine capture fisheries are an underperforming global asset...The difference between the potential and actual net economic benefits from marine fisheries is in the order of $50 billion per year...The principal drivers of the overexploitation in marine capture fisheries and the causes of the dissipation of the resource rents and loss of potential economic benefits are the perverse economic incentives embedded in the fabric of fisheries harvesting regimes, reflecting a failure of fisheries governance...Improved governance of marine fisheries could capture a substantial part of this $50 billion annual economic loss. Reform of the fisheries sector could generate considerable additional economic growth and alternative livelihoods, both in the marine economy and other sectors.

(World Bank/FAO 2008, pp. ix and 39)

globally, for example, is less than a third of this. Fish is also a highly diversified trade product with more than 800 species estimated to be traded worldwide. Further adding to the degree of diversification is the general trend of moving away from the export of raw fish to processed fish products (Ahmed, 2006).

Trade in fisheries products is especially important for developing countries, which generate close to 50 per cent of world fisheries exports – a figure that does not even reflect the significant additional movement of fish products out of developing country waters that results from fisheries access agreements.[1] The rapid expansion of fish exports by developing countries has been due to a variety of factors, including technological advances that have made long-distance marketing of fish products increasingly plausible and affordable. Also contributing to higher trade figures is the increasing internationalization of value chains, with processing often outsourced to Asia and, in some cases, to Central and Eastern Europe. Generally, outsourcing decisions have been taken on the grounds of differences in labour costs (FAO COFI, 2008) and have led to an increase in exports of processed products as well as increased trade in raw-materials (ICTSD, 2006a).

The resulting fish trade has a strong 'south–north' orientation, with the greatest portion (US$32 billion in 2003) taking place between developing and developed countries. Trade among developed countries is also significant (US$23.8 billion), while trade between developing countries is much less significant (US$8.3 billion) and consists mainly of low-value species. Exports from developed to developing countries is only US$4.1 billion. As a result of its south–north character, fish trade is an important source of foreign exchange earnings for many developing countries. Net exports of fisheries products from developing countries have grown significantly over the past decades and now bring in more foreign exchange than exports of coffee, rubber, cocoa, meat and bananas combined (FAO SOFIA, 2008, fig. 32).

The rapid increase in fish trade has created an increasingly complex environment. New actors, such as multinational traders and distant water fishing nations, have emerged, creating new challenges for fisheries managers and policy-makers. And the regulation of fisheries trade itself has emerged as a significant and highly technical field that involves tariffs, quotas, safeguards, anti-dumping measures, standards, health and safety regulations and rules of origin.

Moreover, even as many developing countries have embraced export-led growth in their national fisheries sectors, the ability of fish trade to produce durable benefits has been less than perfectly reliable. While fisheries exports can increase foreign exchange earnings, they can also lead to a decreased availability of affordable fish on local markets, thereby reducing food security (see, for example the case of Senegal explored in Chapter 3, 'Fisheries Subsidies: The Senegalese Experience'). Similarly, even in importing countries, increased trade in fish and fishery products has both negative and positive aspects. On the one hand, increased imports improve food supplies and provide raw materials for domestic industries. On the other hand, cheap imports can undermine local markets and destroy jobs (ICTSD, 2006a). Nor is there any guarantee that the positive effects of trade will reach the poorer parts of society without government measures to guide market forces. Export-led growth can also put significant pressures on fisheries resources, leading to depletion of stocks unless fully effective management systems are in place.

Governance of Global Fisheries

As governments and industry strive to expand their share of the potential economic returns from the fisheries sector, they face a challenge to balance economic growth and sustained employment with the absolute necessity of ensuring long-term resource sustainability. The need for international action to protect the marine environment and its resources started to become apparent at the beginning of the 1970s, as centuries of ever-increasing pollution and demand for marine resources left their mark on the productive capacity of the oceans. Then, in the 1990s, the fisheries crisis itself rose rapidly on the international agenda, producing a series of new international legal instruments. As a result, the international policy and legal framework for fisheries management is dispersed among a variety of treaties, instruments, management schemes and economic policies that operate at international, regional, national and sometimes even local levels.

UNCLOS and its progeny

Historically, governance of the world's oceans was characterized by the principle of 'freedom-of-the-seas', under which coastal nations tenaciously held jurisdiction over the waters directly touching their coastlines while the remainder was open to exploitation by all. This created the context for a classic 'Tragedy of the

Commons' in which individual actors are motivated to exploit but not to husband a limited common resource.

By the mid 20th century, increasing pollution, rival demands for lucrative fish stocks and the rapidly rising value of sea-bed oil and mineral resources combined to throw into sharp relief the inherent weaknesses of the 'freedom-of-the-seas' doctrine. A series of nations asserting absolute right over expanded areas of the sea further eroded the traditional view of the ocean commons and established the need for more cooperative and orderly governance. As a result, in 1982 negotiations were completed for the adoption of the United Nations Convention on the Law of the Sea (UNCLOS), which subsequently entered into force in 1994. To this day, UNCLOS remains the only legally binding global and comprehensive instrument governing use of the seas. It also constitutes the backbone of the international legal regime for fisheries management.[2]

Much of UNCLOS is dedicated to establishing navigational rights, territorial sea limits, economic jurisdiction and the legal status of resources on the seabed. But also among its core articles are provisions relating directly to the use and conservation of living marine resources and the marine environment. In this regard, UNCLOS contains both rights and obligations that are differentiated across three bands of ocean territory. First, UNCLOS grants all coastal states exclusive and sovereign jurisdiction over their 'territorial sea' up to 12 nautical miles from shore (subject to certain rights of passage and safe harbour). Next, UNCLOS establishes a mixed set of rights and obligations over resources, including fish, found within a country's 'exclusive economic zone' (EEZ), which extends from the border of the territorial sea out to 200 nautical miles. Finally, what remains beyond 200 nautical miles is considered the 'high seas' and continues to be governed by customary international law, unless other specific treaty arrangements supersede.

The great majority of the world's commercially valuable marine fishes are found within the EEZs of coastal states and it is with regard to these that the fisheries provisions of UNCLOS are the most influential. UNCLOS grants to each coastal country 'sovereign rights for the purpose of exploring and exploiting, conserving and managing' natural resources within its EEZ, including fish. But UNCLOS also imposes specific duties on coastal states to manage and conserve their EEZ fisheries, including by determining and applying scientifically based catch limits consistent with maintaining 'maximum sustainable yield' of fish stocks.[3] Coastal states are further required to determine their capacity to 'harvest' the fish within their EEZs and to grant access to or sell to foreign fishers any surplus stocks that nationals of a coastal state do not exploit (UNCLOS, Article 62). This last provision provides the basis for access agreements, as discussed in Chapter 5, 'The Special Case of Access Agreements'.

UNCLOS also clearly recognizes that effective fishery management often depends on international cooperation. It includes an annex that lists 'highly migratory species' and contains a general obligation for states to cooperate to achieve the 'conservation and... optimum utilization' of these fish, whether they occur within EEZs or beyond. This provides a legal framework for the creation of 'regional fisheries management organizations' (RFMOs) discussed below. In the 1990s, UNCLOS was supplemented by two further agreements intended to strengthen cooperative fisheries management. The first – the 1993 Agreement to Promote Compliance and Management Measures by Fishing Vessels on the High Seas – specifies certain obligations of 'flag states' to prevent the reflagging of vessels as a means to avoid compliance with international management regimes. The second – the 1995 Agreement Relating to the Conservation of Straddling Fish Stocks and Highly Migratory Fish – establishes detailed minimum international standards for the conservation and management of stocks that straddle EEZ boundaries and of highly migratory fish that swim across boundaries.

The FAO Code of Conduct and its progeny

Alongside the binding legal instruments of UNCLOS and its related agreements, a second major pillar of international fisheries governance is the UN Code of Conduct for Responsible Fishing, along with a series of 'international plans of action' (IPOAs) supplementary to it. The Code and the IPOAs were drafted under the aegis of the UN Food and Agriculture Organization (FAO), which is the leading international body with a global remit to promote the sustainable management and optimum use of fisheries. Examining the state of world fisheries ten years after the signing of UNCLOS, the FAO issued a report in 1993: 'Marine Fisheries and the Law of the Sea: A Decade of Change' (FAO, 1993a). This was one of the first high-level documents to highlight the distressed state of world fisheries. As noted below, this report also helped give birth to the debate over fisheries subsidies.

Around the same time, FAO members began to take concerted action to address the threat of overfishing. An initial declaration adopted in May 1992 by 67 nations and international organizations meeting in Cancun, Mexico, gave impetus towards a more comprehensive and ambitious instrument. Three years later, FAO members adopted the 1995 Code of Conduct for Responsible Fishing (FAO, 1995a).

The Code has manifold objectives and provides a set of principles and standards for the conservation, management and development of all fisheries (see Box 1.3). It also includes standards for the capture, processing and trade of fish and fishery products, fishing operations, fisheries research and the integration of fisheries into coastal zone management.

Although it is formally a non-binding commitment, the Code of Conduct for Responsible Fisheries goes well beyond UNCLOS in its approach to fisheries management. For example, it moves past the single species, MSY-based focus of UNCLOS to begin recognizing the need for a precautionary approach. It also calls for ecosystem-based management which looks at impacts of fishing not only on target stocks but also on non-target species and habitats (IUCN, 2003). While the Code is voluntary, certain parts of it are based on relevant rules of international law, including those reflected in UNCLOS.

Box 1.3 Objectives of the FAO Code of Conduct for Responsible Fisheries (Art. 2)

1. establish principles...for responsible fishing and fisheries activities, taking into account all their relevant biological, technological, economic, social, environmental and commercial aspects;
2. establish principles and criteria for the elaboration and implementation of national policies for responsible conservation of fisheries resources and fisheries management and development;
3. serve as an instrument of reference to help States to establish or to improve the legal and institutional framework required for the exercise of responsible fisheries;
4. provide guidance...in the formulation and implementation of international agreements and other legal instruments, both binding and voluntary;
5. facilitate and promote technical, financial and other cooperation in conservation of fisheries resources and fisheries management and development;
6. promote the contribution of fisheries to food security and food quality;
7. promote protection of living aquatic resources and their environments and coastal areas;
8. promote the trade of fish and fishery products in conformity with relevant international rules;
9. promote research on fisheries as well as on associated ecosystems and relevant environmental factors; and
10. provide standards of conduct for all persons involved in the fisheries sector.

In the hope of moving governments towards concrete and detailed implementation of the principles articulated by the Code, FAO members have also negotiated and adopted International Plans of Action (IPOAs) on four leading subjects:

1. the reduction of sea-bird bycatch in longline fisheries;
2. the conservation and management of sharks;
3. the management of fishing capacity; and
4. the prevention and elimination of illegal, unreported and unregulated (IUU) fishing.

The first three of these IPOAs were formally adopted in February 1999, while the IPOA-IUU was negotiated and adopted somewhat later, in 2001. As with the Code of Conduct itself, the IPOAs are all voluntary instruments. Nevertheless, they contain a series of specific commitments by governments and present a clear roadmap for implementing the Code in several priority areas of policy. The IPOA-Capacity is of particular importance for the fisheries subsidies debate both in light of the direct relationship between inappropriate subsidies and overcapacity and because, as discussed in Chapter 4, the IPOA-Capacity was the first global international instrument to call for fisheries subsidies reform (see extracts in Box 1.4).

Box 1.4 FAO International Plan of Action for the Management of Fishing Capacity

25. When developing their national plans for the management of fishing capacity, States should assess the possible impact of all factors, including subsidies, contributing to overcapacity on the sustainable management of their fisheries, distinguishing between factors, including subsidies, which contribute to overcapacity and unsustainability and those which produce a positive effect or are neutral.

26. States should reduce and progressively eliminate all factors, including subsidies and economic incentives and other factors which contribute, directly or indirectly, to the build-up of excessive fishing capacity thereby undermining the sustainability of marine living resources, giving due regard to the needs of artisanal fisheries.

45. FAO will, as and to the extent directed by its Conference, collect all relevant information and data which might serve as a basis for further analysis aimed at identifying factors contributing to overcapacity such as, *inter alia*, lack of input and output control, unsustainable fishery management methods and subsidies which contribute to overcapacity.

Regional treaties and fisheries management organizations

In operational terms, neither UNCLOS nor the Code of Conduct provides a mechanism for the active management of international fisheries. Instead, this task falls to a growing set of 'regional fisheries bodies' (RFBs) that have been created for that purpose. These RFBs, which vary widely in their geography and authority, now cover most parts of the world's oceans and most fish species. Only some of these have powers to set management measures that are binding on the fishing fleets of their member countries; many have only advisory functions (IUCN, 2003). None has fully-effective enforcement capabilities.

Generally speaking, an RFB is any treaty, commission or organization through which three or more states or international organizations engage in joint management and development of fisheries affairs. RFBs can be categorized as management bodies known as Regional Fisheries Management Organizations (RFMOs), advisory bodies or scientific research bodies (for a full list see Annex 3). Advisory and scientific RFBs are expected to provide management and/or scientific advice to their member states. Their role is principally to translate scientifically-based conservation recommendations into management measures recommended for adoption. Many, but not all, RFBs are gathered under the administrative aegis of the FAO, although the FAO itself has no direct fisheries management authority.

Alongside the RFBs, Regional Seas Conventions (RSCs) are region-specific agreements for preventing marine pollution and developing sustainable use of marine resources. RSCs, which operate under the umbrella of the UN Environment Programme (UNEP), have mandates that can overlap with RFBs, particularly to the extent that the latter are moving towards ecosystem-based management principles that address broader environmental and habitat concerns (UNEP, 2001a). UNEP helps administer 13 regional seas programmes, including through administrative support for the RSCs and the development of action plans. Several of the Regional Seas Programmes refer in their action plans to the need to combat overfishing and unsustainable fishing practices in cooperation with the relevant fisheries bodies.[4]

At the bottom line, however, in the realm of marine fisheries governance it is the RFMOs that play a uniquely critical role in facilitating international cooperation for the conservation and management of fish stocks. These organizations currently represent the only means of governing fish stocks that occur either as straddling or shared stocks between zones of national jurisdiction, between these zones and the high seas, or exclusively on the high seas. Despite efforts over the past decade to improve their management capacity and their images as effective organizations, some RFMOs have failed to achieve their fundamental goal of the sustainable management of stocks. Strengthening RFMOs in order to conserve and manage fish stocks more effectively remains the major challenge facing international fisheries governance.

Subsidies in the Fisheries Sector

Despite almost three decades of international efforts, the rules and mechanisms of fisheries governance have yet to halt the problem of overfishing. While in some specific fisheries a combination of strong regulations and proper economic policies has demonstrated that effective management is possible, such cases remain the exceptions rather than the rule. Clearly, many factors have complicated the search for solutions, including real-world limits on scientific knowledge, on administrative capacity and on political will. But policy-makers and stakeholders have also become increasingly aware that attempts to regulate and end overfishing will probably continue to fail if the underlying economic incentives cut the other way.

There are two fundamental sources of perverse economic signals in the fisheries sector. First, there is the continued prevalence of ‘open access’ fisheries that perpetuate depletion through repeated tragedies of the commons. Second, there is the provision of inappropriate subsidies that support bloated fleets and drive massive overfishing where sheer unprofitability would otherwise reduce or halt them altogether (FAO, 2003e). Regarding the first of these, a growing body of experts is now concluding that fisheries management systems are most likely to succeed if fishermen are given (or sold) secure and enforceable long-term property interests in the fish themselves. Regarding the second, as discussed throughout the remainder of this book, there is broad recognition of an urgent need for widespread subsidies reform.

With a few notable exceptions, the international conventions and codes of fisheries governance rarely mention the problem of subsidies. However, the conditions that gave rise to those instruments – and to some degree the terms of the instruments themselves – paved the way for the discussion of fisheries subsidies to enter the international stage. The above-mentioned FAO report ('Marine Fisheries and the Law of the Sea: A Decade of Change', FAO, 1993a) contains a special chapter examining the economics of the fisheries sector. It found that 'operating costs of the global marine fishing fleet in 1989 were in the order of USD $22 billion greater than the total revenues, with no account being taken of capital costs' (FAO, 1993a, p.20). When capital costs were included, the estimated annual cost-revenue gap rose to US$54 billion. The report went on to state, 'Subsidies are presumed to cover most of this deficit' and concluded that 'a continuation of the high subsidies can only lead to greater and greater economic distress as well as a further depletion of stocks' (FAO, 1993a, p.24).

After the FAO report, as work proceeded towards agreement and then implementation of UNCLOS and the Code of Conduct, international preoccupation with the overcapacity of fishing fleets grew. In the early FAO IPOAs, overcapacity was the first general issue to be addressed, alongside the species-driven IPOAs on seabirds and sharks. The link between subsidies and overcapacity was already well understood and, as noted above, the IPOA-capacity called directly for subsidy reform.

Fisheries subsidies literature and reports by IGOs have also shown how a subsidy affects profits and provides incentives for increased fishery effort, leading to overcapitalization and depletion (Arnason, 1998; FAO, 2000b; Munro, 1999; Nordstrom and Vaughan, 1999; OECD, 2000b, 2003; Stone, 1997; UNEP 2002b, 2004b). This has been confirmed by a number of case studies (e.g. UNEP, 2001b, 2002a, 2004c, 2006a; Chuang and Zhang, 1999; World Bank, 1998; OECD, 2000a) and been echoed in many of the country position papers tabled during the WTO fisheries subsidies negotiations.

The effect on the sustainability of fisheries resources is just one impact of harmful fisheries subsidies. Ill-considered subsidization policies also have simultaneous effects on international trade and regional and national economic growth, not least in developing countries. It is clear that reducing the cost of fishing or enhancing its profitability by increasing revenue allows subsidized producers to reduce prices, gain market share and limit access to common fisheries resources for non-subsidized fleets. This reinforces tendencies to overfish and overinvest. Significant impacts are felt by developing country fisheries through trade distortions that inhibit developing country access to markets of industrialized countries and distant water fishing operations that deprive local people of their fisheries resources. The congruence between environmental and trade effects is well illustrated in Figure 1.1 where the 'fish' represent support measures that are currently being employed by many countries.

No one suggests that fisheries subsidies are employed in a purposefully destructive way. Rather, they can have negative effects that are unintended, unforeseen or unaccounted for in the policy-making process (OECD, 2005a). Moreover, not all fisheries subsidies are likely to be harmful. The kinds of subsidies and the circumstances in which they are employed can have different

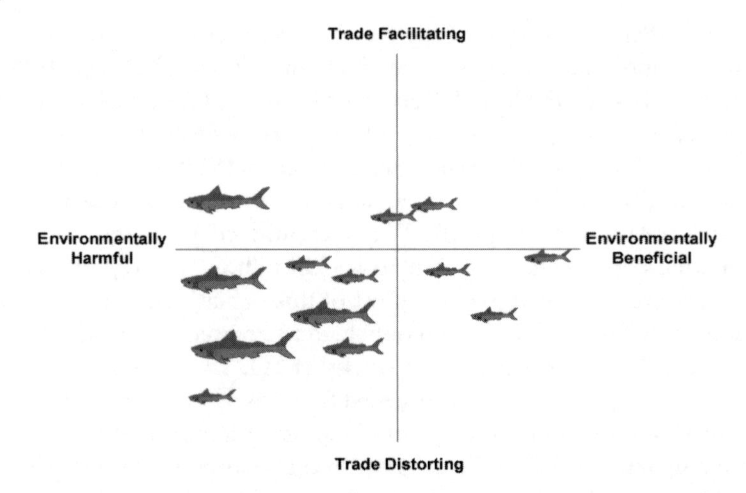

Figure 1.1 *Impact of Fisheries Subsidies on Environment and Trade*

Source: Adapted from Steenblik, (2009)

implications for fishing effort. Subsidies to capital costs and price supports, which can have major impacts on the relative prices in international trade, are particularly likely to distort trade. Subsidies to variable costs and access to foreign fishing waters also directly promote increased fishing pressure and are likely to distort trade. An analysis of different impacts of subsidies on fisheries resources according to their management and stock condition is contained in Chapter 2.

A definition of fisheries subsidies?

To some degree, the debate over fisheries subsidies has been coloured by different views of what precisely is meant by the term 'subsidy'. A subsidy is generally understood to be some type of government support to the private sector, usually serving a public purpose.[5] However, as FAO asserts, not everything the public sector does, or does not do, can be classified as a subsidy. Given the complexity of the fisheries subsidies domain, further qualification of this definition is required (FAO, 2004e).

A key definition in the subsidy debate is that of the 1994 WTO Agreement on Subsidies and Countervailing Measures (ASCM, see Annex 10), which provides the legal definition of a subsidy in international trade law. In the ASCM, any 'financial contribution' by a government (also public or private entities on behalf of the government) is a subsidy, if it is a 'benefit' to specific domestic industry from:

- *transfers of funds* including grants, loans and equity infusions;
- *potential transfers of funds* such as loan guarantees or government insurance;
- *foregone government revenue* from tax exemptions;
- *goods or services* provided to the industry other than general infrastructure;

- *indirect support* through payments to a funding mechanism or privately held body to perform any of the above;
- *price or income support programmes* other than tariffs.

FAO's 'Guide to Identifying, Assessing and Reporting on Subsidies in the Fisheries Sector' offers a somewhat different approach to the definition: 'Fisheries subsidies are government actions or inactions outside of normal practices that modify – by increasing or decreasing – the potential profits by the fisheries industry in the short-, medium- or long-term' (FAO, 2004e, pp.7–8). Other organizations involved in the fisheries subsidies reform debate also prefer relatively broad definitions. For example, the WWF thinks of fisheries subsidies as all government supports to the fishing industry that can effectively reduce the costs of fishing, thus focusing directly on the links between fishing subsidies and unsustainable fishing. The Asia-Pacific Economic Cooperation (APEC) also uses a definition covering all government support measures to the fishing industry (APEC, 2000, pp.3–4).

The OECD does not use the term 'subsidies', but has adopted 'government financial transfers' (GFTs) instead with a broad economic definition. The difference between subsidies and GFTs is that 'subsidies encompass more than just the explicit transfer of money from the public purse to the sector' and also 'include implicit transfers from consumers to the industry', whereas 'GFTs are considered to be a subset of the whole range of subsidies'. They are defined 'as the monetary value of interventions associated with fishery policies, whether they are from central, region or local governments' and 'include both on-budget and off-budget transfers to the fisheries sector' (OECD, 2003). In this context the OECD identifies three categories of GTFs: direct payments, cost-reducing transfers and general services. There is a fourth category, market price support, but it has not been addressed in the OECD study on Transition to Responsible fisheries. There is some controversy over whether general services (e.g. governmental fisheries management activities) should be counted as GTSs, since government reports differ in this matter.

UNEP (2002b, 2004a) employs a composite list of subsidies categories that encompasses the definitions used in other schemes. It categorizes subsidies by their objective rather than by the form of the subsidy. Thus all subsidies that have the same economic effect are grouped together. For example, tax preferences, soft loans and support for vessel modernization or building of new vessels, are categorized as subsidies for capital investment in the fishing industry. Programmes that contribute to the income of both vessel owners and fishermen are also grouped together, as are social or income assistance schemes for fishermen (see Chapter 2).

The Fisheries Centre of the University of British Columbia (UBC) – whose attempt to quantify global fisheries subsidies on a country-by-country basis is discussed below – adopts a rather broad definition of subsidies and classifying them as 'good' (e.g. expenditures for management services), 'bad' (e.g. boat building) and 'ugly' (e.g. vessel buyback programmes) (Sumaila/Pauly, 2006).

All in all, as noted above, the precise definition of 'fisheries subsidy' is context-dependent and can even be a reflection of specific advocacy positions. Generally, however, these distinctions are only significant where detailed technical precision is required, such as in the drafting of proposed WTO law. As noted in the next section, however, the definitional debate is also one source of uncertainty about the precise scale of the subsidies problem.

The scale of the problem

Subsidies have been found to be generally pervasive worldwide (OECD, 2005a) and there is no doubt that they are used on a scale that has a significant impact on the economics of the fisheries sector. However, precise measurement of the level of fisheries subsidies has remained an elusive goal, due in part to definitional uncertainties, but more fundamentally to the continuing lack of transparency in most subsidy programmes.

The earliest global estimate (FAO, 1993a) of US$54 billion was subsequently considered by some fisheries experts probably to be too high. According to the first in-depth study of the extent of fisheries subsidies conducted by the World Bank (World Bank, 1998), subsidies represent nearly 20 per cent of fishing industry revenue and are estimated to total US$11–20 billion per year. A study by a major international accounting firm conducted for the inter-governmental Asia-Pacific Economic Cooperation (APEC, 2000) estimated subsidies to the APEC 'capture fisheries' sector alone at US$8.9 billion. Shortly thereafter, WWF reviewed APEC, OECD and WTO data based on government sources and found that after accounting for likely overlaps 'officially reported' subsidies stood at approximately US$13 billion per year (WWF, 2001). WWF went on to present evidence of widespread under-reporting of subsidies and concluded that the actual global yearly total was 'at the least USD $15 billion, and very possibly higher.'

Figure 1.2 illustrates the breakdown of officially reported fisheries subsidies across OECD countries in 2003.

The most recent and ambitious effort to quantify global fisheries subsidies has been a study by the Fisheries Center at the UBC conducted in 2006, which estimates the worldwide total at US$30–34 billion per year (Sumaila, Pauly 2006). The UBC study further classifies subsidies according to their effects, stating that US$6.6 billion are so-called 'good subsidies', US$3.4 billion so called 'ugly subsidies' and US$16 billion 'bad subsidies'. The data were collected for 141 countries mainly from secondary sources, but also through the publications of intergovernmental organizations and multilateral agencies, such as the FAO or UNEP. The results show that of the 'bad subsidies' 70 per cent are provided in developing countries. Developed countries in turn account for 75 per cent of 'ugly subsidies' and the largest part of 'good subsidies' as well. Meanwhile, a study by Khan et al (2006) in the same year estimates the total magnitude of fishery subsidies in marine capture fisheries at US$25.7 billion for 11 types of subsidies (excluding fuel subsidies). They also suggested that the total estimated

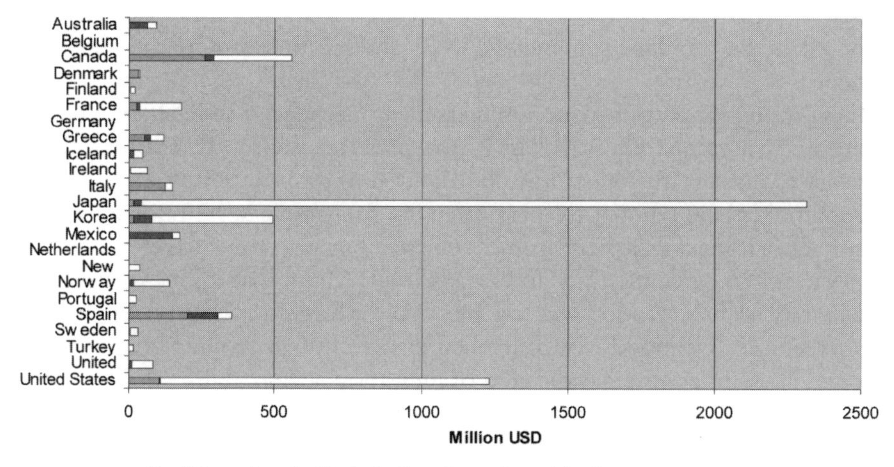

Figure 1.2 *Government Financial Transfers to Marine Capture Fisheries in OECD Countries 2003 (US$m)*

Source: OECD (2008)

non-fuel subsidies are divided nearly evenly between developing countries (US$13 billion) and developed countries (US$12.7 billion). A complementary fuel subsidies study (Sumaila et al, 2006) estimates fuel subsidies at US$6.3 billion, giving a total of US$32 for global fisheries subsidies. Given all these different estimates, today we often talk about subsidies ranging from US$15–34.

Apart from the frequent gaps in public information about subsidies, it appears that many studies have omitted subsidies to aquaculture and government support to the fish processing industry – support which may obviously confer significant (if indirect) benefits on the fishing industry itself. Such subsidies are not insignificant but are also difficult to quantify. APEC (2000) include processing subsidies, estimating them at roughly US$0.7 billion annually. However, the economic or other impacts of these subsidies on fishing activities can be extremely difficult to establish or describe.

Challenges for Reform

Undoubtedly, marine fisheries governance has made significant progress. International cooperation on global fisheries management has been strengthened through the creation of UNCLOS and the establishment of RFBs, which have improved the sophistication and effectiveness of global fisheries management. While transparency on fish production processes is still lacking, more information and more traceability is available than ever before. Improved monitoring has generated the most reliable information to date on fish stock health and fishing fleet capacities. In terms of management standards, the ecosystem approach to fisheries, supported by the precautionary principle, is now being promoted as the standard to which RFMOs and national fisheries management should aspire.

Long-term planning and good management are needed to balance economic growth and sustained employment in the fisheries sector. The reform that is necessary to achieve this has to be coordinated at an international level due to the potential loss in competitiveness affecting first-movers. As reform does not only affect fish stocks and government purses, but also rural development, the process needs to be democratic and transparent. Many developing countries do not only rely on fish for food security, but also for foreign exchange earnings and the provision of livelihoods. Uncontrolled growth in fish production and export is though likely to counteract the goal of long-term sustainability of the fisheries sector. The impacts on poverty, food security and livelihoods need to be taken into account at all stages of the reform process.

In current fisheries management regimes, subsidies to the fisheries industry play a significant role in the depletion of fish stocks. Efforts to reform these subsidies are thus needed to ensure sustainability and sustainable growth of the sector. As is abundantly clear from the material presented in this book, the issue of subsidies reform is a delicate and complex one with no simple solution. In most cases, subsidies are granted on the basis of valid policy goals and legitimate motives, including promoting development and poverty reduction. Efforts to stimulate the development of fisheries – to modernize them, to increase productivity or to support selling fishery products – are generally meant to protect fishermen who are often amongst the poorest workers. And, increasingly, some subsidies are also being aimed at improving fisheries management and ensuring the sustainability of the resource.

The proper motives of governments, however, has not made subsidies substantially less likely to do harm, or substantially more consistent with good management policies. Almost inevitably, subsidies of all kinds enable the industry to operate at lower costs than would otherwise be possible. At its most fundamental, this action causes vital price signals that would otherwise indicate scarcity of fisheries resources to be lost at each point along fisheries supply and consumption chains. Furthermore, if the industry's revenues are sufficiently low, indicating that it has become more expensive to catch fewer fish, some subsidies 'top up' fishing industry revenues to make it economic to continue fishing already overfished resources.

The inherent danger of almost all fisheries subsidies is badly compounded by the almost universal disconnect between subsidies programmes and the fisheries management policies mandated by UNCLOS, the Code of Conduct and their ancillary instruments. In short, fisheries subsidies are too often a classic example of policy incoherence.

While governments become more aware of the need to reform fisheries subsidies, the removal of a harmful subsidy may lead to immediate increases in costs that reduce the competitiveness of national industries. For some, fisheries subsidies removal thus appears a likely descent into uncertainty and further poverty for fishers. This concern is particularly acute for small-scale and artisanal fishing industries. Given the socio-economic stakes discussed above, it is clear that internationally coordinated reform is more likely to succeed politically than 'unilateral disarmament.'

Naturally, however, even governments that seriously seek a cooperative strategy for reforming fisheries subsidies are confronted with a variety of perspectives and priorities that complicate efforts to reach agreement. Furthermore, governments tend to share a dislike for supranational intrusions on their domestic policies. As in any complex international policy negotiation, these factors make it all the more important for cooperative fisheries subsidies reform to be achieved democratically and transparently. This will require even greater emphasis on improving public access to information about subsidies, both to increase the accountability of administrators and to give policy-makers a proper basis to formulate programmes that ensure both resource sustainability and economic development.

Endnotes

1 As discussed in Chapter 5 in 'The Special Case of Access Agreements', fisheries access agreements are the instruments under which foreign fleets can gain access to fisheries within another country's exclusive economic zone. Under the unusual 'rules of origin' normally applied within the fisheries sector, fish products have the nationality of the vessel catching them, no matter where the capture takes place. Thus, fish taken by a European vessel in the waters of Mauritania are not 'imports' when sold in Europe. Such products are generally not counted as having entered international 'trade' but are considered to be produced by the country.

2 It should be noted that a number of countries have still not signed or ratified UNCLOS and may object to some of its provisions. Still, the basic framework of UNCLOS has a significant influence on the organization of oceans governance for all nations.

3 The UNCLOS obligation to manage stocks for 'maximum sustainable yield' is not absolute. Governments are allowed an unspecified degree of deviation from MSY to accommodate 'relevant environmental and economic factors, including the economic needs of coastal fishing communities and the special requirements of developing States, and taking into account fishing patterns, the interdependence of stocks and any generally recommended international minimum standards, whether subregional, regional or global.' (UNCLOS Art. 61.3).

4 The Baltic Marine Environment Protection Commission of the Helsinki Commission, for example, states in its Baltic Sea Action Plan that 'devising long-term plans for the monitoring, protection and sustainable management of coastal fish species' is needed and that this and other related activities 'will be carried out by the competent fisheries authorities in co-operation with the Baltic Sea Regional Advisory Council (RAC) and HELCOM' (Helsinki Commission, n.d., section 4).

5 See, however, the discussion of the OECD's vocabulary, below, which also refers to uncompensated private transfers of value as 'subsidies' and reserves the term 'government financial transfers' for public–private supports.

Chapter 2

Fisheries Subsidies and their Resource Impacts*

Introduction

The purpose of this chapter is to analyse systematically the impacts of major categories of fisheries subsidies on fisheries resources in order to help inform the debate surrounding the disciplining of fisheries subsidies. The analysis takes an approach that ensures systematic coverage of the relevant combinations of subsidy types and relevant characteristics of the fishery and its management regime. It is designed to serve as an analytical basis for designing new or improved disciplines on fisheries subsidies that will help protect fishery resources.

Given the range of interests at stake in the international dialogue over fishing subsidies, the limited scope of this study should be clearly understood. Although the focus of this chapter is on the impacts of subsidies on commercial stocks targeted in a given fishery, it should be borne in mind that other environmental impacts of fisheries subsidies – increased by-catch, marine pollution and damage to habitats from certain fishing gear – also occur as a result of increased fishing effort. Nor does this chapter investigate other non-resource effects of fishing subsidies, such as impacts on trade or on social welfare.

However, impacts of these various kinds are often interdependent and the analysis of this paper can thus be considered relevant to the broader discussions. Moreover, the approach adopted here might be usefully applied to analyses focused specifically on these other impacts. It is worth noting, for instance, that the overexploitation of resources encouraged by some fisheries subsidies is

* This chapter is taken from UNEP (2004a). It has been slightly adapted for the purposes of this book. The original report was prepared by Gareth Porter and has benefited from comments and suggestions from several rounds of review, including a government review in December 2003. It was then discussed at UNEPs Workshop on Fisheries Subsidies and Sustainable Fisheries Management, held in Geneva in April 2004 before its publication (see accompanying CD-ROM).

relevant to the analysis of those same subsidies on social equity and poverty. The assumption often voiced in international discussions that fisheries subsidies may be used for poverty reduction, even if they have environmental impacts, must therefore be carefully analysed in terms of the specific subsidy in question.

The fact that fisheries subsidies can have the effect of increasing fishing effort and thus have negative impacts on the level of fish stocks is universally accepted. Different categories of subsidies, moreover, have different implications for fishing effort. The existing fisheries subsidies literature points to the need for a systematic, case-by-case analysis of the impacts of fisheries subsidies on fishery resources, which will reflect both the different incentive effects of various subsidies and the different circumstances in which subsidies may be provided.

Categories of Fisheries Subsidies

In recent years, a great deal of effort in various forums has been put into defining subsidies and developing frameworks for categorizing subsidies, resulting in a wide variety of definitions and classification frameworks. This variety is primarily a response to the various objectives pursued in the studies as well as the different perspectives taken by authors. In this chapter, a pragmatic approach is taken to the definition and classification of subsidies. This chapter does not attempt to propose either a new definition of fisheries subsidies or a classification system that should be used in international legal or negotiating processes. Rather, the focus is on analysing the effects of different types of subsidies, irrespective of how they might be viewed in the international legal context.

This chapter does not limit the analysis to those subsidies that would clearly qualify under the Subsidies and Countervailing Measures (SCM) Agreement. It analyses the forms of government support that have been most widely used and which have been most prominently discussed in the discourse on reforming fisheries subsidies.

The study does not, therefore, analyse the full range of 'implicit' subsidies that occur as a result of government inaction (such as non-collection of resource rents). While it may be argued that these are indeed subsidies, most of them are peripheral to the main focus of policy discussion at this stage. Moreover, they are perhaps better examined in the context of the overall management of fisheries rather than under the more specific topic of fisheries subsidies.

A study by UNEP (2002b) provides a review of the various classification schemes that have been employed by the Organization for Economic Cooperation and Development (OECD), the Asia Pacific Economic Cooperation (APEC) and the United States. The Food and Agriculture Organization of the United Nations (FAO, 2000b) provides a less detailed scheme that employs broader categories. Categories of subsidies in one categorization scheme often overlap with those used in other schemes. Overlaps occur because some categories are defined by the intended effect of the subsidy, whereas others are based on the form of the subsidy.

This chapter uses a composite list of subsidy categories that encompasses the categories used in previous schemes. It groups subsidies by their objective, rather than by the form of the subsidy. Thus all subsidies that have the same economic effect (in other words, direct funds towards the same objective) are grouped together under one heading, regardless of the form of the subsidy. For example, soft loans, tax preferences, insurance and other risk-reducing programmes, as well as grants for vessel modernization or new vessels, are grouped together as subsidies that channel resources into capital investment in the fishing industry. Price supports and other programmes that contribute to the income of both vessel owners and fishermen, but do not channel the money into any particular kind of investment, are also grouped together, as well as all social or income assistance to fishermen.

The subsidy categories employed are:

- fisheries infrastructure;
- management services and research;
- subsidies for access to foreign countries' waters;
- decommissioning of vessels and license retirement;
- subsidies to capital costs;
- subsidies to variable costs;
- income support and unemployment insurance; and
- price support subsidies.

Some governments and non-governmental actors consider that properly designed subsidies can, in some cases, have positive impacts on the conservation of fish stocks or other aspects of marine ecosystems. Two of the categories – 'management services' and 'decommissioning of vessels and license retirement' – are often justified in these terms. In addition, similar claims are made about some subsidies to capital costs, such as for the acquisition of selective fishing gear. Significant debate surrounds the question whether, or under what circumstances, subsidies of these kinds can have positive impacts on fisheries resources.

Furthermore, as international discussion of fishing subsidies has matured, attention has been increasingly drawn to the differences between large-scale industrialized fisheries and small-scale or artisanal fisheries. This study does not address scale or technological level as significant variables. More empirical work to illuminate the issues surrounding subsidies to small scale or artisanal fisheries is clearly warranted.

Subsidies to fisheries infrastructure

General infrastructure programmes, such as highways and ports, which benefit the general public as well as all industries involved in trade, are excluded from the definition of subsidy under the SCM Agreement. But a fishing port, which provides benefits to the general public, benefits the fishing industry even more. WTO dispute settlement panels have not yet clarified precisely what standard is to be used to determine whether assistance in building fishing ports is a subsidy (World Bank, 1998).

Some trade lawyers have suggested that the criterion for identifying infrastructure subsidies to the fishing industry should be whether the industry pays for fishing infrastructure projects in most countries. By this criterion, fishing port construction would not be considered a subsidy, since financing the construction or modernization of fishing ports has been treated by national and local governments as well as bilateral and multilateral funding agencies as general development projects rather than as a subsidy to the fishing industry. In practice, fishing ports are almost always paid for by the state or by external financing institutions (World Bank, 1998). Applying this criterion would therefore simply legitimize the status quo that results in general subsidization of fishing industries.

A more rigorous set of criteria, in line with the economic definition of infrastructure subsidy, would be whether the fishing industry is clearly the primary beneficiary of the project and whether it pays a reasonable user fee for the services provided by the infrastructure. By this measure, government provision of fishing ports for its fishing industry is always a subsidy, unless the users do in fact pay user fees, because the intention is in every case to benefit local or regional fishing fleets.

Included under this classification are payments for the provision of land-based resources encouraged by government budgetary transfers. Expenditures on fisheries infrastructure include the construction of harbours, the dredging of ports and the improvement of landing installations and equipment. In this classification, financial assistance to institutional infrastructure for fisheries is also included (that is, government transfers to support the activities of producer organizations). Under this classification all payments to infrastructure for a specific fishery that are not offset by user fees may be considered as the subsidy rather than an estimate of the 'economic benefit' to fishers of total infrastructure spending.

Programmes included in fisheries infrastructure are:

- harbour facilities and moorage – provided free or at low rates of moorage for fishing fleets;
- fishing port infrastructure enhancement, such as dredging;
- support to producer organizations – institution infrastructure (except where in the details of the policy the payments are for price support).

Management services and research

Management services usually comprise three functions: administering the existing management system, adjusting management settings within an existing management system and recommending amendments or additions to the existing system. Research is often used as the basis for management decisions and the creation of new management systems. Common examples of research activities include data collection, survey data analysis, stock assessment and risk assessment. Enforcement services typically involve surveillance of compliance with fisheries laws and a role in the prosecution of fishers who do not comply with those laws. Also included in this category is research to develop the fisheries industry and programmes encouraging fish re-stocking and protection of marine areas.

Although management services, including monitoring and surveillance, stock assessments and research on sustainable fishing gear generally benefit both the industry and the general public, they can be considered a subsidy in economic terms to the extent that a reasonable proportion of the costs of those services are not recovered from industry. A number of countries, including New Zealand, Iceland and Australia, have adopted programmes of cost recovery for fisheries management services.

Programmes included in management services and research are:

- stock enhancement programmes, including fish habitat improvements, release of juveniles;
- fisheries management programmes, including monitoring and surveillance;
- fisheries enforcement programmes, including prosecuting of offences;
- programmes to assess fish stocks;
- programmes to identify and develop new fisheries;
- R&D to develop new fisheries technologies;
- protection of marine areas;
- aid for fish re-stocking;
- artificial reefs.

Payments for access to other countries' waters

Some distant water fishing states negotiate fisheries agreements with coastal states that involve the granting of fishing rights conditional on financial payments. Such payments to foreign countries, which may cover a significant part of the effective costs of a distant water fleet's access to a foreign fishery, can have the effect of subsidizing the foreign fleet in question (see discussion on access agreements later in this chapter).

Programmes included in access to other countries' waters are:

- payments of part of the costs of access to foreign fishing waters in conjunction with international fishing access agreements.

Subsidies for decommissioning of vessels and license retirement

Because of massive overcapitalization of fishing fleets, states have used the provision of grants for decommissioning fishing vessels or license retirement as a means of reducing the level of fishing capacity in their fisheries. These schemes involve payments for the permanent removal of fishing vessels or fishing licenses from the fishery or fisheries involved.

Some states have insisted that decommissioning subsidies should be considered as 'good subsidies', because of their objective of reducing fishing capacity. Although this categorization scheme uses the explicit objective of decommissioning in defining this category, it is, however, not meant to imply that its actual *effect* is necessarily to reduce capacity in the fishery. As discussed in the following section, whether a subsidy in this category has a positive or negative impact on fisheries resources depends on the circumstances in which it is provided and the policy conditions attached to it.

It should be noted that, in this classification scheme, this category does not include payments for *temporary* cessations of fishing. Such payments are included under subsidies to income.

Programmes included under decommissioning of vessels and license retirement are:

- payments for the permanent withdrawal of fishing vessels;
- payments for the permanent withdrawal of fishing permits or licenses.

Subsidies to capital costs

Grants or below-market loans to purchase new fishing vessels or to modernize existing vessels reduces a major capital cost. In some countries and at certain times, subsidies to capital costs have represented a large proportion of total new capital investment in the industry, magnifying the impact of such subsidies on both capacity and ultimately fishing effort.

This category includes all government-funded grants for fleet renewal and modernization, loans, loan guarantees and loan restructuring at below commercial lending rates to the fisheries sector. Also included are tax preferences, which reduce the costs of the purchase of capital goods, and risk reduction mechanisms such as loan guarantees. Loan guarantees do not necessarily cost the government anything, and the loans might be made at conventional interest rates, but they still affect investment decisions by reducing the risk of investment.

Programmes included in subsidies to capital costs are:

- grants and below-market loans for fleet renewal and modernization;
- accelerated depreciation that reduces taxation of vessels and fishing gear;
- development grants for fisheries enterprises;
- aid to shipyards which supports fishing boat construction;
- reduction of the financial burden of equipment needed for deep-sea fishing;
- loan guarantees;
- loan restructuring.

Subsidies to variable costs

This category includes policies which reduce fisheries operating costs, including tax concessions, for example, tax rebates on purchases of fuel, vessel insurance programmes provided by the government, payment for damages, bait services, extension services, training and transport subsidies.

Government provision of insurance programmes reduces the rates of insurance paid by the fishing industry thus increasing the numbers involved in the fishery. In such cases, the government, which has a broader ability to bear risk, will provide insurance where private insurers refused to operate because of the high costs of doing business (Schrank, 1998).

Programmes included under subsidies to variable costs are:

- fuel tax exemption or rebates for vessels;
- income tax deferral for fishers;
- vessel insurance and reinsurance programmes;
- subsidies to reduce bait prices;
- support of baiting stations;
- extension services consultant advice for fishers;
- compensation for damaged gear;
- transport subsidies;
- interest deduction for liquidity loans;
- support to energy saving devices onboard fishing vessels.

Subsidies to income

Subsidies included in this classification are income support and unemployment insurance (although, as noted above, some analysts may consider such supports subsidies only if they go beyond the normal unemployment insurance for other economic sectors). Also included are payments to vessel owners for temporary stoppages of fishing, or 'laying up' of vessels.

Programmes included in subsidies to income are:

- payments beyond the norm for other sectors to supplement the incomes of fishers and fisheries workers;
- payments targeted specifically for unemployed fisheries workers and going beyond social insurance in other sectors;
- payments for independent fishermen who are idled by restrictions on fishing;
- payments to vessel owners for temporary cessation of fishing.

Price support subsidies

Price support provided to the fisheries sector is to ensure a minimum price level. Within this category are the budgetary transfers used for maintaining the price for fish products including export subsidies and market intervention programmes, as well as border measures (tariffs and tariff quotas) used to maintain the prices of domestic fish above world levels.

Programmes included in marketing and price support are:

- government support to ensure minimum prices or to keep domestic prices above world prices;
- withdrawal of fish from the market to maintain minimum prices.

Analytical Framework

Whether, and how much, impacts on fish resources are associated with a particular category of subsidies depends on both the circumstances in the fishery and the way in which subsidies are provided. It is widely recognized that some categories of subsidies are more distorting than others. The subsidies that are most distorting are those that are contingent on increases in fishing capacity or effort. Least likely to distort production are those that involve lump-sum transfers unrelated to production, prices or input use.

The extent to which a given subsidy results in increasing fishing capacity or effort in a fishery will be determined in part by the management system governing the fishery in which they operate. A detailed assessment of a particular category of fishing subsidies requires the identification of key characteristics of the regulatory situations. Among the broad variables or parameters that distinguish different management regimes are: (i) whether the fishery is managed with effective catch and effort controls (see Box 2.1); (ii) the presence or absence of incentives for sustainable fishing; and (iii) whether the fishery is experiencing overcapitalization and excess capacity.

The incentives for sustainable management may be either economic or a combination of economic and social. Purely economic incentives are provided by

Box 2.1 Catch and Effort Controls

Fisheries managers in many countries have been using a variety of controls over catch levels, fishing effort and inputs in the hope of reducing, if not eliminating, overexploitation of the fishery. The management tools used for this purpose, as catalogued and analysed by OECD (2000b), are total allowable catch (TACs) quotas for the entire fleet in the fishery and/or vessel; catch limits per trip; restrictions on fishing effort (seasonal and spatial closures, limits on the number of fishing trips or days spent fishing per vessel per year); restrictions on the number of fishers allowed access to the fishery through limited licensing, and limitations on the number or types of vessels and types of fishing gear.

Often restrictions on catch, effort and inputs are used in combination with one another. However, they have been consistently ineffective in preventing or reducing overfishing where the economic incentive for a 'race of fish' has not been eliminated from the fishery. The incentive to catch as many fish in the shortest time possible exists in any fishery that has not allocated individual fishing rights to fishers or managed fisheries though community-based institutions. Under these circumstances, fishermen inevitably exceed catch limits while underreporting catches. This distorts the statistical basis of decisions on catch quotas and leads fisheries managers to set those quotas at levels above the carrying capacity of the resource (FAO, 1983). Effort restrictions, of which limiting the duration of fishing seasons is the most prominent, simply result in more 'capital stuffing' – heavy investment in capital equipment aimed at attaining higher catch rates earlier in the season (Corten, 1996; OECD, 1997a). Restrictions on total number of vessels or certain types of gear usually stimulates growth in other factors that are less well regulated (Townsend, 1985, 1992; Smith and Hanna, 1990).

the allocation of 'property rights' – tradable rights to a share of fishing quotas, often called individual transferable quotas (ITQs). Properly designed and enforced ITQs can provide an economic incentive for fishermen to want quotas to be set at levels that are consistent with the sustainability of stocks and to want the quotas to be fully enforced (see Box 2.2). Territorial Use Rights or community management schemes that involve face-to-face relations among fishers may have a parallel effect, by combining the allocation of fishing rights with social pressures to motivate fishermen to limit fishing effort to agreed biologically sustainable levels (see Box 2.3).

The other broad variable affecting the impact of a given subsidy is whether the fishing fleet is overcapitalized or not – in other words, whether the level of fishing capacity in the fishery exceeds what is required for the desired level of catch (see Box 2.4). The experiences of fisheries management in OECD countries shows that overcapitalized fisheries tend to create powerful pressures on fisheries managers to set catch quotas too high and to impose limits that are not strict enough and are too late to prevent serious resource depletion (OECD, 2000c). The persistence of overcapitalization in many of the world's fisheries remains a cause of concern and is one of the key motivating factors behind the inclusion of fisheries in the Doha Declaration and the WSSD Plan of Implementation. Even if

Box 2.2 Property Rights in Fisheries

Experiences in many OECD fisheries have shown that a particularly effective approach to altering the economic incentive for a 'race of fish' can be the allocation of 'property Rights', in the form of individual transferable fishing quotas (ITQs), to fishers. Such tradable individual quotas represent shares of the total allowable catch and are a longer-term economic asset which gives the fisher an incentive to harvest his individual share at the lowest cost possible and to contribute to conservation of fish stocks.

By the late 1990s, the OECD (2000b) had identified 49 fisheries in 10 OECD countries that had been managed by the use of ITQs, of which more than a third were in New Zealand. Although it does not eliminate the danger of TACs being set too high, and still requires high levels of monitoring and enforcement, the use of ITQs does appear to reduce cheating on quotas and to reduce substantially the incidence of competitive behaviour among fishers (Geen and Nayar, 1988; Clark, 1993; Grafton et al, 1996).

It should be noted, however, that property rights approaches to management have not yet been universally accepted. The effectiveness of property rights regimes may depend on specific cultural, commercial, and economic conditions. Moreover, the means by which property rights regimes are implemented can have significant implications for the social organization of a fishery, including by encouraging concentration of vessel or license ownership.

Box 2.3 Territorial Use Rights and Community-based Management

In relatively small-scale fisheries, rights to fish can be allocated by local communities or cooperatives through systems of Territorial Use Rights in Fisheries (TURFs) or community-based management, rather than by central or state governments. In these alternative forms of management, fishing communities and cooperatives can effectively manage capacity and effort when all the fishermen in the fishery have face-to-face interactions and can bring effective sanctions to bear against violations of locally agreed rules. In such community-based TURF systems, fishing rights are normally not transferable, so that holders are always subject to community pressures.

Once a fishing community or cooperative realizes that overfishing in a well-defined area endangers its livelihood, community-based systems of rights can make necessary adjustments in capacity and effort. Instead of competing for very limited resources, communities can limit access by defining qualifications for membership in the cooperative, reduce and control fishing effort so that it is sustainable by pooling fishing opportunities. Japan has a number of successful examples of TURFs and community-based systems for managing fisheries that have been fully integrated into commercial fish markets, in part because of a cultural heritage of strong community institutions. All fishermen in Japanese coastal communities belong to fishing cooperatives with an average of about 250 members, which are able to make authoritative decisions on fishing that local officials cannot change (FAO, 1993b).

Box 2.4 Defining and Measuring Overcapacity

In the fisheries literature, fishing capacity is generally defined as the maximum available capital stock in a fishery – vessels, engines, gear and equipment – that is fully utilized at the maximum technical efficiency in a given period. In order to measure that capacity against the desirable level of capacity, however, that capital stock must be translated into its potential output – the tonnage of fish that it would catch, assuming constant returns to scale (FAO, 1999d). This is a theoretical level of catch achieved by multiplying the potential catch of the average vessel at maximum efficiency and multiplying it by the total number of vessels. In the real world, of course, if there is too much capacity in the fishery, each boat catches far less than its maximum potential. But establishing this theoretical level of total catching power is necessary to compare actual fishing capacity with the desired level. In almost all cases, the former is much greater than the latter and that difference is overcapacity.

The ideal level of capacity is that which, when fully utilized, would yield a level of catch that maximizes the difference between fleet costs and revenues over time. This level of output is called maximum economic yield (MEY). One measure of overcapacity, therefore, is the ratio of actual capacity to the level of capacity needed to produce MEY. A somewhat higher but still desirable level of output would be the highest level of catch that can be replaced through compensatory changes in the fish population, which is called maximum sustainable yield (MSY). Thus, a second measure of overcapacity is the ratio between the actual level of capacity and the level required for MSY.

Calculating precisely either the theoretical capacity of the fleet or the desirable level of capacity involves a number of technical issues, the most difficult of which is determining what constitutes a biologically sustainable level of catch. Indeed, more often than not catch quotas have been set at levels that turned out to be too high to permit the necessary compensatory changes in fish population, and thus contributed to the depletion of fish stocks. For the purpose of determining the general impact of subsidies on fish stocks, however, a number of gross indicators of the existence of some overcapacity in and overexploitation of a fishery are available that do not require such calculations. Examples of such gross indicators of overcapacity and overexploitation in a given fishery include any evidence of declining biomass, seasonal limits on fishing, decommissioning schemes, laying up subsidies and social schemes for fishermen to compensate for loss of fishing opportunities. Where any one of these indicators is present, it may be reasonably assumed that the fishery has already experienced overcapacity. Very few fishing states in the world have not already reached the point of overcapacity in the exploitation of their commercial fish stocks.

subsidies do not increase capacity or effort in a given fishery, because of severe restrictions on fishing opportunities, they are likely to delay or prevent the decrease in capacity and effort required to achieve sustainability. The present analytical framework therefore includes as a major parameter the level of fishing fleet capacity in relation to fish stocks, which is also referred to as the 'bio-economic conditions' of the fishery.

The combinations of these two broad variables must be considered in order to understand the implications of each category of subsidies for the health of the fishery resources. The analytical framework used in this chapter is based on the matrix approach proposed by the OECD (1993) and by a UNEP study (UNEP, 2002b) to take into account these different combinations of management and bio-conditions of the fishery in analysing the impacts of fishing subsidies on fishery resources. The matrix approach has been further refined by adapting a framework developed for the OECD's study on the liberalization of fisheries markets (OECD, 2003). That study postulated three stylized management regimes – 'open access', 'catch control' and 'effective management' as the basis for evaluating the impacts of fisheries subsidies in general.

In the OECD study, these three ideal types of regimes are assumed to represent management regimes that are effectively enforced. In the analytical framework used for this study, however, this assumption has been relaxed to reflect the fact that in the real world, management systems are not perfectly implemented. Indeed, experiences in a number of OECD countries have shown that in those fisheries which have not allocated property rights and in which fleet capacity already exceeds a level required for a sustainable level of catch, the countries have been unable to prevent widespread cheating on catch quotas and serious overexploitation of fisheries (Commission of the European Communities, 1991; Sutinen et al., 1990; Karagiannakos, 1996; Nielsen and Joker, 1995; Jensen and Vestergaard, 2000).

In this analytical scheme, therefore, the category of 'effective management' assumes that adequate systems of catch and effort controls as well as monitoring and enforcement systems are supplemented by incentives for compliance with fishing regulations. Thus an effective management system in a particular fishery is one in which legal and regulatory systems operate in such a way as to eliminate or radically reduce the 'race for fish' incentives and in which cheating on quotas is practically eliminated. Additionally, 'effective management' assumes that data constraints and scientific uncertainties have been sufficiently overcome or met through application of a precautionary approach.

This chapter assumes that such 'effective management' can only be accomplished in the presence of some kind of property rights regime (see Box 2.2). The definition of 'effective management' used here is based on the reduction or elimination of economic incentives to overfish, regardless of the means by which that result is achieved. The analysis here thus does not go into the details regarding the merits or draw-backs of specific property rights regimes.

In today's world, 'effective management' regimes remain an ideal objective rather than a common reality. Thus, in reviewing the matrices developed throughout this study it is important to recall that the third of each matrix devoted to 'effective management' conditions represents only a tiny fraction of real world circumstances.

Thus the three management regimes types in the OECD study have been translated into three real-world management regimes. The 'open access' regime is one which combines the absence of catch and effort controls and the absence of an effective incentive scheme; the 'catch control' regime combines (imperfect) catch and effort controls and the absence of incentives for sustainable fishing, and the 'effective management' regime combines effective incentives and is relatively more effective (because of the effects of the incentives) catch and effort controls. On this scale of possible management regimes, a few OECD countries have fisheries that approach the 'effective management' regime, but most OECD countries are clustered around the 'catch control' model. Most, but not all, developing country fisheries management systems have few catch and effort controls and little monitoring and surveillance systems or enforcement and are closer to the open access model.

It is generally accepted in the fisheries economics literature that the existence of well-defined and enforced property rights provides a strong incentive for fishers to exploit fisheries resources in an efficient and sustainable manner (OECD, 1997a). Fisheries managed by well designed and fully enforced Individual Transferable Quotas (ITQs) should not experience higher levels of fishing effort and catch as a result of a subsidy unless the subsidy is accompanied by an increase in the Total Allowable Catch (TAC). Without an increase in the TAC, the subsidy would be a straight transfer to the industry (Arnason 1998; Hannesson, 2001). If a set of instruments other than ITQs could effectively control the level of effort at the profit maximizing level, they would also negate the effect of a subsidy in expanding effort and capacity. However, few management systems have demonstrated the ability to do this.

Each major category of subsidy is assessed in the context of each combination of management regime and levels of exploitation, and the results for each of the nine combinations associated with each category are summarized in a matrix. The template for this matrix is shown in Table 2.1. Across the top of the matrix are the three different management parameters, reflecting the presence or absence of property rights and whether catch and effort controls have been imposed. The left hand side of the matrix refers to the bio-economic parameters, indicating whether the fishery is overcapitalized or not fully utilized. Within each cell of the matrix, the effect on fisheries resources of the provision of the particular category of subsidy is assessed as either 'harmful'; 'possibly harmful' or 'not harmful'. These assessments reflect the degree of certainty that can be assigned to the subsidy category under that combination of parameters.

In the case of certain combinations of subsidy type and management and bio-economic parameters, the theoretical case does not arise in the real world, for reasons that can be easily explained. In these cases, the cell in the matrix is labelled 'not applicable'.

In order to translate the effects of a particular category of subsidy on the behaviour of a vessel owner into an impact on fish stocks, this analysis has had to make certain simplified assumptions about the relationship between changes in the level of fishing effort and changes in the biomass of fish stocks. Fish stocks, defined as total fish biomass, fluctuate considerably over time and fluctuations are determined by changes in water temperature and other variables exogenous to the management regime and the level of fleet capacity. There are also cases in which a fishery that has been overexploited and has suffered a collapse of fish stocks and is subject to a moratorium on fishing in order to allow for recovery. However, in

Table 2.1 *Matrix Template*

	Effective Management Catch Controls	*Open Access*
Overcapacity		
Full capacity		
Less than full capacity		

this simplified scheme, the assumption is made that in a fishery that already has overcapacity, any increase in fishing effort induced by the addition of a fishing subsidy will have some negative impact on fish stocks. It is believed that this assumption reflects the reality for most fisheries most of the time.

Analysis of Impacts by Category

Subsidies to fisheries infrastructure

The construction and maintenance of fishing ports provides a service to the fishing industry for which it would have to pay in the absence of government funding of the costs. Because the fishing industry has seldom had to spend any money on fishing ports, infrastructure subsidies represent a large hidden cost of fishing that has been avoided. That means that the cost curve for each level of fishing effort is shifted upward, so a higher level of fishing is profitable in most cases.

It can be argued that fishing ports have some public goods elements, particularly because they are more likely to protect public health and landings. Government administration of such facilities, therefore, is better for the public than private industry administration. The issue is not whether governments build and run the ports, but whether industry pays its fair share – presumably the largest share – of the costs of construction and maintenance, as is already the case in some countries.

To analyse the effects of infrastructure subsidies to the fisheries sector in the context of different management and bio-economic parameters, an assumption must be made about how much the fishing industry would have spent on fishing infrastructure in the absence of the state subsidy. It need not be assumed that the industry would have spent the full amount. Some fraction of the infrastructure spending may exceed what the industry would need to be profitable. In the Philippines, an analysis of fishing ports (Israel and Roque, 2000) concluded that most of them represented 'excess capacity' for fish landing and processing, partly because such projects are funded on a political basis, and partly because of unanticipated declines in fish catch. The following analysis assumes, therefore, that the industry would have had to spend most, but not all, of the amount spent by the state on fishing ports and related infrastructure.

Perhaps the most ambitious such programme has been financed by Japan's Ministry of Agriculture, Forestry and Fisheries (MAFF). The MAFF announced in 1994 that its Long-Term Plan for Fishing Ports, running from financial year

(FY) 1994 through FY 1999, would require US$5 billion annually (Japan MAFF, 1994). The total commercial value of Japan's fisheries production in 1996 was approximately US$18 billion (OECD, 2001). Assuming that the allocation of fishing ports reflects political influence as well as the needs of the fishing industry, however, this figure inflates the actual cost savings to industry.

Many other countries have relatively ambitious programmes of building fishing ports for their fishing industries, although the costs would not represent such a high proportion of total fishing costs or of the commercial value of the catch. Indonesia is in the process of expanding two existing ports, with funding from Japan, at a cost of US$45 million (Australian Trade Commission, 2003). The Philippines is still constructing regional fishing ports in new areas (Israel and Roque, 2000).

A distant water fleet vessel owner may find deep sea fishing in a given pelagic fishery more attractive because of access to a fishing port. It is also possible that the same vessel owner would not enter into the pelagic fishery if access to the port were to cost. This effect of infrastructure subsidies on fishing decisions would affect only those with a particular cost structure, rather than all those interested in deep sea fishing.

The effect of infrastructure subsidies on fishing capacity and the effort in a given fishery depends on the management and bio-economic conditions in which the fishery operates. It is also affected by the type of fishery (in-shore, coastal zone or deep sea). The differences in effect given different combinations of these conditions are examined below.

Effective management

Fisheries that are managed with individual tradable fishing quotas have no economic reason to subsidize the fishing industry by providing fishing ports to them free of charge. However, a state may be motivated to provide fishing ports or other infrastructure projects to a fishing industry as an inducement to accept a new management system that includes individual transferable quotas.

As long as the fishermen are assured of a total catch level, they will have no incentive to increase fishing effort. Instead, they will take advantage of the subsidy to fishing infrastructure to minimize their costs and increase their profit at the existing level of effort (Arnason, 1998; Munro, G. and Sumaila, U.R., 1999; Hannesson, 2001). The same is true of a fishery that is under community-based management.

Catch controls

Reducing the cost of infrastructure to the fishing industry would not alter the producer's plans for fishing capacity and effort. The incentive for increasing capacity or effort would remain the same, because the profitability of a given resource allocation decision would be no different from the profitability of the same decision without the subsidy. Therefore subsidies to fishing infrastructure do not have any impact on fishing effort or capacity for vessel owners already in the fishery.

Table 2.2 *Expected Impact of Subsidies to Fisheries Infrastructure on Fish Stocks*

	Effective Management	*Catch Controls*	*Open Access*
Overcapacity	Not Harmful	Harmful	Harmful
Full capacity	Not Harmful	Harmful	Harmful
Less than full capacity	Not Harmful	Not Harmful	Not Harmful

However, subsidies to infrastructure projects make fishing more profitable regardless of the existing level of effort and therefore provide an incentive for vessel owners to remain in the fishing industry, if they had been below the level at which continued fishing is financially viable in the absence of the subsidy. If the fishery is already at full exploitation or overexploitation, the incentive effect of the subsidy will be to discourage disinvestment that otherwise would have reduced overexploitation. For a fishery that is still underexploited but has catch and effort controls, the impact of the subsidy would not be harmful to fish stocks.

Open access

In the vast majority of fisheries in which the catch is not controlled, fishing capacity is already well above the level needed for a biologically sustainable catch. In an open-access fishery that is already fully exploited, and which is likely to experience continued growth in capacity in the future, infrastructure subsidies would not have an immediate impact on stocks but would be a future liability with regard to efforts to reduce capacity. Therefore, it can be considered harmful to fish stocks in the longer term. If the fishery is not fully exploited, the eventual impact would depend on whether the dynamic in the fishery is undergoing relatively rapid increases in capacity or not. If the fishery is clearly moving toward full exploitation and beyond, it could be considered as potentially harmful. If not, the risk to fish stocks would be small.

Management services and research

Management services to the fisheries sector include setting fishing regulations, surveillance and enforcement of those regulations, stock assessments, and a wide range of research on fish habitats, fishing technology, market issues and other topics. Together, these services constitute an estimated 36 per cent of the total financial transfers from OECD governments to the fisheries sector (Flaaten and Wallis, 2000). Whether the costs of these services should be borne by the fishing industry and the uncompensated provision of the services considered as a subsidy has been the subject of vigorous debate. To the extent that these services provide conditions under which output can be increased over time, costs per unit of effort reduced and return per unit of output maximized, they benefit the fishing industry but they also benefit consumers more generally from the sustainability of fish supplies over time (Hayes et al, 1986).

Under these circumstances, it is argued that the public interest requires government intervention, thus distinguishing the issue from one of discretionary service to the industry (Hannesson, 1991). Based on this view, most countries have not demanded that commercial fishing interests pay any of the costs of management services. It can also be legitimately argued, however, that the fishing industry is the main beneficiary of management services. Australia, New Zealand and Canada have already adopted for a cost recovery policy in regard to management services in the fisheries sector based on that rationale and in Canada user fees collected from license holders pay for part of the costs of those management services (Hatcher and Pascoe, 1998; Pascoe et al, 2002). A potential danger of the cost recovery approach, however, is that the industry may seek greater influence over the management regime if it is paying its full costs and may obtain increased opportunities for rent-seeking (Hatcher and Pascoe, 1998).

Failing to recover the costs of management services that are in the public interest is not a direct subsidy to industry. Failure of producers or consumers to internalize the externalities of the production of a particular product, however, has been considered by some economists to be an 'implicit subsidy' (OECD, 1996; Reijnders, 1990). Thus, it may be argued that failure of producers to pay the full social costs of sustainable fisheries management is an implicit subsidy.

Some fisheries research programmes are clearly for the benefit of the industry. These include programmes aimed at improving the marketing of fish products and identifying new and unexploited fisheries. Norway conducts research on fishery technology and on market issues that are of particular interest to the fishing industry (Myrstad, 1996). Similarly, the United States funds some research and feasibility studies that would otherwise have been paid for by the industry (OECD, 2000a). Even research on new, more selective fishing technologies that are clearly in the public interest, could also be viewed as directly affecting the fishery resources on which the future of the industry depends. Canada has set aside up to 5 per cent of the coast-wide total allowable catch on its Pacific coast to pay for selective fisheries projects (Fisheries and Oceans Canada, 2001). However, it is difficult to distinguish in principle between such research programmes and stock assessments because both are also clearly in the public interest.

It is not always easy to determine whether fisheries research programmes are subsidies to the industry or whether the industry would have borne the costs in the absence of government payment for them – and the answer might change over time. Until very recently, the fishing industry in OECD countries probably would not have been willing to finance research on selective gear, but that may be changing.

Most fisheries management services are considered, for the purpose of this analysis, to have no negative impacts on fisheries resources, on the grounds that government-funded services have, on balance, at least the potential for greater restraint on exploitation of fisheries than the financing of those same services by the industry. This generalization applies across all bio-economic parameters and all management parameters. Research programmes that are not necessary for the

Table 2.3 *Expected Impact of Provision of Management Services on Fish Stocks*

	Effective Management	*Catch Controls*	*Open Access*
Overcapacity	Not Harmful	Not Harmful	Not Harmful
Full capacity	Not Harmful	Not Harmful	Not Harmful
Less than full capacity	Not Harmful	Not Harmful	Not Harmful

regulation of fishing and directly benefit the industry, however, are considered to be similar in impact to government financing of fisheries infrastructure, albeit on a smaller scale. The matrix in Table 2.3, therefore, should be understood to refer to dominant components of fisheries management services.

Subsidies for access to foreign countries' waters

A subsidy provided to a fishing fleet through an international fishing access agreement operates in the same manner as a subsidy to any other variable cost, reducing the costs of fishing per unit of effort for the distant water fishing fleet being subsidized.[1] Unless other factors prevent it, such a subsidy will provide an incentive for the distant water fleet to invest in greater fishing effort in that country's waters. Beyond the incentive to the distant fleet of reduced fishing costs, the subsidy may cause increased fishing in those waters by giving the coastal state government an incentive to allow more distant water fishing capacity than would have been the case in the absence of the subsidy. The actual impact of the subsidy on fish stocks will depend on the management and bio-economic conditions in the country that has given access to the foreign fleet.

Effective management

International fishing access agreements normally cover relatively short time periods (usually four or five years). At present, no state allows foreign fleets to obtain domestic fishing quotas. It is theoretically possible, however, for coastal states entering into such agreements with distant water states to offer property rights, such as tradable individual catch quotas or other effective means of eliminating economic incentives to overfish, to distant water fleets. Under such ideal circumstances, subsidizing the costs of access for distant water fleets would not provide an incentive for increased fishing effort. It would not have any impact on fish stocks, regardless of the previous level of exploitation of stocks. In reality, no such case of a coastal state with effective management of a fishery signing a fishing agreement with subsidies to access for a foreign fleet is likely to arise.

Catch controls

In the ideal situation of a completely effective system of catch controls, in which biologically safe catch quotas are imposed on the fishery and are effectively enforced on all fishers, the incentive effects of the subsidy to fishing access would not result in any increase in catch beyond a sustainable level. In the real world, however, no states signing bilateral or multilateral fishing agreements in which subsidies are embedded have had management systems with effective catch controls, even though catch quotas are sometimes imposed on the distant water fleet. Even if catch controls were as effective as they are in OECD countries, however, the risk that subsidies to distant water fleets would cause additional harm to fishery resources would be very high. And in the real word, catch controls suggest that the fishery is already at or above full exploitation.

Open access

Without a system of catch and effort controls, backed by adequate monitoring, surveillance and enforcement capabilities to ensure that the catch by distant water fishing fleets remains within biologically sustainable limits, subsidies to access foreign countries' fishing waters will induce a higher level of catch than would have been the case in the absence of the subsidy. This might take the form of increased effort by subsidized vessels that were already in the fishery, a larger number of vessels from the subsidizing distant water state's fleet entering the coastal state's waters, or both. If the level of total capacity in the fishery is still well below the level sufficient for the maximum sustainable yield, the increased fishing effort caused by the subsidy will not have any short-run impact on the health of stocks. However, if the fishery is already fully exploited or overexploited, the subsidy will result in some harm to fish stocks.

This combination of parameters describes the management system in the countries that have actually entered into international fishery agreements that provided subsidized access to their fishing zones. The African states that have allowed EU vessels access to their fishing zones under EU-African fishing agreements have not set catch quotas based on the known state of the fish stocks (WWF, 1997). Systems of monitoring and enforcement, moreover, have been virtually nonexistent, and African coastal states have obtained little data on the level of catches and even of fishing effort of distant water fleets (Sub-Regional Fisheries Commission, 1997; Acheampong, 1997).

In most of the fishing zones covered by EU bilateral fishing agreements with West African coastal states, total fishing fleet capacity was already excessive in the 1980s (Fishery Committee for the Eastern Central Atlantic, 1992). Studies of exploitable biomass of demersal fish off the coast of Mauritania and Senegal indicate severe and continuous reductions between the early 1970s, the early 1980s and the early 1990s (Johnston, 1996; Bonfil et al, 1998; UNEP, 2002c). The fisheries affected by these agreements have therefore been overexploited for many years. Distant water fleet fishing that has been permitted under those agreements has continued to deplete the fishery resources of African coastal countries. Nevertheless, because of reduced fishing costs, distant water fleets can continue to fish profitably for a longer period, in spite of the depletion of the resources, than would have been the case in the absence of the access subsidies (WWF, 1997).

The Pacific Island states that signed a fishing access agreement with the United States in 1987 have not imposed either catch or effort limits on its distant

Table 2.4 *Expected Impact of Subsidies for Access to Foreign Countries' Waters on Fish Stocks*

	Effective Management	*Catch Controls*	*Open Access*
Overcapacity	Not Harmful	Harmful	Harmful
Full capacity	Not Harmful	Harmful	Harmful
Less than full capacity	Not Harmful	Not Harmful	Not Harmful

water fleets (Aqorau and Bergin, 1997, 1998; Hunt, 1997; Petersen, 2001). The subsidy to the US tuna seiner fleet's access presumably caused an increase in the fleet's capacity and effort in the fishery relative to an unsubsidized scenario, by increasing its attractiveness in relation to the other option available to the fleet: the Eastern Tropical Pacific tuna fishery. By reducing costs at each level of effort, it also provides an incentive for more trips and therefore higher catch levels than would have been the case in an unsubsidized scenario.

However, the US and other distant water tuna seiner fleets in the Pacific Island region are targeted on skipjack and yellowfin tuna. Thus far, no evidence of generalized overexploitation of those species has come to light, but serious overcapitalization had already been reported in Fiji's fishing zone and improvements in statistical coverage in some critical regions are needed (Anonymous, 2001, Silbert, 1999; Coan et al, 2000). It can no longer be assumed automatically that subsidies to Pacific Island States' tuna fisheries are not harmful.

Subsidies to decommissioning of vessels and license retirement

Decommissioning schemes have represented a significant share of government transfers to the fishing industry for many years, as nearly every fishing state has financed programmes to buy back vessels, and sometimes licenses, in order to reduce overcapacity in their commercial fishing fleets, increase the efficiency of their fisheries and transfer income to the fishing industry. Vessel and license buy-back programmes represented the second largest category of subsidies in the OECD countries in 1996–97 after fisheries infrastructure (OECD, 2000b).

Decommissioning and license withdrawal programmes influence the economic behaviour of the fishing industry on multiple levels. Although their objective is to reduce the level of fishing capacity in a fishery, decommissioning schemes also have unintended impacts on industry behaviour that undermine that objective. The mechanism on which the programmes rely is a cash transfer to vessel and/or license owners for withdrawing either a vessel or license or both. This in turn creates quasi-rents in the fishery by increasing catch per vessel and per unit of effort. Given the 'race for fish' in any but an ideally managed fishery, however, this rent in fishery will spur efforts to capture more of the rent through 'capital stuffing' by the vessel owners remaining in the fishery (Townsend, 1985).

The removal of some active capacity from the fishery also increases demand for idle vessels in the fishery (Gates et al, 1997a; Holland and Sutinen, 1998).

The vessel owner who receives the decommissioning premium has an economic incentive to use the additional capital to reinvest in the same fishery or another that is well regulated. If the decommissioning scheme gives vessel owners the discretion to resell the vessels to be withdrawn, rather than scrapping them, it further increases the incentive to invest in more additional fishing capacity.

The differences in technological capabilities between newer and older vessels of the same tonnage and engine power is so great that decommissioning programmes are virtually certain to increase the level of fleet capacity if they target the oldest or least productive vessels in the fishery, and then allow a lesser number of new vessels to enter the fishery based on a formula using similar physical characteristics (De Wilde, 1999; Coglan et al, 2000; Eggert, 2001; Pascoe et al, 2002).

Beyond these direct effects of the availability of additional capital and the temporary increase in rents, moreover, it has been widely observed that the existence of vessel buy-back programmes encourages vessel owners and potential investors to believe that the risk of additional capital investments in fishing is significantly reduced, even if stocks have been or are being depleted. This belief would tend to increase investment in the fishery or to discourage disinvestment from it (Gates et al, 1997a; Arnason, 1998; Munro, 1999; Jorgensen and Jensen, 1999; Munro and Sumaila, 1999; OECD, 2000b). However, no statistical methodology exists to estimate such an indirect effect. Unless the management regime discourages additional capacity through ITQs or community-based management, or by tight controls over technological improvement and increased effort, over time decommissioning subsidies will not prevent and will even contribute to the replacement of all the withdrawn capacity and the addition of more capacity. Even a programme that ostensibly purchases destructive fishing technologies, such as the Indonesian buy-back of trawlers, would pose the problem of decommissioning premiums being used for reinvestment in another overexploited fishery.

It cannot be safely assumed, therefore, that the net impact of decommissioning subsidies on a fishery will be positive. Case study data are available on 11 different decommissioning schemes in 9 different countries. Table 2.5 summarizes the data on these 11 case studies. Some of these case studies do not provide quantitative estimates of net results of the programme in terms of changes in capacity or effort in the fishery, much less the impact on the health of commercial fish stocks. But for each case study at least some data are available on the major policy conditions associated with the programme and on the general situation regarding capacity and effort in the fishery or fisheries affected. The discussion of the implications of different management systems for the effects of decommissioning subsidies on economic behaviour and on fish stocks draws on this body of case study evidence.

The analysis of impacts under different bio-economic and management parameters below does not consider the possibility of anything other than overcapacity in the fishery. Vessel and license buy-backs are invariably a sign of overcapacity in the fishery, because the incentive for decommissioning schemes

Table 2.5 *11 Decommissioning Schemes: Policy Conditions and Results*

Decommissioning Programme	*Payment based on vessel catch or revenues*	*Retirement of vessel from fishing required*	*Reduction of total capacity achieved*	*Reduction in total effort achieved*	*Increased capacity/effort in specific fisheries encouraged*
Denmark	No	Yes	Yes	No	No
Norwegian Purse Seine	No	Incentives provided	Yes	No	No
United Kingdom	No	No	No	No	Yes
Chinese Taipei	No	Yes	No	No	Yes
Netherlands	No	No	No	No	Yes
US NE Multi-Species Groundfish	Yes	Yes	No	No	Yes
Canada Atl. Groundfish	Yes	No	No	No	Unclear
Canada Inshore Lobster	Yes	No	No	No	Probably
France Scallop	No	No	Yes	No	Unclear
Japan Akita	No	Yes	Yes	Yes	No
Japan Shimane	No	Yes	Probably not	Probably not	Probably

Sources: Denmark: Lindebo, 2000; Danish Directorate of Fisheries, 2001; Gates et al, 1997b; Salz, 1991. Norwegian Purse Seine: OECD, 2000a; Trondsen, 1999; Asche et al, 1998; Hannesson, 1992. United Kingdom: Nautilus Consultants, 1997; Banks, 1999; Gates et al, 1997b; Greboval and Munro, 1999. Chinese Taipei: Chuang, Zhang, 1999; WWF, 1998b. Netherlands: Court of Auditors, 1994; De Wilde and van Beek, 1996; Frost et al, 1996; De Wilde, 1999; Gates et al, 1997b. United States NE Multispecies Groundfish: Thunberg, 2000; Kitts et al, 1999; Walden and Kirkley, 2000; Gates et al, 1997b. Canada Atlantic Groundfish: Auditor General of Canada, 1997; Gates et al, 1997b. Canada Inshore Lobster: OECD, 2000a; Gates et al, 1997b. France Scallop: Guyader et al, 2000; Daures and Guyader, 2000. Japan: OECD, 2000b.

would not exist in the absence of such severe overcapacity in the fishery. Therefore it is assumed that all fisheries in which decommissioning schemes are implemented have already experienced significant overcapacity. The economic dynamics of decommissioning subsidies depend on whether the economic incentive to 'race for fish' remains imbedded in the management system of the fishery, whether effective control over catch levels are exercised and on the degree of overcapacity in the fisheries.

Effective management

If the management system for a fishery includes both effective overall controls on catch and effective incentive systems, a vessel decommissioning or license withdrawal programme should be successful in substantially reducing capacity and effort in the fishery. In this case, neither the additional capital obtained by the industry nor the creation of rent in the fishery by removing excessive capacity would cause vessel owners to make additional investment or to increase effort in the fishery.

Owners who are assured of a definite share of a specified catch level and who know all other owners are similarly allocated a definite share, would be able to allocate fishing effort more evenly through the entire fishing season. Less capacity would be required, therefore, to catch a given level of fish. Costs per unit of effort would thus be reduced and overall profit would increase. Without a 'race for fish', vessel owners would have no incentive for 'input stuffing'. Thus, the capacity withdrawn from the fishery by a decommissioning programme would not be replaced by idle capacity or by new capacity financed by the decommissioning premiums.

If property rights or other economic disincentives to overfish have been established in a fishery management system, however, it would not be necessary to use decommissioning payments to reduce fleet capacity in the fishery to the level required to catch the quota at the least cost. Under an Individual Transferable Quota (ITQ) system, the adjustment in capacity level would take place even without a vessel-buy back programme. Because the incentive is to catch the quota level at the least cost, excess capacity would be withdrawn voluntarily, as has been shown in a number of fisheries that have adopted ITQs (Geen and Nayar, 1988; Gauvin et al, 1994; Grafton, 1996; Wang and Tang, 1996; OECD, 1997a; Trondsen, 1999; Runolfsson and Arnason, 2001).

Although ITQ-based fisheries have made major adjustments in capacity in as little as a year or two in several cases, it could take several years to remove excess capital from a fishery under certain circumstances (Squires and Kirkley, 1991). A decommissioning scheme may help to speed up the process of capacity adjustment in a rights-based fishery, but it is not necessary for such adjustment.

The only two documented examples of a decommissioning subsidy programme implemented in a rights-based fishery are the Norwegian Purse Seine decommissioning scheme and the US vessel buy-back programme for the Bering Sea Pollock Fishery. In both cases the decommissioning scheme was either completed or mostly completed when the rights-based system of quotas became fully effective.

In the case of the Norwegian Purse Seine Fleet, the first step toward property rights, taken in 1973, was to require licenses that entitled the owner to a certain level of catch depending on the cargo capacity of the vessel. Initially, however, the licenses were not transferable. By the mid-1980s, there was an actual market in such licenses, providing an incentive for vessel owners to further reduce the number of vessels in the system (Flaaten et al, 1995). In 1990, the transfer of such licenses was legally accepted for the first time (Asche et al, 1998).

The decommissioning programme, which began in 1979, reduced the number of purse seine vessels from 202 to 90 and the licensed catch capacity of the purse seine fleet from 1.3 hectolitres of fish in 1973 to 0.8 hectolitres from 1978 to 1990 (Asche et al, 1998; Hannesson, 1992). From 1990 to 1992, the number of purse seine vessels fell still further from 90 to 63 (Asche et al, 1998). The effectiveness of the decommissioning programme was undoubtedly enhanced by the existence of a license transfer system that allowed owners to buy up catch rights and achieve more catch with less capacity and cost. The number of vessels withdrawn was also boosted by export subsidies for vessels (Flaaten et al, 1995).

The US Bering Sea Pollock Fishery was among the most overcapitalized fisheries in the United States in 1998, with more than five times the fishing capacity needed to catch the allowable quota for the fishery in 1998 (WWF,

1998b). The US buyback programme for the fishery purchased 9 of the 29 factory trawlers operating in the fishery in 1998, but they accounted for only 10 per cent of the fleet's catch capacity (US GAO, 2000), which would have left the fishery still vastly overcapitalized and would not have reduced effort. This programme is distinguished by the fact that most of the costs of the buy-out were borne by the companies themselves through a fee on the inshore Pollock sector (US GAO, 2000; APA, 1999).

However, the legislation mandating the programme also established a fishing cooperative by the remaining 20 factory trawler owners, which allocated individual catch quotas to its members and thus ended the race for fish.[2] In the following season, without the pressure of competition for the catch, only 16 of the 20 remaining factory trawlers in the fishery fished during the winter/spring season and only 14 fished in the summer/fall season. More importantly, the average daily catch rates for the fleet during the winter/spring season were only one-third of the catch rate of the fleet during the same season during the previous year (APA, 1999). The combination of decommissioning and property rights thus achieved a major reform of fishing effort that should have been beneficial to Pollock stocks. Without the creation of individual catch quotas, the short-run capacity reduction achieved by the decommissioning subsidy would have been much smaller and would have had little or no impact on fish stocks.

Catch controls

In a perfect system of catch controls, with biologically safe catch quotas and completely effective enforcement, a decommissioning programme would not contribute to additional pressure on the stocks, even without effective management to curb the race to fish. It could not guarantee that the programme would actually reduce the level of capacity in the fishery, however, unless it included an effective system to prohibit additional capacity being introduced into the fishery through improvements in vessels and gear.

In the real world, in fisheries with systems of catch control but no property rights or community management, catch quotas and enforcement systems have not prevented stocks from being overfished. Under these circumstances, the race for fish has continued, as vessel owners seek to maximize their share of the catch. Limited access, in the form of caps on licenses and controls on effort through limits on days at sea and seasonal closures, has come only after excessive capacity and overfishing has caused stock collapses. In this combination of bio-economic and management conditions, decommissioning schemes are unlikely to help solve the overcapacity problem and could make it worse. If the fishery is fully exploited but not clearly overexploited, there is normally little reason for a decommissioning scheme, but the result is likely to be increased rather than reduced capacity, since the incentive for new investment in the fishery would be even greater under such a combination of management and bio-economic conditions.

When rent is created in a fishery without property rights by the withdrawal of some capital, extremely tight controls over increased effort and technological improvements are required to prevent the temporary gain from being nullified by the race for fish. That task becomes virtually impossible if there is a very large overhang of idle capacity licensed for the fishery or fisheries, or if the programme is geared towards the withdrawal of older and less efficient vessels. When large overcapacity exists, most vessels in the fishery are operating at low levels of efficiency and increased effort can easily swallow up any short-term effort reduction achieved by the programme (Lindebo, 1999; Eggert, 2001; Pascoe et al, 2002).

The danger of a programme that permits some replacement of capacity is even greater when subsidies for capital cost are allocated across an entire national fleet without regard to the impact on fleet capacity in each specific fishery. In that case, the replacement capacity is most likely to flow into those fisheries that are already most highly exploited and where the danger of accelerated depletion of fish stocks is greatest (Nielsen, 1992; Banks, 1999).

When a new vessel, or an older vessel with modern gear, replaces an older vessel withdrawn under a buy-back programme, it always represents an increase in capacity. A study of the English Channel beam and otter trawl fisheries found that the capacity of newer vessels was 1.6 per cent greater for every additional year of difference in vessel age (Pascoe and Coglan, 2002). Thus a new vessel would have 40 per cent more fishing capacity on average than a 25-year old vessel with the same tonnage and engine power.

Table 2.5 shows that only a few of the 11 decommissioning programmes in fisheries without property rights (or other means for eliminating economic incentives to overfish) achieved any net reduction in capacity in the fishery, and that only 1 (Japan Akita Province) achieved any reduction in the level of fishing effort. These programmes failed to prevent the erosion of the initial capacity reduction achieved through vessel and license buy-backs, because of the combination of incentive effects, additional capital introduced into the fisheries by the decommissioning premiums and the insufficiency of controls over increased capacity or effort.

Of these 11 programmes, only 3 based the purchase of licenses or vessels on actual catch history or historical revenues in order to ensure the removal of the greatest capacity for the money spent. The others were biased in favour of withdrawal of older vessels that had been less active or less efficient and would have been most likely to exit from the industry within a relatively short time even without a decommissioning programme.

Except for the programmes in Chinese Taipei and Japan, all these programmes allowed the vessel owner to resell the vessel outside the fishery or nearby fisheries, thus adding to the effective premium obtained for withdrawal. In many cases, the decommissioning scheme led to the transfer of capacity into fisheries that were less regulated, often on the high seas (OECD, 2000b; Gates et al, 1997a). The likelihood that a decommissioning scheme will not succeed in achieving any net reduction in active capacity, or even a reduction in total capacity licensed for the fishery, increases if the programme purchases only the withdrawal of the least active and productive vessels (Pascoe et al, 2002). Vessel buy-backs that are biased towards retiring the oldest and least efficient vessels provide a windfall profit to those owners who might otherwise be considering withdrawing from the fishing industry, so they are particularly prone to slowing the normal rate of disinvestment from the fishery, as illustrated in studies of a range of fisheries in Japan and France (OECD, 2000b; Guyader et al, 2000; Daures and Gayuder, 2000; Giguelay and Piot-Lepetit, 2000).

Unless it is otherwise forbidden, vessel owners who divest themselves of old and inefficient vessels may then use their decommissioning premiums to buy a new vessel, invest in more modern gear or even reconfigure another existing vessel to increase its catch capability. Meanwhile, any temporary reduction in total capacity creates an incentive for others to upgrade existing vessels and increase their effort in the fishery. The net result may be even more capacity than when the programme was introduced.

Three basic factors cause levels of capacity and effort in the industry to be higher than would be the case without the decommissioning subsidies: the introduction into the fishing industry of additional capital that can be used for capital investment, the incentive for investors to take more risks investing in the fishery and the incentive for vessel owners to stay in the industry longer than would otherwise be the case. The first can be limited or even blocked through the most stringent capacity and effort controls. The latter two cannot be reduced through management measures except for adopting a strong prohibition against vessel decommissioning subsidies.

The UK decommissioning programme provides an example of a programme that purchased the withdrawal of the least productive vessels in the fishery. Most of the vessels leaving the fisheries under that scheme had about half the catch rates and days at sea of the average vessels in the fisheries targeted. Furthermore, the programme included a major loophole that allowed vessel owners to use their decommissioning premium to buy a new vessel (Nautilus Consultants, 1997). The result was a failure to achieve a net capacity reduction. Of the three biggest segments of the UK fleet (beam trawl/seiners, demersal trawl/seiners and nephrops trawl), representing 76 per cent of the total decommissioning premiums and 65 per cent of total capacity, only the nephrops trawl segment did not actually increase its catching power during the 1992–1996 period, despite the decommissioning programme (Banks, 1999).

The US decommissioning programme for its Northeast Multi-species Groundfish Fishery was unusual in basing its bidding system on actual catch history, so it avoided one weakness of most such programmes. However, it also illustrates the problem of huge latent capacity waiting to take advantage of any rents created by a decommissioning or license retirement programme. Latent capacity consists of both vessels that are completely idle and vessels that are actively fishing but using only a part of the days at sea allocated to them. One indication of the latent capacity in the fishery is that, of the 832 otter trawl vessels licensed to operate in the fishery, only 128 or 15 per cent of the total would have been required to take the entire cod quota for 1998 (Walden and Kirkley, 2000).

Another estimate of the overall level of unused capacity in the fishery is the gap between the number of fishing days of effort in the fishery estimated to be required to catch the quota, on one hand, and the number of fishing days allocated to those with licenses, on the other. The estimate of fishing days needed to catch the quota for all commercial species in 1989 was 49,000 days. The number of fishing days of effort actually used was 75,000 days, meaning that only one-third of the vessels actually fishing were needed to catch the quota. At the time the decommissioning programme was launched, the groundfish fleet as a whole was allocated a total of 249,000 days at sea. Thus, less than one-fifth of the capacity in the licensed fleet was needed to catch the quota, since groundfish quotas in 1996 were lower than they had been in 1989 (Kitts et al, 1999).

The Northeast Multi-species Groundfish Fishery vessel buy-out that began in 1996 removed the equivalent of roughly 16.8 per cent of fishing days actually used during the 1996 fishing year, but only 4.9 per cent of the total allocated fishing days (Kitts et al, 1999). In terms of reducing total capacity, including latent capacity that could become active if controls over effort were relaxed, therefore, the decommissioning programme was meaningless. In the absence of unusually stringent effort and capacity controls in the fishery, moreover, it could have worsened the overexploitation of stocks.

The New England Fisheries Management Council has imposed deep reductions in days at sea and even a ban on the use of pair trawlers on the fishery to limit further overexploitation of groundfish resources (WWF: Porter, 1998b). Little or none of the previously idle capacity in the fleet, therefore, could be introduced into the fishery in response to the capacity reduction. However, some of that idle capacity has been diverted to *other* fisheries where effort restrictions were less stringent (Thunberg, 2000). Increased capital available for investment in those other fisheries because of the buy-back programme has contributed to some of the increase in exploitation experienced in those other fisheries.

Meanwhile, cod stocks in the Northwest Groundfish Fishery are still being overexploited, despite warnings that fishing mortality must be 'substantially reduced' (Mayo and O'Brien, 2000). Even if the programme has not contributed directly to a worsening of stock depletion in the fishery by encouraging vessel owners to remain in the fishery and wait for stocks to recover, it has slowed the rate of exit from the fishery and reduced pressures for a more fundamental solution to the problem of the race for fish.

In the absence of property rights (or other means for eliminating economic incentives to overfish) a vessel or license retirement subsidy is bound to fail to reduce both capacity and effort significantly and is likely to make the overcapacity situation worse unless very stringent and well-enforced regulations are in place to prevent increased capacity and effort. Unless very strong attempts are made to control effort, the perverse incentive built into the fishery management system for a race for fish will cause additional increments of capital from the programme to be used to increase pressure on fish stocks. Even with such stringent controls, moreover, such decommissioning programmes are likely to discourage normal exits from the fishery.

Open access

If the fishery into which a decommissioning scheme is introduced lacks either a system of control over catch levels or individual quota rights for fishers, the likelihood of actually reducing either capacity or fishing effort through such a programme is very low and the danger of worsening the state of the stocks is correspondingly greater. Since the race for fish is not blocked by meaningful limits on catch, a decommissioning premium would both provide an inducement to capacity increases by creating rent in the fishery and would also serve, in effect, as a fund for modernizing the fleet.

Assuming that the fishery is already seriously overcapitalized, or even fully capitalized, the result of a decommissioning programme would almost certainly be further depletion of fish stocks. Even prohibiting the entry of new vessels would not prevent the improvement of existing vessels that had been idle or that were already active in the fishery. Efforts to limit new capacity by restricting gear types would be ineffective, given the limited resources for regulating the fishery implied in the lack of catch controls. The kind of strict controls over capacity and effort required to prevent further damage to stocks would not be a realistic possibility.

Table 2.6 *Expected Impact of Subsidies to Decommissioning of Vessels and License Retirement on Fish Stocks*

	Effective Management	*Catch Controls*	*Open Access*
Overcapacity	Not Harmful	Probably Harmful	Harmful
Full capacity	Not Harmful	Probably Harmful	Probably Harmful
Less than full capacity	—	—	—

Subsidies to capital costs

Subsidies to capital costs, whether by grants, loans at below-market rates, loan guarantees or tax benefits, have a direct impact on the economic behaviour of fishers. By definition, these subsidies are used only for the purchase of fishing vessels, new fishing gear, motors or engines or other fixed cost investments. Unlike other categories of subsidies which have an indirect effect on fishing capacity or effort, subsidies to capital costs cause increases in fishing capacity. How significant that contribution to total fleet capacity is and how it affects fish stocks, however, depends on the size of the subsidy and how much of the total costs it covers.

In the worst case, these subsidies are directly responsible for leaps in fishing fleet capacity that lead quickly to overexploitation and fisheries depletion. Even in the best combination of management and bio-economic conditions, however, large-scale capital subsidies can have serious economic and biological consequences.

In theory, capital subsidies could be given solely for the purchase of selective fishing gear, although no state has used subsidies for that objective thus far. If the subsidy were limited to gear that had been proven to reduce buy-catch or capture of juvenile fish, it would not be harmful, regardless of the management and bio-economic conditions of the fishery.

Effective management

Because this combination of management parameters is always associated with fisheries that are already fully capitalized or overcapitalized, a fishery that is not already at least fully exploited can be dismissed as irrelevant to this exercise. Similarly, a fishery with both effective catch controls and property rights, community-based management or some other means for eliminating economic incentives to overfish would not remain seriously overcapitalized, because the fishers would have a strong incentive to reduce overall capacity. Therefore, this combination would exist for a relatively short time. The strong tendency in the fishery would be towards a reduction in capacity to a level compatible with maximum profitability.

That leaves the case of a fishery that does have full capacity and the combination of effective controls and property rights or community-based management. In that case, a subsidy to capital costs would not have any impact on the level of capacity or effort, according to economic theory (Arnason, 1998). If overall quotas are set at a level of maximum economic yield (MEY), and each vessel has an individual quota assured, the fishing industry would be motivated to replace existing capacity, making the fleet more efficient but not increasing total effort. As profit-maximizing producers, vessel owners would not want to add to total capacity, because it would reduce the catch per unit of effort. Hence, the subsidy to capital costs would increase profits but not output. This combination would also imply that fisheries managers would apply very stringent controls to prevent an increase in overall capacity.

In the real world, however, no fisheries managers who have already assigned property rights to fishers would be motivated to seek or to continue subsidies to capital costs. As Arnason (1998) notes, it would not be 'deemed politically appropriate for a well managed ITQ fishery'. There are no cases of fisheries with this combination of management conditions that have provided subsidies to capital costs. Therefore, this case is of theoretical rather than actual policy significance.

Catch controls

If the management system of a fishery includes catch controls but no assignment of individual catch quotas to fishers or community-based management, subsidies to capital costs would have a very strong tendency to increase capacity as well as fishing effort. A system of catch controls implies that full exploitation of fishery resources has already been reached or surpassed. Although an ideal catch control system would prevent increased fishing effort induced by subsidies to capital investment, in practice the effectiveness of the system of catch controls would be affected by the level of capacity in the fishery.

Because an incentive for a race for fish would strongly influence fishing industry behaviour, vessel owners would take advantage of the subsidies to increase their own fishing capacity and effort, because they could do so at reduced cost. This motivation would prevail regardless of the level of capacity in the fishery in relation to resources. Even in a grossly overcapitalized and overexploited fishery, where additional capital investment by the industry and additional effort would reduce overall profitability in the fishery, the individual owner would have to focus on the relative gains to be obtained from acquiring increased capacity at a reduced cost.

The motivation for governments to provide subsidies to capital costs is either to increase capacity in particular fleets or to make the fishing industry more profitable by modernizing it, or both. Subsidies for the purpose of expanding fleet capacity are often extended to the industry when fishing fleets are believed to be at less than full capacity. Under these circumstances, the risk that the subsidies will lead to excessive fleet capacity is particularly great. Estimates of the biomass and spawning stocks of the fish resources in question may be significantly overstated, making any increase in capacity dangerous to fish stocks. But even assuming that the estimates are accurate, subsidies for vessel construction or vessel modification almost inevitably lead to capacity overshoot in the relevant fisheries (De Wilde, 1999). These subsidies not only increase the number of vessels in the fleets but speed up the introduction of new fishing vessel technologies and gear. When one group of owners decides to invest in the new technology because of the reduced costs, it induces others in the industry to do the same in order to remain competitive. The likely result is that the catching power of the fleet will quickly grow far beyond a level needed for sustainable catch.

Since the premise of the policy was that fleet capacity could be safely increased without damage to the health of fish stocks, such leaps in capacity inevitably translate into substantially increased fish mortality and will very likely result in serious depletion of resources.

When fishing fleets already have significant overcapacity in relation to the fishery resources they target, the motivation for subsidies to capital costs is normally to make the fleet more efficient and more profitable without increasing capacity. Subsidies to capital costs only for the purpose of modernization are usually combined with decommissioning schemes in order to maintain or reduce the net level of fleet capacity. Such schemes for subsidized replacement of older vessels with more modern vessels are almost certain to fail to prevent overall increases in capacity.

Capital investment subsidies combined with a system that allows the replacement of vessels on the basis of formulae involving tonnage or engine power have two major weaknesses that make them very unlikely to achieve their goal. First, they fail to take into account the fact that newer vessels replacing older ones have much higher catching capability than is indicated by similar tonnage and engine power. Thus, they hide actual substantial increases in capacity and risk undermining the controls over even nominal additions to capacity (De Wilde, 1999).

Second, such programmes are likely to allocate subsidies to capital costs to entire national fleets without regard to the impact on fleet capacity in each specific fishery. The result of such a loosely drawn 'fleet adjustment' programme is that the reductions in capacity will be concentrated in fisheries that are less profitable, allowing those fisheries that are more profitable to remain highly overcapitalized and worsening the effect on stocks in these fisheries (Banks, 1999; Nielsen, 1992).

The likely impact of subsidies on capital costs of fishing effort in a 'catch control' management system varies according to the bio-economic conditions of the fishery. A fishery with stocks that have already been severely depleted, and that has stringent limits on days at sea, and seasonal closures is likely to experience less net increase in fishing effort because of capital subsidies than a fishery that already has full capacity but has not yet imposed such severe effort restrictions. On the other hand, the greater the level of overcapitalization, the more difficult it is to avoid the setting of overall quotas at levels well above biologically sustainable levels and to prevent cheating on quotas. These increased pressures on the management system in a highly overcapitalized system limit the effectiveness of effort restrictions in protecting against even higher levels of fish mortality in already overexploited stocks.

Open access

If neither effective catch and effort controls nor appropriate incentives exist in a fishery, the purpose of subsidies to capital costs must be to increase the level of fleet capacity. In the absence of catch controls, even if the fishery is less than fully exploited, such subsidies are likely to cause an overshoot in fleet capacity well beyond a biologically sustainable level. Subsidies to capital costs are likely to encourage the adoption of much more powerful fishing technologies, causing the capacity level to grow so much that the fishery shifts from less than fully exploited to seriously overexploited. Thus, subsidies to capital costs can cause unintended damage to fish stocks in either or both of these two situations.

If a foreign fleet is also operating in the same fishery, the difference between the undercapitalization and overcapitalization of a national fleet obviously will be much smaller than would otherwise be the case. The Canadian subsidies for capital construction of its Northwest Atlantic Fleet from 1954 to 1968 illustrate the dynamics of such subsidies under this combination of management and bio-economic conditions. Canada wanted to increase its fleet in order to compete for the catch more effectively with European fleets operating in the same waters off its coast. Over this 14-year period, Canada provided grants and low interest loans to its Northwest Atlantic Fleet that increased its catching capacity by as much as 18 times, according to one estimate. This enormous leap in fishing capacity was spurred mainly by encouraging the general adoption by the fleet of not one, but two successive generations of fishing vessel technology: the side trawler and then the much more powerful stern trawler. By 1970, the Canadian government estimated that the fleet already had twice as much capacity as needed for a sustainable level of catch, taking into account the increased capacity of the European fleet also fishing in the same waters (WWF, 1998b).

Senegal's programme of subsidizing the purchase of engines for fishing vessels in the small-scale fishing sector illustrates the use of subsidies to capital costs in a developing country's fishery under these conditions. Beginning in the 1960s, the state sold engines to the owners of pirogues tax-free and perhaps at below market prices as well. This programme continued on a much larger scale in the 1970s and 1980s and is still apparently in operation today (UNEP, 2002a).

The effect of this subsidy has been to set off and accelerate a technological revolution in the Senegalese small-scale fishing fleet. The number of fleets that were motorized increased from none in the 1960s to nearly 2000 in 1970, and doubled to 4000 by 1974. This engine subsidy also made possible the introduction and generalized use of purse seine nets in the 1970s and 1980s (UNEP, 2002a). The Senegalese domestic fleet increased its share of the catch in the national economic zone (as against the share of distant water fishing fleets) from a very small fraction in the 1960s to one-third by 1970 and to three-quarters by 1975 (Bonfil et al, 1998). Although most of this development probably would have occurred even in the absence of the engine subsidy, it certainly increased the speed and total impact of motorization (UNEP, 2002a).

In the absence of any limitations on access or catch, the engine subsidy has exacerbated the depletion of fishery resources that followed the modernization of the Senegalese fleet. The sharp reductions in stocks of most major commercial species, indicated by data on catch per unit of effort, quickly followed the trajectory of modernization. The decline in biomass has brought, in turn, a rise in destructive non-selective fishing methods (UNEP, 2002a).

Table 2.7 *Expected Impact of Subsidies to Capital Costs on Fish Stocks*

	Effective Management	*Catch Controls*	*Open Access*
Overcapacity	Not Harmful	Harmful	Harmful
Full capacity	Not Harmful	Harmful	Harmful
Less than full capacity	Not Harmful	Harmful	Harmful

Subsidies to variable costs

As noted above, subsidies to variable costs include all those that reduce actual operating costs of fishing vessel owners. Those that have been provided in the past have reduced the costs of a wide range of items, including income taxes, vessel insurance and re-insurance, bait services, damaged gear and energy-saving devices onboard. The most popular subsidies to variable costs, however, have been tax rebates or other special arrangements to reduce the costs of fuel. As fisheries have become increasingly overexploited, the costs of fuel per unit of catch have continued to increase, so fuel costs have become a large proportion of total operating costs.

In general, reducing these costs through a subsidy makes each fishing trip less expensive and tends to increase fishing effort. Although subsidies to variable costs generally have less impact on fish resources than subsidies to capital costs, subsidies to fuel use may actually affect the level of fishing capacity indirectly through their technology effects. They provide an incentive for vessel owners to use more powerful and fuel-consuming engines (FAO, 1983; McGoodwin, 1990). They also induce more use of refrigeration on vessels by making it more profitable. Both effects of fuel subsidies give vessel owners greater incentives to extend fishing trips in time and space, implying large increases in catch. The seriousness of the impact of subsidies to variable costs on fish resources thus depends on the size of the subsidy, the degree to which it has a technology effect and the combination of management and bio-economic parameters in the fishery.

Effective management

Assuming effective catch controls as well as the existence of individual fishing quotas, subsidies to variable costs should not increase fishing effort or capacity. As discussed in relation to subsidies to infrastructure, which would otherwise be a variable cost, the assurance of a proportion of the total allowable catch will take away the incentive to subsidize fuel costs or other variable costs to increase capacity, even if it can be done at less cost per unit of additional capacity. Rather, the recipients of such subsidies would choose to simply minimize costs and increase their profit, while maintaining the existing level of effort (Arnason, 1998; UBC, 1999; Hannesson, 2001). As noted in the discussion of other categories of subsidies, however, the assumption of property rights makes the choice of subsidies to variable costs by fisheries managers most unlikely.

Catch controls

As previously discussed, in the absence of property rights, community-based management, or other means for eliminating economic incentives to overfish, catch control systems involving catch quotas, monitoring and surveillance and time and geographical restrictions on fishing effort have not been able to prevent significant quota violations in fisheries which are overexploited. The failure of such control systems is likely to be greater as fleet overcapacity in the fishery increases.

Therefore, the effect of subsidies to variable costs, which is to induce marginally higher fishing effort, and in the case of fuel subsidies, to encourage greater use of more fuel intensive fishing vessels, is only partially mitigated by the existence of catch control systems. If fleet capacity is already greater than required to exploit the resources at a sustainable level, a subsidy to variable costs will carry a high risk of degrading fish stocks.

Open access

In fisheries with no effective controls over catch and no means for eliminating economic incentives to overfish, subsidies to variable costs, and particularly to fuel, will certainly increase capacity and effort. Although few empirical studies of this case are available, they show that the result of such subsidies is to promote technological change towards the use of more powerful engines that consume more fuel. The case study of Senegal's fuel subsidies shows that it provided an incentive for boat owners to acquire engines that consumed more fuel and which could take fishermen further out to sea, as well as sustain longer trips. Thus it contributed to opening up new fishing areas and to the development of a purse seine industry. The result was significantly increased catches (UNEP, 2002a).

Under the bio-economic conditions of fleet overcapacity in either commercial or inshore small-scale fisheries, such capacity and effort increases induced by subsidies to fuel use cause further depletion of the fish resources. Thus the impact of subsidies to variable costs is particularly damaging to resources when they have indirect impacts on technology choices and occur in fisheries that are already fully or nearly fully exploited.

In a clearly underexploited fishery, this category of subsidy would not necessarily harm the resources. Many developing countries have deep-sea fisheries that are not fully exploited, for example, and the fish stocks would not be damaged in the short-run. The potential for a 'technology effect' of such a subsidy, however, raises the possibility that it could induce a rapid increase in capacity and cause an overshoot of the sustainable level of capacity and effort. In an open access fishery, such an overshoot can occur within a very short time span, as was the case in Senegal. The capacity of the inshore, small-scale fleet was clearly below the level required for full exploitation when the motorization of the fleet began in 1970, but by the early 1980s stock biomass for demersal fish had already been reduced by half, and it was halved again by the early 1990s (Bonfil et al, 1998). A direct relationship unquestionably exists in the Senegalese case between fuel subsidies, steeply rising fleet capacity, overexploitation and enormous losses of biomass. Subsidies to variable costs pose potential dangers, therefore, over a 5–10 year period, which would need to be addressed in a broader fisheries policy.

Most countries with no catch controls and no property rights already have fishing fleets that are either close to or well above full capacity. Under these circumstances, subsidies to variable costs in fisheries are certain to be harmful to the resource.

Table 2.8 *Expected Impact of Subsidies to Variable Costs*

	Effective Management	*Catch Controls*	*Open Access*
Overcapacity	Not Harmful	Harmful	Harmful
Full capacity	Not Harmful	Probably Harmful	Harmful
Less than full capacity	Not Harmful	Probably Harmful	Probably Harmful

Subsidies to income

The category of subsidies to income includes two distinct sub-categories that are different in their relationship to fishing effort. The larger and more prevalent sub-category is income support and unemployment insurance for fishermen. The smaller sub-category is subsidies to the income of vessel owners, primarily through payments for temporary cessation of fishing.

OECD countries have provided income support to fishermen through a range of social insurance programmes, including minimum basic wage levels for fishermen, unemployment insurance and compensation for days lost at sea (for a certain minimum number of days in the month) because of bad weather. A key distinction among different income support plans is whether the plan provides income only for wage-earning fishermen or for both wage-earners and self-employed fishermen. Where social insurance for wage-earning fishermen is in line with that available to wage-earners in other sectors, theoretically it will not constitute a particular incentive for employment in the fishing sector. Similar benefits for self-employed fishermen on the other hand, would in theory alter the attractiveness of fishing in relation to other economic activities.

The question of whether income support or unemployment insurance for fishermen provides an incentive for increased fishing effort, or otherwise causes an increase in fishing effort, must be broken down into four sub-issues. First, do such programmes draw additional fishermen into the industry? Second, do they discourage exit from a fishing industry that has collapsed or is in decline because of resource degradation? Third, do they provide an incentive for fishermen already active in the fishery to increase their fishing effort, and fourth, do they increase pressure on fisheries managers to relax controls on catch or effort?

The main form of subsidy to the income of a vessel owner is a 'laying up' grant to compensate for the temporary withdrawal of a vessel from active fishing for a given period of time. Such subsidies are normally prompted by a moratorium on fishing for certain depleted species and are therefore subsidies in support of idle capacity. The effect of the subsidy to idle capacity is therefore to discourage exit by vessel owners who might otherwise consider retiring from an industry that has been seriously affected by overfishing (Garrod and Whitmarsh, 1991). In the late 1980s and early 1990s, the European Union (EU) carried out a significant programme of laying up subsidies. Finland, Germany and Italy still report the use of such subsidies to the OECD (OECD, 2003).

Based on the EU experience, laying up grants lend themselves to fraud and are unlikely to result in substantive reductions in fishing compared with the baseline situation. An investigation of the EU laying up grants programme from 1987 to 1990 found that it permitted vessel owners to be compensated for periods in which fishing was traditionally reduced in any case and failed to dissuade them for fishing for the most profitable species threatened by overfishing (European Court of Auditors, 1992). Laying up is thus likely to represent a financial transfer to the vessel owners.

Effective management

In a rights-based or community-based management fishery, income subsidies would not provide an incentive to increase fishing effort, which is determined by the individual quota. The only exception would be if the eligibility requirement were set in terms of weeks fished or income that would require a catch level higher than the quota.

The issues of adjustment and political influence of unemployed fishermen on fisheries management decisions only arise in bio-economic conditions of overexploitation. In a fishery with property rights (or other means for eliminating economic incentives to overfish) and some catch controls, self-employed fishermen would also have to be those with fishing rights. Under those circumstances, they should have no effect on adjustment to bio-economic conditions by the fishing industry, since those decisions would be made on the basis of fishing rights and anticipated future fishing opportunities.

The use of laying up grants in a fishery managed with property rights is very unlikely, since excess capacity would be eliminated by owners' adjustments, who would maintain only enough vessels in the fisheries to catch the level to which they have transferable quotas. In the unlikely event that laying up grants were offered, however, they would have no impact on the level of capacity or effort in the fishery, because the vessel owner would have no incentive to increase the level of effort.

The laying up premium issue also arises only in conditions of overexploitation of a fishery. Assuming that catch quotas are set at a sustainable level, vessel owners would have much less reason to want to increase effort beyond the level needed to catch the amount of fish to which it has been allocated rights. Laying up premiums would thus be redundant and highly unlikely to be adopted as a policy option, and would have no discernible effect on fishing effort even if they were adopted.

Catch controls

The theoretical proposition that an income subsidy will provide incentives for expanded fishing effort is based on the assumption of an open access fishery (Poole, 2000). The question of whether an income subsidy has an effect in a fishery that imposes controls on the catch and effort must be also broken into the four sub-issues mentioned above.

Research on the impact of income support programmes on fishermen has been focused on Canada's Unemployment Insurance (UI) programme in Atlantic fisheries and particularly in Newfoundland. One study (Ferris and Plourde, 1982) argued on the basis of an econometric analysis that the UI did in fact contribute to the growth of the inshore fleet in Newfoundland. In the Canadian Atlantic provinces, self-employed fishing families received roughly 30 per cent of their income from UI benefits and data for Newfoundland does show a marked increase in fishermen after the start of the programme (Poole, 2000).

However, Canada only established catch controls after 1977 and both the data for the next decade and their interpretation are ambiguous. The number of fishermen increased dramatically from 1977 to 1980 (Schrank, 1998), but that period coincided with the creation of Canada's 200-mile exclusive fishing zone, the expulsion of European fleets from Canadian waters and a new programme of subsidized vessel construction and modernization (WWF, 1998b). Furthermore, the number of fishermen in Newfoundland's inshore fleet fell by 17 per cent during the 1980s, presumably reflecting a sharp decline in profitability as catch per unit of effort dropped off (Schrank, 1998).

The evidence from Newfoundland does not really help answer the question of whether the UI programme has had an attraction effect. Since income support programmes are established in large part because of major economic shocks affecting fleet size, it seems likely that similar results would be found in other fisheries.

The issue of whether income programmes discourage exit from a fishery has not been the subject of research and analysis. The data on Newfoundland could be read as supporting such an effect, because the percentage decline in profitability caused by changes in bio-economic conditions was significantly greater than the percentage decline in numbers of fishermen.

Poole (2000) examines the question of whether unemployment insurance provides an incentive for increased fishing effort or political efforts by fishermen to influence managers' decisions on catch and effort controls. In his model of a regulated, restricted-access fishery, if the UI payments are a function of income, they will cause an expansion of fishing effort; if it is a lump-sum income transfer with a qualifying threshold measured in weeks fished, and benefits diminish at higher income levels, the marginal incentive to fish will be minimal, unless its minimum requirements for weeks fished are greater than the cap on effort in the fishery. If the minimum requirements for weeks fished are above the effort cap, fishermen will be strongly motivated to use their political influence to loosen the restrictions on effort; on the other hand, if the requirements are below the effort cap, impacts on marginal incentives to fish will be minimal.

Apart from the relationship between an effort cap and minimum fishing period requirements, income support programmes may have an indirect political effect on fishing when a fishery has experienced serious fleet overcapacity leading to stock collapse, and income support programmes are justified as a means of tiding over the industry until 'the fish return'. It is under these circumstances that income support programmes are most likely to result in increased pressures on fisheries managers to reopen the fisheries prematurely (McCleod, 1996; Schrank, 1997; OECD, 2000a). Pressures for maximizing fishing opportunities would exist even in the absence of social insurance. It has been hypothesized that the effectiveness of fishermen as a pressure group in seeking to reduce catch and effort limits is directly proportional to their numbers (Poole, 2000), but such a scenario would occur only in a fishery that is already overcapitalized.

In a fishery in which some catch and effort controls have been imposed, but in which the race for fish is still a factor, a grant to the vessel owner unlinked to specific inputs or outputs does not have the same impact on industry behaviour as production or input-linked subsidies. Neither does it cause fishermen to increase the level of fishing capacity or effort. At the same time, it has been shown to be almost completely ineffective in inducing owners to reduce the level of effort below what it would otherwise be. The ways in which the system can be manipulated to obtain payments without any actual reduction in fishing are too many and too easy.

Laying up subsidies may provide some incentive to remain in the fishery rather than to disinvest, because it increases the expected level of profitability (or reduces the level of expected loss) of the fishing industry. That means that some marginal reduction of capacity may not occur because of the provision of laying up premiums.

Open access

In an open access fishery, an income support programme would, in theory, encourage additional entry until all rents are dissipated, and if the income subsidy is a function of monthly income, it would induce increased effort as well (Poole, 2000).

In the absence of catch and effort controls, there would be no rationale for laying up subsidies (LU), which are supposed to be compensation for foregone fishing opportunities. Therefore, this combination of parameters and subsidy is of no theoretical or practical significance.

Table 2.9 *Expected Impact of Subsidies to Income*

	Effective Management	*Catch Controls*	*Open Access*
Overcapacity	UI: Not Harmful LU: Not Harmful	UI: Possibly Harmful LU: Harmful	UI:Harmful LU: Harmful
Full capacity	UI: Not Harmful LU: Not Harmful	UI: Possibly Harmful LU: Harmful	UI:Harmful LU: Harmful
Less than full capacity	UI: Not Harmful LU: Not Harmful	UI: Possibly Harmful LU: Harmful	UI:Harmful LU: —

Price support subsidies

Price support subsidies are mechanisms for increasing fishing industry revenues, with the bulk of the benefits going to vessel owners. Price support programmes use different mechanisms and provide different levels of compensation to the industry. Some compensate for any market price reduction below a target level that is well above the world price, whereas others only compensate for serious falls from the world market price. The EU price intervention regime is of the latter type. It has several intervention mechanisms, including withdrawal prices, carry-over aid and private storage aid. Withdrawal price, limited to 24 specific products, is the most widely used (OECD, 2000a).

In economic theory, a subsidy to producers through price intervention will cause fish production to rise above the level that would have been produced in the absence of the subsidy. The actual effect of either a price support subsidy, however, depends on the management and bio-economic conditions of the fishery. The analysis that follows focuses on price support subsidies.

Effective management

In a fishery with both property rights or community-based management and an effective system of catch and effort controls, the effect of price support on fishing effort should be negligible if the fishing fleet is already overcapitalized or at full capacity. Vessel owners would be motivated to reduce fleet capacity to the level necessary for MEY or to keep it at that level, despite artificially higher prices. Fishing effort would be restrained by the desire to ensure future sustainability as well as by the catch quotas and monitoring and surveillance systems. Under these circumstances, the subsidy would be a straight resource transfer from taxpayers or consumers to the fishing industry.

If the fishery had these management characteristics, but the fleet was below the capacity necessary for MEY, price supports would provide the incentive for fishing effort to rise to the level of MEY. However, it is highly unlikely that both catch controls and property rights would be established in a fishery that is not already fully exploited.

Catch controls

In a fishery without individual fishing rights, price supports provide an incentive to increase production. Although catch controls should in theory constrain this incentive effect, in reality the constraint is limited. As has been noted earlier, at high levels of fleet overcapacity in a fishery without allocation of fishing rights, catch quotas and systems of monitoring and surveillance provide only a partial defence against the effect of price support policies in inducing higher levels of fishing effort and catch. Fishermen are motivated and able to use their political influence to pressure fisheries managers to compromise the scientific integrity of catch quota decisions and to flout the quotas.

In a fishery in which fishing effort is at or above a biologically sustainable level, price support measures harm fish stocks, depending on the degree of reduction of biomass already taking place. The clearest case study of the impacts of price supports as the dominant form of subsidy to the fishing industry is in Norway, where price support represented as much as 75 per cent of total transfers in the late 1970s and 40–50 per cent in most of the 1980s. Annual average price supports for cod in the north and northwest of Norway increased fivefold from 0.2 NKr/kg in 1978 to 1.08 NKr/kg in 1985, which represented an increase from 5.5 per cent of the total price of cod to 16 per cent. The level of the subsidy was set in relation to landed value, so it provided a powerful incentive for both increased investment in the industry and increased fishing effort (OECD, 2000b).

Norway's price support subsidies allowed the industry to ignore the signals that Norwegian fisheries were already seriously overcapitalized. Thus gross tonnage of the Norwegian fleet increased by roughly 30 per cent between 1973 and 1987, even though its purse seine and herring fleets were already seriously overcapitalized. This caused depletion of stocks and a 28 per cent decline in catch from 1978 to 1984 (Salz, 1991; OECD, 2000b; Flaaten and Wallis, 2000). New investment in the fisheries continued, as the financial effects of this decline were masked by the increase in price support subsidies (OECD, 2000b; Flaaten and Wallis, 2000). If the fishery is exploited at levels well below a biologically sustainable level, the incentive effect of price support measures would not necessarily result in harmful impacts on stocks, but it could cause the level of effort to increase to a point above the biologically safe level of exploitation.

Open access

In the absence of catch controls and incentives for sustainable fishing, the impacts of price support interventions are quite similar to those in a fishery with catch controls, but they will be of greater magnitude, given the same bio-economic circumstances and rates of price support. Instead of impaired effectiveness of catch and effort controls, the fishery will have no limitation on the incentive effects of price supports. If a fishery's resources are being overexploited, price support subsidies will certainly induce increased fishing effort and cause stocks to decline.

In theory, price support in a fishery that is clearly at less than full exploitation may not harm stocks immediately, but will certainly speed up the transition to overexploitation and may do so very rapidly. As noted above, in practice, such subsidies are likely to be used only under conditions of overexploitation.

Table 2.10 *Expected Impact of Price Support Subsidies on Fish Stocks*

	Effective Management	*Catch Controls*	*Open Access*
Overcapacity	Not Harmful	Harmful	Harmful
Full capacity	Not Harmful	Harmful	Harmful
Less than full capacity	Not Harmful	Possibly Harmful	Harmful

Summary

Using an analytical framework that takes into account different combinations of management parameters and the degree of exploitation in a fishery, this chapter has examined the impact of the following eight categories of fishing subsidies on fishery resources:

1. subsidies to fishing infrastructure (e.g., construction of port facilities);
2. management services (e.g., monitoring and surveillance, management-related research);
3. subsidies to securing fishing access (e.g., government-to-government payments that cover significant portions of the cost of access to foreign fishing grounds);
4. subsidies to decommissioning of vessels (e.g., vessel or license retirement);
5. subsidies to capital costs (e.g., grants, loan guarantees, or tax incentives encouraging fleet renewal or modernization);
6. subsidies to variable costs (e.g., subsidies on fuel, bait, insurance, or other operating costs);
7. income supports (e.g., special unemployment insurance or 'lay up' payments); and
8. price supports (e.g., government market interventions to guarantee a minimum price on fish products).

Each of these subsidy types is investigated within a matrix that reflects two major variables: level of fleet capacity (expressed as 'overcapacity', 'full capacity', or 'less than full capacity') and type of management regime (categorized broadly as 'open access', 'catch control' and 'effective management' regimes). For purposes of analysis, the management regime categories are necessarily presented as stylized ideal types. 'Open access' regimes are defined as lacking any legal or regulatory framework – a condition that has been considered a root-cause of fisheries over-exploitation. 'Catch control' regimes are those that use a variety of controls over catch levels, fishing effort and input, such as 'total allowable catch' quotas, catch limits per day or trip, area and seasonal closures, restrictions on certain vessel types or fishing gear and limitations on fishing licenses.

The assumption is made that, even in a management regime with catch controls and no property rights, the combination of overcapacity, inappropriate quotas, imperfect monitoring and surveillance and the continued existence of a race for fish leads to some degree of overexploitation of the fishery. An 'effective' management regime, therefore, is one that combines scientifically-based catch and effort controls, adequate monitoring and surveillance measures and socioeconomic incentives for sustainable fishing. It is important to recall that the category of 'effective' management regimes remains extremely rare in the real world. The vast majority of the world's fisheries conform to the conditions specified in either the 'catch control' or 'open access' management regimes described.

Finally, the analysis assumes that the total biomass of commercial fish stocks in an overexploited fishery is not increasing at the time the subsidy is applied. Therefore, under this simplified model of the impacts of fisheries subsidies, an increase in effort will have a negative impact on fish stocks in a fishery that is already overexploited.

As illustrated in Table 2.11, the results of the impact analysis for each of the eight categories can be summarized as follows:

- **Subsidies for fisheries infrastructure** are expected to be harmful to fisheries resources except where incentives ending the race for fish are provided by an effective management system or where the fishery is clearly less than fully exploited. While subsidies to infrastructure might not cause actual increases in fishing effort within 'catch control' regimes, they would provide disincentives to reduce capacity, even in an overexploited fishery. As in the case of all fisheries subsidies other than management services, subsidies to infrastructure will always be harmful in an open access regime that is fully exploited, over-exploited, or capitalized to or beyond full capacity.
- **Subsidies to management services and research** have not proved harmful to fishery resources. Subsidies to research that clearly benefits only the fishing industry and is not in the general public interest are an exception, although these are likely to be marginal in their impact.

- **Subsidies for access to foreign countries' waters** could theoretically be beneficial in the presence of effective management. However, such subsidies are expected to be harmful to fisheries resources, unless the fisheries covered by the agreement are clearly undercapitalized. Unfortunately, bilateral access agreements in the real world have almost universally involved host country fisheries in which capacity or exploitation levels are already high and/or in which management controls are absent or weak.
- **Subsidies to decommissioning of vessels and license retirement** are provided only in fisheries that are already overcapitalized and are usually also overexploited. In the presence of effective management controls, including property rights, a vessel decommissioning or license withdrawal programme can often be successful in substantially reducing capacity and effort in the fishery. However, effective property rights regimes should be able to reduce or prevent overcapacity without resort to decommissioning subsidies. Historically, in fisheries with systems of catch control but no property rights or community management, catch quotas and enforcement systems have not prevented stocks from being overfished. In some such cases, decommissioning subsidies have increased rather than reduced capacity. In an open access fishery, the likelihood of actually reducing either capacity or effort through a decommissioning scheme is low, and the danger of worsening the state of the stock is significantly greater.
- **Subsidies to capital costs** are expected to be harmful in all circumstances unless the fisheries management system provides for property rights, community-based management or other means for eliminating economic incentives to overfish. They can be harmful even in fisheries that are less than fully exploited, where subsidies to capital costs encourage the adoption of much more powerful fishing technologies, potentially causing an overshoot in fleet capacity well beyond a biologically sustainable level. Only under the extremely rare ideal circumstances of an effective management regime, could such subsidies be benign.
- **Subsidies to variable costs** provide an incentive for vessel owners to use more powerful and fuel-consuming engines and are expected to be harmful unless 'effective management' exists or the fishery is less than fully exploited. These subsidies are similar to subsidies to capital in their potential for harm, although in 'catch control' fisheries that do not suffer from overcapacity or overexploitation, the likelihood of harm is considered 'probable'.
- **Subsidies to income**, particularly for vessel owners, could be harmful if the fishery is fully or overexploited and lacks economic incentives to eliminate the 'race for fish', or when open access prevails. Income subsidies in the form of 'laying up' subsidies are likely to have the effect of discouraging reductions in capacity that would otherwise be financially more attractive. Where catch control fisheries are at or beyond full capacity, the impacts of unemployment insurance schemes are likely to be less harmful than 'laying up' subsidies.

- **Price support subsidies** are expected to be harmful in all circumstances unless the fishery has appropriate economic incentives to eliminate incentives for overfishing, such as property rights or community-based management. Price supports have had a clear impact on levels of fishing effort and can speed up the transition from a condition of less than full exploitation to overexploitation.

Viewed synoptically, the analysis shows that most subsidies have the potential to be harmful to fish stocks under the management and bio-economic conditions found in the vast majority of fisheries today. While all fisheries subsidies would theoretically be 'not harmful' in the presence of truly effective management, very few if any fisheries today are subject to management systems that are sufficiently 'effective' to ensure that fisheries subsidies will not harm fisheries resources.

More specifically, five categories of fisheries subsidies (subsidies to fisheries infrastructure, subsidies for access to foreign waters, subsidies to capital costs, subsidies to variable costs and price support subsidies) can generally be considered harmful to fisheries resources under most real-world conditions. Subsidies that contribute directly to increased fishing capacity or effort, such as subsidies to capital or operating costs, are among the most harmful. Of the remaining three categories, two (decommissioning and income supports) are considered 'possibly or probably' harmful. Only management services can generally be considered 'not harmful'.

Properly designed fisheries subsidies, including government programmes for the reduction of fishing capacity and the improvement of fishing techniques, can contribute to the achievement of sustainable fisheries, provided that effective safeguards are put in place. Subsidies for decommissioning deserve particular attention. They are likely to be harmful to fisheries resources under the conditions found in the fisheries where they have been and are most likely to be used. Stringent policy conditions are necessary to accompany such a subsidy programme in order to avoid altered incentives to enter or exit the industry or to invest in the modernization or purchasing of new vessels. These safeguards could include mandatory physical scrapping of vessels, prohibition of introducing new vessels and commitment to time limits of decommissioning programmes.

This analysis has used the 'matrix approach' to assess the resource impacts of different types of subsidies under a number of management and bio-economic conditions. This approach could usefully be developed further to include other environmental, social and economic impacts, aquaculture and the fishing processing industry. It could also address the dynamics of subsidies and fisheries management systems, elaborate further on the differences of scale of subsidies and their impact taking into account the differences of subsidies to small-scale and artisanal fishing and those to large-scale and industrial fishing.

Table 2.11 *Impact of Eight Categories of Fisheries Subsidies on Fish Stocks*[3]

	Catch Controls			*Open Access*		
	Over-capacity	*Full capacity*	*Less than full*	*Over-capacity*	*Full capacity*	*Less than full*
Fisheries Infrastructure	H	H	NH	H	H	NH
Management Services	NH	NH	NH	NH	NH	NH
Access to Foreign Waters	H	H	NH	H	H	NH
Decommissioning	PH	PH	—	H	PH	—
Capital Costs	H	H	H	H	H	H
Variable Costs	H	PH	PH	H	H	PH
Income Supports	PH	PH	PH	H	H	PH
Price Supports	H	H	PH	H	H	PH

NH = Not Harmful
PH = Possibly or Probably Harmful
H = Harmful
— = Not Applicable

Endnotes

1 For a detailed discussion of the conditions under which access agreements count as subsidies, please see Chapter 5, 'The Special Case of Access Agreements'.

2 In fact, these individual quotas were not yet fully protected use rights, since the system was to expire in 2004 unless renewed (testimony of Mike Hyde, President, American Seafoods, Senate Commerce Committee, Magnuson-Stevens Hearings, 18 January 2000).

3 Note that the 'effective management' category has been omitted from this table since these regimes are very rare in the real world. Note also that the term 'not harmful' here is narrowly defined to mean that a given subsidy type under given conditions cannot create a direct economic incentive to overfish. This definition assumes the subsidies in question will not be able to encourage cheating on catch limits or other regulations. It also ignores possible political impacts of subsidies, such as the creation or maintenance of enterprises which may have a vested interest in relaxed management controls.

Chapter 3

National Experiences with Subsidies, their Impacts and Reform Processes

Introduction

The international debate over fisheries subsidies has always had as its ultimate focus the practical experiences of governments and fishing industries at the national or even sub-national level. Empirical studies of these national experiences therefore form an important part of the fisheries subsidies literature. This chapter presents three sections illustrating recent country study work.

UNEP has been one of the leading sponsors of fisheries subsidies country studies, having commissioned papers examining the realities on the ground in Argentina, Bangladesh, Ecuador, Senegal and Vietnam. Another set of national studies has been brought forward by the OECD, which has published papers on 'government financial transfers' to the fishery sectors in Canada, the EU, Japan, New Zealand, Norway and the United States (OECD, 2000a). Still other studies have been sponsored by NGOs such as WWF (Argentina, the EU, the Baltic) and The Nature Conservancy (Indonesia).

Considering that fisheries subsidies are employed by dozens of countries around the globe, these case studies are still relatively few in number. The studies themselves also vary widely in methodology, data coverage and analytic depth. Taken together they have all provided important insights into the diverse fisheries subsidies programmes.

It is also important to note that in some cases the country studies themselves have been vehicles for engaging local stakeholders in unprecedented dialogues about the subsidies that affect their lives. UNEP's case studies in particular have always involved stakeholder roundtables at which environmental, industrial and governmental perspectives are represented. In the case of Senegal, for instance, UNEP commissioned a series of papers in which the later investigations reconvened stakeholders to develop concrete mechanisms for implementing earlier recommendations. Another leading example of stakeholder engagement through the country study process is the paper on Indonesian fishery subsidies recently published by The Nature Conservancy (included on the accompanying CD-ROM).

The following three sections have been chosen for the different perspectives they offer on the fishery subsidies problem. The Senegal paper, like the UNEP-sponsored Senegalese studies that preceded it, directly examines some of the difficulties experienced by a fisheries-dependent developing country with its own domestic use of fisheries subsidies. It identifies the main subsidies and their related environmental impacts. The Ecuador study, while also examining Ecuador's own subsidies, is focused mainly on the negative commercial impacts on Ecuador's tuna export industry as a result of foreign fishery subsidies. The Norway study, based on work undertaken by the OECD, also focuses on domestic programmes, this time in a major developed country fishing power that has largely phased out its use of fisheries subsidies in recent years. It shows the reform process and provides details on how market-based management instruments were used to help the industry adjust to the reduction in subsidies and contributed both to a reduction in capacity and longer-term sustainability.

While the studies of Senegal, Ecuador and Norway are obviously different, the three sections reveal certain common themes that are also often repeated throughout the fisheries subsidies case literature. First, the studies suggest that governments often share similar intentions in choosing to subsidize their fishing sectors. Modernization, efficiency and the maintenance of healthy small-scale fishing communities are frequently among the leading motives. But the studies also generally reveal that there is a gap between what governments intend to accomplish through fisheries subsidies and what they actually achieve. As will be discussed briefly in the conclusion to this chapter, these gaps between intentions and results can often be traced to the failure to fully assess their impacts and integrate subsidies programmes into sustainable resource management policies.

Fisheries Subsidies: The Senegalese Experience*

Introduction

In Senegal, fisheries are vital to the lives and livelihoods of more than 12 million people. Located on the west coast of Africa and adjacent to one of the most biologically productive ocean currents on Earth, Senegal depends on its fisheries for the majority of its edible protein, for many of the jobs in its coastal communities and for its most valuable export products. But today, Senegal's rich fisheries are being depleted, its domestic food supply has become less secure and its efforts to increase export earnings have faltered. Overfishing is beginning to take its toll.

The causes of overfishing in Senegal are multiple and complex. Inadequate fisheries management is, of course, at the root. But among the contributing factors have been fisheries development policies and government subsidy programmes that have sometimes failed to meet their stated goals of contributing to Senegal's sustainable development. This section provides an overview of some of the main uses of fisheries subsidies in Senegal over the past decades, exploring their links to Senegal's general strategy for its fisheries sector and discussing their apparent impacts on sustainability.

As is often true with country studies of fisheries subsidies, it is difficult if not impossible in the case of Senegal to draw simple connections between subsidies practices and specific environmental or economic impacts. The factors that determine levels of fishing effort and the health of fish stocks are simply too complex for a reductionist analysis. However, the basic facts can be displayed and correlated. Accordingly, the first section provides a brief overview of the state of Senegalese fisheries, noting some worrying trends and their implications for Senegalese society. The second section then reviews the main uses of fisheries subsidies in Senegal and discusses the types of impacts they have had or seem likely to have had. The third section briefly discusses policies that have more recently been put in place in an as yet unsuccessful effort to reduce the chronic overcapacity of Senegal's fishing fleets. Finally, a number of basic conclusions are drawn and put in relation to the results of a survey conducted on Senegalese stakeholders regarding attitudes towards the subsidies issue.

---

* This section was written by Moustapha Dème (Centre de Recherches Océanographiques de Dakar-Thiaroye (CRODT/ISRA), Dakar, Senegal), Pierre Failler and Thomas Binet (Centre for Economics and Management of Aquatic Resources, Department of Economy, University of Portsmouth, UK), in collaboration with UNEP staff. It is an update based on UNEP's 2004 publication 'Fisheries Subsidies and Marine Resource Management. Lessons Learned from Studies in Argentina and Senegal' and UNEP's 2002 publication 'Integrated Assessment of Trade Liberalization and Trade-Related Policies. A Country Study on the Fisheries Sector in Senegal'. Data and facts were updated and a questionnaire was submitted to policy-makers, researchers, fishermen, traders, processors and NGOs for further information. The chapter benefited from research carried out within the EU Research and Cooperation ECOST (Ecosystems, Societies, Consilience, Precautionary principle: Development of an assessment method of the societal cost for best fishing practices and efficient public policies), 2005–2010.

As discussed in the final section, the results of this review confirm the continuing need to reform Senegalese fisheries subsidies practices and to increase the attention of the government and other stakeholders to fisheries subsidies policies.

The Senegalese fishery context

Socio-economic importance of fishing in Senegal

The Senegalese economy once depended heavily on phosphates, groundnuts and other cash crops. Since a major drought in the 1970s and a crisis that took place in the mining and agricultural sectors, however, fishing has become a vital sector of the economy. Thus, today about 15 per cent of the working population of Senegal is directly or indirectly involved in fishing.[1] With 60,000 fishermen,[2] 40,000 processors and 100,000 handlers, fish mongers, wholesalers and retailers, the fishery sector largely contributes to the absorption of unemployment in the country.

The fisheries sector also receives special attention from the government as a major export sector and a potential source of hard currency earnings to restore Senegal's trade balance. For the period 1996–2008, the fishery sector accounted for about one-third of the value of exports, followed by oil products (17.5 per cent), phosphoric acid (11 per cent), mineral fertilizers and chemicals (5 per cent) and groundnut products (3 per cent). It generates a value added[3] estimated at 195,000 billion CFA francs (US$390 billion), which represents 5 per cent of the gross national product (GNP) (WWF, 2007b).

With an annual production of about 400,000 tonnes (of which only 80,000 tonnes are exported), the fishery sector also significantly contributes to local food security. Fish products represent up to 75 per cent of Senegalese total animal protein intake, annual consumption reaching as high at 30kg per capita.[4] The supply of the domestic market has been defined as one of the six goals identified by the government's strategy for sustainable development of fisheries and aquaculture, as adopted by the Ministry of Fisheries in 2001.[5] Given its central role in food security, fisheries also play a crucial role in poverty reduction – a central objective of government policy recalled in 2003. Small pelagic fish provide affordable, low-cost fresh and processed products that play a strategic role in the diet of both rural and urban populations. The overwhelming presence of women in the artisanal processing of fish products is also perceived as a favourable factor for reducing poverty, as incomes generated from activities conducted by women are generally used for basic family needs. Especially in the context of the current world economic turmoil, the fisheries sector is often perceived as an opportunity for improving living conditions of Senegal's most vulnerable groups.

Finally, fishing communities in rural coastal areas also contribute to reducing the rural exodus towards major urban centres such as Dakar or Saint-Louis. In some cases, fishing activities are undertaken as a complement to agricultural activity, especially during the dry season. For instance, the collection of marine invertebrates – which is not as demanding work as going to sea – is often conducted by women and children between rice harvests. In other cases, fishing constitutes a sole occupation and is often characterized by seasonal migrations along the Senegalese coast or to waters of neighbouring countries such as Mauritania to the north and Guinea Bissau to the south.

In short, it is clear that fishing plays an important role in the economic and social organization of coastal communities and in the Senegalese economy overall. The desire of the government to promote the development of the fishery sector is thus obvious. Similarly, failure to maintain a healthy fisheries resource base could spell significant hardship for Senegalese society.

State of Senegalese fisheries

Senegalese seas and their maritime resources are characterized by a variety of species. The most targeted species include four groups with different bio-ecological characteristics and socio-economic importance: (i) high-sea pelagic resources; (ii) small pelagic resources; (iii) coastal demersal resources and (iv) deep sea demersal species.

Generally, only a few species are underexploited with level of catches below 'maximum sustainable yield' (MSY). Most of the assessed species are fully exploited or overexploited. The threatened conditions of many of Senegal's fish stocks are specifically obvious with regards to declines in 'catch per unit effort' and average size of catches. Without attempting a complete review of Senegalese stocks, the next two sections illustrate the situation for the important in-shore demersal species, most of which are overexploited.

Decline in catch per unit of effort (CPUE)

Trends in 'catch per unit of effort' (CPUE) are a basic indicator of stock health. As stocks are depleted, the amount of effort necessary to locate and catch fish increases, with obvious impacts on the cost/revenue ratios for the fishermen. For many of Senegal's shallow demersal stocks, the CPUE data (measured in kilograms of fish taken per hour of fishing) have been strongly negative since the 1970s, as illustrated by the following data:

- for the Mottled grouper (*Mycteroperca rubra*), CPUE was less than 10kg/h in 1998, compared with 50kg/h in the 1970s;
- for all species of rouget (*Pseudolithus* spp), CPUE was less than 10kg/h in 1998, while at the end of the 1970s it was over 2 tonnes per hour;
- for red sea bream, CPUE had fallen to 50kg/hr by 1998, compared to more than 300kg/hr in 1975;
- for seabream (*Pagellus bellott*), CPUE was over 1000kg/hr until the early 1980s, but then declined sharply to 200–400 kg/h since 1990 (it has recently shown a slight upward trend, which may be due to the fact that this trade category also includes species found on the edge of the continental shelf or on the continental slopes);
- for sea catfishes, or mâchoiron, (*Arius* spp), the pattern has been the same as that of *pageot*, with CPUE falling sharply in the second half of the 1980s from over 4000kg/hr in 1981 to approximately 100kg/hr in the early 1990s;

- for the Lesser African threadfin (*Galeoides decadactylus*), CPUE has declined from over 1000kg/hr in 1981 to around 130kg/hr in 1995;
- for Rubberlip grunt (*Plectorhinchus mediterraneus*), CPUE had been over 140kg/hr in 1977 and fell to less than 20kg/hr by 1998;
- for shrimp (*Penaeus notialis*), CPUE had fallen to 60kg/hr in 1998, as against 140kg/hr in the early 1970s; and
- for grouper (*Epinephelus aeneus*), CPUE was less than 10kg/hr in 1998, as against 140kg/hr in the early 1970s.

Decline in fish size

The evolution of the average size of a species is a good illustration of the status of fish stocks and potential overexploitation. A decrease in average size generally marks the decline of the relative abundance of a stock and indicates that fish are being captured at a steadily younger average age. To ensure the renewal of the stock, the minimum size of individual fish captured must be greater than the size of first sexual maturity (spawning size). Here again, a sampling of data reveals significant negative trends, with several major species now experiencing average catch sizes close to the minimum sizes permitted. For example from 1995 to 2000:

- the average size of sea bream fell from 16.3cm to only 12.2cm;
- the average size of Lesser African threadfin, or capitaine, fell from 38.5cm to 25.4cm; and
- the average size of red snapper (*Lutjanus agennes*) fell from 17.2cm to 12cm.

For all three of these commercially important fish, average catch size is now below the size of first sexual maturity, jeopardizing the renewal of fish stocks and the sustainability of fishery activities on vital demersal resources. In fact, the growing consumption of young and immature demersal fish (mostly capitaines) used in the commercial category called ‘frying’ is a clear symptom of the overexploitation of resources.

Catch levels beyond MSY

A review of data for five of the main species targeted by Senegalese fishers, for which calculations of MSY are also available provide direct evidence of overfishing. Based on information for 2001, four of the five species are experiencing fishing significantly above MSY levels, as follows (see Table 3.1):

Trends in Senegal’s fish exports

As noted above, exports of fish products are vital to the Senegalese economy and for many years government policy has sought to promote such exports. The history of Senegalese fish exports since 1990, however, does not suggest that these policies have succeeded. As indicated in Table 3.2, after the devaluation of the CFA franc in 1994[6] Senegalese fish exports increased but quickly reached a ceiling at slightly over 100,000 tonnes per year up to 1998. In 1999, total fish exports reached an

Table 3.1 *Catches versus MSY for Five Major Species*

Species	*2001 MSY (tonnes)*	*2001 Catch (tonnes)*	*Actual Catch as % of MSY*
Seabream	4996	10720	215%
Grouper	1302	3620	278%
Mottled grouper	2836	5650	199%
Lesser African threadfin	3400	4470	131%
Mullet	1900	1400	74%

exceptional level of 125,000, due to an unusual increase in exports of cephalopods that year. The quantities exported then fell steadily to a lower level of about 87,500 tonnes in 2002 – this marked a drop of 26 per cent compared to the figures of 1997. There was some recovery in exports in 2003 and 2004, only to be followed by another decline, down to only 74,000 tonnes in 2006 – the lowest level of exports of fish products in Senegal over the past 20 years.

Fortunately for Senegal's trade balance, the stagnation of fish exports in terms of quantity has been softened by a simultaneous increase in the commercial value of fish products, which have escalated by 30 per cent to 50 per cent in value, depending on species. The rise in export prices seems to be attributable to the scarcity of fish in the domestic market coupled with a very strong external demand. Thus, even this 'good' news suggests problems with fish supply that seem related to problems with domestic supply, as discussed below.

Also, one can note that the breakdown of Senegal's fish exports have shown an increase of exports of unprocessed products while classic development theory

Table 3.2 *Export Trends by Product 1990–2006 (tonnes)*

Year	*Fish*	*Crustaceans*	*Molluscs*	*Processed Products*	*Total*
1990	79,233	5433	16,146	23,860	124,672
1991	62,232	5182	25,918	25,523	118,855
1992	46,371	3925	12,775	21,040	84,111
1993	33,969	4883	12,188	32,762	83,802
1994	44,054	4632	14,947	30,041	93,674
1995	43,327	5677	13,271	41,188	103,463
1996	53,558	5993	12,924	34,547	107,022
1997	57,698	6239	11,327	36,893	112,157
1998	51,490	8438	14,650	34,910	109,488
1999	44,990	7111	46,626	26,611	125,338
2000	45,276	6860	13,014	22,870	88,020
2001	53,170	7757	9885	16,220	87,032
2002	33,764	8246	22,056	23,498	87,564
2003	51,642	6524	20,631	16,878	95,675
2004	52,207	7196	19,963	13,129	92,494
2005	53,565	2824	15,583	11,132	83,104
2006	47,736	4646	13,125	8517	74,023

Source : Direction de l'Océanographie et des Pêches Maritimes (DOPM) (2007)

Table 3.3 *Fish Exports in 2006 According to the Nature of the Products*

	Fish	*Crustaceans*	*Molluscs*	*Processed Products*	*Total*	*%*
Raw products	41,412	4064	9957	0	55,433	74.9
Processed	6323	582	3168	8517	18,590	25.1
Total	47,735	4646	13,125	8517	74,023	100.0

Source: DOPM (2007)

would encourage the increase of exports of processed goods in order to increase the domestic value added of exports. Between 1993, the year preceding devaluation, and 1999, the share of whole unprocessed products rose from 60 per cent to 80 per cent of exports. In 2006, the figures were 75 per cent for whole fish and 25 per cent for processed products (Table 3.3). In short, Senegalese exports of fish products have remained dominated by unprocessed products.

This situation is unsatisfactory both from the environmental and economic point of view. Exports of unprocessed products mean that volumes are in effect valued more highly by producers than margins and that increased export revenue only comes from increased fishing pressure. From the angle of economics, the low value added means less profit earned in-country.

Senegalese fish processing consists mainly of canning tuna. In comparison, preparation of fillets, fish steak and shrimp peeling are not significant. But Senegal's canning factories are now facing an acute crisis. Only one out of the three owned by Senegal operates regularly. This crisis is firstly due to the shock resulting from upgrading to international technical standards. In order to comply with European directives and to be eligible for export permits, canning factories invested 4 billion CFA francs in 1995. However, this investment was not followed by significant productivity gains. Among the other factors affecting production are continuing difficulties securing raw fish inputs.

The Senegalese fish exports will now have to evolve in a renewed international context which revolves around an overall erosion of trade preferences and the setting up of stronger barriers to trade:

- Economic Partnership Agreements (EPA) between the European Union and ACP countries, agreements that have been in place since 2008;
- an increased globalization of markets that open the regional and sub-regional markets to international competition;
- market access requirements based on sanitary and phytosanitary measures (SPS), measures on the rules of origin and the upcoming required certification of legal catches from the EU;[7] and
- potential new requirements of the World Trade Organization in terms of new rules on fisheries subsidies.

In order to meet these challenges, new strategic directions may be required in relation to fisheries resource use.

Trends in Senegal's small-scale fishing sector

Two basic trends in the small-scale sector have been particularly important. First, there has been an appreciable shift of 'artisanal' fishing effort away from small pelagic species and towards coastal demersal species. Today, canoes using demersal fishing gear represent 94 per cent of the small-scale fleet, with only 6 per cent now equipped with pelagic gear. This intensification of demersal fishing has clearly contributed to the stock depletion of demersal species. As discussed below, this shift was largely driven by incentives for small-scale fishers to shift from supplying the domestic market with small pelagics to supplying 'higher value' demersal species to the export market. As a result, the small-scale fishery supplied up to 60 per cent of the export markets in 2009.

Second, there has been a large increase in fishing capacity in small-scale fleets. This is largely due to the motorization of pirogues, but also to the overall enlargement of artisanal pirogues which can reach up to 22 metres. Also, the adoption of more efficient fishing gear has contributed to an enhanced fishing efficiency: the emergence of fishing lines and purse seines in the 1980s and the use of combined gear which enable fishing in all seasons and target various species. Importantly, the use of ice on board has allowed fishers to sail further and longer. These combined changes in fishing technologies have led to a massive increase in fishing effort on demersal species. This has added to the already intensive fishing by national and foreign industrial trawling fleets. The national industrial fishing fleet is also marked by the emergence of more powerful ships with increased average gross registered tonnage.

Trends in Senegal's fisheries management policies

The foregoing discussion reveals significant and fundamental challenges to the sustainability of the Senegalese fishing sector with regards to ecological, economical and social perspectives. Importantly, this question of sustainability can be addressed through sound governance and decision-making. Fortunately, over the course of the past decade there has been a growing recognition of the problem and of the need to address sustainability within the Senegalese fisheries management framework. In 2000 for the first time, all major stakeholders came together in a concerted national action that led to a revised framework for fisheries policy. Another major consultation, convened by UNEP/ENDA Diaopol/CRODT[8] in 2003, helped define regulatory perspectives for the small-scale fishery (UNEP, 2004e).

In 2004, the government initiated a review of progress achieved since the announcement of the 2000 policy reforms. The review concluded that, despite the promulgation of some technical measures, the overall situation had worsened since 2000. Following this review, the government has noted its intention to improve policies towards:

- reduction of overcapacity;[9]
- control of access to resources, particularly through establishment of fishing permits in the artisanal sector and through the implementation of a system of

concessions for access rights to fisheries resources;
- control of fishing effort (for example, through temporary closures, creation of artificial reefs and marine protected areas);
- strengthening of fishermen's responsibilities through co-management and the creation of local fisheries committees;
- introduction and upgrading of monitoring, control and surveillance mechanisms; and
- upgrading of hygiene and sanitary conditions in the fisheries sector.

Implementation of these policies has advanced further in the industrial sector than with regards to the artisanal fisheries. A new administrative unit has been created and charged with industrial restructuring of the fishery and development of an industrial restructuring plan. A focal element of these policies in the industrial sector is the introduction of fishery management plans to identify possible and necessary improvements in the current system of concessions of fishing rights (licenses), taking into account the objectives and management measures established. The progressive implementation of fisheries management plans, starting with the most threatened stocks, is conceived by the fishing authorities as the most appropriate way to ensure sustainability of the resource and the fishing activity (biological, economic and environmental sustainability). These plans will form the cornerstone of the adjustment of fishing capacity and measures to control access in the industrial fisheries. Management plans for coastal demersal and deep sea shrimps were under development by the end of 2009. Overcapacity reduction would therefore begin to be effective in 2010.

Although among the stated priorities of the government for sustainable management and restoration of fishery resources is the general implementation and improved effectiveness of the artisanal fishing permit system, the implementation of a system of concessions for access rights to fisheries resources and the decentralization of fisheries policy through co-management[10] implementation in these areas is not as well advanced. This is largely due to the priority given to industrial fisheries management. The question of capacity management is also a priority on the agenda. This is planned to be addressed through buyback programmes, but such programmes are unfortunately far from being established at the moment.

Finally, on top of fisheries management policies and in an attempt to counterbalance overexploitation of marine resources and the inherent threat to coastal livelihoods and food security, the Senegalese authorities aim to foster domestic aquaculture in the country. This has been initiated with the creation of a national agency and will lead to further actions in the near-future.

Altogether, it is important to emphasize that the fisheries sector remains a key contributor to the national economy, although many important commercial fish stocks are clearly overfished, the development of the export sector has lagged behind hopes and food security has been affected by a substantial shift in national fishing effort away from supplying local markets towards selling into export markets. Also, in spite of the failure of past management policies to address such issues, the revised management framework recently released holds the potential for a significant improvement in Senegalese fisheries management.

Based on this short overview of the context for Senegalese fisheries sector, the following will examine the role subsidies have played in leading to such a worrying situation. The section will also analyse the role they may be able to play in moving Senegal's fisheries towards a more sustainable and economically rewarding future.

Direct and indirect subsidies to Senegal's fishing sector

State financial support for the Senegalese fisheries sector began shortly after the country's independence in 1960 and can be divided into several main periods. Subsidies until the 1980s largely focused on industrial fishing. Although in theory such subsidies would be likely to have a direct negative effect on the equilibrium of stocks, it turned out that they have had only a small effect (as discussed in the 2004 UNEP study on which this section is partly based). Despite very limited government support, small-scale fisheries remained competitive enough to slow down the industrial sub-sector's development. Additionally, the projects implemented under this policy were not sufficiently long-lived to have major effects on social or environmental issues. They had one economic effect, though: the inappropriate allocation of investments at the national level.

Starting in the 1980s, government support to the fisheries sector was heavily oriented towards encouraging exports and went hand-in-hand with other pro-export policies and changes (including the 1994 devaluation of the CFA franc). Unlike subsidies in the earlier period, these had mostly negative effects on fish stocks. In more recent years, an emphasis has been put on improving the situation of small-scale fishers. In spite of greater concern of national authorities for sustainable fish exploitation, the 1980s and the 1990s were a period of maximum growth of small-scale fishing and closer analysis reveals that not all the anticipated results in terms of fisheries' sustainability have been achieved. The assistance to small-scale fisheries did not manage to offset the problems arising from the increase in exports. In fact, many small-scale fishers have turned their attention to export oriented high-value species.

Within the latter period, the Senegalese government has been providing subsidies to encourage modernization of the fisheries sector, such as tax cuts on fishing equipment (e.g. out-board motors, fishing gear), subsidies on fuel, soft loans for the small-scale fishery and infrastructure building. Special support has also been given to small-scale and environmentally-friendly fisheries and processing through low lending rates, micro credits and other measures. The export industry has been subsidized since 1980 to facilitate market penetration and to meet foreign standards, while access agreements allow foreign fleets to fish in Senegalese waters. A number of specific subsidy types are discussed in the following sub-sections.

Free taxes on outboard motors and fishing gear

Annual tax reductions granted in connection with the purchase of outboard engines rose from 2 billion CFA francs in 1986 to over 6 billion CFA francs in 1994 (Kebe and Dème, 1996). In 2005 these subsidies were reduced by 0.5 billion CFA francs (Programme de gestion intégrée des ressources marines et côtières/Integrated Marine and Coastal Resource Management Project (GIRMaC), 2007).

These subsidies could only have accelerated the considerable impacts of motorization, from both the technical and the economic points of view. The use of engines significantly extended the capacity of small-scale fishing vessels by enabling them to reach previously inaccessible distant fishing areas. It greatly reduced travel times and substantially extended the time available for actual fishing operations. It encouraged migration of Senegalese small-scale fishermen along the coasts of the West African sub-region and the development of distant fishing. There can be no doubt that the introduction of outboard engines in small-scale fishing has been the main factor in promoting the enlargement of pirogues, thereby facilitating their adaptation to new fishing techniques such as purse seine.

Today, almost 90 per cent of the Senegalese pirogues are motorized. Following tests conducted with purse seine gear in the early 1970s with the assistance of the FAO, purse seining also became widespread from 1973 onwards. This was, after motorization, the second major technical break-through in pirogue fishing since 1960. Its consequences have been enormous, with an unprecedented growth of landings, resulting in the development of fresh fish marketing and of the small-scale braising industry, and technical effects of the construction of large pirogues capable of carrying large catches (up to 20 tonnes). In 2005, tax reductions granted to replace or buy new purse seine gears amounted to only 0.1 billion CFA francs (GIRMaC, 2007).

It could be peremptory to assert that withdrawal of subsidies would have prevented most of the operators from being able to self-finance motorization and other activities. Perhaps only the least profitable ones might have been forced out of the sector. And as the least profitable are mostly those targeting the small pelagic stocks for sale into domestic markets, it is possible that removal of subsidies might have had a negative impact on the country's food security (Dème, 2007).

Nevertheless, it is very likely that subsidization at least accelerated and intensified the motorization and expansion of the small scale fleet (Chauveau, 1988). Moreover, these fishing units are subject to sharp and continuous decline of catches due to decreased fish stocks and lower value of targeted species. This poses additional economic problems for the small-scale fishermen targeting pelagic fish for the domestic market. It is not easy under such circumstances to identify a simple and appropriate support policy. It does appear, however, that the programmes administered so far have used a 'one type of subsidy fits all subsectors' approach that needs to be reconsidered (see 'Lessons Learned' below).

Table 3.4 *Evolution of the Consumption of Fuel by the Small-scale Fishery and the Level of Subsidies*

Year	*Consumption (million litres)*	*Commercial value (billions FCFA)*	*Annual subsidies (billions FCFA)*	*Subsidies per litre (FCFA)*
1992	24.5	4.2	2.2	87.8
1993	24.9	6.5	3.3	132.6
1994	27.7	7.2	3.5	124.8
1995	27.9	6.6	3.2	112.8
1996	31.9	8.3	4.0	124.7
2001	39.5	12.0	4.5	114.8
2002	46.0	13.4	5.1	110.8
2003	51.3	15.1	5.7	111.9
2004	45.3	14.9	5.5	121.2
2005	41.6	16.2	5.8	138.0
2006	40.6	18.3	5.7	136.5

Source: DOPM (2007)

Fuel subsidy

The artisanal fleet consumes 40 to 50 million litres of fuel yearly. Through subsidization, the price paid by fishermen is less than that paid by the general consumer. A fixed subsidy of 87.7 CFA francs (€0.13) per litre was prevalent up to 1992. Since then the subsidy has ranged from 110 to 138 CFA francs per litre. The total fuel subsidies granted to the small-scale fishery each year, initially estimated to be less than 2 billion CFA francs (€3 million), reached almost 4 billion CFA francs (€6.1 million) in 2006. This increase has been allegedly caused by the sharp increase in the number of canoes, the use of more powerful engines and the exploration of new remote fishing areas as resources become scare.

As with the subsidies for motorization and gear, there is no differentiation between small-scale and industrial fishing concerning fuel subsidy grants, which has resulted in the industrial fleet also benefiting from reduced fuel prices. Over the period 2001–2006, the industrial fleet consumption has ranged from 60 to 85 million litres of fuel per year. With an average subsidy of 229 CFA francs (€0.35) per litre consumed, the industry benefited yearly from a total average subsidy of around 17.7 billion CFA francs (€27 million) (Table 3.5) resulting in a total of 106 billion CFA francs (€163 million) during the period. For the artisanal and industrial sub-sectors together, the cumulative fuel subsidy from 2001 to 2006 reached almost 140 billion CFA francs (€215 million), of which more than 76 per cent went to the industrial fleet.

The fuel subsidy in the small-scale sub-sector has been a decisive factor in the modernization of the small-scale fishing equipment, facilitating the use of more powerful engines, contributing to the enlargement of pirogues and helping to prolong sea trips and to open up new fishing areas. It considerably reduces the operating costs of fishing units, thereby keeping the price of fish caught by small-scale fishermen at levels compatible with the purchasing power of the Senegalese

Table 3.5 *Evolution of Annual Subsidies to the Industrial Fishery*

Year	*Annual subsidies (billions FCFA)*	*Subsidies per litre (CFAF/Litre)*
2001	16.5	213.4
2002	18.3	237.4
2003	17.7	216.2
2004	18.9	225.2
2005	14.7	231.7
2006	20.0	250.8
Average	17.7	229.1

Source: DOPM (2007)

population, but also increases fishing efforts by industrial fishing fleets.

There can be no doubt that the fuel subsidy has had a significant impact in terms of extending the length of sea trips of icebox pirogues and has led to an intensification of the demersal fishing effort and increased the pressure on many fish stocks. On the other hand, a simple suppression of fuel subsidies for the artisanal pelagic fishing units could exacerbate existing trends towards migration of effort to demersal targets for export, further reducing the availability of affordable fish for the local populations and accentuating the animal protein deficit already evident in the country. For the largely profitable demersal fishing units, because of the export markets targeted, there is a need to reconsider the fuel subsidy, in terms of reduction or even suppression. Eventually, the subsidy could be limited to the purse seine and the surrounding fishing units as they are targeting species to supply the home market for food security needs.

Below market loans and micro-credit for small-scale fishers

Insufficient financial resources has long been a major impediment to the development of the artisanal fishery sub-sector. In response to this constraint, in the early 1990s several professional associations (with support from local NGOs) set up a wide programme of mutual funding and networking to help fishers get access to credit under acceptable conditions. The Senegalese authorities, with assistance from development partners (e.g., African Bank of Development, Canadian Agency of International Development, French Agency for Development), set out several projects to create mechanisms such as mutual funds that provide an alternative to the traditional credit system[11] (high interest rates on loans) and formal banking[12] (far from caring for the needs of the fishing communities). These projects aimed to provide funds to meet the equipment needs of fisherman as well as other actors indirectly involved in fishing activities (e.g., wholesalers, artisanal fish products processors, carpenters, outboard engines mechanics). Other objectives of these development projects were related to the building of wharfs, the setting up and equipment of fish processing areas, the improvement and the introduction of new fishing technologies and investment in community infrastructure. These projects by the government and development partners mobilized a total investment of more than 9 billion CFA francs (€13 million).

Formal banking initiatives, on behalf of small-scale fisheries, have suffered from serious shortcomings reflected in the smallness of the fund's portfolio, which has not risen above 3.2 billion CFA francs (€4.9 million) in 10 years of intervention in the sector. The difficulties encountered range from restrictive conditions of access to credit to a dissuasive interest rate (12.5 per cent), as well as difficulties of loan recovery due to producers' insolvency and the lack of permanent guarantees. Following this programme interest rates are expected to drop to between 4 and 6 per cent with no guarantee or financial contribution to the investment required from beneficiaries.

Currently, financing schemes known as Mutual of Saving and Credit (MSC) exist to support artisanal fishery. Through the acceptance of small deposits, the MSC gives the most vulnerable families access to credit. The MSC has thus managed to set up a flexible policy of credit in which the short-term commitments are important and more adapted to the needs of the members, which has considerably increased the productivity of the beneficiaries.

The setting up of MSC has contributed to the fight against the practice of usury that has prevailed in fishing and other economic activities in Senegal. The decentralized financial system has played an important part in the proximity approach with the introduction of several MSC funds in relatively poor rural areas where a 'credit culture' was lacking. The availability of credit has greatly increased the productivity of beneficiaries. The MSC has created a culture of credit in many professional fishers enabling access to credit that did not exist before. Moreover, the diversification of activities related to credit has brought greater security and less dependence of fishermen on middlemen. In many localities, the MSC has enabled individuals and families living in a precarious financial situation to gain greater security by making an income from self-employment. A culture of savings has been developed among stakeholders as MSC's small deposits policy offered to the most vulnerable fish actors faster and easier access to credit. The credit has a real impact on community development as it affects activities as varied as fishing, small-scale processing, local fish trading, development of the local economy and social protection. The implantation of MSC in remote areas has contributed to a diversification of local economies and keeps people in rural areas. MSC is now widely supported by grassroots organizations. They are strongly rooted in the local economy.

Subsidized infrastructure for landing, processing and marketing

Until the 1980s, most landing centres for small-scale fisheries were quite unsanitary. They constituted domestic waste dumps and latrines for neighbouring populations. Even at the beginning of the 1990s, sanitary conditions in Senegal's major landing sites were far from satisfactory. Catches were deposited on the sand to await buyers, so the risk of contamination was significant. No parking lots or packaging areas were available to wholesalers. In the absence of storage infrastructure on the landing sites, fishermen were in a weak position as traders and were obliged to sell at unsatisfactory prices.

In order to improve product quality for sales a hundred miles away from the coastline, the Senegalese fish centres of Kayar, Joal and Rufisque with a production capacity of 40 tonnes per day were built in the early 1980s. Ice factories have been also set up across the country for a better distribution of fresh fish to inland regions. Despite these efforts, support to local marketing of seafood products has remained insufficient and marketing support is still more focused on export promotion than on service to local consumers.

Until 1992, rudimentary fish markets existed in all major landing centres and around the country. Important losses of fish were registered during peak production periods due to lack of means of conservation. To address these problems, two central markets were built, one in Dakar and another one in Kaolack.

The central Fish Market of Dakar was built in 1992 and became operational in 1993. It was financed jointly by Japan, the Government of Senegal and The Urban Community of Dakar and cost 3.1 billion CFA francs (€4.7 million). The Central Fish Market of Dakar ensures the preservation of unsold products and provides ice to fish traders at competitive prices. As the country's first central fish market it facilitates deliveries to secondary markets and helps to regulate the fish supply in the Dakar area.

The central fish market of Kaolack was built in 2003. It has played a positive role in improving the quality of marketed products in the Kaolack region and had helped to regulate the fish supply in the neighbouring cities and villages.

The construction of fishing wharfs is likely to reduce post-capture losses and has a positive environmental and sanitary impact. Therefore, such infrastructure development is likely to be beneficial both for public health and resource sustainability. The Senegalese Fishery Department is currently seeking to increase the construction of such infrastructures along the Senegalese coastline. Yet, the fact remains that the support to marketing has consistently focused on export promotion and local marketing of sea products is still receiving relatively poor treatment.

Support to small-scale fish processing

Small-scale processing stabilizes the fresh fish market as it constitutes an important and secure outlet for the fishermen during periods of overproduction. It makes use of products rejected by wholesalers. This sector also has a very important social function, in terms of employment (mostly women) and contribution to animal protein intake especially in the inner parts of the country where fresh fish is rarely available. The outputs of small-scale processing plants are also exported to countries of the West African sub-region (Mali, Côte d'Ivoire, Ghana, Burkina Faso, Nigeria etc.) thus contributing to the expansion of regional trade.

Government interventions in this sub-sector have historically been very limited and have focused mainly on the introduction of new products or new processing techniques (Institut de Technologie Alimentaire (Sénégal)/Institute of

Food Technology (ITA), 1986). Until the 1990s most processing procedures were carried out on the ground, causing production losses and unsatisfactory product quality. In order to improve the quality of processed products, to extend their period of conservation and to develop new products, public authorities, with support from development partners, have experimented with breeze block ovens along the Senegalese coastline. The first experiments date back to the mid 1980s and continued until the late 1990s. The government provided training costs and built ovens which were used for training. These ovens were then returned to processors.

In addition, at the beginning of the 1990s the construction of stores and warehouses in most processing areas reflected the strategic function of storage as an element in marketing processed products. It allowed traders to anticipate variations in demand, to face up to fluctuations in producer prices and to avoid paying for costly transport to the processing centres. Sheds and shelters were built to improve the working conditions of processors even during rainy seasons. The main fish landing centres benefitted from such investments. Other measures taken to remedy shortcomings linked to the small-scale processing sector included the PAPA-SUD (programme to support the development of artisanal fisheries in the southern region) technological dissemination programme designed in 2001 to increase output and to improve hygiene in fishery operations.

Export subsidies

Most of the subsidies mentioned above directly or indirectly benefit the export industry. In addition, there have been substantial subsidies specifically to export activities, dating from 1980.

The granting of export subsidies was part of a general policy of encouraging exports to international markets. The strategy aimed to increase competitiveness and offset certain local costs. It was motivated by a desire to contain Senegal's trade deficit after a period of import growth and poor performance of traditional exports (groundnuts and phosphates). Initially set at 10 per cent free on board (FOB) value (price at the departure of Dakar), it was raised to 15 per cent in 1983. At first, this export promotion programme was limited to agricultural products, but it was soon extended to tuna and to all fisheries products in 1986. The same year, the subsidy rate was raised from 15 per cent to 25 per cent and a more accurate definition was given to the subsidy base by introducing the criterion of 'national industrial value added incorporated in the final product' (Law 86–37 of 4 August 1986). These subsidies – which, for example, resulted in support of approximately 1.2 billion CFA francs (€1.8 million) to exports based on demersal species during the fiscal year 1991/1992 – allowed exporters to be more competitive in the international markets. This mechanism was, however, cancelled after the 1994 devaluation of the CFA franc in view of emerging possibilities of recovery of the sea fishing industry.

After the cancellation of the direct export subsidy, however, other programmes to support exports were implemented. These were more indirect aid to export, but still aimed at sustaining exports through aligning to importing countries' standards, mostly with regards to sanitary measures. Thus, in 1995, as part of the 'support to the restructuring of the fishing industry' project, the Senegalese authorities with the help of the French government started a policy aiming to align export companies and industrial fishing vessels (freezing ships) to European standards. The European market being the main destination for Senegalese sea products, this was meaningful assistance. A subsidy of 1.7 billion CFA francs (€2.6 million) was granted to some companies in a bid to finance up to 30 per cent of their investments in order to comply with health and sanitary requirements in the European market.

This project focused on export platforms and the industrial frozen trawlers and tuna ships that supplied them. However, in 2004 the EU health and consumer protection authorities visited some landing sites for the small-scale fishery and noted the need to establish acceptable standards for the artisanal fishery as well, mainly with regards to conditions on the pirogues, at the landing sites and in the vehicles transporting fish products. Following their recommendation, eight landing sites were identified along the coastline from north to south and a process to bring them into compliance with European sanitary standards started, beginning with an evaluation of weaknesses and infrastructure needs. In return for such support, Senegal agreed to keep supplying fish to the European markets. This arrangement is crucial for the fishery industry as the European market absorbs up to 70 per cent of the Senegalese fish exports.

Subsidies for sanitary improvements, ice plants and freezing and refrigerated rooms for use by the small-scale fisheries do not as such have direct adverse impacts on fish stocks. However, it remains the case that the benefits of these programmes are largely restricted to export companies. Moreover, the large extent of subsidization available to any export-oriented enterprise has encouraged many newcomers who lack independent financial or technical capacity to enter the sector. Such new entrants tend to employ very basic handling and processing techniques, resulting in a shift of export structures towards fresh and frozen products instead of towards more advanced processing with its additional value added and higher profit margins.

If subsidies did help Senegalese exporters to consolidate their position in the international market, they have undoubtedly also generated indirect negative environmental impacts. First, along with other programmes aiming at raising the level of exports, they have steadily contributed to increasing the pressure on stocks of high-value species. They have caused a shift of fishing effort targeting a whole range of species to a focused effort on coastal demersal species, which in turn has led to stock depletion and the risk of biological collapse to these stocks. And by failing to encourage value-added processing, they have sought to raise export incomes in a fashion that depends directly on increased fishing pressure.

Trade liberalization and export-oriented policies

Beyond the direct subsidies to encourage exports, there have been a number of trade-related policies that create values and incentives with effects that can be similar to export subsidies. The Lomé Convention, the devaluation of the CFA franc and the establishment of free export zones are three such mechanisms that have artificially lowered the export price for fish and therefore stimulated exports, with effects paralleling those discussed in the previous section.

Lomé Convention and Cotonou Agreement

In 1975, the European Communities concluded the trade and aid agreement known as the Lomé Convention with a group of African, Caribbean and Pacific (ACP) countries who were former European colonies. The Lomé Convention (now replaced by the Cotonou Agreement) put in place a regime granting almost all ACP products access to the European market free of any tariffs or quotas imposed on other supplying countries.

By instituting a duty-free regime, the Lomé Convention largely contributed to stronger competitiveness of Senegalese fish products in the European market. In the years following the adoption of Lomé, Senegalese exports to Europe kept growing. Between 1982 and 1991, exports rose from 90,000 tonnes to 125,000 tonnes. As noted above, fish exports since the mid 1990s have been declining for a number of reasons, including depletion of fish stocks, growing competition from other suppliers to the European market and the need to comply with new rules of market access. Nevertheless, Europe remains by far the main destination for Senegalese fish exports, accounting for about two-thirds of total exports. While Senegalese exports to other African countries account for most of the remaining one-third of export volumes, the African markets weigh only marginally in value terms. From that standpoint, Europe continues to be an essential market as it absorbs most of the high market-value exports. This situation holds true for many other African countries, with Europe as the destination market for 80 per cent of fish product exports from Africa generally.

Since demand from Europe is mostly focused on species with a high market value, the trade advantages under the Cotonou Agreement probably contributed to an increase in fishing pressure on these (already endangered) species, such as demersals, crustaceans and cephalopods.

Devaluation of the CFA franc

On 11 January 1994, the CFA franc used by 14 African countries was devalued by 50 per cent of its value against the French franc,[13] following a process of structural adjustment policies for several African countries under the auspices of the International Monetary Fund. The expectations were that devaluation would enhance export competitiveness and restore macro-economic creditability. In this context, the fisheries sector received special attention from donors as it was seen as important both for food security and for increased exports.

The devaluation of the CFA franc had a significant impact on the sector. While exports had sharply declined between 1991 and 1993, especially exports of frozen products to Europe, the devaluation immediately improved their competitiveness. It restored operating margins and boosted exports up to 125,000 tonnes in 1999 from 80,000 tonnes in 1993, an increase of about 56 per cent. Although a large part of this increase was driven by an exceptional volume of exports of octopus in 1999 the volumes of 100,000 tonnes and 110,000 tonnes achieved in previous years do constitute a high average in view of stock limitations (see Table 3.2 for export data 1990–2006).

Following the devaluation, strong external demand further inflated export prices. These price increases meant that even modest growth in export quantities resulted in significantly higher earnings. For example, between 1993 and 1994 exports increased only by 10,000 tonnes (from 80,000 to 90,000 tonnes), export receipts soared from about 50 to 83 billion CFA francs (77 and €127 million). They later continued their upward trend and reached 174 billion CFA francs (€277 million) in 1998 whereas volumes stagnated between 100,000 and 110,000 tonnes. This is a perfect illustration of a situation in which devaluation in a context of some production constraints induced a price effect rather than a volume effect. This effect not only led to the reopening of various companies which had been shut down, but it also attracted new investors. In about a year the number of enterprises in operation thus rose from about 40 units to almost 80.

Following devaluation, the exports of frozen products to Europe increased significantly from 21,000 tonnes in 1993 to 58,000 tonnes in 1999 – thus accounting for 37,000 out of the total 41,500 tonne increase in exports between those two years. This is evidence that devaluation tended to encourage export of raw products rather than helping to valorize products and to develop processing industries. Since then a decrease can be seen in Senegalese seafood exports. The same tendency has been noticed in 2009 as unprocessed products continue to make up more than 80 per cent of exports of fish products.

As with the direct subsidies to exports, the devaluation of the CFA franc had undeniable environmental impacts. Lower export prices led to an increased profitability of export companies which was reflected in an intensification of fishing effort with harmful impacts on fish stocks. The devaluation also led to an increase in import prices, but this did not restrain the process of stock depletion. Still, a distinction must be drawn between export companies and those supplying the domestic market. In the case of the former, the growth of external demand more than compensated for rising input costs, whereas the latter faced deteriorated operating accounts after devaluation, further eroding the economics of fishing for the domestic market and giving rise to fears that domestic demand might have to face serious problems of supply.

Free Zone and Free Exporting Enterprise status

The Dakar Industrial Free Zone (IFZ) was established in 1974. This zone was expected to provide an attractive framework for encouraging foreign investors to come and establish export-oriented and labour-intensive industries. Therefore, authorized enterprises were granted a variety of tax and customs reductions as incentives to investment. Later, Law 91–30 of 13 April 1991, extended these incentives to exporting industries operating outside the IFZ. In 1995, the scope of application of Law 91–30 was further extended to all agricultural enterprises operating on the national territory and exporting at least 80 per cent of their production (fishing being included in agriculture). The main objective was to boost the development of Senegalese exports to reduce the gap in the balance of trade through foreign exchange earnings and the creation of local added value. Other objectives were to encourage employment and to speed up the country's industrialization.

Duty-free exporting enterprises benefit from many customs, tax, financial, social and economic advantages, ranging from the exemption from duties and taxes (such as those levied on capital goods, equipment, commodities and finished or semi-finished goods entering or leaving the country) to the payment of reduced company tax on net annual benefits of 15 per cent instead of 33 per cent.

Such incentives have attracted sea product packaging/processing companies eager to take advantage of the growing demand for seafood products worldwide and especially in the developed countries. However, these are mostly concentrated on preparing the raw material in order to export. Accordingly, the presence of large numbers of such companies is exerting strong demand for inputs into exportable raw products rather than giving opportunities for access to value for Senegalese products. It therefore creates an additional threat to stocks of demersals, crustaceans and cephalopods without creating added value.

Foreign financial flows

In addition to the subsidies discussed above, additional financial flows come from outside the country. These flows cannot be considered as 'direct subsidies', but they are entering the national fisheries sector and thus deserve some consideration here.

Fisheries access agreements

Fisheries access agreements are an important dimension of the international economics of fishing and fish trade (see Chapter 5, 'The Special Case of Access Agreements'). They have been a significant feature of Senegal's fishing sector for decades. The contemporary era of Senegalese access agreements began following the creation of the European Economic Community, when EU policy dictated that from January 1977 EU member countries would negotiate fishery agreements under the control of the European Commission. That put an end to the era of bilateral accords concluded individually by European countries. The framework convention signed between Senegal and the European Union in 1979 made it possible to reconsider previous fishery agreements which are now based on the payment of a financial contribution, jointly negotiated and agreed. Since that year, nine protocols have been renewed every two years with the exception of the 1997 protocol and that of 2002 signed for four years each.

Financial transfers from the EU to the government of Senegal have averaged between 8 and 9 billion CFA francs (€12–14 million) per year (1994–2001) for the two agreements. For the fisheries agreement 1997–2001, financial flows (financial compensation, fees on fishing licenses and various taxes) have been estimated at 36.7 billions CFA francs (€56.5 million). Economic flows comprising local expenditures of the European fleet, intermediation port expenses, and maintenance and repairs of ships are estimated at an additional 11.3 billions CFA francs (€17.3 million). The social effects (various direct and indirect jobs generated) amounted to 1.5 billion CFA francs. In total, the flows generated by the fisheries agreement Senegal/European Union over the period 1997–2001 amounted to 49.5 billion CFA francs.

The total financial compensation of the fishery agreement Senegal/European Union covering the period from 01/07/2002 to 31/06/2006 was €64 million. Of the total, approximately €12 million were earmarked for the implementation of targeted actions aimed at promoting resource conservation and sustainable development of fisheries. Activities funded included the monitoring of the resource (€500,000), strengthening the security of the artisanal fisheries (€500,000), control and surveillance of fishing activities (€700,000), institutional support for the establishment of sustainable fisheries (€500,000), strengthening of human capacity (€700,000) and evaluation and audit of partnership actions (€100,000).

The EU-Senegal agreements are, in fact, a form of disguised subsidy – not to the fishers of Senegal, but rather to their competitors in Europe's distant water fleets. The major part of the access fee is borne by the European flag state authorities. The part paid by the European ship-owners represents only about 10 per cent of the access fee, the remainder corresponding to counterpart funds paid by the European Commission. This situation allows fleets which otherwise would probably have been forced to withdraw from a highly competitive sector in Europe, to capitalize on their fishing equipment in African waters. It is also unclear whether even the total access fee reflects a good rate of return for Senegal on the value of the resources to which access is granted – raising additional questions about the true value of the subsidy that the access represents (ICTSD, 2006d, UNEP, 2008a).

Since 2006, there has been no fishery agreement between Senegal and the European Union as a consequence of disagreements over target species, the amount of compensation and the employment of Senegalese fishermen in European vessels. With the overexploitation of the coastal demersal resources and the need to reserve pelagic fish for local food supply, the Senegalese authorities offered European vessels access to tuna and the deep sea demersal fisheries. The European Union, however, is seeking access to coastal demersal fisheries for its Spanish vessels. Since 2006, joint venture fleets (French, Spanish and Greek) have been granted access on a bilateral basis to operate on tuna and deep sea demersal resources.

The prospects of concluding new fishery agreements with the European Union are currently poor, as the local fishing industry, environmental groups and other stakeholders are becoming more aware of the state of overexploitation of resources and are expressing opposition to the signing of new fishery access agreements by Senegal. The more advanced technologies, higher catch per unit of effort, better spatial occupation of the fishing grounds and more sustained fishing effort of foreign vessels give them an advantage in terms of productive efficiency with regard to the local fleet. This results in serious competition between fishing products landed by the European fleets operating in the Senegalese waters and Senegalese products on the exports markets. Resources like demersal fish stocks located in deep waters varying from 150 to 1000 meters require the use of very powerful fishing units with specific rigs. Such fishing units, are very expensive and are not affordable for the Senegalese operators. This explains the relative exclusive presence of European industrials in the deep dermersal fisheries. These obvious technological advantages of the European Union boats improve their fishing power by enlarging fishing zones, by ameliorating their catch per unit of effort, their storage capacities and their autonomy.

Fishing agreements are believed by local fishermen and NGOs to be one of the main causes of the overexploitation of marine resources in Senegal. By lowering the production costs of fishing efforts units, fishing agreements encourage them to increase the export level and fish beyond the economic optimum compatible with sustainable resources management. Facing constraints of conservation capacities, vessels also do not hesitate to engage in environmentally destructive 'high grading' – the practice of throwing back overboard low-value catches or catches that are not of the desired size in order to maximize the revenues of fishing trips. Furthermore, in the absence of proper means and equipment the Senegalese authorities are unable to monitor and enforce foreign fleet compliance with Senegal's fishing regulations.

Official Development Assistance

An analysis of the investment programmes listed in the Senegalese Consolidated Investment Budget over the period 2001–2005 reveals that Official Development Assistance (ODA) contributions to support development of Senegal's fisheries sector exceeded supports from national government resources. Over that period, the national budget for fisheries budget was 22.4 billion CFA francs,[14] while donor countries contributed a total of 29 billion, as follows: Spain (12 billion CFA francs), Japan (10.7 billion CFA francs), the European Union (3.4 billion CFA francs) and France (2.9 billion CFA francs). Since 2005, ODA has also surpassed investment taken directly from the national budget in Senegalese fisheries.

Development assistance projects have focused on infrastructure development (10.1 billion CFA francs, over the period 2001–2005), followed by projects providing institutional support to the fisheries sector (6.9 billion CFA francs). Projects are also undertaken in the area of resource assessment, support to professionals (especially women) and conservation of the marine environment. All these projects share the common objectives of enhancing export competitiveness, improving sanitary standards of industrial and artisanal fisheries and strengthening the management of fisheries resources.

There are several other sectoral programmes that have a positive interface with fishery policies. These include the national programme for the development of rural infrastructure, the national programme of sectoral transports, the programme to fight against poverty, the national programme of integrated development in the health sector, the programme of rural electrification, the programme of agricultural services and support to producer organizations and the national programme to protect the environment.

Subsidized transfers of fishing capacity from Europe

An additional point worthy of note here is the transfer of fishing capacity from the European Community to West Africa that occurred up until 2004, in line with the development of joint fishing ventures in Senegal. The fishing vessels in Europe which benefited from a decommissioning scheme were able to transfer their vessels to a third country (most often a country in West Africa, such as Senegal). A vessel could then be decommissioned in Europe and transferred to West Africa to continue fishing there, while the owner of the boat was subsidized for leaving the activity in European waters, but built up a joint venture for fishing in Senegal. This mechanism of capacity transfer as part of a decommissioning scheme was banned in 2004. However, it can certainly be considered as an external subsidy to the Senegalese fisheries sector that is likely to have impacted the West African fish stocks very heavily from the 1990s to 2004.

Proposed subsidies for reducing overcapacity

As noted above, the problem of overcapacity in Senegalese fisheries has been recognized and the government has indicated that adjustment of capacity is one of the high priority goals for improved fisheries management. While the foregoing discussion makes clear that subsidies have been among the causes of the current overcapacity, there is also room for subsidies to be part of the solution. To this end, the government has already initiated policies to use subsidies as one tool to reduce overcapacity and to reach a sustainable level of fisheries.

The government's proposed capacity reduction programme takes consideration of both small-scale and industrial fisheries. An evaluation of the cost of such a programme was conducted in 2006 by the fishery department (MEMTM, 2006). For artisanal fisheries, an excess capacity of 4000 small-scale fishing units is estimated out of 10,000 fishing units that are specifically targeting heavily overexploited coastal demersal resources. For industrial fleets, one-third of the 145 boats should be withdrawn to adjust overcapacity and reach a sustainable fishing effort level (MEM, 2004). The proposed approach considers three possible scenarios for payment of a premium for destroying ships that are at least 35 years old in 2008, or to deploy boat owners in activities other than fishing.

Unfortunately, slow progress has been made towards implementing these plans, as negotiations with stakeholders are still ongoing over the conditions of withdrawal and the necessary funds have not yet been found. However, it is worth reviewing some of the details of the proposal.

Table 3.6 *Proposed Capacity Reduction Programme – Projected Costs*

Lines of the funds	*Millions of CFA*	*% total cost*
Buying back the artisanal fishing units	9094	36.5
Buying back the industrial fishing units	6460	25.9
Grant for wholesalers and processors	1555	06.2
Grant for the conversion of fishermen to other activities generating revenues	2601	10.5
Credit from national financial institutions	5198	20.9
Total	24,908	100.0

Source: MEMTM (2006)

As it attempts to move ahead, the financial effort of the state will focus on creating a fund for the adjustment of fishing capacity. The fund will finance the buying back of the artisanal fishing units and supporting the loss of revenues of owners and crew members to be withdrawn from the fishery. It will also fund the repurchasing of fishing industrial units and payment of the loss in investment and income for owners and crews, plus a grant for wholesalers and processors and a grant to finance the retraining of fishermen. Table 3.6 above summarizes the projected cost of the programme.

If successful, the programme will have both environmental and economic benefits. By reducing excess fishing pressure, the programme will improve fisheries productivity, thus allowing fishermen still active in the fishery to profit from a more abundant natural resource. Based on the fall of 30 per cent of the total catch of demersal species over the past 15 years and the natural productivity/resilience of some species (Gascuel et al, 2004), effective capacity reduction could allow progressive restoration of the average productivity of the affected fisheries of 1–2 per cent per year. The resulting economic rate of return would be in the order of 18 per cent over 15 years. Thus, the cost of adjustment should be offset by the gain in the value of production over a decade, while fishers enjoy significant improvement in their profitability.

Through this programme, the state should also gain significant revenues from the fishery sector through increased customs duties on imported fishing equipment, increased revenues from the special fuel tax for fishermen and improving revenue from fish licensing. If capacity reduction were sufficient to create long-term profitability in the industry, the state could pass from a position of providing financial support to fisheries (subsidy on fuel, tax cuts in fishing equipment) to a situation of collecting receipts for the public treasury.

It will, however, require more than just funding and stakeholder agreement to make such a buy-back programme work. Implementation will also depend on more careful management of foreign participation in the industrial fishery sector and the identification and registration of the artisanal fleet. As noted above, efforts to implement such measures are ongoing.

Conclusions and way forward

The impacts of fisheries subsidies in Senegal

Among the key problems facing Senegal's fishing sector, there are four that have been repeatedly mentioned above:

1 The overcapacity of Senegal's fishing fleets is causing excess fishing effort and is sharply reducing profits and returns on investments for many fishers.
2 Generalized overfishing (and particularly on high-value shallow demersal stocks) is threatening many of Senegal's most valuable fisheries with biological collapse.
3 Policies that have created strong and undifferentiated incentives to pursue export markets have shifted a significant portion of national fishing effort away from small pelagics and on to shallow demersals, simultaneously depleting an environmentally sensitive resource and endangering Senegal's food security through reductions in domestic supply.
4 Despite the overemphasis on policies to promote exports, Senegal's fish exports have failed to grow substantially while exports of unprocessed products with low value-added have increased rather than decreased.

While clearly a variety of factors has contributed to these negative aspects of Senegal's status quo, there is no doubt that subsidies have played a role in creating all four of these circumstances. Overcapacity and overfishing have been directly encouraged by subsidies to capital and operating costs, such as subsidies for motorization, gear acquisition and fuel. A heavy emphasis on export promotion since the 1980s (and particularly intensified around the time of the CFA devaluation in 1994) included a number of direct and indirect subsidies, such as direct financial rewards for export sales, the targeting of subsidies to export-oriented fishing capital and port infrastructure and the creation of Industrial Free Zones and other tax and duty havens for export-oriented enterprises.

The impact of these subsidies on Senegal's food security merits particular attention. The prospects of better profits in export markets through subsidies and cheaper availability of fishing gear and fuel attracted many newcomers to the sector, but fish stocks are not infinite. This situation was reflected in a sharp rise in commodity prices. The extra demand from the export units resulted in an intensification of the fishing effort at a time when demersal captures were already clearly on the wane. Additionally coastal demersal species targeted by the European fleet are those heavily exploited by national fleets and exported in high quantities to the European Union and to Japan in the case of cephalopods.

In effect, Senegal's national vessels increasingly operate on behalf of the European consumer, not on behalf of the African consumer. Analysis of the different national fleets and their national fish market supply shows a progressive decline in supply due to increasingly higher incomes through sales to export markets. On local markets, only those products with a low commercial value are available. The scarcity of high value species on the one hand and the high level of exports on the other have led to a compression of national market supply (Dahou and Dème, 2002; Failler et al, 2004).

Consequently, exported species are seldom available on local markets, and even when they are available, they are frequently sold at prices that are not affordable to local households. The absence or low level of landings (except for small pelagic, which are currently in a period of abundance) have a number of effects:

- A decrease in the availability of fish on the domestic market and a rise in prices, which in turn leads to a decrease in the purchasing power of local households, the majority of which already live below the poverty line.
- Traditionally consumed species are being replaced by species that were either not consumed, or rarely so, a decade ago. Demand for low value species results in increased consumption of species that were previously left for lower-income households.
- The substitution of poultry for fish occurs because of lower prices for white meat than for fish (poultry that is to some extent imported from Europe in particular thanks to European export subsidies).
- The opposition of Senegalese public opinion to the conclusion of fishing agreements.

Overall, it appears that the high level of national exports and fishing agreements limit the supply for the domestic market with negative consequences on food security in Senegal.

The foregoing does not mean, of course, that all of Senegal's fisheries subsidies have had negative impacts. As noted, loans and other mechanisms to increase access to capital for fishers have tended to create a culture of savings and investment and have often helped some of the most vulnerable fishermen. Similarly, subsidies to port infrastructure and processing can help improve product quality and value, which result in higher profits for fishermen without necessarily intensifying fishing pressure. And, of course, if subsidies can be effectively applied to help reduce and adjust capacity levels, this will have significant environmental and economic benefits.

Lessons learned

From the discussion above it is possible to draw a few conclusions that may be of use to policy-makers and stakeholders:

- Granting subsidies that increase capacity and fishing pressure can directly contribute to overfishing if done in the absence of proper regulatory control.
- An over-emphasis on export promotion can have unintended consequences for the organization of production within the sector and for product flows, with implications both for stock health and for food security.
- To function properly, subsidies must be designed and implemented in ways that carefully target the kinds of activities and actors to be subsidized.

This last point is an important recurring theme of the discussion above and it deserves some further elaboration. Fuel subsidies, for example, have significantly different implications for the demersal and small pelagic sectors and yet no distinction between them is made by current Senegalese policy. It could well make sense for fuel subsidies to be phased out of the demersal fisheries, while being carefully maintained in the small pelagic fisheries (subject to appropriate management conditions).

Similarly, Senegal's experience with export subsidies also suffered from insufficiently precise targeting. Broadly available rewards for anyone able to fish for export resulted in a wave of new entrants into the demersal fisheries who had neither the necessary skill nor financial means to push the industry towards higher value-added activities. If instead export subsidies had only been selectively available to fishers, and had more specifically targeted value-added processors, export revenues might have been higher, with lower pressure on stocks. Moreover, no steps were taken to offset the negative financial impacts of the 1994 devaluation on fishers focused on domestic sales, who saw increases in both investment costs and operating costs when the CFA franc was devalued. Rather than use subsidies to help ensure continued supplies to the domestic market, untargeted subsidies directly reduced Senegal's food security.

Perhaps most fundamental from an environmental perspective is that none of the subsidies mentioned here were targeted on the basis of conditions in specific fisheries. European market demand is mainly for high-value, 'noble' species. But these shallow demersal fish stocks are far more environmentally brittle than the small pelagics that live in the rich upwelling off Senegal's shoreline. But the biological sensitivity to overfishing of particular target species was not considered in forming subsidies policies. Nor have policies been made responsive to evidence of overfishing. On the contrary, subsidies have helped offset declines in catch per unit effort (CPUE) by making it financially possible to continue fishing where low returns would otherwise cause fishing to stop.

Subsidies on the agenda of government and stakeholders

At the level of small-scale fishing, interviews with public authorities and industry representatives and professionals in 2009 have highlighted a strong desire for construction of additional infrastructure and for enhancing fish production and allowing professionals to operate in better working conditions. State authorities have insisted on the fact that, from now on, new infrastructure should meet strict standards to ensure the quality and hygiene of products and in the medium-term, their traceability. Particular emphasis will be placed on the sanitary condition of pirogues and the landing sites. Future investments by the state will also be devoted to the rehabilitation and upgrading of existing infrastructure.

With respect to industrial fishing, ship-owners and the public authorities expressed similar concerns about maintaining the important achievements on the policy side and on the upgrading of enterprises. A broad consensus among stakeholders is to focus in the short and medium term on the implementation of the programme of safeguarding the national agreement to get access to

European markets by strengthening local institutional capacities, the upgrading of production units (plants, freezer vessels) and the restructuring of the processing industry. The establishment of a fund for the restoration of the fishery industry, highly demanded by professionals in the sector is largely supported by the fisheries administration. The Senegalese government has expressed a commitment to mobilize resources for the fishery sector, including through international financial cooperation to support the process. These investments need to be accompanied by effective management measures of fishery resources in order to avoid a continuing race for fish which would increase overcapacity and overfishing.

The way forward and pathways for improvement

Fishery resources are necessary for long-term poverty alleviation and for securing economic growth. Therefore subsidies should not be addressed only in terms of their short-term benefits to the fisheries population and industry. Rather, they should be examined from a broader perspective that considers their direct and indirect effects in inducing overcapacity and overfishing.

The discussion above specifically indicates the need to turn away from subsidies and other policies that seek to pursue development based on an increased exploitation of natural resources while marketing them without advanced processing. It also plainly calls for policies that give greater priority to protecting domestic fish supplies and to addressing the needs of Senegal's most vulnerable communities.

In short, subsidies that provide an incentive for overfishing have to be addressed with urgency and reform has to be initiated based on a balancing of interests between different stakeholders.

In addition to a fundamental reorientation of Senegal's subsidies away from encouraging overfishing and towards capacity reduction, enhancing value and improving management, additional effort will be needed to ensure that subsidies are more carefully targeted and are more responsive to realities both in the fisheries themselves and in fishing communities. There is a range of economic, social, biological and environmental issues that are important to consider when discussing fisheries in Senegal, namely:

- the overcapacity of the fishing fleets operating in national and international waters (Gascuel et al. , 2004);
- the loss of biodiversity and biomasses that has occurred over the last 50 years;
- increased poverty, either at the local level (in fishing communities) or at the national level, as revealed by an increase in external indebtedness (Failler and Kane, 2003);
- the introduction of fishing management practices designed to address the problem of overfishing and non-sustainability, but which, in doing so, raises the cost of fishing activities.

In addition, more detailed studies on subsidies and their impacts in both Senegalese and adjacent waters is recommended to set up strong bases for discussion and the design of sound policies. This work cannot be performed without taking into account all other factors that generate externalities and lead to the depletion of marine resources and the crisis of the fishing sector.

Finally, the interviews conducted clearly indicate that there is still incomplete knowledge of the amounts and forms of subsidies in the fishing sector and their effects are even less known. Information on subsidies and their effects should be made widely available in Senegal as well as in the context of WTO current and forthcoming negotiations on subsidies.

The Impact of Fisheries Subsidies on Tuna Sustainability and Trade in Ecuador*

The following section identifies and evaluates Ecuadorian fisheries subsidies affecting tuna and seeks to estimate and evaluate the main subsidies in other tuna fishing countries affecting Ecuador's tuna industry. It assesses the impact of such subsidies on trade and the environment. Furthermore, it discusses subsidies reform within the context of national and regional fisheries management policies, the WTO negotiations on subsidies reform, as well as market-driven reform through certification and eco-labeling schemes.

Introduction to the fisheries subsidies problem

Fisheries subsidies present an issue of great importance due to the connection between the support received by the fishing industry and the overexploitation of marine resources. Fisheries subsidies, besides causing irreparable damage to the environment, can also generate unfair competition and distort trade, mainly affecting developing countries.

Tuna fisheries are especially affected by this situation. Worldwide, approximately 8 per cent of tuna have been depleted, while 50 per cent are fully exploited and run the risk of overexploitation. The risk to tuna stocks by subsidization is therefore rather high. (FAO, 2006c)

The negative impacts of fisheries subsidies have given rise to efforts by governments, non-governmental organizations and civil society, geared towards reducing and preventing their negative consequences. At the initiative of multiple countries, work is ongoing in both the WTO and the FAO towards improving the subsidies discipline policies and marine resource sustainability.

In 2001, at the launch of the WTO Doha Round, it was agreed to 'clarify and improve' the WTO disciplines on fisheries subsidies. This discussion has gained strength in the past few years. Presently, within the WTO framework, a text for new disciplines is being debated according to which a large number of fisheries subsidies would be banned.

Despite the difficulties inherent in this type of negotiation, there is basic agreement that the reduction in fisheries subsidies is necessary, particularly of those related to the enhancement of fishing capacity and fishing effort. Beyond an outright ban on certain subsidies, however, there remains a question of the kinds of measures or criteria to be adopted in order to discipline those subsidies that continue to be permitted. Some countries and international organizations have suggested that criteria be established with which countries would have to comply

* This section is based on a summary of the UNEP-CPPS-Ecuadorian government study 'Impacto de los subsidios pesqueros en la sustenabilidad y el comercio del atun en el Ecuador'. The author of the study and its summary is Ivan Prieto. The project was managed by Alfonso Jalil, CPPS, in collaboration with UNEP. Important contributions to the project were made by Jimmy Anastacio, Martin Velasco, Rafael Trujillo, David Schorr, Pablo Guerrero and Ramón Montaño.

Table 3.7 *Subsidies for Fishery Activity in Ecuador*

Subsidy/Help Programmes for the Fishery Sector	*Estimated annual amount (thousands of US$)*	*Relevant to the industrial tuna fleet*
Tariff deferment for the importation of capital goods, supplies and raw materials for the fishery sector in general	2000	Yes
Loans to the fishery sector in general	170	Yes
Subsidy for the installation of Satellite Monitoring Devices for minor vessels.	524	No
Fuel subsidy for artisanal fisheries	12,000	No
Infrastructure development programme for the artisanal fisheries sector	50,000	No

Source: UNEP-CPPS (2009)

as a precondition to the use of subsidies in accordance with WTO norms or any other relevant norms required by international law and existing agreements.

Two previous UNEP publications are of particular relevance to this case study. First, 'Analyzing the Resource Impact of Fisheries Subsidies: A Matrix Approach' demonstrates that the potential damage caused by fisheries subsidies depends on the conditions in which they are granted and in particular on fishery management (see Chapter 2). The Matrix approach provides a rubric for analysing the risks of the subsidies examined in this study. Second, 'Sustainability Criteria for Fisheries Subsidies', suggests a range of sustainability criteria to be considered by governments and other relevant forums (WTO, FAO, etc) as preconditions to subsidizing a fishery (see Chapter 5, 'Sustainability Criteria for Fisheries Subsidies'). The proposed criteria have a three-fold focus: the health of the fish stock, fleet capacity and the quality of the fisheries management system. These serve as an important input into the context evaluation and recommendations in this study.

Main fisheries subsidies in Ecuador

The tuna industry is one of Ecuador's traditional industries; nearly 200,000 Ecuadorians depend on it. The tuna fleet captures between 25 per cent and 30 per cent of all the tuna-like species catches in the Eastern Pacific Ocean (EPO), with purse seining representing the main fishing method used (www.iattc.org).

Both national vessels as well as those sailing under a foreign flag, which operate in Ecuador under association contracts, supply the Ecuadorian tuna processing industry. It can be estimated that the associated foreign fleets supply between 40 per cent and 50 per cent of the raw material.

As can be seen from Table 3.7 above, fishing subsidies in Ecuador are mostly destined to strengthen and advance the development of the artisanal fishing sector and the vulnerable coastal communities in which they operate, with fisheries located mainly within the 'Exclusive Economic Zone' (EEZ). Of the programmes listed in the tables, the recent tariff deferment for capital goods and raw materials imports, as well as the credit lines that offer certain flexibility in

Table 3.8 *Tuna Catching Countries in the EPO*

Country	*Average EPO Catch Participation 2000–2007*
Ecuador	27.31%
Mexico	20.00%
Venezuela	11.95%
Panama	8.14%
Japan	4.94%
Spain	2.81%
Vanuatu	2.34%
Korea	2.01%
Colombia	1.79%
Taiwan	1.66%
United States	1.32%
Nicaragua	0.72%
China	0.56%
Honduras	0.59%
El Salvador	0.52%
Costa Rica	0.18%
Belize	0.17%
French Polynesia	0.18%
Perú	0.16%
Chile	0.01%
Guatemala	0.00%
Canada	0.00%
Other	12.63%

Source: IATTC, last accessed January 2009 (www.iattc.org)

loan repayment schedules for the fishing sector, would be measures favouring the national fishing industry and, particularly, the tuna fishing industry.

The price of diesel in Ecuador, fixed by the monopoly of the public company, has frequently remained below international prices. On other occasions the reverse situation has occurred, when local diesel oil prices were above international levels without evident effects on fishing effort level. Although it might be argued that the fuel subsidy in Ecuador is not specific to the fishing sector, and thus might technically fall outside some definitions of 'subsidy',[15] the price of diesel for fisheries under the conditions given in Ecuador has at times provided a benefit for fishing vessel owners.

Main subsidies in other tuna fishing countries in the EPO

Due to the concentration of the Ecuadorian fleet in the EPO, this region is of particular interest to Ecuador (Table 3.8).

Most countries that operate in the EPO maintain significant fisheries subsidies. Among Latin American countries, the principal ones are Panama, Venezuela, Colombia and Mexico, which grant (i) fuel subsidies, (ii) rural fishery development projects, where significant amounts of money are invested for fortifying artisanal fishing and (iii) tax reduction on fishery activity, to name the most important.

Table 3.9 *Subsidies in Important Fishing Regions (US$ thousands)*

Types of fishery subsidies	*Japan*	*Korea*	*Philippines*	*Indonesia*	*China*	*Spain*	*Thailand*	*Taiwan*
Vessel construction, refitting and modernization	37,491	63,681	194,485	–	24,200	64,071	67,254	79,400
Fisher development projects and assistance services	142,172	–	454,665	36,677	125	63,658	–	–
Fishery port construction and renovation	–	160,977	–	91,200	64,625	11,765	–	126,514
Programmes for storage and processing infrastructure and trade assistance, including price maintenance programmes	43,008	4422	2180	98,200	100	59,500	66,960	6300
Tax exemption programmes	50,958	4157	350,482	–	149,990	–	39,173	
Payment for access to other territorial waters: programmes with other countries be they for technological transfer, market access, including monetary transfers	200,000	43,606	–	–	193,418	111,047	–	21,098
Fuel subsidies	1114,750	331,380	168,300	218,890	1,815,660	119,943	241,280	119,610
Assistance programmes for fishers, income maintenance programmes	512,333	–	–	–	1,508,002	248	–	14,500
Fishing vessel buyback programmes	19,010	29,453	–	–	–	88,094	–	–
Fishery management programmes and services	2,807,057	48,558	94,593	84,549	11,124	22,672	24,625	21,794
Fishery development and investigation programmes	–	4953	47,942	–	875	6715	2379	4081
Maintenance of protected marine areas	33,046	2250	14,182	18,100	4136	13,780	3517	3214
Total	4,959,825	693,439	1,326,829	547,613	3,772,255	561,493	445,188	396,511

Source: FERU – UBC (January 2009)

Other countries operating in the EPO tuna fisheries are major distant water actors such as Spain and Japan. These countries are known for the significant amounts of subsidies that they give to their respective fishery sectors. Their impact on Ecuador's tuna industry is considered in the more global context of the next section.

Subsidies in tuna fishing countries in other fishing regions

The Western and Central Pacific Ocean (WCPO) is the fishing region where the most tuna is caught. In this fishing region particularly, tuna fleets (from some of the countries considered by leading studies to be among the largest providers of fisheries subsidies) are active. According to a recent University of British Columbia (UBC) study, these same countries – Japan, China, Indonesia, Philippines, Thailand, Republic of Korea and Spain – add up to over 35 per cent of the total amount of fisheries subsidies on a global scale (see Table 3.9). They also play a dominant role in tuna fishing and tuna commerce, making up more than 50 per cent of tuna catches worldwide.

The combination of high tuna catch levels and high levels of subsidization makes the aforementioned countries of particular interest to this study. Fishing assistance in these countries was accordingly reviewed, principally using information published by the Fisheries Economic Research Unit of the UBC. The most noteworthy and relevant data are as follows:

- Japan has been identified by other studies as the largest fishing activity subsidizer. However, half of the funds granted have been identified as going towards programmes directed at the strengthening of fisheries management in this country. The data observed from 2000–2006 show no significant increase in fishing effort (measured in days at sea), but did reveal a significant increase in 'catch per unit effort', resulting in an increase in Japanese tuna catches. It can be assumed that the increment in productivity may somehow have been facilitated by the fisheries subsidies which stimulated technological development especially considering the important amount destined for vessel modernization.
- China is another main fishing subsidizer whose fuel subsidies and income sustainability programmes for fishermen stand out. The data observed for the tuna fishery of this fleet during the 2001– 2006 period reveal growth levels which reflect the country's considerable development of this activity.
- Republic of Korea's fuel subsidies and subsidies destined for construction and renovation of fishing ports are also noteworthy. Tuna fishing statistics show that this activity has expanded during the past few years, with an increase in fleet catches, as a result of an increase in fishing effort.
- Taiwan has allocated subsidies for the development of fishing port infrastructure, as well as for fuel and fleet construction, renovation and modernization. As far as tuna fishing is concerned, relatively stable tuna catches have been observed since 2003, although effort has tended to drop slightly.

- In the Philippines the main subsidies are those allocated for the development of fishing projects and tax exemptions. Fishing effort statistics show a period of slight decrease in effort during 1999–2003 and recovering in 2004. As far as catches are concerned, these have been on the rise since 2001 and – while fishing effort increased in 2004 – fleet productivity decreased notably. This means that the fleet may have needed to increase its fishing effort in order to maintain its catch volume.
- Spain is one of the main countries to receive government help geared towards the fishing sector. Subsidies allocated for fuel, for the repurchase of fishing vessels and for access to other territorial waters are noteworthy. Statistics reveal that the Spanish tuna fleet is especially active in the Indian Ocean, where they obtain more than 50 per cent of the Spanish tuna catch.
- Other countries that stand out in tuna commerce and capture are Thailand, which has an important tuna processing industry, and Indonesia, which like other previously mentioned countries is one of the main suppliers of raw tuna for the Thai market.

Fishery subsidies impact on the environment

To assess the environmental impact of fisheries subsidies in the focal countries mentioned above, information was applied both from previously published sources and from data gathered during the course of this study. For the assessment of subsidy impact on fishing resources, the methodology adopted was based on the 'matrix' approach proposed in Chapter 2.

According to UNEP's matrix analysis, the effect of fisheries subsidies on fish stocks and the environment depends on the conditions under which the subsidies are granted, on the effectiveness of the management system and on the bio-economic conditions of the fisheries sector.

In the case of Ecuador, the study showed that the fisheries subsidies currently employed in existing management and industrial conditions would be categorized by the UNEP matrix approach as for the most part 'probably harmful' to fish stocks. They are evaluated on the premise that the fisheries are being exploited at full capacity and that the current fisheries management system is a 'catch control' system.

Each subsidy granted by countries fishing in the EPO, as reported by UBC, was classified according to the matrix approach categories under 'catch control' and 'full capacity' scenarios. This matrix evaluation of the environmental impact of fisheries subsidies in the EPO reveals that many fisheries subsidies to other Latin American fishing industries were, like Ecuador's, in the majority, 'probably harmful' to the resource. Again, it remains unclear what effect these subsidies have had on tuna fisheries in particular. Similarly, UBC data placed into the UNEP matrix under the aforementioned stated scenario suggest that for Spain and Taiwan 'harmful' subsidies to fish stock are above 50 per cent of the total subsidies granted (see Figure 3.1.)

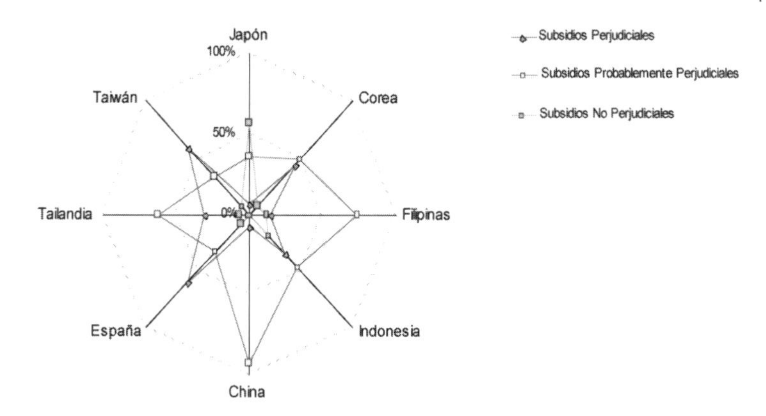

Impact	*Japan*	*Korea*	*Philippines*	*Indonesia*	*China*	*Spain*	*Thailand*	*Taiwan*
Harmful	6.0%	43.6%	14.8%	34.6%	7.5%	59.6%	30.1%	58.8%
Possibly Harmful	36.7%	48.4%	73.4%	46.7%	92.1%	32.7%	63.0%	33.8%
Not Harmful	57.3%	8.0%	11.8%	18.7%	0.4%	7.7%	6.9%	7.3%

Figure 3.1 *Impact of Fisheries Subsidies on Fishery Stocks (Japan, China, Taiwan, Thailand, Indonesia, Philippines, Spain)*

Source: Elaborated based on data presented by the UBC's FERU (available at www.seaaroundus.org/) 2009

Effects of fishery subsidies on trade

Fisheries subsidies, besides affecting marine resource populations, can influence the commerce of fisheries products and affect the interests of competing countries. The following matrix provides a general idea of the effects of different types of fisheries subsidies on the economy and trade of the fishing industry (Table 3.10).

As can be seen from Table 3.10, some subsidies allow fishing fleets to reduce their costs of operation, which decreases the marginal cost of fishing effort assumed by ship owners (UNEP, 2004a). This results in larger catches and a greater supply of raw tuna, with the consequent reduction in international prices.

The fall in the price of raw tuna leads to reduced prices of canned tuna for the processing industry and thereby to a consequent rise in the derivative demand and incentives to increase the production of canned tuna. This in turn gives rise to an increase in the supply of processed tuna, such as in the case of raw tuna, in which increases in the supply lead to decreases in the price of both products.[16]

The reduced prices of raw material result in an increase in the supply of exports of the country that receives subsidized raw material. This leads to a new equilibrium condition with reduced international prices and a shift in trade from the country with lower capacity to subsidize the fishery industry to the countries with important fisheries subsidies.

Table 3.10 *Effects of Subsidies on the Economy and Trade of the Fishing Industry*

Type of Fishery Subsidy	*Effects*	
	Direct	*Indirect*
Vessel construction, refitting and modernization	Increase in transport capacity, better efficiency	Lower capital cost
Fishery development projects and assistance services	Generate incentives to enter into the industry	Could encourage an increase in capacity or effort in the medium or long term
Fishery port construction and renovation	Better efficiency in unloading	Makes the fishing industry more profitable, creates an incentive to pursue fishing activities
Programmes for trade assistance and for storage and processing infrastructure, including price maintenance programmes	Increases the income generated in the industry	Encourages increases in the production level
Tax exemption programmes	Variable cost reduction	Can encourage increases in fishing effort
Payment for access to other territorial waters: programmes with other countries be they for technological transfer or market access, including monetary transfers	This subsidy reduces the fishing cost per unit of effort of the vessels in distant waters	Can encourage an increase in the fishing effort in distant fishing zones
Fuel subsidies	Reduction in variable operation costs	Increase in fishing effort Can have technological effects since it can encourage the use of stronger and more powerful engines or refrigerating systems
Assistance programmes for fishers. Income maintenance programmes	Reduces the expectations of losses in the industry	Encourages vessel owners to stay in the industry. Discourages disinvestment
Fishing vessel rebuy programmes	The intention of these programmes is to reduce fishing capacity, however, its effectiveness is questioned	Under a system in which incentives to increase catches exist, these programmes may or may not be entirely effective since they can generate capacity replacement, that is, allow the increase in more modern vessels, with better capacity, power and better storage systems
Fishery management programmes and services	Sustainability of the resource as a way of procuring the sustainability of the industry	Reduction in costs of some services that are not paid by the industry or which are partially paid
Fishery investigation and development programmes	Investment in investigation, covered by public institutions, such as in the development of relevant public information of a fishery	Possible improvements in efficiency, for instance in the development of better catching or processing methods

Source: UNEP (2004a); Munro and Sumaila (1999); FAO (2004e); WWF (2003); APEC (2000)

The supply of additional fish resulting from subsidies of industrialized countries causes lower prices compared to those that would prevail in the international markets in the absence of subsidies. This means a decline in export earnings for developing countries. In the medium term, the lower prices may thus discourage the fishing effort of developing countries and thereby cause a shift of the activity to industrialized countries.

However, the possibility that low prices will give rise to a decrease in fishing effort undertaken by ship owners in the short term deserves to be studied in depth. In fact, in view of the lower prices, there is the possibility that the ship owners will also increase the fishing effort in order to compensate the income deficit caused by the low prices. This necessity to intensify the effort could in some cases also discourage the compliance with certain conservation measures.

The Impact of Foreign Subsidies on Competitiveness of Ecuador's Tuna Exports

As noted above, Ecuador's main competitors for international tuna markets are among the heaviest fisheries subsidizers in the world. This suggests that Ecuador's tuna industry may suffer a competitive disadvantage as a result of foreign subsidies. While data limits prevent a direct empirical test of this proposition, application of an inferential method strongly suggests such a competitive impact occurs. The method applied can be summarized as follows:

- Estimation of the likely value of subsidies to the tuna industries of each principal competitor country, assuming that the tuna industry receives a share of national fisheries subsidies roughly in proportion to the share of tuna catches in the overall fisheries production of each country.
- Estimation of likely quantities of subsidized tuna canned and exported from principal competitor countries.
- Correlation of Ecuadorian market share for canned tuna sold in the United States with value of subsidies and subsidized tuna exports from competitor countries.

An analysis based on this inferential method reveals a striking correlation between estimated levels of foreign tuna subsidies and the displacement of Ecuadorian tuna exports into the US market. Table 3.11 shows that in 2003 Thailand exported 395,428 tonnes of canned tuna, of which 27 per cent went to the US–American market. The calculations also show that 36 per cent of the processed tuna was produced with subsidized raw material, coming from the most important suppliers of frozen tuna to the Thai market, during 2003. This means that 36 per cent of the total value of the exports of canned tuna from Thailand to the United States could have benefited from subsidies.

It is important to consider the effect that the values incorporated as subsidies into the production of supplier countries could have had in this case. In fact, these subsidies could have diminished the costs and/or maintained the incomes of

Table 3.11 *Determination of the Volume of Subsidized Processed Tuna in the Thai Market*

Production of canned tuna that benefited from subsidies which have a commercial impact	*2001*	*2002*	*2003*	*2004*	*2005*	*2006*
Volume of fisheries introduced to the Thai market that benefited from subsidies and have a possible commercial impact (tonnes)	138,851	152,515	269,547	234,366	287,963	331,374
Processed tuna benefited through subsidies with possible commercial impact (tonnes)	72,318	79,435	140,389	122,065	149,980	172,591
Volume of exports of canned tuna	298,417	320,241	395,428	377,519	453,517	501,443
Percentage of canned tuna benefited from subsidies with possible commercial impact	24%	25%	36%	32%	33%	34%
Participation of Thailand in the US–American market	41%	38%	42%	40%	42%	44%

Sources: USTRADE (www.fas.usda.gov); THAI CUSTOM (www.customs.go.th); FERU (www.seaaroundus.org/sponsor/feru.aspx, Processed by author

the suppliers shifting the supply of these countries and increasing the fishing effort and the level of captures. The higher quantity of catch increases the offer of raw material for the processors in Thailand while diminishing the prices and increasing the derived demand of the processors.

The following two figures on page 105 summarize the results, showing correlations between price decreases and subsidy intensity (Figure 3.2) and between the growth in heavily subsidized Thai market share and declining Ecuadorian market share for US canned tuna sales (Figure 3.3). The participation of the Ecuadorian products in the US–American market decreased significantly during the last years, declining from 24 per cent in 2002 to only 7 per cent in 2008. During the period ranging from 2002 to 2008, the participation of Ecuador in this market decreased by 71 per cent, but the participation of Thailand increased by 21 per cent during the same period.

Therefore the negative impact of foreign subsidies on Ecuador's tuna industry appears to be significant.

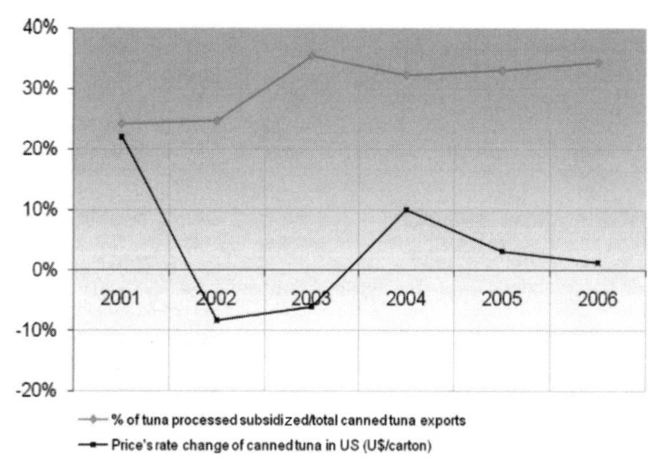

Figure 3.2 *Processed Tuna with Subsidized Raw Material versus Range of the Price of Canned Tuna in the US–American Market (2001–2006)*

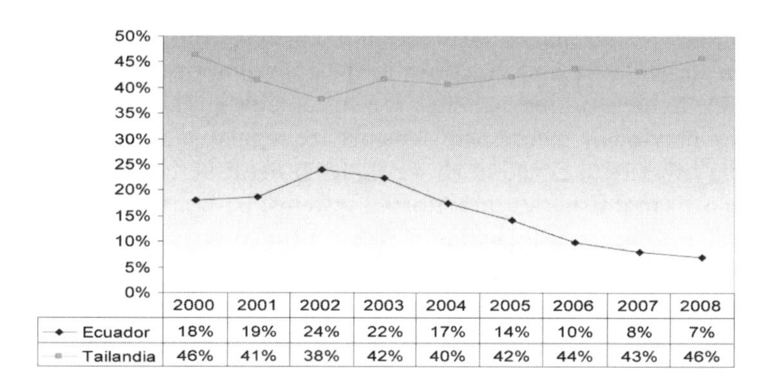

	2000	2001	2002	2003	2004	2005	2006	2007	2008
Ecuador	18%	19%	24%	22%	17%	14%	10%	8%	7%
Tailandia	46%	41%	38%	42%	40%	42%	44%	43%	46%

Figure 3.3 *Market Participation of the Processed Tuna from Thailand and Ecuador in the US–American Market (2000–2008)*

Source: US TRADE (available at www.fas.usda.gov/ustrade/usthome.asp – last accessed in January 2009)

The Ecuadorian fishery management system and its capacity to prevent the negative effects of fishery subsidies

The effect of fisheries subsidies on stock levels depends, among other conditions, on the effectiveness of the management system.

In developing countries, limited budget capacity is one of the problems that fishery management systems face, since the implementation of effective controls and surveillance systems implies incurring costs that exceed the fishery management's budget possibilities.

Apart from a basic management infrastructure, vessel registration systems, vessel-monitoring systems, fishing licenses, catch records and other mechanisms are needed to enable an effective fishery management system. Figure 3.4 shows

Figure 3.4 *Elements of an Effective Fishery Management System*

Source: Based on UNEP–WWF (2007b)

that there is a need for permanent evaluation of capacity and fishery resource stock, but also surveillance and sanctions in the case of noncompliance. Only then is it possible to have fishing activities at a sustainable level.

Most of the previously mentioned elements are regulated and enforced for purse seine tuna fisheries in Ecuador. However, for other fisheries in the country, particularly in artisanal fisheries, monitoring systems, on board observers and adequate catch records are still lacking. Table 3.12 incorporates the measures implemented by the fisheries legislation system of Ecuador, making a comparison based on the international requirements of the evaluation criteria established by UNEP and WWF and measures adopted by the Ecuadorian fisheries legislation system.

Fishery management systems and their capacity to prevent the negative effects of Fishery Subsidies

Tuna resource management in Ecuador is a responsibility shared by the national fishery management system and the Inter American Tropical Tuna Commission (IATTC). Both of these components must be taken into consideration when evaluating the fishery management system because tuna is a migratory resource, temporarily passing through the EEZ, but mostly staying in international waters on the high seas.

The IATTC has implemented a series of control measures, some of which have been deemed responsible for programmes with effective achievements carried out on a regional level (Morón, 2006). The IATTC's principal management measures are:

- a closed season for purse seine tuna fishing vessels;
- resolution for the restriction of fishery capacity growth in the EPO;

Table 3.12 *Comparison of the Evaluation Criteria of UNEP-WWF and Measures Adopted by the Ecuadorian Fisheries Legislation System*

Fishery management related criteria	*Minimum International Requirements*	*Fishery Management System (FMS) Ecuador*	*Observations*
Resource Evaluation	Scientific stock evaluations based on catch or effort data	• Scientific evaluations of species subject to exploitation within the EEZ are carried out by the National Fishing Institute • As far as the tuna resource in the EPO, Ecuador's FMS uses, among others, regional evaluations of the resource carried out by the IATTC's (Inter American Tropical Tuna Commission) scientific team for management making decisions such as additional closed seasons to those that have been established on a regional level	The Ecuadorian Fishery Management System's (FMS) limited budget capacity is the main restriction to carrying out scientific evaluations on fishery resources
Capacity Evaluation	Scientific capacity evaluations with quantitative estimates of fleet capacity and trends	• As for fishery capacity, an FAO mission by means of an expert is elaborating an evaluation of the fishing capacity state of Ecuador's main fisheries (2009), with the aim of identifying the outline for the elaboration of an Action Plan for Fishery Capacity Management in Ecuador. This is mainly for other fisheries as purse seine fishery capacity was limited by an IATTC Resolution	Technical assistance from the FAO with this plan's elaboration would strengthen management carried out by Ecuador's Fishery Management System, above all for decision making and the creation of policies concerning fishery capacity
Controls	Formal management plans, including the FAOs IPOA-capacity fishery management plan Limit Reference Points for resource condition as well as capacity based on scientific evaluations, taking the MSY as the maximum acceptable limit for resource biomass	• On a national level is the Fishery and Aquaculture Management Plan of Ecuador • The FAO, by means of a fishery expert, is elaborating a fishery capacity management plan, which would serve additionally as a pilot programme for the rest of the Comisión Permanente del Pacífico Sur (CPPS) member countries. The programme's goal is to avoid fishery overcapitalization and overexploitation. • On a national level there are specific regulations for territorial waters, for example, a ban on whale and marine turtle fishing, a closed season for lesser pelagics, molluscs among them. Besides regulations for preventing incidental fishing, like the establishment of TEDs for turtles and a special screen on purse seine vessels	Ecuador intends to comply with international regulations and standards with regard to controls to its fisheries activities. This is evident and fully in place for its export products such as tuna Bans are in place and enforced for many fisheries. Results are positive in many cases. Recently fisheries authorities have increased budget and personnel for control nationwide Evaluation of results of control and other management activities might be improved

Table 3.12 *continued*

Fishery management related criteria	*Minimum International Requirements*	*Fishery Management System (FMS) Ecuador*	*Observations*
	Pre-established regulations that serve as an answer when fishery reference points have been exceeded	• Ecuador's SAP works in accordance with the IATTC's regional regulations regarding tuna resources. The IATTC's principal regulations are: – A closed season for tuna fishing with purse seine fishing vessels – IATTC resolution for restricting fishery capacity growth in the EPO – Agreement on the International Dolphin Conservation Program (AIDCP) in the EPO – IATTC resolution banning high seas transfer of tuna These regulations are recognized by Ecuador's Subsecretaria Regional de Pesca (SRP) and controlled by the Dirección General de Pesca (DGP). In addition to the IATTC's regulations; Ecuador has also implemented additional closed seasons of its own initiative. Another initiative by Ecuador's SAP is the mandatory excluder screen for juvenile fish on class six tuna vessels	
Compliance/ Enforcement	Procedures to allow effective and reasonable preventive actions and against illegal fishing in fisheries Mandatory Devolution of subsidies granted to any vessel that has partaken in illegal fishing activities	• The Scientific Research Permit (SRP) requires fishery inspectors to consolidate compliance with management measures The registry of actions performed by the fishery control team is public and available on the SRP's website. On a national level there are operations for closed season and incidental fishing tracking and control • The National Directorate of Aquatic Spaces (DIRNEA) carry out operations against illegal fishing by marking the vessels' motors in strategic places, the authorities can identify the vessels. Any unidentified vessel will be detained and auctioned off and the fisherman will be placed at the disposal of the district attorney's office • To combat illegal fishing, satellite monitoring devices have been installed on the vessels. Port and fishery management authorities	Coordination among different institutions empowered by Ecuadorian legislation has been able to prevent and detect most illegal fisheries activities of industrial fleets. Enforcement of all bans to the artisanal fleet has been difficult because of the large number of small vessels. Efforts for improvement are being put in place including satellite monitoring devices to the artisanal fleet with partially subsidized government funds

<table>
<tr><td></td><td></td><td>work together through the control and surveillance system with the goal of counteracting illegal fishing
• The SRP, through regulations against illegal fishing, has permanently banned discharges, commercial transactions, transfers, and all imports of fish caught by vessels implicated in illegal, unreported and unregulated (IUU) fishing activities, as well as those declared IUU by the IATTC.</td><td></td></tr>
<tr><td>Monitoring, Control and Surveillance Infrastructure</td><td>Mandatory registration of all fishing vessels, supplying the mandatory information required by the HSVAR database.
Mandatory licenses for all vessels and fishers, detailing fishing authorization and license information contained in a public registry of licenses.
Mandatory Catch and Landing report.</td><td>• Mandatory Registry of Active Vessels. The SRP publishes a registry of industrial as well as artisanal vessels on its website, which partially complies with the information required by the FAO's HSVAR. On a regional level, Ecuadorian tuna fishing vessels that operate in the EPO are also registered and the information is published by the IATTC
• Mandatory catch and landing reports
• Fishing permits/licenses for vessels and shipowners
• Vessel satellite tracking system
• Inspectors for fishery port controls
1 Closed season monitoring and control
2 Vessel inspections
3 Plant inspections
4 Maritime patrol
5 Port Control
6 Beach Inspections
7 Turtle excluder device inspections
8 Etc....
• Onboard inspectors for class IV, V and VI vessels based on the AIDCP agreement for dolphin protection and monitoring of the IATTC's other resolutions</td><td>The registry of active vessels by Ecuador's Fishery Management System partially covers the requirements for the High Seas Vessels Authorization Record (HSVAR) format of the FAO's database
The existence of fishing permits/licenses, the mandatory catch and landing reports, as well as other existing mechanisms in Ecuador's SAP are used for fishery control and regulation compliance. Fishery controls such as patrols are carried out together with the DIGMER and the DIRNEA, executing control actions against illegal fishing among others
Limited budget assignations impede the assignment of inspectors for all fisheries, especially for artisanal fisheries that catch a certain quantity of large pelagics. This limits the inspection of certain fisheries by vessel type and size</td></tr>
</table>

Source: Subsecretaria de Recursos Pesqueros del Ecuador (www.srp.gov.ec); Instituto Nacional de Pesca del Ecuador (www.inp.gov.ec); and Inter-American Tropical Tuna Comission (www.iattc.org) last accessed in January 2009; Processed by author

- the Agreement on the International Dolphin Conservation Program (AIDCP) in the EPO;
- resolution banning high-seas transshipment of tuna.

The effectiveness of Regional Fisheries Management Organizations (RFMOs) to manage has been questioned by some experts, NGOs and others. The problem most frequently noted is the consensus-based governance structure under which any single member can veto rules proposed for adoption by the organization.

The RFMOs, conscious of their important role in bio-aquatic resource conservation, drew up a list of criteria to undertake their auto-evaluation. Some of these organizations have initiated this process with a view to improving their performance and improving the management of tuna resources (IATTC, 2008). This is an endeavour that should be undertaken widely.

Proposed auto-evaluation criteria focus on are the adopted measures of conservation and management, as well as surveillance and control systems. They propose that stock evaluation be based on specific data and criteria and that they establish stock protection measures, and monitoring, control and surveillance systems for fishery activity. However, the effectiveness of RFMO management systems is still an issue. Although the IATTC's control measures in particular could be viewed as superior to those executed by other RFMOs, it is still necessary to evaluate this organization's effectiveness in light of the results concerning the state of the resource.

Towards fishery subsidies reforms

The WTO is debating the possibility of banning capacity and effort enhancing fisheries subsidies. This ban would have important exceptions within the framework of special and differential treatment for developing countries, potentially including large nations such as China, India, Brazil and Mexico, which have significantly developed fishery industries. Due to the size of these economies, exceptions to fishery subsidy grants could have an impact on resource conservation and cause significant harm in economic terms to other developing nations, given their weight in international aggregate supply and consequently in the trade of other developing nations.

Regarding this possible scenario, it would be in the best interests of smaller developing countries, which participate in international trade, for the special and differential treatment extension to be limited, in order to guarantee that any exception allowed has as its only objective the development of artisanal fisheries, aimed at the internal markets of the country where the subsidy is granted. The exceptions should be allowed only if the fisheries fulfill the minimum sustainability requirements described in this document.

Another recommendation is to implement mechanisms, over a fixed period of time, that guarantee the establishment of minimum international management standards in countries that grant fisheries subsidies, under an international surveillance plan that involves the respective RFMO and other relevant international bodies, for high-seas fisheries.

In the case of the Ecuadorian fishery management system, the information reviewed leads to the recommendation that for new subsidies to be granted to artisanal fisheries, effective management systems should be a prerequisite. These should include scientific evaluations, data collection and high seas observation, among other proposals, as minimum international requirements.

The new subsidy disciplines that will be approved by the WTO will probably not ban some assistance for export-oriented fishing activities in developing countries. These activities are common in some developed countries and have been identified as causing the greatest impact on the environment and trade. Since such practices may not be fully disciplined by new WTO rules, market mechanisms – such as ecolabels – should be explored to help correct the imbalances that might otherwise persist.

The current trend is to use eco-labels, which focus on providing information that allows actors in the marketplace to distinguish between fishery products on the basis of the environmental impacts of their production process. In fact, environmental and management quality certifications could be linked to informing the consumer and distribution channel that the fishery employs sustainable methods and that it has not received subsidies. Or, in the case of having received subsidies allowed by the WTO, they would reveal that these have been granted within minimum sustainability standards, including healthy stock levels and effective management systems.

In order to help to resolve the fisheries subsidies problem, it would be interesting to consider eco-labeling systems to incorporate restrictions as related to subsidization. It could thus be assessed whether the subsidy would be permitted under the WTO and whether – under the prevailing conditions in the fishery – would be considered harmful or possibly harmful.

It is therefore necessary to establish the implementation of sustainability criteria in the fishery management systems of countries as a prerequisite for the use of subsidies and perhaps also certification. Furthermore, there is a need to agree on clear definitions concerning fisheries subsidies and impact identification criteria that differentiate between subsidies that are harmful and those that would favour conservation activities.

Conclusion

Many fisheries subsidies have negative effects on the environment and trade of developing countries. The extent of the negative impact of fisheries subsidies depends on the subsidy type, the condition of the resource, the existing level of fleet capacity and the fishery management system.

Fishing assistance granted by industrialized nations can be detrimental to the commerce of developing nations, especially to those of lesser size, which have a smaller capacity to subsidize a specific economic activity and less weight in the international aggregate supply.

As far as the impact on sustainability is concerned, the evaluations carried out in this study on fisheries subsidies of the countries with the largest catches indicate that the impact of fisheries subsidies on tuna stocks should at a minimum be classified as 'probably harmful' – given the conditions under which tuna is currently managed. However, the percentage of subsidies that should be classified as simply 'harmful' is considerably higher in the substantial number of fisheries already at full exploitation.

The possibility of a harmful impact can be decreased by implementing improvements in the fishery management systems of the corresponding countries and RFMOs. The improvements must coincide with the evaluation criteria of the RFMOs. In other words, they must ensure that stock evaluation is reasonable and based on scientific data. Additionally, mechanisms that allow the implementation of measures that protect stocks need to be put in place, and control, monitoring and surveillance systems for fishery activities have to be implemented.

However, RFMOs have financial, operational and legal limitations that restrict the effectiveness of their actions. New management systems that have been proposed in the international arena such as individual transferable fishing rights are promising, but it will probably be difficult to find a consensus in the short term. Since effective RFMO reform based on property rights is not likely to be widespread in the near term, fisheries subsidies will continue to pose serious risks to tuna stocks and to the competitive possibilities for developing countries in the market for tuna products.

Therefore, it is vital to find subsidy solutions in the framework of the WTO, FAO and other forums, considering the necessity of conservation and equitable trade among nations at different stages of development. Also, the search for market solutions should be intensified. These include the use of eco-labeling, for which precise definitions of subsidies need to be adopted, and the ones with a greater negative environmental impact need to be singled out.

For a fundamental reform of fisheries subsidies to be possible and for better fishery management systems to be implemented, nations must reconsider the major aspects of international relations with an integral vision that contemplates environmental and equitable development aspects.

International assistance is fundamental for those countries with less fiscal resources to be able to comply with the minimum sustainability criteria proposed for effective fishery management.

In the context of globalization and increased trade, any new fishery subsidy disciplines need to promote better management regimes in order to allow for the maintenance of sustainable capacity and viability of long term fishery activities.

Fisheries Subsidy Reform in Norway*

Introduction

The story of Norway's fisheries subsidies provides a salutary lesson to the world's fishing sector. For many decades, the development of the Norwegian fisheries sector was supported by significant subsidies. Subsidies reached a peak of NOK 1.3 billion (approximately US$227 million[17]) in the early 1980s. However, a combination of resource crisis, poor fleet profitability and political backlash prompted major reforms to the sector. Subsidies declined to less than NOK 200 million by 1994 and to only around NOK 50 million in 2008. At the same time, significant changes were made during the 1990s to the management regimes used to govern the fisheries sector, with an increased emphasis on the use of market-based management instruments. The package of reform measures has resulted, over time, in a profitable and sustainable industry capable of being self-reliant, flexible and which is a major contributor to the Norwegian economy.

This section reviews the process of reform in the Norwegian fisheries sector, covering the offshore and coastal fleets, engaged in catching primarily cod, saithe, herring, capelin and mackerel. The reforms encompassed the package of policy changes that resulted in the drastic reduction of subsidies and the introduction of market-based mechanisms to manage the sector. The central lesson from this review is that it was the combination of these two policy shifts that has been the key to the development of a sustainable and profitable industry over the last couple of decades. The next section provides some background on the Norwegian fishing sector. The rise and fall of the subsidy regime is then described. The last section provides a number of policy insights of relevance to other efforts to reform fisheries subsidies.

Background

Norway is the tenth largest fisheries producer in the world, with catches of around 2.4 million tonnes in 2006. The bulk of the catch is based on relatively few species, with almost 90 per cent of the volume and value of the catch covered by just 10 species (see Table 3.13). In addition, 90 per cent of the catch comes from stocks that are shared with other countries including the Russian Federation, European Union, Iceland, Faro Islands and Greenland. The fishing sector makes a relatively small contribution to the GDP of Norway (0.7 per cent in 2006) but is a major export sector, accounting for 5.1 per cent of total Norwegian exports in 2006 (third behind oil and gas and metals); Norway is also the second largest exporter of fisheries products in the world (behind China).

---

* This section was written by Anthony Cox, based on work undertaken while he was in the Fisheries Policies Division, Trade and Agriculture Directorate, OECD, and reported in OECD (2010). The views expressed in the paper do not necessarily reflect those of the OECD or its member countries.

Table 3.13 *Wild capture fisheries production, 2008*

Species	*Catch (tonnes)*	*Value (NOK million)*
Atlantic cod	215,221	3497
Atlantic herring	1,025,493	2796
Northeast Atlantic mackerel	121,492	1290
Saithe	227,513	1266
Haddock	74,617	646
Northern prawn	30,180	581
Blue whiting	418,289	424
Greenland halibut	9058	182
Red king crab	5199	135
Other	306,054	1284
Total	2,433,116	12,101

Source: Fiskerdirektoratet (Norwegian Directorate of Fisheries, 2009)

The numbers of both fishers and vessels has been declining steadily in Norway over the past decades as a result of increases in technical efficiency and government policies to reduce the fleet size. The number of full and part time fishers was just under 13,500 in 2007, down from around 30,000 in 1985. The number of vessels has declined from over 25,000 in the early 1980s to 7,041 in 2007. Around 60 per cent of the vessels are less than 10m in length and this vessel class has experienced the greatest proportional decline over the last decade (down from 10,601 vessels in 1996 to 4056 vessels in 2007, although many of these were inactive) (Fiskerdirektoratet, 2009). There is a strong regional pattern to the fishing fleet. In the northern region of Norway, the fleet is dominated by small fishing vessels operating close to shore, while the larger fishing vessels operating in the Barents Sea and the North Sea mainly belong to owners on the west coast, apart from the cod trawlers which are mostly owned by a few large processing companies.

The regional pattern is also reflected in the important role that fishing, processing and, more recently, aquaculture have played in rural development. The sector has been essential in securing basic employment in a large number of coastal communities. The alternative employment opportunities in some of these communities are relatively limited, especially in the north of the country. The regional concentration of fisheries has meant that they play an important role in the politics of Norway as the sector plays an important role in the Norwegian government's overall policy to maintain the settlement structure in coastal communities, especially in the northern part of Norway.

The objectives of Norway's current fishery policy therefore reflect two policy priorities. Fisheries management is intended to maximize the profits of the sector through an economically efficient use of the resources, while at the same time ensuring socio-economic optimization with respect to the returns to communities along Norway's coast. Reconciling these two priorities has been a consistent feature of the fisheries policy reform debate in Norway over the past few decades. For example, opposition from the fishery industry was a decisive factor behind Norway's decision not to join the European Community in 1972 and again in 1994. Fisheries will probably continue to be one of the main issues in the discussion on a possible membership of the European Union in the future as well.

The political importance of the fisheries sector is reflected in the organization of the sector and the institutional design for decision-making (Hersoug, 2005). The political interests of the fishers are represented primarily by the Norwegian Fishermen's Association (NFA), established in 1926, and the Norwegian Coastal Fishers' Association (NCFA), which was set up in 1990 to cater for the smaller scale coastal fishers. The NFA is, in essence, a labour union which has a strong political influence and takes part in discussions and decisions on fishery management, including quota decisions. Norway has a long tradition of user-participation and the fishery policy has been formed taking into account the views of the fishers, represented by the NFA and the NCFA.

The Ministry of Fisheries is responsible for overall policy in the fishing sector while the implementation of management measures is carried out by the Directorate of Fisheries. In the national regulation process, the Directorate makes a proposal on how the Norwegian share of internationally agreed Total Allowable Catch (TAC) should be shared amongst the various fleets and regions. This proposal is then presented in an open meeting where representatives from the NFA, NFCA, Federation of Norwegian Food and Allied Workers Union, The Sami Parliament and others who are interested are invited to participate. Based on discussions in the meeting the Directorate gives its recommendations to the Ministry. The Ministry then evaluates the recommendations and the Minister of Fisheries decides how fishing should be arranged for the following year. Throughout the whole regulatory process, a great deal of importance is attached to cooperation between the authorities and the various fishers' representative organizations (OECD, 2006).

The rise of fisheries subsidies in Norway

The origins of Norway's fisheries subsidies can be traced back to the 1920s and 1930s when the fisheries sector went through a number of crises as markets failed and prices fell. The industry's problems were increasingly regarded as a matter of public concern and a number of policy measures were implemented to try and address the fluctuating fortunes of the industry. The first sales organizations were established around this time and a model of close cooperation between industry and government in managing the fisheries was formulated. Industrialization of the fishing industry became the policy priority and measures were introduced to improve and modernize the technological capacity of the industry and increase economic efficiency and profitability. Subsidies were a key part of this process and entailed a range of measures including support for vessel construction and modernization, infrastructure, expansion into processing and price support.

Until the beginning of the 1960s, the level of state subsidies was decided by negotiations between the fishers' sales organizations and the government. However, in an effort to simplify the negotiation process, a 'General Agreement' was established in 1964 whereby the level and utilization of subsidies was settled through annual negotiations between the state and the Fishermen's Union.[18] The General Agreement specified the participants, goals, means and rules of the negotiations and served to institutionalize the strong and close relationship between the industry and the government in the area of subsidies. The aim of the General Agreement was to provide support in order to increase the efficiency of harvesting, processing and marketing and ensure a rational development within the entire industry. Significantly, it was also intended to be temporary, with support being withdrawn once the industry was able to function competitively and became economically self-sufficient (Jentoft and Mikalsen, 1987). Over time, the General Agreement also encompassed an additional goal of preserving settlement patterns and strengthening the economic viability of marginal regions.

The subsidies provided under the General Agreement included price support, income support, subsidies for inputs and structural measures. The main principle underlying the General Agreement was that subsidies were required when the profitability of the fleet was too low to secure the average fisher an acceptable income (where an acceptable income was defined as the average income in the manufacturing sector). In addition to this agreement, there were generous subsidies for vessel construction and favourable loans available from the Fishermen's Bank.

Figure 3.5 provides an overview of the level and types of state subsidies provided for the period 1980–2008. From a level of around NOK 150 million in 1964, subsidies reached a peak of NOK 1333 million in 1981 before declining to less than half that level by 1988. After a brief surge in the late 1980s and early 1990s, the level of subsidies fell dramatically during the 1990s. Several points are worth highlighting. First, the level of total subsidies fluctuated significantly from year to year. The amount of subsidies was influenced by the general economic situation prevailing at the time, as well as by the fisheries prices in any given year. Second, price support accounted for around half of the total subsidies. Price subsidies were used to cushion the domestic industry from fluctuations in world fish prices and were based on the volume of fish landed.

The decline and fall of the subsidy regime

The subsidy regime established under the General Agreement ultimately became a self-defeating policy, working against the objectives for which it was originally established. The initial idea of the subsidies being temporary was quickly forgotten as the industry became ever more dependent on the support provided by the state. While there was political pressure to rein in the subsidies, this was rebuffed by the strongly linked corporate model of governance in the sector and by the over-riding regional development objectives (Hersoug, 2005, pp.176, 250–1). The subsidies were considered a necessary price to pay for preserving the prevail-

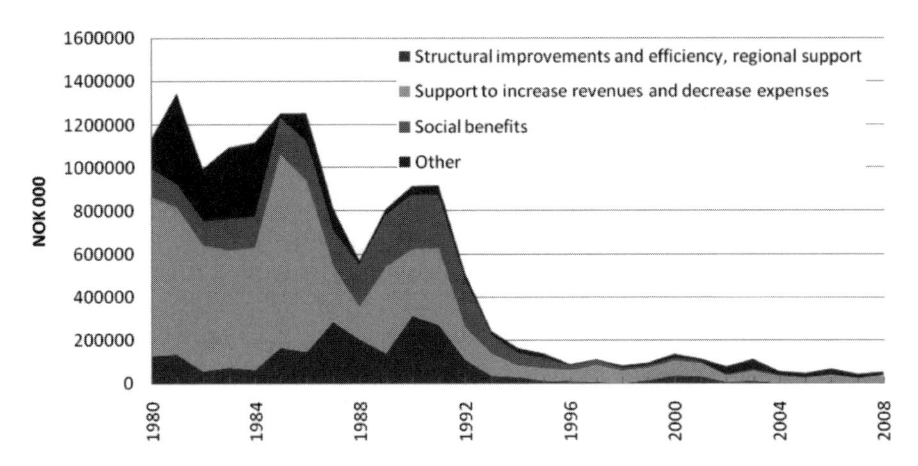

Figure 3.5 *Government support to the fishing fleet 1980–2008 Fiskeridirektorat (Norwegian Directorate of Fisheries, 2009)*

ing settlement structure (Jentoft and Mikalsen, 1987). However, there were a number of powerful interconnected factors that served to highlight or exacerbate the perverse incentives that were generated by the General Agreement and it was these that resulted in the Agreement being abandoned in 1995.

The first of these factors was the incentives provided to fishers by the nature of the Norwegian fisheries management prevailing at the time that the subsidies were provided. From the 1930s, the Norwegian industry had gradually been regulated by a series of licensing regimes for the different fleet segments (trawlers, purse seiners, etc). However, the industry remained an effectively open access fisheries with no effective limits on catches or capacity. It is well established that this type of management regime will inevitably result, over time, in overcapacity, reduced catches and stock levels and reduced resource rent and profitability (OECD, 2006). While there may be limits on entry, there is a strong incentive for fishers to over-invest in capacity in order to compete in the 'race to fish' against other fishers. Adding subsidies to this situation further alters the incentives facing fishers, encouraging further investment in capacity and shielding fishers from price signals that may provide incentives to adjust their operations and investments. This was the case in Norway where overcapacity resulting from subsidies provided in a limited access regime led to a series of resource crises and poor profitability. In effect, the industry was modernized with subsidies for vessel construction, supported in its operations with price subsidies and then restructured using government support for scrapping vessels. It was not until the late 1980s that efforts were made to 'close the commons' and address many of the perverse incentives associated with the effective open access fisheries.

A second factor was the effect of technological progress. The rapid modernization of the Norwegian industry meant the uptake of new technology in the construction of fishing vessels which significantly increased the technical efficiency and catching capacity of the fleet. This further exacerbated the effect of the subsidies for vessel construction and operations.

A third factor was the perverse incentive created by the form of price subsidies used to support the industry. The price subsidies were provided whenever the export prices fell below an 'acceptable' level. With no constraints on the industry about how much it was allowed to catch, the price subsidies effectively rewarded increased catches and provided an incentive to expand effort to 'harvest the subsidies'. As a result, the price supports tended to favour large-scale, capital-intensive operators at the expense of small-scale fishers. Furthermore, the price subsidies tended to increase income differences in the fisheries between fleet segments and between different regions (due to geographical differences in fleet structure) (Jentoft and Mikalsen, 1987).

Fourth, it became clear that what had become one of the main rationales for the subsidy regime – regional objectives – were not being met. The logic of the price subsidies meant that the largest companies were receiving the largest share of the subsidies and these companies were generally not located in the remote regions that had been the intended target of regional support objectives. During the early 1980s, the counties in western Norway received on average more support per fisher per year than the counties in northern Norway (Jentoft and Mikalsen, 1987). Western Norway is the centre of much of the oil-related economic activity and the cost of labour is relatively high compared to the northern regions where alternatives to fishing are much fewer (Hersoug, 2005).

Finally, the combination of these factors culminated in a series of resource crises that led to poor economic returns, pressure for further support and strong support for major institutional change to alter the cycle of subsidy dependence. At the end of the 1960s, there was a total collapse in the Norwegian herring fisheries. The impact on fishers was masked to a large extent by the availability of subsidies to fishers to compensate for lack of profitability. However, it became increasingly evident that the fleet had to be reduced as overcapacity was hampering economic recovery. In fact, overcapacity had been a recurrent theme in official communications for many years and successive governments had stressed the need for continuous adjustment of fishing capacity, since there was little scope for increasing total catches. The standard means of adjusting capacity at that time was the use of scrapping schemes, both in the offshore fleet and in the coastal fisheries. A series of government-financed scrapping schemes was instituted through the General Agreement with some 3500 vessels being scrapped since the 1960s. With respect to the offshore fleet, 393 vessels were decommissioned over the period 1960–1993 at a cost of NOK 592 million. In the period 1998–2002, nearly NOK 200 million was used to scrap coastal vessels (Danielsen, 2004). However, the conflicting signals provided by the coexistence of scrapping schemes with subsidies to fleet operations and regulated open access fisheries meant that the industry had entered a vicious cycle of subsidy dependence.

In 1989, matters came to a head with another looming cod crisis. This created a strong incentive for both government and many fishers to engage in debate over how to improve the profitability and adaptability of the industry.

The resource crises and accompanying low profitability and excess capacity in many fleets made it clear that it was necessary to change the open access nature of fisheries that were the norm in Norway at the time. One by one, Norway's key fisheries were shifted from open access to closed access, primarily through the installation of limited entry based on a tight licensing regime. These events were accompanied by a massive reduction in subsidies to the sector, from a peak of over NOK 1.3 billion in the early 1980s to less than NOK 200 million by 1994 and only NOK 50 million in 2006. This required the sector to be more self-reliant and flexible in generating profits, rather than relying on government transfers to carry them through fluctuating fortunes. The General Agreement was officially terminated in 2004.

Shifting to a market-based approach to managing fisheries

While the resource crises were a major trigger for reform in the sector, it was the shift from open access fisheries to closed access that generated the most heated debate and ultimately led to the introduction of market-based instruments to manage the chronic over-capacity in the sector (Hersoug, 2005). Closing the commons had significant implications for the distribution of wealth within the sector. It was necessary for the government to develop flexible management systems that would generate resource rent, while at the same time addressing the distributional concerns brought about by the changes in access arrangements and the radical reduction in subsidies. Market-based instruments have helped to create a profitable and sustainable sector, but the process has not always been smooth and has been marked by an evolution of the design and implementation of management instruments to meet different policy challenges. The rest of this section discusses the reform process.

As noted above, it was evident in 1998/99 that a cod crisis was looming. The issue of over-capacity was again put on the agenda, although this time it mostly affected the coastal fleet. The issue of establishing an Individual Transferable Quota system (ITQ) was introduced through a report from a working group on the structure of the harvesting sector.[19] The group comprised representatives from the Norwegian Ministry of Fisheries, the Directorate of Fisheries and the NFA. The original idea was to introduce enterprise allocations to the offshore fleet, thereby making it possible for companies with two or more vessels to rationalize harvesting capacity and, in addition, making it possible for two or more companies to cooperate in reducing effort. This was, according to most fishers and politicians, seen as more or less the same as ITQs. The proposal created a heated debate, with strong opposition from the coastal fishers and politicians. The proposal was not acceptable to the fishers because the fisheries policy was perceived as a regional policy contributing to the maintenance of the established settlement pattern.

Faced with opposition to the proposal, the Ministry of Fisheries prepared a white paper in 1992 for the Parliament. The report described the existing ITQ schemes in Australia, New Zealand, Iceland and Canada, and the group presented an overview of different forms of ITQ. The group ended up by recommending different forms of ITQ systems with strong geographical limitations on transferability. The report's preferred version (relating to vessels longer than 8 metres) was based on TAC allocation to various groups (vessels and regions) based on historical catch. Individual quotas, defined as a share of the TAC, would be allocated for a limited period of time (five years) and be subject to an annual resource fee to be paid to the government. Quotas would be tradable within groups and regions, while transfers across vessel groups and regions would require permission from the ministry.

By taking the demand for larger flexibility and the need for regional stability into consideration, the ministry thought the proposal would meet acceptance, not only by the fishers, but by regional politicians as well. However, '[t]he overwhelming majority of those consulted were strongly against ITQs, even in the modified version suggested in the draft' (Apostle et al, 1998). The main reason for the scepticism was the fear of privatization of the commons. While it was generally agreed that it was necessary to have TACs and closed access, and that the exclusive right to fish is distributed to a limited number of fishers based on tradition, it was not considered legitimate that someone should be given an exclusive right to trade and make profit from the fisheries resource, without actually fishing. The pure forms of an ITQ system therefore did not favour in political discussions. In the 1992 report from the parliament's standing committee on fisheries, the majority rejected an ITQ option. This effectively scuttled future debate on the use of ITQs in Norway.

However, an alternative to an ITQ system, the Individual Vessel Quota (IVQ) system, was established and implemented in most of the Norwegian fishing fleet. Most attention was focused on the northeast Atlantic cod stock that was in a serious state in the late 1980s. Due to a sudden and unexpected decline in the size of the cod stock, the TAC was set to 340,000 tonnes in 1989, down from 630,000 tonnes the previous year. In 1989, the coastal fisheries were closed after only three and a half months. Because of this, an individual quota system was established during the fall of 1989 and implemented for the 1990 season in the coastal fleet.

The IVQ system was a two-tiered system. The most active vessels, as measured by the quantity of cod landed in the 1987–1989 period, were put under a vessel quota regime (the priority Group I, most active vessels). These quotas were exclusive, so that the vessel owner had full discretion to decide when or where to fish. On the other hand, the less active vessels were allowed to fish competitively under a group quota (Group II, less active vessels). There were no restrictions on participation to this fishery, as long as the fisher fulfilled the requirements of being a registered fisher. However, the allocation to this group was about 10 per cent of the quota given to the coastal vessels in Group I.

When the IVQ system was implemented, the dramatic condition of the cod stock made it inevitable that steps needed to be taken to improve the situation for the full-time fishers. However, the IVQ regime was initiated as a response to the resource crisis and initially the idea was to abolish the system once the situation returned to normal. This may have been the main reason why the IVQ system was adopted so quickly, with relatively few objections. Even though both the fisheries authorities and the Fisherman's Association regarded the IVQ system as transitional, the IVQ system became permanent when the crisis passed. The major reason for the shift was that the owners of vessels in priority group I discovered the benefit of being inside a closed group (Holm et al, 1996; Hersoug, 2005). The exclusion of some 4000 vessels from full quota rights meant that there was a dramatic improvement in wealth and profit for the approximately 3500 remaining rights holders in Group 1. It allowed the rights holders to fish their quota when it suited them, or to sell the vessel with the additional quota value. The Norwegian Fishermen's Association therefore began to work to protect the value of rights created by the IVQ and was supportive thereafter of measures to improve the profitability of the sector through the further refinement of the quota-based management instruments (including transferability).

Following this shift in position, a series of further variations on market-based instruments were introduced throughout the fleet over a number of years. In summary, these were as follows:

- *Unit Quota System (UQS)*: this quota transfer system is an adjunct to the IVQ system and is intended to reduce the number of vessels in the offshore fleet. The system allows the owner of two vessels to transfer the quota of one vessel to the other vessel. If the surplus vessel is sold or scrapped, the vessel owner can hold the additional quota for 18 years (13 years if the vessel is sold rather than scrapped). The time limit was abolished in 2005 when the Structural Quota System (SQS) was introduced to the offshore fleet as well (see below).
- *Structural Quota System (SQS)*: this quota transfer system was introduced in 2004 to encourage capacity rationalization in the coastal fleet. It enables vessels between 15 and 21 metres and between 21 and 28 metres to transfer quota from one vessel to another vessel if one vessel is scrapped. 20 per cent of the quota attached to the scrapped vessel is returned to the regulation group while the remaining 80 per cent is held in perpetuity by the vessel owner. To avoid geographical concentration, the SQS is subject to certain limitations. The government decided (in spring 2007) that the structural quota would be given with a predetermined time limit (20 years). The SQS for the coastal fleet would also be extended to cover the group of vessels with a quota length of 11–15 metres (or 13–15 metres in the case of the coastal group of mackerel fishing)
- *Quota Exchange System*: this was introduced in selected coastal counties in 2004 as a temporary measure for the coastal fleet. The system allows two vessel owners within one vessel group to team up and fish both quotas on one vessel for a period of three out of five years. This system was settled by the end of quota year 2007.

Table 3.14 *Timeline for Norwegian fisheries management changes*

Year	*Event*
1960	First decommissioning scheme undertaken
1984	Quota transfer system (early version of Unit Quota System) introduced in part of the cod trawler fleet
1990	Debate opened on individual transferable quotas Individual vessel quotas introduced in Norwegian cod fisheries Unit Quota System introduced in cod trawler fleet
1994	Unit Quota System introduced in Greenland shrimp trawler fleet
1996	Unit Quota System introduced in purse seiner fleet
2000	Unit Quota System introduced in long-liner fleet
2001	Unit Quota System introduced in saithe trawler fleet
2002	Unit Quota System introduced in industrial trawlers
2004	Structural Quota System introduced to coastal fleet
2004	Quota Exchange System introduced for coastal fleet as temporary measure
2005	Unit quota system for offshore fleet redesigned to mirror structural quota system Moratorium on further use of market based instruments and review of fleet structure policy
2007	Structural Quota System re-introduced, with some modifications.
2008	Quota Exchange System for coastal fleet (introduced in 2004) was settled.

Source: OECD (2006); Hersoug (2005)

The remaining Norwegian fisheries were gradually closed and brought under these types of market-based management regimes (see Table 3.14). Allocation of catch shares was, naturally, a very contentious issue and it was settled when the 'trawler ladder' fixed allocation key that provided the sharing formula between the offshore fleet and the coastal fleet was finally decided in 1989, following extensive consultations between the Ministry and stakeholders (primarily the fishers associations). Based on the five-year allocation key, imposed and followed up by the Ministry of Fisheries, the task of scaling down the trawler fleet could start. The 'Unit Quota System' was introduced in 1990 to accomplish this.

In 2001, the allocation keys were up for discussion. This time the threats of breaking up and leaving the organization were clearly pronounced by the subordinate organization of the offshore fleet, centering on even minimal changes to the previous allocation keys. The case had been thoroughly prepared through a large committee, consisting of fishers from most fleet segments. The committee managed to obtain a compromise and after days of negotiation, a similar compromise was reached in the NFA to binding the allocation until 2007 when it was further renewed with little controversy. As part of the agreement, the Ministry was requested to 'close' a number of fisheries, that is, to limit the participation according to certain criteria. It was also a request that led to the so-called 'Finnmark model', where the coastal fleet is divided into four length-groups, each group being allocated a quota according to a historical share. These requests followed the line of the Ministry of Fisheries' work to take further steps to reduce capacity in the coastal fleet. The 'Finnmark-model' was implemented from 2002, as well as the new allocation keys.

During 2002, the Ministry of Fisheries introduced three important changes to the Norwegian access regime. The first was a green paper presented in the spring of 2002, proposing to close the saithe, haddock and herring open access fisheries. In the northern part of the country, the saithe and haddock fisheries had remained open, even when the most important demersal fishery, the cod fishery, was closed in 1990. The Ministry argued that no further restructuring in the coastal fleet could take place before access was closed to all the fishers and a white paper was presented to parliament in 2003.

The proposal introduced a dual structure, as already implemented in the cod fisheries, giving the priority Group I vessels of the cod fisheries an individual vessel quota on saithe and haddock according to vessel size. Vessels not holding a permit in the cod fisheries had to qualify according to their historical catch of saithe or/and haddock or/and cod to be given a permit to the new priority Group I, now enlarged to include cod, saithe and haddock. Vessels not qualified were given access in the open fisheries of cod, saithe and haddock in the so-called Group II. During 2002, the last open fishery of mackerel was closed, as well as the coastal fisheries of Norwegian Spring Spawning herring.

The second proposal concerned a decommissioning scheme for the coastal fleet partly financed by a fee on first-hand sales of fish. The idea was to build up a 'Structural Fund' over a five year period aimed at buying out and scrapping coastal vessels of less than 15 metres. This is contrary to the traditional approach in Norway, where the Government has financed all decommissioning schemes. The principle of 'all pay, some receive' was chosen to establish a fund big enough to have an effect, while keeping the fee as low as possible. This proposal was also supported by NFA, on the condition that the government contributed matching funds to the Structural Fund. The government did that in 2003 and 2004, but gave no guarantee for further contributions. The government has, in fact, given some contributions for the years after 2004.

The third proposal concerned the actual restructuring of the coastal fleet, considered as having substantial overcapacity. From the Ministry's point of view, it was important to offer the coastal fleet an option that they would choose to use, depending on its individual situation. In the summer of 2002, after a period of consultation with stakeholders, the ministry presented a paper suggesting two main directions of policy: either cooperation through the exchange of quotas between various vessels, the Quota Exchange System or a more permanent restructuring through the merging of vessels that each holds a fishing permit (the Structural Quota System).

After thorough consideration, parliament went along with all the proposals with only one minor change during the spring of 2003. This led to the implementation of the jointly funded decommissioning scheme from the summer of 2003, the Structural Quota System and a trial period for the Quota Exchange System from 2004 (this system was settled by the end of quota year 2007).

In October 2005 there was a cabinet reshuffle in Norway and the new government decided to stop all the market-like instruments pending a review of the fleet structural policy. The SQS was suspended and a committee was set up to

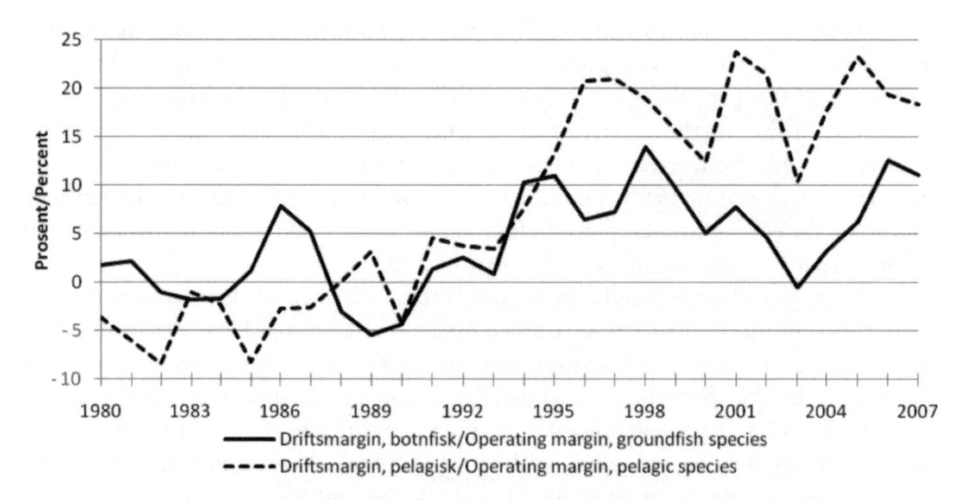

Figure 3.6 *Average operating margin and total operating revenues for vessels 8 metres and above 1980–2007*

Note: Operating margin = (operating result/operating revenues) × 100

assess the impact of established structural measures in relation to the government's objectives of: securing fish resources as commons, securing a fishing fleet that would serve to sustain activity the length of the Norwegian coastline and ensuring that such a fleet would be modern, diversified and profitable. The committee was a diversified group including fishers and vessel owners, representatives from the fish processing industry, stakeholder organizations, the sales organizations, researchers and political representatives. The authorities were only observers to the committee.

The key point to emerge from the series of papers and public hearings in this process was that there was strong agreement for the need for structural policy instruments. The government decided to continue providing the decommissioning schemes for the coastal fleet, at least up until July 2008, when it would be reviewed. It was also decided to continue with the SQS, although with some modifications.[20] This would be evaluated at the end of 2009. There were also changes in the rules on quota ceilings in the SQS for the coastal fleet in order to meet the government's objective of regional distribution of fishing resources and revenues.

In summary, the combination of reforms to the subsidy regime, and the management regimes, have marked a significant change in the Norwegian fisheries sector. Profitability has increased (see Figure 3.6) while fleet size and employment have declined (see Figure 3.7). There has been a degree of concentration in the industry, which is an inevitable outcome of the restructuring process as marginal operators exit the industry. A major innovation has been the use of decommissioning schemes that are largely industry funded (with some government support through soft loans and top-ups), indicating a shift in the outlook by both industry and government towards a more rational 'beneficiary

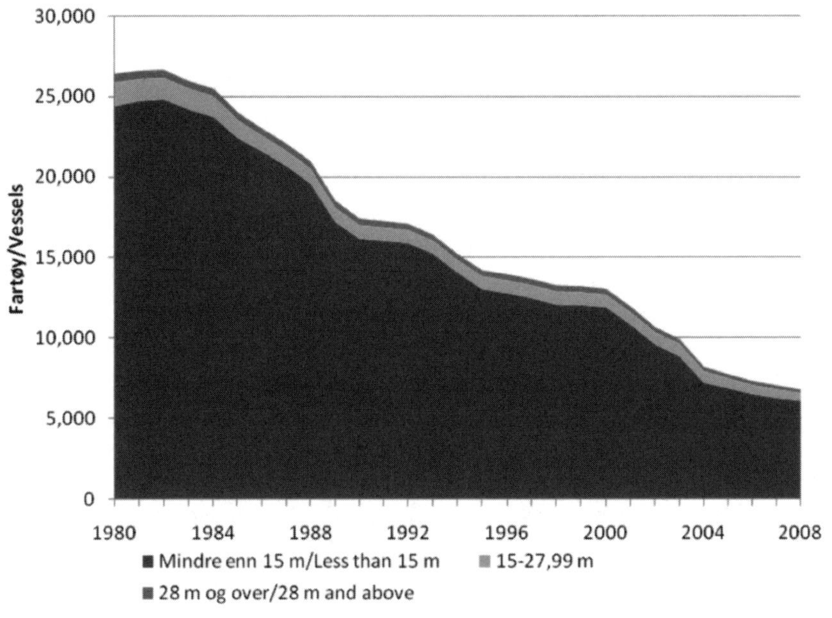

Figure 3.7 *Vessel numbers 1980–2008*

Source: Fiskeridirektorat (Norwegian Directorate of Fisheries, 2009)

pays' principle governing the design of decommissioning schemes (OECD, 2009). These involved the use of soft loans and industry contributions to fund the decommissioning of vessels.

Conclusion

The Norwegian fishing industry has grown from a heavily subsidized and overcapitalized industry to a competitive one that stands on its own two feet. This has been achieved through comprehensive structural adaptation and efficiency improvements. Reshaping the industry has resulted in reduced employment and increased concentration. There is also increased resource rent being generated for fishers through higher sales prices and a high level of profitability.

The focus now is much more on profitability within each vessel group. Approximately 95 per cent of the catch value comes from access-regulated fisheries. The TACs are distributed to the various vessel groups through fixed allocation keys and are further allocated as Individual Vessel Quotas. Different quota-transfer systems have been developed to meet the challenge of an increasing overcapacity due to technical development in vessels, gear and equipment. The management instruments implemented leave the responsibility for adjusting the fishing capacity to the available resources to the industry and thus secure higher profitability. However, it was a long and slow process and required a step-by-step approach to reform to ensure stability and cooperation amongst stakeholders.

Economic crisis drove reforms

As is the case in many other countries, it was the economic crisis that provided the impetus for reform. While there had been an increasing degree of resource pressure on stocks, it was not until this was translated into severe impacts on the profitability of key fleet segments that the inertia to undertake significant reform was overcome. The temptation to keep subsidizing losses in the industry, and defer politically difficult reform, was significant, particularly as Norway enjoyed significant oil-based wealth. However, the economic crisis arrived at a time of change in the approach of the government of the day towards more self-sufficient industries in general. This was assisted by the powerful political voice of the NFA. Many fishers, especially the larger vessel owners, recognized the potential benefits in terms of increasing profits from closing access and reducing the number of participants in the industry.

Sequential reform, continuous improvement and the demonstration effect

The Norwegian reform process reflected a gradual approach to reform with sequential changes to the fisheries management system providing a smooth introduction of reforms over time. This was in marked contrast to the relatively rapid reduction in fisheries subsidies, which declined from around NOK 900 million in 1992 to less than NOK 200 million just two years later. The management reforms effectively provided a means for the industry to restructure and become self-sufficient, but they were introduced over a longer period of time. The management reforms tackled the relatively 'easier' fleet segments, and those facing the more immediate economic crises, before turning to the more politically difficult fleet segments after years of experiences with the various systems. In this way, the reforms introduced a series of tailor-made schemes, which demonstrated an adaptive policy response but kept the basic policy principles intact: primarily by closing the commons and, through the use of market-based mechanisms, creating better-aligned incentives for fishers to effectively manage capacity and support sustainable fisheries. This reflects a philosophy of continuous improvement that provided a demonstration effect as successive fisheries moved from open access to closed access. This enabled support for further reform to be developed and then sustained. Such an approach is supported by all stakeholders through the extensive consultation process which is an integral part of the Norwegian fisheries management system.

Strong stakeholder involvement

The varied fishing fleet in Norway has created challenges in trying to reduce subsidies and to design efficient instruments to manage overcapacity. The key to overcome the management challenges has been continuous cooperation between the authorities and the stakeholders in the fishing industry and other affected organizations/institutions. Norwegian fisheries management has been characterized as 'a system of centralized consultation' (Mikalsen and Jentoft, 2003). While the central government retains the ultimate authority to manage fisheries, there is a significant element of power-sharing through the institutional arrangements facilitating participation in decision-making. When it became clear that the subsidy regime was going to be significantly reduced, the focus of government/industry discussions turned to designing and implementing mechanisms that would lead to a profitable and sustainable industry. For example, the distribution of resources between vessel groups is largely determined by the NFA, with the allocation keys proposed by the NFA largely being adhered to by the government. This is in the interests of ensuring long-term stability within and across vessel groups so that the benefits of the structural systems accrue to the fishers within the vessel groups. There is also strong stakeholder involvement in policy reform. For example, the development of the 2006 white paper was based on a year-long committee process which comprised key stakeholders, with the government acting as agents of change to help move the process along. This is not to say that the process has been easy or smooth at all times. However, the long history of institutionalized cooperation provided a strong basis for discussing issues and resolving disputes.

Compensation to achieve support for reforms

A holistic approach was taken to the reform process in Norway. Compensation to those who exited the industry was an essential part of the reforms and was largely effective in ensuring ongoing support for policy changes. Decommissioning schemes have been used as a means not only of assisting the transition to a lower level of capacity, but also as a means of compensation and buying support for reform from affected groups of fishers. The use of decommissioning schemes in this way needs to be carefully managed, otherwise they may themselves create perverse incentives. These schemes need to be well-targeted, temporary, involve an element of industry contribution and be introduced in conjunction with management reforms that encourage automatic adjustment of capacity in the future (OECD, 2009).

Support for the reforms was also achieved by ensuring that there was stability in the distribution of resources between the various vessel groups. The benefits of improved management were distributed among the groups through long term allocation keys that had been agreed within the industry (although not without dissent, particularly between the coastal fleet and the offshore fleet, as represented by the NCFA and the NFA, respectively). Such stability, transparency and inclusiveness helped to reduce the perceived threats from policy reforms targeting capacity reductions and improved efficiency.

Meeting regional development objectives

Norway has a varied and technologically advanced fishing fleet, encompassing both small coastal vessels and large off-shore trawlers and purse seiners. A fleet composed of a variety of sizes has been seen to be vitally important to keep up both employment and livelihood in many coastal communities, but also because a varied fleet of smaller and larger vessels has the advantage of being able to exploit all parts of the fish-stocks, in-shore as well as off-shore, in a rational fuel- and cost-efficient way.

The government has faced an ongoing challenge in meeting the twin objectives of modernizing the Norwegian fishing industry through the use of market-based instruments and reduced subsidies, and maintaining coastal or regional development. As noted above, a feature of the reforms has been the sequential nature of policy changes which has allowed management instruments to be successively fine-tuned in response to real-world experience. The latest round of policy changes for the coastal fleet represents a complex development of the type of market-based instrument used elsewhere in Norway. It is too early to tell if this will result in the intended outcomes as envisaged by the government.

Common lessons from Senegal, Ecuador and Norway cases

Without attempting to summarize the three sections that have been presented in this chapter, it is worth considering briefly some of the common lessons that they teach. If a single theme unites them all, it would seem to be that fisheries subsidies have often had unintended negative consequences. What has been the basic nature of these consequences and how can they be explained? The answer seems to lie in the fundamental fact that fisheries subsidies are routinely used as instruments of industrial policy without adequate attention to resource impacts or to real socio-economic dynamics.

The Senegalese case is perhaps the most distressing because it reveals how unintended consequences prevented a developing country from reaping real social and economic benefits from subsidies programmes that the government was scarcely able to afford. By offering subsidies preferentially to export-oriented industrial scale fishing, the policies effectively stimulated the depletion of biologically sensitive pelagic stocks while simultaneously driving up the local price of fish central to the Senegalese diet. At the same time, the rapid motorization of Senegal's pirogue-based fisheries put sudden and excessive pressure on critical in-shore stocks. Moreover, even in purely 'industrial' terms, many of the subsidies failed to take account of the comparative advantages enjoyed by Senegal's efficient small-scale fishers – creating a commercial handicap for a subsector that was potentially highly competitive and traditionally responsible for significant levels of local employment.

For Ecuador's tuna industry, the case study is distinct. While fisheries subsidies do exist in Ecuador, they have been mainly oriented towards the artisanal sector and have not greatly benefited the domestic tuna fleets. Even so, the study argues that Ecuadorian fisheries subsidies need to be more carefully aligned with policies for sustainable fisheries management. But the main conclusions of the study focus on the reduction in Ecuador's tuna exports to the US market as a result of the subsidies enjoyed by competitive tuna fleets from countries such as Japan, Republic of Korea and Spain. It might be said that in this case the subsidies provided by such countries did have at least some of their intended impact – obtaining commercial advantages for their national industries. But the case study also clearly demonstrates that many of the subsidies provided by these countries are in the highest category of risk for causing the depletion of fish stocks, a fact that clearly raises questions about the long-term durability of the national advantages such subsidies secure.

In the Norwegian case, subsidies were also employed with the aim of increasing efficiency and competitiveness and of supporting traditional coastal fishing communities. It was considered necessary to promote modernization and to protect fishermen from large fluctuations in fish prices that occurred from year to year. And it was even believed that subsidies could be used only on a short-term temporary basis, to help the industry 'adjust' and then stand on its own. But things did not work out as intended. 'Temporary' subsidies became a permanent fixture of Norwegian fisheries policy over the course of more than a generation. Subsidized modernization and generous price supports, when applied in the context of open access fisheries, rapidly resulted in fleet overcapacity and the deterioration of stocks. And policies meant to favour remote rural communities instead concentrated their benefits on large-scale industrial enterprises. Then it was the impacts on resource stocks and profitability that provided the impetus for reform. Rapid reduction of subsidies was accompanied by gradual improvements to the fisheries management system through constant stakeholder cooperation.

In all these cases – as in many of the other country studies mentioned in the introduction to this chapter – it is clear that domestic subsidies are routinely maintained without serious thought for the impact on fisheries resources and often without a proper understanding of how the subsidies will affect the economics of the sector they are meant to benefit. Repeatedly, subsidies that are divorced from fisheries management considerations turn out to be not only bad for the fish but also for fishing communities and the entire economies. These studies show why the stakes are so high in the fisheries subsidies debate and why the essential thrust of reform must be towards more careful and better integrated policy-making.

Endnotes

1 Respectively 200,000 and 400,000.

2 Of these, about 57,000 are involved in small-scale fishing and 3000 in industrial fishing. Over 80 per cent of fishermen are active in the coastal marine fishery, the other 20 per cent in inland fishery.

3 The value added can be defined as the measure of wealth created by an economic activity. For any economic agent, the value added created corresponds to the difference between the value of output produced and the value of intermediate consumption (inputs) used for this purpose.

4 For comparison, the per capita consumption of meat is about 13kg (chicken, beef, lamb, etc.).

5 The five other goals are:
   1 To ensure sustainable management of fisheries and aquaculture;
   2 To improve and modernize operations in artisanal fishery;
   3 To increase added-value to fishery products;
   4 To develop a sustainable system for financing fishing and aquaculture activities;
   5 To reinforce international cooperation in fisheries and aquaculture.

6 The CFA franc is the currency used by Senegal and the 13 other African countries of the 'Franc Zone' (mostly former French colonies). Since 1945, the CFA franc or its predecessors have enjoyed a fixed exchange rate against the French Franc (and now against the Euro, by agreement with France). In 1994, the exchange rate was changed for the first time since 1960, when the value of the CFA franc was halved in order to help boost African exports. The value of the CFA franc has not been adjusted since. See website of the Musée de la Banque Centrale des Etats de l'Afrique de l'Ouest (BCEAO), www.bceao.int/internet/bcweb.nsf/pages/musee (last accessed 24 November 2009); see also http://ec.europa.eu/economy_finance/the_euro/euro_in_world9371_en.htm (last accessed 24 November 2009).
7 In 2008 (CEC, 2008) the European Union adopted a regulation for the implementation of a plan to fight against Illegal, Unreported and Unregulated fishing (IUU); this plan provides for the implementation of catches certificates to all fish products entering the European Community.
8 It followed the 2001 UNEP work on trade liberalization in the Senegalese fishery that highlighted the fact that fishery cannot be managed without taking into account trade issues. This was the first time that trade was associated with the management of fishery resources in Senegal.
9 Following a 2004 FAO group on overcapacity in fishery.
10 To follow other policies decentralized in 1990 (Failler et al, 2002).
11 A practice known as 'tontines', friendly associations in which members contribute money that is provided to one of them on a revolving (daily, weekly or monthly) basis were seen as a way to have working capital or to accumulate enough capital to invest in fisheries. However, the risks of misuse of funds were very high especially in big tontines.
12 Formal financing for the artisanal fishery in general is deeply insufficient despite the efforts deployed by public authorities. Among the difficulties identified are the limited access to credit (high fluctuation in production capacities, high interest rate), difficulties in loan reimbursements (producers' insolvency), absence of collateral, lack of banking culture among actors of the sector and their non affiliation to insurance companies (rapid depreciation of the fishing equipment, high risk of accidents and thefts, various frauds).
13 The CFA franc went from CFA 50 to CFA 100 = FF1 (French franc), a measure virtually imposed by France, the IMF and the World Bank.
14 The exchange rate has been fixed to 1 CFA-Franc = 0.152449 euro.
15 See WTO definition on subsidies, as contained in Article 1 of the WTO Agreement on Subsidies and Countervailing Measures (Annex 10 to this book).
16 An analysis of relation between catches, prices and exports in tuna market explained in Catarci (2004) illustrate the kind of effect of changes in fishing effort.
17 At the average 1981 exchange rate of 1 US$ = 5.734 NOK.
18 The General Agreement is also sometimes referred to as the 'Main Agreement'.
19 While details vary across fisheries where they are applied, the system of ITQs basically involves the division of a total allowable catch among a number of fishers who then have the right to trade their individual quotas amongst themselves, or sometimes with new entrants to the fishery.
20 For example, the SQS for the coastal fleet would be extended to cover the group of vessels with a quota length of 11–15m (or 13–15m in the case of the coastal group for mackerel fishing).

Chapter 4

Emergence of an International Issue: History of Fisheries Subsidies in the WTO

Introduction

Fisheries subsidies have been the subject of the first global effort to use the laws and institutions of international trade to promote sustainable development in a key natural resource sector. The history of the issue – how it developed politically and technically – is an important chapter in the history of international environmental policy and of international economic relations generally. An account of that history may be of use to current WTO delegates – who often work in four-year rotations – as well as to other stakeholders ready to contribute to this major effort of developing international rules on fisheries subsidies. It can also provide a valuable context for better understanding the technical questions covered in more detail in Chapter 5 of this book.

A confluence of intellectual, institutional and political factors has shaped the emergence of the fisheries subsidies issue over the course of a history that can be viewed in five overlapping phases, all highly influenced by the overall context of trade negotiations:[1]

- Phase I (1992–1998): empirical and analytical studies carried out by academics and civil servants open the debate and lead to early attention by international policy-makers.
- Phase II (1998–2001): rising public controversy over the phenomenon of 'globalization' helps activists raise the political profile of the issue and moves fisheries subsidies on to the core agenda of the WTO at the 2001 launch of the Doha Round of global trade negotiations.
- Phase III (2002-2005): the form and content of the WTO negotiations take shape as both key issues and key stakeholder groups emerge.

- Phase IV (2006–2008): a high level of ambition is set for the outcome of the WTO talks as domestic policy-makers pay increased attention to the need for reform of national fisheries subsidies practices; simultaneously, technical and political obstacles to reform become more apparent.
- Phase V (2008–): with a draft text finally on the table, the talks become more political and technical, and then enter a period of uncertainty as the 'Doha Round' of comprehensive trade talks reaches a long impasse.

As suggested by the inconclusive fifth phase, the fisheries subsidies issue approaches the start of its third decade without a clear indication of how the story will turn out.

Unfortunately for fishermen and fish alike, factors largely external to the debate have prevented the swift adoption of binding new international rules. At the time of this writing, the future direction and influence of the WTO as a global institution remain more uncertain than at any time since the birth of the multilateral trade system at the close of World War II. Simultaneously the international response to climate change has risen to the top of global environmental and economic concerns, introducing new complexities, priorities and opportunities into the global policy arena.

In this context – and with the crisis of excess capacity and overfishing still far from resolved – the problem of fisheries subsidies remains as urgent as ever. The story has not yet reached its endpoint. But, as the following history makes clear, significant steps have at least been initiated towards integrated environmental, social and economic policies.[2]

Phase I: Early Analysis and Preliminary International Action

> As the opportunities for an increased catch from fishery resources have declined considerably, a continuation of the high subsidies can only lead to greater and greater economic distress as well as a further depletion of stocks.
>
> (FAO, 1993d)

Fisheries subsidies are not a new phenomenon, and have been the subject of technical examination and discussion by international bodies such as the FAO and the OECD since at least the 1960s. It was not until the 1990s, however, that the issue began to gain the attention of policy-makers and stakeholders in civil society. Between 1992 and 1998, a series of studies conducted by the FAO, the OECD, UNEP, WWF and the World Bank, among others, revealed a significant level of subsidization to the fisheries sector globally and suggested strong links to fisheries depletion.

The most influential of these early studies was the 1993 FAO publication *Marine Fisheries and the Law of the Sea: A Decade of Change* (FAO, 1993a)[3] –

one of the first major international publications to sound a clear warning of the growing fisheries depletion crisis (see Chapter 1). While the phenomenon of fisheries subsidies had been known and discussed among experts for at least the previous quarter century, the FAO study was among the first to emphasize subsidies among the underlying causes of declining fish stocks. It was also the first to offer an initial (albeit indirect) estimate of the scale of the fisheries subsidies problem. After reviewing available data on the costs and revenues of commercial fishing worldwide, the FAO found that industry-wide costs appeared to exceed revenues by an estimated US$54 billion per year and concluded that '[s]ubsidies are presumed to cover most of this deficit.' The $54 billion figure was all the more remarkable in comparison with annual industry non-subsidy revenues, which the report estimated at approximately $70 billion. Although by the end of the decade most experts agreed that the ratio of subsidies to revenues was not quite so dramatic, the FAO study placed the fact of massive and harmful fisheries subsidies squarely before the international community.

Around the same time, the Organization for Economic Co-operation and Development (OECD) undertook a review and discussion of 'assistance instruments in the fishing industry' in the major industrialized economies (OECD, 1993b). The work examined border adjustments (tariffs, quotas, etc.) as well as direct 'domestic supports', focusing almost exclusively on their potentially distorting effects on international export markets for fisheries products. Unlike later OECD work on fisheries subsidies, it did not consider the problem of fisheries depletion. The OECD activity did, however, contribute to the growing international technical discussion of subsidies and the international economics of fishing.

As a practical matter, the influence of these first studies was restricted largely to the world of fisheries policy experts. But several factors would soon combine to begin raising the visibility of the fisheries subsidies problem to a broader set of stakeholders and policy-makers.

First, in the span of just a few years in the mid 1990s, a series of international declarations and agreements reflected the rapidly increasing priority of the fisheries crisis as a global issue. These new international instruments included the Cancun Declaration on Responsible Fishing (1992), the Rio Declaration on Environment and Development (Earth Summit, 1992) and the Rome Consensus on World Fisheries (1995). In October 1995, this trend culminated with the adoption of the UN Code of Conduct for Responsible Fishing – a text that remains the leading codification of international fisheries norms (see Chapter 1).

The declarations and codes of the early to mid 1990s were heavily focused on the need to improve fisheries management through science-based regulations and international cooperation. The texts – and the meetings that generated them – devoted little or no time to the problem of fisheries subsidies themselves. But the adoption of the UN Code of Conduct marked a turning point after which international attention turned increasingly from the definition of general principles to their implementation through specific obligations and norms.

Among the first topics to be addressed in this implementation process was the problem of excess fishing capacity. It had been known for some time that on an aggregated global basis – the world's fishing fleets were far larger and more powerful than could be sustainably utilized at full production. Put simply: there were (and still are) too many boats chasing too few fish. Accordingly, the FAO's Committee on Fisheries began work in 1997 towards an International Plan of Action for the Management of Fishing (IPOA) Capacity. From the outset, the consultations and negotiations over the IPOA Capacity included the question of subsidies and their relation to excess fishing capacity.

Another trend that helped focus attention on fisheries subsidies was the rapid rise in the international trade in fish products. From 1985–1995, the quantity of fish products entering international trade rose by more than 50 per cent, with trade in valuable oceanic species more than doubling (FAO SOFIA, 2002). Naturally, this led policy-makers to take a growing interest in the terms of trade affecting the fisheries sector. One result was the inclusion of fisheries products in an ambitious 'early voluntary sectoral liberalization' initiative undertaken by the Asia-Pacific Economic Cooperation (APEC) trade ministers in 1997 to reduce tariff and other barriers to trade in fish products.[4] While the APEC Early Voluntary Sectoral Liberalization did not ultimately succeed, the initiative paralleled long-standing (and still ongoing) efforts by the United States to achieve the total elimination of tariffs on fish products and later developed into a formal proposal to WTO members during preparations for the 1999 Ministerial in Seattle. The APEC work programme also led to the commissioning of a major survey of fisheries subsidies among APEC members (APEC, 2000).

As attention to fisheries management, fleet overcapacity and fisheries trade grew through the mid 1990s, so did the activities of academics, environmental NGOs, governments and intergovernmental bodies aimed directly at reforming fisheries subsidies practices. 1997 turned out to be a critical year during which,

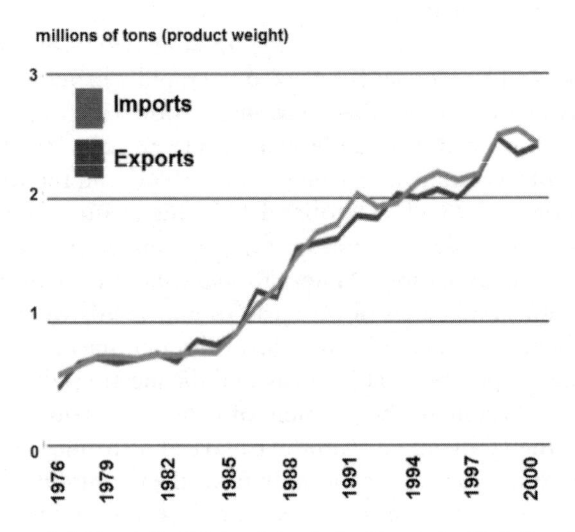

Figure 4.1 *Rapidly Rising Trade in Oceanic Fish Products*

Source: FAO SOFIA, 1992, Fig. 11

among other activities, UNEP co-hosted the first international conference ever dedicated to the fisheries subsidies problem. The conference, held in Geneva, was produced in collaboration with WWF and marked the start of serious and coordinated environmental advocacy in favour of fisheries subsidies reforms. Around the same time, the United States and New Zealand became the first governments to raise the issue of fisheries subsidies at the WTO, tabling separate papers before the WTO's Committee on Trade and Environment (CTE) and calling for attention to the potential role the WTO could play in addressing the problem.[5]

These activities contributed to the consideration given to fisheries subsidies by the UN General Assembly when it met in 'special session' in mid June 1997 to mark the fifth anniversary of the 1992 Earth Summit. A principal outcome of that meeting was the adoption of a 'Programme for the Further Implementation of Agenda 21,' paragraph 36(f) which called on governments:

> to consider the positive and negative impact of subsidies on the conservation and management of fisheries through national, regional and appropriate international organizations and, based on these analyses, to consider appropriate action,[6]

This language in a UN General Assembly resolution – 18 months before the completion of the FAO IPOA Capacity – constituted the first formal call by a global intergovernmental body for cooperative international action to reform the use of fisheries subsidies.

The momentum that started to gather in 1997 received additional energy in 1998. While negotiations over the IPOA Capacity were actively underway in Rome, a series of new studies and policy analyses began to appear. The most influential of these was a technical paper published by the World Bank that offered the first direct empirical effort to quantify fisheries subsidies on a global basis (World Bank, 1998). The study drew on a combination of publicly available data and information provided to the author through US embassy sources around the world. This data, along with some extrapolation, led its author (Matteo Milazzo) to estimate that 'environmentally harmful' (in other words, capacity-enhancing) subsidies to commercial fishing worldwide totaled US$14–20 billion per year – or roughly 20–25 per cent of global fishing industry revenue. Although now more than a decade old, Milazzo's paper remains one of the few authoritative catalogues of global fisheries subsidies practices.

1998 was also the year in which environmental activists first began to push directly for WTO action on fisheries subsidies. The groundwork had been laid in 1997 both by the US and New Zealand papers tabled at the WTO and by early academic analyses of the potential application of international trade law as a tool for restricting fisheries subsidies.[7] Legal and policy experts had recognized that existing WTO rules already imposed restrictions on fisheries subsidies to the extent they caused harmful distortions in export markets. However, the existing WTO rules would need to be changed if they were to address the linkage between subsidies and the rapid depletion of fish stocks.

Building on these concepts, WWF soon became the first major non-governmental organization to issue a direct call for new WTO rules on fisheries subsidies, launching its initiative during a 'High Level Symposium' at the WTO in March 1998. Later that year, separate publications by both UNEP and WWF laid out the technical case for WTO action, initiating what would become an important and ongoing dialogue with trade technocrats and policy-makers (UNEP, 1998, WWF 1998c).

Meanwhile, in the last months of 1998 the FAO negotiations over the IPOA Capacity produced a final text. Among its provisions was a clause that called on governments to:

> reduce and progressively eliminate all factors, including subsidies and economic incentives and other factors which contribute, directly or indirectly, to the build-up of excessive fishing capacity thereby undermining the sustainability of marine living resources. (FAO, 1999c).

When the text was formally adopted by FAO members in February 1999, the IPOA Capacity became the first normative international instrument to address the fisheries subsidies problem.

The IPOA Capacity was, however, voluntary and hortatory in its legal effect. With stock depletion continuing unabated, stronger action to reduce subsidized fishing pressure was clearly needed. But it was not until fisheries subsidies became linked to the heated politics of globalization that the issue gained a serious foothold on the agendas of world leaders.

Phase II: Globalization and the Shift of Focus to the WTO

'Trade and environment' in the globalization debate

The phenomenon of globalization had been stirring passions for years before the birth of the fisheries subsidies issue. Dissatisfaction with new laws and institutions promoting 'free market' international economic relations had been growing for a long time and in many different national circumstances. For decades, the IMF and the World Bank had been encouraging government budget cuts and the rapid removal of market controls in developing countries. In more than a few cases, these policies led to intermittent but sometimes serious public disturbances. Similarly, in the United States and in Europe post-war moves towards trade liberalization had at times been met with fierce resistance from organized labour and special domestic industrial interests.

By the early 1990s, public fears over the impacts of globalization were growing broader and deeper. In the United States in particular, negotiations over the North American Free Trade Agreement (NAFTA) and the creation of the WTO prompted an unprecedented level of popular concern. Among the main preoccupations of US activists was the fear that trade liberalization would cause a deregulatory 'race to the bottom' that would eviscerate national environmental standards. Similar concerns eventually spread in Europe and even among some activists in developing countries. The result was the birth of 'trade and environment' as a focus of activism, policy and academic debate.

At its inception, the trade and environment movement was widely viewed as a predominantly 'northern' phenomenon. The movement received its first major impetus from the 'tuna–dolphin' dispute, at the WTO, in which Mexico and other Latin American countries successfully challenged a US dolphin protection law as contrary to the General Agreement on Tariffs and Trade.[8] The law in question imposed a trade ban on any tuna caught in a manner that US law did not deem 'dolphin safe.' As a result, tuna caught by Mexican fishermen was excluded from the US market even though Mexico was fully complying with an international agreement (to which the US was also a party) that established strict limits on dolphin mortality from tuna fishing. A GATT dispute resolution panel found that the US import ban was illegal under international trade law.

For many citizens, including thousands of schoolchildren who had petitioned the US Congress in favour of dolphin protections, the idea that an international trade treaty could limit the ability of the US to protect dolphins came as a disturbing surprise. Environmental groups and citizens in some European countries had also joined the 'dolphin-safe' tuna movement, and similarly viewed the GATT case as a threat to national and international environmental policies. For Mexican fishermen (and many other observers across the developing world), however, the GATT ruling was seen as an appropriate rebuke to 'green protectionism' and to the illegitimate imposition of US environmental standards on foreign industries.

Even though it was never implemented,[9] the tuna–dolphin ruling generated a public backlash that opened a new chapter in the debate over globalization. At the same time, the United States, Mexico and Canada were negotiating NAFTA – a major regional free trade agreement, with an environmental 'side agreement'.

The NAFTA side agreement set a precedent for including environmental issues as a component of international trade agreements. When the negotiations to create the WTO concluded just a few years later, environmental issues were once more brought to the fore, again largely at the insistence of the United States. When it was formally born on 1 January 1995, the WTO was mandated by the first paragraph of its charter to pursue trade relations that promote 'the optimal use of the world's resources in accordance with the objective of sustainable development, seeking both to protect and preserve the environment and to enhance the means for doing so...'. The Committee on Trade and Environment (CTE) was established among the standing bodies of WTO governance and was given a broad mandate 'to enhance positive interaction between trade and environmental measures, for the promotion of sustainable development, with special consideration to the needs of developing countries, in particular those of the least developed among them.'[10]

Such moves towards integrating environmental concerns into international trade relations did not, however, dampen the rapid emergence of an anti-globalization movement around the world. When WTO members announced their intention to open a new round of trade liberalizing negotiations – to be formally launched at a ministerial meeting in Seattle, Washington, at the end of 1999 – the debate over globalization intensified dramatically. Grass-roots groups in the US and elsewhere began planning massive street demonstrations to accompany the Seattle ministerial meeting. Meanwhile, in many developing countries, citizens groups and governments alike were voicing increased resistance to 'northern' demands for further market liberalization.

These developments culminated in the historic explosion of riots at the WTO meeting in Seattle in December 1999. Those riots – the first of a long series of violent international protests against the WTO, the G8 and global financial institutions such as the IMF and the World Bank – put the anti-globalization movement on front pages and TV screens around the world. Meanwhile, in the barricaded meeting rooms of Seattle's convention center, trade ministers were finding it impossible to reach common ground on a new negotiating agenda, with the split particularly dividing developed and developing countries. In the end, the Seattle ministerial collapsed in physical and diplomatic shambles, with broken glass in the streets and a shattered consensus for launching the 'Seattle Round'.

The dramatic failure of the Seattle meeting, however, masked an important fact: if the meeting had not failed, the agenda for the Seattle Round would have included a mandate on fisheries subsidies.

The politics of fish from Seattle to Doha

The path fisheries subsidies took to Seattle was shaped by the political, ideological and commercial interests of a variety of governments. President Clinton, who had come to power in the US in 1992, repeatedly promised that market liberalization could go hand in hand with environmental and social improvements. However, there was significant disquiet among developing country WTO members who increasingly feared the impacts of further market liberalization on weaker and less developed economies. In response, the Clinton administration ramped up its pro-trade advocacy by promising a new phase of public-spirited policies that would achieve 'globalization with a human face'.

In this context, the fisheries subsidies issue was primed for top-level attention from US policy-makers. From a commercial perspective, the US ranks far behind other major powers such as the EU and Japan in the level of subsidies it provides to its fishing industry, leaving US fleets at a competitive disadvantage. It was thus no surprise when environmental groups pushing for US leadership on the issue were joined by the National Fisheries Institute, a leading lobbying group for some US fishing interests. Meanwhile, at governmental level, the US was already engaged in technical and diplomatic work through the discussions of fisheries subsidies in the FAO and the United Nations Commission on Sustainable Development (UNCSD), with significant 'buy-in' from key policy-makers in the US Department of State and the National Marine Fisheries Service within the Department of Commerce. With stakeholders, bureaucrats and politicians all behind the issue, fisheries subsidies quickly became part of the US package for Seattle.[11]

Several other leading developed countries with strong and largely unsubsidized fishing industries – such as Iceland, Australia, New Zealand and Norway – had both ideological and commercial motives to join the fight. In addition, a number of developing countries – such as Peru and the Philippines – had clearly recognized the negative impacts of foreign subsidies on their own fishing sectors. At the initiative of the United States, five of these countries came together in the spring of 1999 during the WTO's annual meeting of civil society stakeholders to announce their joint intention to seek a negotiating mandate on fisheries subsidies as part of the anticipated Seattle Round (Crutsinger, 1999). A few weeks later, all seven formalized their position in a series of proposals tabled as part of the WTO's preparatory process for the ministerial meeting.[12] A number of other countries, including Argentina, Brazil, Canada and Chile also quickly voiced support for the move (Williams, 1999).

> *The sovereign right of countries to manage their own natural resources must be respected. Yet the global community must cooperate in ensuring the conservation and sustainable use of marine living resources. No one single action could bring about such positive results towards achieving sustainable development in fisheries as would the elimination of government subsidies.*
>
> (WT/CTE/W/103, 25 January 1999, Submission by Iceland)

But support for bringing fisheries subsidies into the WTO was not universal. Europe and Japan had been reluctant to support international action to reform subsidies from the outset of the IPOA Capacity discussions in the FAO. When the US, Iceland and New Zealand began promoting attention to fisheries subsidies in the WTO, Europe and Japan resisted with even greater vigor, arguing that the link between fisheries subsidies and depletion had not been well established and that questions of fisheries management were beyond the remit of the WTO and should be left to bodies such as the FAO.[13]

The approach of the EU and Japan was, however, different in subtle but important ways. The EU argued that its own subsidy regime was environmentally positive in character, claiming that the 'main aim' of EU subsidies was 'to adjust fishing activity and restructure the fisheries sector'.[14] Further, the EU noted that it was doing a much better job than other WTO members at complying with WTO subsidy notification rules and called for improved transparency from other governments. So while it clearly resisted bringing the fisheries subsidies issue into the WTO – and repeatedly argued that management, not subsidies, was the main fisheries problem – the EU's insistence on its own environmental credentials and above-average transparency reflected domestic political and legal realities that would continue to shape its posture in the WTO fisheries subsidies talks for many years.

For its part, Japan was vehement in arguing that no link between subsidies and depletion had ever been established and that the WTO was the wrong forum to address issues related to fisheries depletion. Moreover, Japan's principal 'pre-

Seattle' concern in the fisheries sector was to avoid the tariff reductions proposed by the US, Iceland and others. Japan had repeatedly raised environmental arguments against these tariff cuts and had received some support for their position from environmental groups. In what may have been partly a tactical move, Japan called for treatment of all issues relating to fisheries and forest products in separate sectoral discussions within any new WTO round, stressing the peculiar characteristics of sectors based on trade in exhaustible natural resources.[15] Around this time Japan was also among the strongest supporters of using 'trade related environmental measures' (in other words, trade sanctions) as instruments to help enforce the rules promulgated by regional fisheries management organizations. These and similar trade measures in other multilateral environmental agreements (MEAs) were known to be in potential conflict with WTO rules – one of the concerns frequently raised by NGOs active in the 'trade and environment' debate.

In short, even as it opposed WTO action on fisheries subsidies, Japan was taking positions that, at least at face value, proposed directly linking fisheries conservation and trade policies. Although Japan's 'green' motives were questioned by some of its trade partners, it is nevertheless interesting to consider what might have been possible if WTO members had heeded Japan's call for 'comprehensive' treatment of trade in fisheries and forest products, 'giving due consideration to the global environmental issues and to the resource conservation and management issues, in order to ensure a sustainable utilization of resources'.[16] In any case, the US and others were not open to such an approach and within a few years Japan itself had shifted to arguing that there was no basis for giving special treatment to the fisheries sector within WTO subsidies rules.[17]

Whatever the mixture of motives and interests among the governments gathering in Seattle, the issue of fisheries subsidies gained substantial momentum in the weeks preceding the meeting, due in part to vigorous diplomatic leadership by Iceland and the United States and in part to an aggressive media campaign waged by WWF that specifically targeted the EU. By the time the ministerial opened, fisheries subsidies had become a prominent issue on the meeting's agenda and preliminary drafts of the ministerial declaration contained language launching work on fisheries subsidies within the anticipated new round. When countries such as Morocco – which had not spoken to the issue previously – used the opening speech of its trade minister to highlight the need for action on fisheries subsidies,[18] it was clear that a genuine 'north–south' coalition in favour of the agenda item was emerging. A high level press conference in support of the issue – attended by ministers from Argentina, Australia, Iceland, New Zealand, Peru, the Philippines and the United States – gave the issue additional political prominence.

The 'Seattle Round' was, of course, destined to be no more than a historical footnote of what might have been. But just hours before the ministerial meeting was declared a failure, a final draft version of the ministerial declaration was circulated to delegates. Among the paragraphs that had effectively been accepted up to that time was the following:

> **Subsidies and countervailing measures:** the rules shall be reviewed, and where necessary amended, on the basis of proposals by participants, taking into account, inter alia, the important role that subsidies may play in the economic development of developing countries, and the effects of subsidies on trade. In the context of these negotiations, the areas to be considered shall include, inter alia, certain subsidies that may contribute to over-capacity in fisheries and over-fishing or cause other adverse effects to the interests of Members. The work on fisheries subsidies shall be carried out in cooperation with the FAO and drawing also on relevant work underway within other intergovernmental bodies, including regional fisheries management organizations. It shall consist of (i) the identification and examination of subsidies which contribute to over-capacity in fisheries and over-fishing, or have trade distorting effects, and (ii) the clarification and strengthening, as appropriate, of disciplines under the [Subsidies and Countervailing Measures (SCM)] Agreement with respect to such subsidies. (Ibsen, 2000)[19]

The major strength of this language was that it called directly for the 'clarification and strengthening' of WTO subsidies disciplines with regard to fisheries subsidies, and explicitly focused the mandate on subsidies that contribute to overcapacity and overfishing. Moreover, the location of the paragraph in the draft ministerial text made clear that fisheries subsidies were to be addressed in the context of core negotiations on possible changes to the WTO rule system. The explicit mention of a cooperative role for the FAO and Regional Fisheries Management Organizations (RFMOs) was also a progressive achievement. Indeed, the final version of the fisheries subsidies mandate was substantially stronger than the proposal that Iceland and others had brought into the earlier draft declarations – due to the elimination of language that would have delayed the onset of actual negotiations until after a preliminary period of study that was to last until ministers met again to confirm the need to alter existing WTO rules.

After the collapse of the Seattle meeting, it was widely understood that the major economic powers would again try to initiate a new round of WTO talks within a few years. Accordingly, both governments and stakeholders remained active, promoting attention to fisheries subsidies and laying the groundwork for the anticipated negotiations. In the CTE, a series of papers tabled by Iceland, New Zealand and the United States – along with technical notes drafted by the WTO Secretariat – kept the formal WTO process alive. In fact, the continued activity in the CTE prompted Japan, now joined by the Republic of Korea, to keep actively resisting progress towards new WTO fisheries subsidies rules.

Meanwhile, outside of the formal WTO process, other intergovernmental organizations and environmental stakeholders remained very active. In the period after Seattle and prior to the WTO Doha ministerial in November 2001, contributions to the fisheries subsidies dialogue included:

- publication of the PriceWaterhouseCoopers study of fisheries subsidies in APEC countries (October 2000);
- publication by the OECD of its major empirical and analytical study, *Transition to Responsible Fisheries: Economic and Policy Implications* (September 2000);
- ICTSD-IUCN-UNEP dialogue, 'Ensuring Trade Rules Affecting Fisheries are Supportive of Sustainable Development' (Geneva, October 2000);
- WWF's 'Fishing in the Dark' conference (Brussels 28–29 November 2000) and communications initiative to highlight the lack of transparency and public accountability in fisheries subsidies programmes;
- UNEP's Fisheries Subsidies Workshop (Geneva, 12 February 2001), including the presentation of drafts of two leading cases studies on fisheries subsidies in Senegal and Argentina;
- publication of WWF's 'Hard Facts, Hidden Problems' (October 2001), creating a comprehensive review of publicly available data on fisheries subsidies.

These and other activities also led to the growing visibility of fisheries subsidies in popular press outlets – a trend that was reinforced by continuing bouts of violence at major global economic meetings.

When trade ministers ultimately gathered in Doha, Qatar, in November 2001 to launch a new round, the political and diplomatic context for action on fisheries subsidies remained very favourable. The still-new administration of George W. Bush was signalling its intention to keep fisheries subsidies on the 'short list' of US demands for the new round, while the EU had emerged as the most aggressive proponent of addressing 'trade and environment' issues, such as the relationship between WTO rules and trade measures in environmental treaties. Europe's environmental posture in Doha made outright opposition to a fisheries subsidies mandate impossible. As a fall-back, the EU hoped to downgrade the topic by placing it in the 'environmental' negotiating basket. But the *demandeurs* prevailed, and fisheries subsidies emerged from Doha as a key item in the mainstream negotiations over WTO rules. The relevant language of the Doha ministerial declaration read:

> WTO RULES
>
> 28. In the light of experience and of the increasing application of these instruments by Members, we agree to negotiations aimed at clarifying and improving disciplines under the Agreements on Implementation of Article VI of the GATT 1994 and on Subsidies and Countervailing Measures, while preserving the basic concepts, principles and effectiveness of these Agreements and their instruments and objectives, and taking into account the needs of developing and least-developed participants. In the initial phase of the negotiations, participants will indicate the provisions, including disciplines on trade distorting practices, that they seek to clarify and improve in the subsequent phase. *In the context of these negotia-*

> *tions, participants shall also aim to clarify and improve WTO disciplines on fisheries subsidies, taking into account the importance of this sector to developing countries. We note that fisheries subsidies are also referred to in paragraph 31.* [Author's emphasis][20]

With these words, the link between trade rules and natural resource husbandry made its first appearance on the formal negotiating agenda of the global trade law system. It would, however, be several more years before the character and true direction of the WTO fisheries subsidies negotiations would be fully known.

Phase III: The WTO Negotiations Take Shape

Overcoming early resistance to the mandate

The Doha negotiating mandate on fisheries subsidies was a clear victory for the 'Friends of Fish' and environmental stakeholders, the importance of which was underlined several months later when heads of state and world leaders meeting at the World Summit on Sustainable Development in Johannesburg included international action on fisheries subsidies among the top priorities for achieving sustainable fisheries. The Johannesburg Plan of Implementation went on to call specifically for a successful conclusion to the newly-launched WTO fisheries subsidies negotiations (see Annex 6).

The language of the Doha mandate had several strengths. First, through the unusual mutual cross references with Paragraph 31, it clearly established the environmental orientation of the talks. Second, it also emphasized the importance of the fisheries sector to developing countries, and the need for any new disciplines to take that importance into account. And third, it explicitly located fisheries subsidies within the 'rules' basket of negotiating topics, along with other key elements of the WTO rules system, including rules on 'antidumping' and on subsidies generally. When the Doha talks were organized administratively it meant that fisheries subsidies were placed into the Negotiating Group on Rules (NGR) while the 'trade and environment' issues were relegated to 'special sessions' of the WTO's CTE. As a result, fisheries subsidies were to be negotiated together with other topics whose commercial relevance gave them high political priority within the round. This helped reinforce fisheries subsidies as a serious negotiating subject and created possibilities for trade-offs between fisheries subsidies and other core issues during the bargaining process. As discussed below, this positioning within the structure of the round would create perils alongside the obvious opportunities for fisheries subsidies.

Despite its strengths, however, the Doha mandate was just a starting point. The language of the declaration itself was terse and somewhat ambiguous. Unlike the language that had appeared in the pre-crash draft of the Seattle declaration, it did not explicitly refer to the problem of overfishing, or to the involvement of the FAO or RFMOs.

The vague instructions 'to clarify and improve' WTO disciplines left open the nature of any changes that might be required, or even whether the fundamental character of existing rules needed to be changed at all.

The first years of fisheries subsidies negotiations within the NGR reflected these strengths and weaknesses of the Doha mandate. In one of the earliest working papers submitted to the NGR, the 'Friends of Fish' coalition[21] submitted a strong call for new WTO rules to reflect the special value and characteristics of the fisheries sector and to help offset the 'production distortions' caused by the subsidized depletion of an exhaustible natural resource.[22] At the same time, however, the EU, Japan, Republic of Korea and Chinese Taipei sought to resist advancing the negotiations towards any substantive result, arguing that fisheries subsidies were not responsible for depletion and that there was no reason for the WTO to craft special subsidies rules for the fisheries sector.

This situation could have led to an unproductive stalemate had not several factors come into play that gradually altered the dynamics. First, environmental stakeholders led by WWF and UNEP kept up steady efforts to educate governments and the general public about the fisheries subsidies problem and its potential solutions. WWF initiated a series of intensive consultations with trade experts in Geneva that, over the course of more than two years, resulted in a dialogue of genuine mutual education. In the end, this process allowed WWF to produce a comprehensive technical proposal for WTO rules that could be both plausible within the institutional context of the WTO and effective in disciplining fisheries subsidies.

UNEP was also very active in the first years of the negotiations, bringing forward case studies and policy papers and convening a series of open consultations and workshops in Geneva. Over the course of 2002–2004, UNEP published studies examining fisheries subsidies in Argentina, Bangladesh, Mauritania and Senegal.[23] These studies, which in some cases produced multiple papers moving from problem identification to proposed solutions, provided an unusual source of detail about real world experiences with fisheries subsidies in developing countries. UNEP also built upon these case studies with a series of more policy-oriented papers, including a seminal analysis of the risks posed by subsidies under various fisheries management conditions (see Chapter 2).

Also influential were UNEP-sponsored workshops – such as those held in March 2002, July 2003 and April 2004 – which quickly became key elements of an ongoing multi-stakeholder dialogue that unfolded in parallel with the formal discussions inside the WTO. These workshops – along with occasional briefings and other seminars sponsored by UNEP or other stakeholders such as WWF or the International Centre for Trade and Sustainable Development (ICTSD) – provided rare opportunities for trade and fisheries experts to interact and for WTO negotiators to share ideas among themselves in a non-negotiating environment. Meanwhile, in the FAO a series of technical consultations on fisheries subsidies was underway as part of implementing the IPOA Capacity. FAO also organized a number of inter-agency consultations to ensure coordination on work on fisheries subsidies between the different IGOs. All of these processes helped move the fisheries subsidies dialogue towards a much greater level of technical maturity and began to give comfort to trade technocrats who had approached the issue fearing that the talks would lead to the dilution of WTO rules with inappropriate environmental considerations.

A second factor in the first years of the negotiations was the early emergence of developing country issues as a vital focus of the debate. The Doha mandate had emphasized the need to address developing country issues in the fisheries subsidies talks. Moreover, the Doha Round itself was being billed as the 'development' round – a reflection, in part, of the increased influence developing countries had begun to enjoy within the multilateral trade system.

Although developing countries were not especially active in tabling papers in the first years of the fisheries subsidies talks, the parallel processes hosted by UNEP, WWF and others quickly revealed the sensitivity of developing country issues. The discussion of these issues in parallel fora helped broaden the set of active governments beyond the established actors within the 'Friends of Fish' and their antagonists.

A third factor that had significant consequences for the WTO fisheries subsidies process was the internal dialogue taking place within the European Union over its own fisheries policies. The backbone of EU fisheries regulation was (and still is) the Common Fisheries Policy (CFP). The CFP that was in place at the start of the Doha Round was due to expire at the end of 2002, leading to the first major review of EU fisheries policy since the mid 1990s. With the stark facts of the global fisheries crisis ever more evident, the 2002 expiration of the CFP became the spur to substantial reforms – the greening, some would say, of European fisheries policy. Among the key elements of reform was a recognition of the problem of overcapacity within EU fleets, and of the need for restructuring the industry.

In parallel with the CFP reform, one of the main baskets of EU fisheries subsidies programmes – then known as the Financial Instrument for Fisheries Guidance (FIFG) – was due for a 'mid-term' review in 2003, just following the expiry of the CFP. The FIFG, and now its successor (known as the European Fisheries Fund (EFF), operated on a six-year cycle that required the adoption at the EC level of an overall regulatory framework for fisheries subsidies. As with the CFP reform itself, the review of the FIFG created pressures towards reforming EU fisheries subsidies policy, mainly in the direction of reducing subsidies that contribute to increased fishing capacity.

The CFP and FIFG reform processes opened the way for the EU to change its posture at the WTO fisheries subsidies talks. In April 2003, it submitted a formal paper that for the first time embraced the Doha negotiating mandate on fish and announced, 'The purpose of this paper...is to take the process in the Rules Group further forward.'[24] The paper went on to propose the prohibition of certain categories of fisheries subsidies, as well as the creation of a category of 'permitted' or 'non-actionable' subsidies. Contrary to the EU's earlier efforts to deny a link between subsidies and depletion, the paper stated: 'We have no doubt that introducing such rules on subsidies will eventually lead to a reduction in overcapacity, and therefore also to reduction in over fishing.' The paper also stressed the need to provide special treatment for developing countries and proposed improvements to WTO transparency rules applicable to fisheries subsidies.

Although the European change of heart would turn out to be limited and in some measure transient, its impact on the negotiations was real and lasting. While still bolstered by allies in the Republic of Korea and Chinese Taipei, Japan was left as the only major developed country opposing the Doha fisheries subsidies mandate – a position of isolation that was deepened when, just a week after the EU, China tabled a paper that also signalled its acceptance of the mandate and its readiness to negotiate within the 'traffic light' framework proposed by the US and the 'Friends of Fish', according to which fisheries subsidies were to be classified into 'prohibited' (red light), 'actionable' (amber light) and 'non-actionable' (green light) categories (Williams, 2004).

It is worth noting that Chinese acceptance of the fisheries subsidies mandate was accompanied by signals that China, by far the world's leading producer of aquaculture products, would strongly oppose including subsidies to aquaculture within the scope of new fisheries subsidies disciplines. Nor did China turn out to be alone in this regard, which became clear when Brazil came forward as a leading voice in the talks in 2005. The question of aquaculture was always delicate. Since the late 1980s, expanded aquaculture has been the only source of growth in the overall production of fish products, while total catches from wild capture fisheries have remained essentially flat. Today, aquaculture provides approximately a third of all fish products and aquaculture products are heavily traded. Moreover, in physical and biological terms, aquaculture and wild capture fishing are inextricably linked. Almost 25 per cent of marine fish caught in the wild are processed into oil and fishmeal, much of which becomes feedstock consumed by aquaculture operations. More recent practices, such as 'tuna ranching', depend on the capture of juvenile fish which are then grown to market weight in captivity.

In both industrial and environmental terms, therefore, aquaculture is an important factor affecting marine fisheries. Accordingly, some governments within the 'Friends of Fish' coalition tended to feel that aquaculture should be included within WTO fisheries subsidies rules. But strong political opposition from China and others, coupled with real differences in the industrial dynamics of aquaculture and wild capture production, made it appear likely from early in the negotiations that government support for aquaculture would largely escape the new fisheries subsidies disciplines.

The year following the acceptance by the EU and China of the basic negotiating mandate was a slow period in the negotiations during which delegations were heavily focused on a mid-round ministerial meeting scheduled for Cancun in September 2003. The meeting ended with an embarrassing stalemate among WTO members over how to proceed with the overall Doha Round. But as negotiations got going again in Geneva in the spring of 2004, the fisheries subsidies talks took another step forward: Japan abandoned its blanket opposition to WTO fisheries subsidies disciplines and began arguing instead for a 'balanced discussion' based on what it called a 'bottom-up' approach.[25] Rather than opposing new rules outright, Japan rejected proposals that had been tabled by the US and New Zealand calling for a broadly-worded ban on fisheries subsidies and proposed instead that fisheries subsidies should only be prohibited on a case-by-case basis.

The Japanese shift came amidst other signs of gathering momentum at the negotiating table and in the parallel process. In the preceding months, some 'Friends of Fish' delegations had individually tabled papers beginning to outline the components of a high ambition outcome to the talks. A paper from a group of small vulnerable coastal states signalled their deep concern that new fisheries subsidies rules would diminish their rights to support their impoverished artisanal fishing communities or to receive payments from other governments for access to their EEZ fisheries.[26]

Meanwhile, the two-day workshop sponsored by UNEP in April 2004 was heavily attended and raised the parallel process to a new level of visibility and intensity. As noted in the Chair's Summary to the UNEP workshop,[27] the negotiations had entered a new phase focused not on *whether* but on *how* new WTO rules should address the problem of fisheries subsidies.

WWF then released a detailed proposal for new WTO rules – backed by 150 pages of technical analysis –developed through WWF's two-year consultation process with trade officials and fisheries experts.

In short, by mid 2004, the political and intellectual obstacles to progress had been substantially reduced.

The 'top-down versus bottom-up' debate and its effects

Clarifying the substantive issues

With these developments it became clear that the WTO process was actively moving towards adopting new rules on fisheries subsidies and that those rules would probably include a prohibition on at least some kinds of subsidy programmes. But what still remained fundamentally in doubt was the so-called 'level of ambition' for the talks – in other words, whether new rules would be weak or strong, and the degree to which they would require governments to alter their behavior from the status quo.

For the next year (and with many echoes after that), the fight over the level of ambition was waged in terms of the so-called 'top-down versus bottom-up' debate. 'Friends of Fish' continued to submit papers and make oral arguments at NGR meetings in favour of a broadly-worded prohibition on most (or even all) fisheries subsidies, to be circumscribed only by a carefully articulated 'negative list' of exceptions. Japan continued to lead the charge in favour of a 'positive list' approach, insisting that only a few subsidies could be reliably associated with fisheries depletion and pushing for a 'bottom-up' prohibition limited to a narrowly drawn list of specific subsidy types. The EU was also sympathetic to the 'bottom-up' modality, although its formal position remained more ambiguous. Although it was often argued that the final result of both approaches could be very similar, the choice of negotiating method would have implications on the scope of the prohibitions, notification requirements, burden of proof as well as the level of transparency achieved in the negotiations (von Moltke, 2007).

Despite its repetitive form, the year-long 'top-down versus bottom-up' debate altered the dynamics of the negotiations both substantively and politically. At a substantive level, the tabling of positive and negative lists by the two camps began to clarify the categories of subsidies that required priority attention and also advanced the discussion of the kinds of conditions and limits for subsidies that remained permitted. For example, in November 2004 a number of 'Friends of Fish' delegations proposed that fisheries subsidies not to be prohibited might include:

- government expenditures for management frameworks, including those relating to surveillance, monitoring, enforcement and associated research;
- government expenditures for general infrastructure;
- certain fisheries-related social insurance programmes (e.g. job training to assist the transition out of the industry);
- government expenditures for access;
- appropriately-structured decommissioning subsidies.[28]

Since this was the proposal of the most ambitious *demandeurs*, an instantaneous consensus effectively emerged that such subsidies would not be prohibited under new rules, regardless of the form of an eventual ban. But the *demandeurs*, as well as participants in the parallel discussions organized by UNEP and WWF, also pointed out the need for limits and conditions on even the most positive sounding subsidies. A good example was the detailed US discussion of vessel buy-back programmes.[29] Echoing similar observations in UNEP's 'Matrix' paper (2004a, see Chapter 2), the US pointed out that although buy-back programmes are aimed at reducing fishing capacity, problems in their administration 'can lead to the return of the removed overcapacity or even an increase'. Accordingly, the US proposed that buy-back subsidies be permitted, but subject to 'appropriate programme conditions' such as rules to guarantee permanent removal of capacity and accompanying management measures to limit entry and effort in the fishery.

Meanwhile, Japan was ready to agree that subsidies directly increasing fishing capacity (such as for vessel construction) 'are at the centre of the problem', but went on to argue that these subsidies pose dangers only in poorly managed fisheries.[30] In 'properly managed' fisheries, they proposed, such subsidies should be allowed. This view led the Japanese to present some preliminary ideas for how a 'properly managed' fishery could be defined. Japan also signaled the importance it attached to maintaining subsidies for the infrastructure of fishing operations and fisheries communities (such as port facilities). The Japanese proposals thus helped bring additional focus to the talks, even as they clearly indicated Japan's desire to exempt most of its current subsidies from any eventual ban.

In short, although preliminary in character and lacking in legal detail, the substantive proposals from various perspectives helped identify issues and define the parameters of the negotiations.

A political focus on 'Special and Differential Treatment' (S&DT)

In addition to its substantive influence, the 'top-down versus bottom-up' debate had important political consequences. To understand these, it is first necessary to recognize the political state of play then prevailing within the talks. By the time the 'top-down versus bottom-up' debate began, several distinct 'camps' had begun to emerge within the negotiations:

- First, there was the 'Friends of Fish' *demandeur* group, whose membership was never very precise but which clearly included the US, New Zealand, Australia, Iceland, Peru, Argentina, Chile and (at least in the first years) the Philippines, among others.
- Second, there were the clear antagonists of an ambitious outcome to the talks, Japan, Republic of Korea and Chinese Taipei. Also in this 'low ambition' group was the European Union, but the EU's posture and interests were sufficiently distinct that it might best be considered a 'camp' unto itself.
- Third, a group of 'small vulnerable economies' (SVEs) and small island states had emerged as an economically weak but politically significant special interest bloc, under the leadership of Barbados and the Pacific Islands Forum.[31] As noted above, this group approached the negotiations with deep defensive concerns around the issues of 'access agreement' payments and artisanal fisheries.

Beyond these more or less well-defined camps, many delegations – the majority, in fact – had yet to take a clear stand on the level of ambition for the talks. Most of these were developing countries which had played a purely reactive part in the debate up to that point. In general, these delegations shared a basic concern with the terms of the 'special and differential treatment' (S&DT) to which they would be entitled under new rules.[32] But within this non-aligned group were developing countries with very different situations and interests, ranging from tiny island states such as Tonga to global economic giants such as China.

As in other areas of WTO rule-making, the question was growing whether large economically advanced developing countries should benefit from the same S&DT as much smaller and weaker economies. Outside the political boundaries of the negotiations, few policy-makers would argue that 'one size fits all' S&DT could really make sense. But at the negotiating table the economically most powerful developing countries, such as China, India and Brazil, were vehemently opposed to any approach to S&DT that included 'differentiation' among developing countries. This was a fundamental concern that cut across a broad range of issues under discussion in the Doha Round, and remained a constant factor in the background of the fisheries subsidies talks.

In addition to the majority of developing countries, the 'not clearly aligned' delegations included a few important developed countries, such as Canada and Norway. Both of these countries had tended to group themselves with the *demandeurs* and Norway in particular has at times joined in actions of 'Friends of Fish'. But both Canada and Norway also had economic and ideological perspectives that were not in accord with the approach taken by the core *demandeur* group. Both, for example, were extremely sceptical of bringing questions related to fisheries management into the WTO. And both also had what they considered 'small scale' or even 'artisanal' fishing communities that were or might become the object of subsidies.

The political impact of the 'top-down versus bottom-up' debate began with the way it raised the importance of the non-aligned delegations as potential 'swing votes' in the debate over the level of ambition for the talks. As long as the EU and Japan were directly resisting the negotiating mandate, there was little space for manoeuvre by countries who were sceptical of the talks but unwilling to take strong positions. To oppose 'Friends of Fish' would have required either joining Japan and the EU in directly resisting the ministerial mandate, or taking some leadership in articulating a 'third way'. But with the end of European and Japanese rejectionism – and with the 'bottom-up' approach offering a negotiating modality that could be embraced without taking a clear position on the ultimate level of ambition – a plausible option for non-hardline sceptics was on the table. Thus, the contest for the allegiance of the unaligned became a significant factor in the tactics of the more active delegations. Since the bulk of the non-aligned were developing countries, a direct consequence was the rapid emergence of S&DT as a primary negotiating topic.

The emergence of the S&DT issue was, of course, not a surprise; the special mention of developing country concerns in the Doha fisheries subsidies mandate reflected some important realities. As outlined in Chapter 1, hundreds of millions of people directly depend on fishing for some or all of their livelihoods – 90 per cent of them in developing countries. For many developing countries, fish is one of the primary sources of protein for human consumption. And in recent decades trade in fisheries products has become one of the leading sources of export earnings for developing countries. Developing countries provide nearly half of all products entering fisheries trade – a figure that does not even include the fish taken from developing country EEZs under access agreements with developed countries (FAO, 2006).

Moreover, and of fundamental importance in the fisheries subsidies debate, the fishing industries of some developing countries are still at relatively low levels of capacity. Even in a world where aggregate fishing capacity is far above sustainable levels, these countries claim the right to grow their fleets and expand their fishing – to have, in effect, their fair share of the oceanic pie. Although obviously complicating global efforts to combat overfishing, these claims are widely viewed as equitable as well as consistent with the rights of countries under the UN Convention on the Law of the Sea (UNCLOS) to exploit fisheries and other resources within their 'exclusive economic zones'.[33]

In short, there has never been any doubt that developing countries would be important stakeholders in the fisheries subsidies debate. In the context of the 'development' orientation, or at least the 'development' rhetoric, surrounding the Doha Round, the potential weight of the unaligned developing countries in the fisheries subsidies debate was still stronger.

Moreover, the dynamics of the fisheries subsidies issue itself inevitably led to a key role for the S&DT issue. It was widely assumed that the great bulk of fisheries subsidies were provided by a handful of developed countries and that developing countries were at an inherent competitive disadvantage in any subsidized race for fish or for export markets.[34] Moreover, while fleet overcapacity was and is a global problem, it was also clear that the fisheries sectors of many developing countries remained underdeveloped – and that some even lacked the industrial capacity to fish their own waters up to maximum sustainable (or maximum economic) yields.

Very few developed and developing governments are ideologically opposed to the use of subsidies. Furthermore, many developing countries correctly argued that subsidies had been used to obtain existing competitive advantages by a number of major first world fleets. It was therefore widely understood that new WTO rules would have to maintain substantial freedom for developing countries to continue using fisheries subsidies, including subsidies in many of the categories to be prohibited for developed countries. How to accomplish this in a way that would be both equitable and consistent with sustainable fisheries management was known to be one of the central problems negotiators would eventually face.

The prominence of the S&DT question was quickly evident in the actions of leading 'Friends of Fish'. For example, in September 2004 New Zealand hosted a two-day workshop in Geneva for 'Friends of Fish' and a handful of 'non-'Friends of Fish' developing countries, such as Brazil, Mexico, Venezuela and the Pacific Islands bloc. The structure of the workshop focused on categories of subsidies that might be excluded from an eventual ban, with an emphasis on some subsidy types of high-priority to developing countries (for example, subsidies related to access payments). Two months later, in an important and unusual development, a joint 'Friends of Fish' paper explicitly called for 'early' and 'parallel' discussion of S&DT.[35]

The 'Friends of Fish' offer was unusual because developed countries have generally shown great reluctance to discuss the terms of S&DT until the last phase of a trade negotiation, when leading governments have arrived at basic consensus on the primary positive obligations from which S&DT are to derogate. New Zealand had explicitly pushed for this traditional order of negotiations on fisheries subsidies in the weeks before the Japanese acceptance of the mandate.[36] Thus, the tactical shift by New Zealand in September 2004 was an important invitation – and one that leading developing countries would soon accept in a way that significantly altered the course of the negotiations.

S&DT: The technical challenges

The political importance of the S&DT issue brought with it a new set of technical challenges that soon came to dominate the activities within the parallel process unrolling alongside the formal negotiations. UNEP, which had already focused substantial attention on developing country issues,[37] commissioned three new papers dealing directly with topics of central concern: artisanal fishing, technical aspects of the application of S&DT to fisheries subsidies and foreign access agreements.[38] The first two of these were the focus of a full-day workshop hosted by UNEP in June 2005.[39] Similarly, other leading parallel process actors such as WWF and ICTSD engaged in activities directly focused on developing country concerns.[40]

Substantively, a balanced approach to S&DT would mean accommodating the desire of developing countries to maintain 'policy space' for pro-development subsidies while avoiding rules that encourage subsidized overfishing or provoke significant trade distortions. Achieving this balance would require answers to three basic questions:

1. What should be the *scope* of S&DT? What kinds of subsidies for what purposes or for what types of fishing should be included?
2. What *conditions* should developing countries be required to fulfill in order to benefit from S&DT to help avoid subsidized overfishing or trade distortions? What would be reasonable and effective 'sustainability criteria' for allowing the use of otherwise prohibited subsidies under WTO rules?
3. Should there be *differentiation* among classes of developing countries within S&DT? Should the most powerful developing countries be subject to the same S&DT rules as smaller and economically more vulnerable countries?

So long as the ultimate position of many developing countries remained unclear, leaving unresolved a key signal about the level of ambition for the fisheries subsidies talks was left unresolved.

This was precisely the situation as the year 2005 got underway, and as preparations began for a crucial WTO ministerial in Hong Kong in December 2005. The Hong Kong meeting, which had been formally announced in October 2004, was to be critical for the struggling Doha Round as a whole, with ministers meant to settle controversial issues related to agricultural subsidies and industrial tariffs that could determine the level of ambition for the entire Round. As Hong Kong loomed on the horizon, therefore, the question of ambition on fisheries subsidies was also very much in play.

Bringing environmental issues to the fore

If the 'top-down versus bottom-up' debate had remained only between the 'Friends of Fish' and Japan et al, it is unlikely ministers in Hong Kong would have been able to achieve much progress on fisheries subsidies. But a remarkable series of proposals tabled by the government of Brazil starting in March 2005 helped clear a path to a much more significant result.[41]

The essential character of the Brazilian intervention was threefold. First, it directly embraced the top-down approach by endorsing a broad ban on fisheries subsidies, so long as such a ban were combined with clear and effective S&DT for developing countries. Second, it proposed that capacity enhancing subsidies be allowed under S&DT only where a lack of capacity was preventing developing countries from the full exploitation of the fisheries resources to which they were entitled (this concept would later sometimes be called the 'room to grow' concept). And third, it proposed that the right to subsidize in accordance with S&DT be conditioned on factors relating to the sustainability of the fishing to be subsidized, including the health of the target stock and the presence of an effective fisheries management regime.

The truly novel and forward-looking nature of the Brazilian proposals was significant not only for the fisheries subsidies negotiations but for the history and development of the multilateral trading system itself. As noted above, when the 'trade and environment' movement was born it faced significant resistance from developing countries who feared green protectionism and the quasi-imperialist imposition of 'northern' environmental standards on their struggling economies. Moreover, there was broad scepticism among developing and developed country governments alike regarding the WTO's involvement in 'environmental' issues such as fisheries management. While many of the negotiating papers tabled prior to the Brazilian proposals had accepted the relevance of environmental factors in determining the risks and impacts of fisheries subsidies, none had directly proposed that such factors be given legal relevance within new WTO fisheries subsidies rules.

Brazil not only embraced this idea, but went on to make detailed technical proposals for how this kind of environmental conditionality could be appropriately accomplished. Brazil thus became the first government to propose limiting significant trade-law rights of developing countries in order to ensure the long-term sustainable use of natural resources. The proposal strongly reinforced the environmental orientation of the negotiations and opened the first serious technical discussions for how environmental considerations could be properly accommodated within WTO rules. While elements of the Brazilian proposals engendered significant debate, the conversation was embraced by most delegations as appropriate and necessary. The level of ambition and the environmental focus of the talks thus advanced together.

The breakthrough in Hong Kong

The Brazilian proposals, along with elements collected from the papers tabled by Japan, the EU and 'Friends of Fish', helped prepare for the Hong Kong ministerial meeting in December 2005 to take a real step forward on fisheries subsidies. In the weeks before Hong Kong, support for a high ambition result was further strengthened when 'Friends of Fish' were able to table a paper that was joined by both Brazil and Pakistan – key developing countries that were not members of the coalition.[42] Moreover, the politics of Hong Kong favoured productive attention to the fisheries subsidies issue. With overall success for the Doha Round still very much in doubt, and with a high level of attention and protest activity anticipated from civil society groups, a genuine 'win-win-win' issue with broad north–south support was again an attractive package for many governments.

And once again, leading parallel process actors such as UNEP and WWF were able to play a significant role. A key activity that occurred in Hong Kong itself was a 'high level' press conference sponsored jointly by UNEP and WWF, at which ministers from seven countries (including the US, the EU and Brazil) made statements in support of WTO action on fisheries subsidies. The preparations for this public event, which began months in advance, undoubtedly helped raise the level of political commitment among key ministers for a positive result on fisheries subsidies in Hong Kong.[43]

After the failure of the ministerial meeting in Cancun two years earlier, expectations for Hong Kong were relatively low. Still, it was understood that some signs of progress on core issues such as agricultural subsidies or industrial tariffs were needed or the Round would appear effectively dead. As the Doha negotiations were the first major project of the still young WTO, the consequences of a failed Round for the multilateral trade system were considered potentially severe. Meanwhile, anti-globalization violence in the streets of Hong Kong was again anticipated and the global media spotlight shone on a WTO meeting once more.

In the end – after the requisite brinkmanship indoors and teargas outdoors – ministers were able to eke out a face-saving result, based in part on the reluctant acceptance by the EU of a date (far in the future) for ending its agricultural export subsidies. And so, unlike in Seattle and Cancun, the underlying 'north–south' split within the WTO did not result in a diplomatic crash. The Doha Round continued forward, but still without the clear 'modalities' that were considered necessary to enter the final and most technical phase of the talks.

For 'Friends of Fish' and environmental stakeholders in the fisheries subsidies talks, however, the Hong Kong ministerial meeting was a success. Shortly after the coordinated public statements of leading ministers at the UNEP-WWF high-level event, the WTO membership as a whole agreed to include language in the ministerial declaration committing themselves to negotiate a ban on fisheries subsidies that contribute to overcapacity and overfishing. The language went on to stress the need for transparency and enforceability, and – in detailed language originally proposed by the SVE group – for effective S&DT focused on livelihoods, food security and poverty reduction. The full text of the Hong Kong Declaration language on fisheries subsidies reads:

> [We ministers] recall our commitment at Doha to enhancing the mutual supportiveness of trade and environment, note that there is broad agreement that the Group should strengthen disciplines on subsidies in the fisheries sector, including through *the prohibition of certain forms of fisheries subsidies that contribute to overcapacity and over-fishing*, and call on Participants promptly to undertake

> further detailed work to, inter alia, establish the nature and extent of those disciplines, including *transparency and enforceability*. Appropriate and effective special and differential *treatment for developing and least-developed Members* should be an integral part of the fisheries subsidies negotiations, taking into account the importance of this sector to development priorities, poverty reduction, and livelihood and food security concerns; [author's emphasis].[44]

Although this commitment was only an 'agreement to agree' – and lacked the reference to 'sustainability' that had been removed at the insistence of the Indian delegation – it represented a genuine political breakthrough that newspapers around the world reported prominently as an agreement by WTO officials to do away with subsidies that drive overfishing (Bradsher, 2005). The fact that it was one of the few clear success stories to emerge from the meeting further increased its political weight. For the first time, fisheries subsidies were not only in the news, but were making top-story headlines on page one of international newspapers (see Figure 4.2).

Phase IV: Towards a High-Ambition Draft

The strengthened Hong Kong mandate fundamentally altered the dynamics of the fisheries subsidies talks. In political terms, Hong Kong solidified the developments of the previous year, which had shifted the political center of the talks towards a high ambition result based on a formula that would combine a broad prohibition with significant S&DT but with no 'blank cheque' for permitted subsidies.

Technically, Hong Kong opened a period of negotiations focused more directly on legal language, with actors on all sides of the debate submitting proposals on particular issues of concern. By the early spring of 2006, it was anticipated that the next milestone in the negotiating process would be the tabling of a proposed draft text by the chairman of the NGR. The timing of the anticipated draft was linked to the progress of the broader Doha Round negotiations. While ministers in Hong Kong had been unable to adopt modalities on agricultural subsidies and industrial tariffs, they had agreed instead that modalities should be adopted no later than 30 April 2006. It was generally understood that a draft text on rules would be expected from the chair within days after agricultural and industrial tariff modalities were in place.

Like so many other Doha deadlines, however, the 30 April date came and went without result. Instead, a 'mini ministerial' meeting was planned for midsummer in Geneva, in the hopes that ministers could make the hard political choices that had evaded them in Cancun and Hong Kong. Once again, the Geneva meeting was advertised as a 'make or break' moment for the Round. Within the Rules Group, negotiations proceeded on the assumption a chair's draft would issue on the heels of the mini-ministerial meeting.

Herald Tribune
THE NEW YORK TIMES

Collective stance at WTO
Activists ally with nations on fishing aid

Trade Ministers at W.T.O. Talks Agree to Limit Fishing Subsidies

All parties at least appear to agree on fishing subsidies

Sideline deals end in smiles all round

Figure 4.2 *The breakthrough in Hong Kong was hailed in newspapers around the world*

This was a nail-biting moment for the 'Friends of Fish' and their supporters among environmental stakeholders. It was widely believed that the level of ambition contained in the chair's draft could set the maximum possible level of ambition for the eventual outcome of the talks – a low ambition draft would all but doom the talks to a weak result. And with a text likely to issue in just a few months, there was still no general consensus in favour of a high ambition outcome. Despite the momentary consensus in Hong Kong, and the support of a number of developing countries for a strong prohibition, the traditional opponents of strong fisheries subsidies rules remained firm.

A wave of proposals, but little convergence on the scope of the ban

Indeed, in the period after Hong Kong the EU swung noticeably away from the greener stand it had adopted during 2002–2003 and turned back towards resisting a high ambition text. This was likely due to several changes in the domestic European context. First, by early 2006 the high-level, future-oriented policy debates that had typified the CFP reform process had been replaced by more concrete, immediate and politically sensitive internal negotiations over specific elements of member state fishing practices. In addition, the FIFG basket of fisheries subsidies was expiring in 2006 and a new EC regulation was needed to create its successor, the European Fisheries Fund (EFF), to provide basic authority and guidance for EU fisheries subsidies for the period 2007–2013 (WWF, 2006a).

The hard work of implementing the promises of the reformed CFP was also taking place in a more challenging economic and political environment. World oil prices – which had hovered roughly at US$25–30 per barrel in the years 2000–2003 – began a steady and then accelerating rise from the start of 2004 to the middle of 2006. By the time negotiations over the new EFF were underway in the first months of 2006, oil had reached then-historic highs of more than US$60 per barrel (US Energy Information Administration). Fishers worldwide were significantly affected by this more-than-doubling of fuel costs. The European fishing sector – with its gas-guzzling industrial fleets, distant water operations and significant reliance on trawling – was hit relatively hard and fishers were soon crying out for publicly funded relief.

In addition, a domestic political backlash was brewing against the CFP's gentle hostility to capacity-enhancing subsidies more generally. The CFP reform had recognized the need to replace the subsidized expansion of capacity with publicly funded restructuring of the sector. That reform had received strong support from powerful and economically conservative member states such as Germany, the United Kingdom, Sweden and The Netherlands. But, in opposition, there appeared a coalition of member states who wished to see a return to more liberal use of capacity-enhancing public supports. This coalition included not only traditional members – such as Spain, Portugal, Italy and France – but also Estonia and Poland, which had been among the ten new member states joining the EU as of 1 May 2004.[45]

Although the EU did not wholly abandon its commitment to CFP reforms, the political and economic realities in 2006 drove Brussels to adopt a series of official acts that clearly undermined the CFP's purported policy against subsidized expansion of fishing capacity.[46]

The context that produced these acts also led the EU to table a paper at the WTO in April of 2006 that aligned it closely with Japan's hard-line positions against robust new fisheries subsidies rules. The EU proposed narrow language for a prohibition that would have permitted directly effort-enhancing subsidies (e.g., fuel subsidies, payments to offset operating costs or price supports) and proposed only a very loose prohibition on subsidies that increase fishing capacity.[47] Further, in what appeared to be a bid to bring developing countries back towards a low ambition posture, the EU suggested rules for S&DT that came very close to offering a 'blank cheque' and in any case had much fewer sustainability-related limits and conditions than had been suggested by leading developing countries themselves.

Japan, Republic of Korea and Chinese Taipei likewise hardened their positions, tabling legal language that permitted not only subsidies to operating costs but also a wide range of capacity-enhancing subsidies upon the fulfillment of minimalistic management conditions.[48] The paper also took a clear stand against granting S&DT for subsidies to fishing beyond the EEZ limits of developing countries.

But the momentum after Hong Kong also had strong encouraging elements: New Zealand tabled proposed language for a strong prohibition and also sought to move the negotiating ball forward by including a significant list of 'non-prohibited' subsidies.[49] In an important gesture towards Japan, New Zealand included on its 'non-prohibited' list subsidies to the infrastructure of fishing communities and port facilities. With this paper, New Zealand continued to take the lead in setting out the basic parameters of the 'broad ban' approach. Similarly, the United States returned to the table with a proposal that began to address important concepts that would contribute to the 'sustainability criteria' discussion. In a paper tabled in April 2006, the US included careful environmentally-oriented limits on a proposed exception for subsidies to 'buy-back' programmes, and further focused on the need for new institutional mechanisms to involve fisheries experts in the administration of eventual WTO fisheries subsidies rules.[50]

The S&DT dialogue continues to mature

The papers tabled by the major players in the months following Hong Kong did little to resolve the stalemate over the desired scope of an eventual prohibition. In the area of S&DT, unanswered technical questions remained a stumbling block to clarifying and resolving some of the underlying political issues. With developing country support still an important prerequisite to a high ambition result, the first half of 2006 saw another flurry of activity focused on S&DT, both at the negotiating table and in the parallel process.

The quick failure of first efforts at 'differentiation'

A basic issue that rose momentarily to the top of the S&DT agenda involved the underlying question of 'differentiation'. As noted above, the question had been constantly lurking in the background whether S&DT rules should apply equally to all developing countries, or should differentiate among them according to their situations. The main argument in favour of differentiation was commercial and economic: in terms of fishing power, the industries of China and Peru, for example, are among the most 'developed' in the world. If differentiation could be applied on the basis of fishing power, S&DT could be made available to a large number of developing countries but still cover only a small percentage of world fisheries production. Arguably, this would sharply limit both the potential economic and environmental impacts of fisheries subsidies and potentially do away with the need for stringent sustainability criteria.[51]

Although differentiation was known to be opposed strongly (and probably fatally) by the large developing countries such as China, Brazil and India,[52] the idea was formally proposed by both Japan and New Zealand shortly after the Hong Kong ministerial meeting.[53] Both suggested what is technically known as a '*de minimis*' approach, according to which developing countries with fishing industries falling below a certain minimal size should simply be exempted from most new WTO fisheries subsidies rules. New Zealand was the first to table this idea formally in a March 2006 paper stating that a new WTO prohibition 'shall not apply to fisheries subsidies provided by a developing country Member where such subsidies do not exceed the *de minimis* level for that Member.'[54]

While New Zealand's formal paper did not elaborate how *de minimis* would be defined, it immediately put forward details during an informal one-day workshop that it sponsored for selected WTO delegations. In a 'non-paper' distributed at that workshop, New Zealand suggested exempting any developing country whose industry was responsible for less than 0.1 per cent of global marine wild-capture fisheries production.[55]

Based on FAO data from 1999–2003, such a rule would grant broad S&DT to the majority of developing countries. However, more than 40 developing countries would fall above the 0.1 per cent line, including countries with relatively small and clearly underdeveloped economies, such as Papua New Guinea, Angola, Bangladesh and Senegal, among others – while even a *de minimis* line set an order of magnitude higher would still deny S&DT to a significant set of developing countries.

Reaction among developing countries against the informal New Zealand proposal was swift and unanimous. Even if it were plausible to split the developing country bloc against a handful of biggest fishing powers among them, the data made clear that no line could be drawn that did not either catch too many developing countries above it, or leave too many significant fishing powers below it.[56] Japan and its allies formally tabled a similar proposal one month later.[57] Later, some smaller developing countries also showed interest in a limited use of the *de minimis* concept as a way of ensuring broad S&DT for themselves.[58] But as the fundamental basis limiting S&DT, the *de minimis* idea soon faded from the proposals of the main protagonists.

Progress on S&DT and the link to sustainability

The *de minimis* discussion reinforced the continued importance of the S&DT issue to the overall debate. To begin with, the parallel process in this period was especially active and almost wholly focused on developing country issues. The FAO and the UN Conference on Trade and Sustainable Development (UNCTAD) hosted an unusual two-day workshop in Geneva in late March on 'The WTO and Fisheries', while in both April and May 2006 UNEP joined with ICTSD and WWF to produce events focused on S&DT. The first of these events was a briefing for developing country trade bureaucrats who had come to Geneva for the WTO's annual week-long capacity-building seminar.[59] The second was a full day workshop focused on artisanal fishing, access agreements and S&DT that again brought together fisheries, trade and development experts and government representatives from the Geneva-based diplomatic corps and from developing country capitals.[60]

Similarly, in the burst of diplomatic activities after Hong Kong, S&DT figured prominently. India, for example, made its first written WTO submission on fisheries subsidies to emphasize the needs of its enormous small-scale sector.[61]

But the centre of the S&DT debate focused on the series of proposals Brazil continued to bring forward, soon to be supplemented by complementary proposals from Argentina.[62] These papers sparked the first detailed and continuous discussion of sustainability as a core element of trade rules ever to take place within a formal WTO negotiating process.

Taken all together, the diplomatic and parallel process activities focused the fisheries subsidies talks directly at the junction of sustainability and development. Moreover, the detailed and practical level of debate significantly advanced the negotiations.

A leading example of this was the dialogue over the treatment of subsidies related to fisheries access agreements. As noted above, small island developing states (SIDS) and other 'small and vulnerable economies' had indicated they would strongly oppose any result on fisheries subsidies that threatened the access payments upon which many of them depended. These delegations wanted access agreements simply written out of the WTO fisheries subsidies rules and some 'Friends of Fish' delegations were tempted to accept this as the price of neutralizing this politically sensitive issue. But environmental stakeholders were adamant that inappropriate and highly subsidized access agreements were a real part of the overfishing problem.

Access agreements became one of the core topics discussed at the FAO/UNCTAD and UNEP-ICTSD-WWF parallel process events[63] and the subject of a number of substantial research papers.[64] UNEP brought out a very influential paper looking at the WTO definition of subsidies in the context of access agreements and examining how access agreements could be dealt with under new fisheries subsidies rules (see Chapter 5, 'The Special Case of Access Agreements'). The resulting discussions clarified several facts:

- the core sensitivity of SIDS/SVEs was regarding the government-to-government fees they received, while the core environmental focus was on the financial benefits to the industrial fleets that gained subsidized access to foreign EEZs;
- the access deals themselves are very often inequitable, with developing countries receiving only a small fraction of the value of the fish extracted from their waters;
- access agreements are often negotiated without serious regard for the conservation of fish stocks, and consequently have a history of causing depletion; and
- access agreements are often negotiated in secret, with first world distant water fleet nations wielding significant effective power over the much smaller and politically disunified SIDS/SVEs.

From these observations, a potential compromise began to emerge. 'Fisheries subsidy' could be defined to exclude government-to-government payments while including the onward transfer of the access rights from a developed country government to its private industry.

Access subsidies thus defined would need to be allowed under WTO rules, even if they were in essence first world subsidies with capacity- and/or effort-enhancing effects. But, SIDS/SVEs might have something to gain from WTO rules that increased the transparency of access agreements and the environmental responsibilities of access-purchasing governments.

The interplay between the Brazilian papers and the parallel process was immediately evident. The version of Brazil's paper tabled in April 2006 directly targeted government-to-government payments as fisheries subsidies, allowing them under certain conditions.[65] By June, Brazil had adopted the concepts that had emerged from the UNEP-ICTSD-WWF workshop and altered their approach to exclude government-to-government payments while treating as subsidies 'the onward transfer' of access rights.[66] The interplay between the negotiating and the parallel processes in the spring of 2006 had turned a likely deal-breaker into a productive search for common ground on an important issue.[67]

The dialectic between the parallel and formal negotiating processes was also apparent on other key S&DT issues, although without such immediately tangible results. For example, questions such as the treatment of 'small scale' and 'artisanal' fishing, the need to demonstrate 'room to grow' and the potential application of management system requirements were all subject to simultaneous informal study and diplomatic discussion. In each case, an examination of the negotiating record clearly reveals the influence of the parallel process debate.

The progress on S&DT was, however, far from achieving solutions to all the key questions. While the debate over the Brazilian and Argentine papers clearly advanced the dialogue, many gaps and difficulties remained. Nor were all of the developments in the formal proposals positive from the perspective of environmental stakeholders.[68] And the relative silence of 'Friends of Fish' such as New Zealand and the US with regard to the details of a workable approach to sustainability criteria further reinforced the sense that more work would be needed before a widely acceptable approach was agreed. Nevertheless, the formal and informal conversations in the spring of 2006 clearly established the necessity of sustainability criteria as part of the WTO fisheries subsidies rule system and helped set out what some leading governments considered essential elements of a sustainability criteria approach.

Uncertainty in the NGR; Collapse in the TNC

As the July 2006 mini-ministerial meeting approached, the incomplete convergence of views on fisheries subsidies posed a significant challenge to the chair of the Negotiating Group on Rules (NGR), who was busily preparing his first draft text. Since the chair had indicated that he would issue an 'un-bracketed' text – that is, one that unambiguously proposed new rules rather than presenting a set of options – the stakes were high all around.

After reviewing the various legal proposals, WWF concluded that 'the proposals now on the table give cause for both hope and fear. Read generously and in combination, they contain the seeds of robust and effective new disciplines; read narrowly and separately, they lead towards rules so full of gaps and weaknesses that they would appear designed to fail' (WWF, 2006b).

This, then, was the state of play on fisheries subsidies as WTO ministers gathered in Geneva in July 2006, hoping to achieve the negotiating modalities needed for the Doha Round to enter its final stage. But in a pattern that was now revealing a deepening crisis in the post-WWII trading system, the ministerial meeting of 2006 added itself to the list of failures that included Seattle, Cancun and (albeit less overtly) Hong Kong. Once again, the key obstacles were the unwillingness of major northern powers to undertake aggressive reforms of their agricultural subsidies and the reluctance of southern market countries to make further cuts in their industrial tariffs. And once again the WTO had called ministers to a high profile meeting without any assurance that compromise was plausible. On 24 July, after a series of acrimonious and inconclusive meetings among the leading powers, WTO Director General Pascal Lamy formally suspended the Doha Round talks, 'pending the resumption of the negotiations when the negotiating environment is right' (WTO, 2006b).

Although hardly unexpected, the crash of the ministerial meeting and the suspension of the Round created an unprecedented level of uncertainty surrounding not only the Doha talks, but the future of the WTO itself. However, the six-month hiatus that began in late July 2006 would turn out to be part of another fruitful stage of work towards new WTO fisheries subsidies rules.

A lull in the round, sustainability criteria issues come into focus

The suspension of the Doha Round created room for further work on fisheries subsidies and particularly on sustainability criteria. Since formal negotiations were impossible, political pressures were somewhat reduced and the NGR chair was free to convene delegates in 'informal' technical sessions. One such session that materially advanced the talks was a session called by the chair in October 2006 for WTO delegates to meet with senior technical staff from the FAO Fisheries and Aquaculture Department.

The meeting with FAO fisheries officials was significant. It had been known since early in the negotiations that the FAO could potentially play a significant role in the administration of fisheries subsidies disciplines. Several delegations had formally called for the involvement of FAO and other experts.[69] Others, however, were highly sceptical of bringing the FAO in as a partner organization to the WTO. And it was not known whether the FAO itself wanted to get involved. The meeting with FAO officials signalled that the fisheries subsidies talks had ripened to the point that direct dialogue with the FAO was timely and important. It also helped highlight sustainability criteria as a crucial element of the talks.

The meeting itself, which took place behind closed doors, had several results. First, it gave many delegations some assurances that the FAO had available the norms and data necessary for the administration of WTO sustainability criteria on fisheries subsidies, addressing the biological, industrial and regulatory requirements. It also left delegates with a sense that the FAO was willing and able to play an appropriate role on fisheries subsidies in liaison with the WTO, if it were asked to do so by its membership. But it was also clear that the FAO would not act as 'policeman' on WTO fisheries subsidies rules or make voluntary FAO norms into enforceable international law. Still, on balance, the meeting with FAO officials contributed momentum to the talks and reinforced the plausibility of fisheries-specific sustainability criteria within WTO rules.

Around the same time, the NGR chair issued a 'non-paper' outlining five pages of very detailed questions about how sustainability criteria in new WTO rules could work clearly indicating the rapid maturing of the technical debate.

The fall of 2006 also saw another significant development in the academic literature on fisheries subsidies. The Fisheries Centre at the University of British Columbia (UBC) published a major new empirical study titled 'Catching More Bait: A Bottom-Up Re-Estimation of Global Fisheries Subsidies' (Sumaila and Pauly, 2006). The paper built upon the earlier catalogues of fisheries subsidies compiled by the World Bank, the OECD and APEC, supplementing existing data with new country-specific research and filling in gaps with estimates based on methods of extrapolation devised by UBC. True to UBC's reputation as a source of provocative fisheries research, the study reached a number of stimulating conclusions, including:

- on a global basis, fisheries subsidies were conservatively estimated at US$30–34 billion per year for the period from 1995 to 2005 – less than the FAO's original indirect estimates, but far more than previous data collections had shown;
- developing countries were estimated to provide 40–45 per cent of global fisheries subsidies – a much higher share of the total than previously reported (although still far lower than developed countries on a per-country basis);
- more than half of the subsidies catalogued or estimated were considered likely to contribute to overcapacity or resource depletion...and the majority of these were found in developing countries, where the ratio of 'bad' subsidies to 'good' was substantially higher than in developed countries; and
- fuel subsidies were roughly 20 per cent of the total and approximately 30 per cent of the capacity- or effort-enhancing subsidies.

These findings, although considered controversial since a significant portion was based on extrapolated data, were another wake-up call that helped maintain momentum at the WTO negotiating table, despite the suspension of formal talks. The UBC data on developing country subsidies also reinforced the importance of finding workable and effective sustainability criteria to define the scope of and condition S&DT. Not only did the findings suggest that significant environmental impacts could flow from developing country fisheries subsidies, but they also clearly indicated the potential for significant 'south–south' competitive subsidization.

With the chair's anticipated text on hold, and with many of the chair's technical questions still unanswered, UNEP and WWF launched a year-long phase in the parallel process aimed at facilitating the sustainability criteria discussion at a deep technical level. The process, which eventually produced the joint UNEP-WWF publication *Sustainability Criteria for Fisheries Subsidies: Options for the WTO and Beyond* (UNEP-WWF, 2007b) began with the formation of an expert group and initiation of research in the fall of 2006. A working draft of the UNEP-WWF paper was presented at a jointly sponsored symposium in Geneva in March 2007.[70] The two-day symposium brought together over 120 government officials and experts, representing more than 50 countries.

Among the speakers were senior technical staff from both the FAO and WTO as well as stakeholders representing small-scale fisheries, who spoke directly to the technical plausibility of the emerging UNEP-WWF recommendations. The final paper would ultimately be presented at another jointly-sponsored workshop, in September 2007.[71]

The Round resumed, the NGR moves towards a chair's draft

On 7 February 2007, Pascal Lamy announced the resumption of formal Doha Round negotiations.[72] The resumption represented both a hope and a gamble by the Round's leading protagonists. Months of quiet, high-level bilateral and small-group discussions on agriculture and industrial tariffs had failed to produce the negotiating modalities whose elusiveness had caused the July 2006 collapse. But a further delay in resuming the talks risked an increasing series of obstacles to completing the Round any time soon. The situation in the United States was a prime concern, since at the end of June 2007 the President's so-called 'fast track' legislative authority to conduct the Doha negotiations was set to expire, only to be followed by the opening of the US presidential campaign season in the autumn of 2007. It was widely assumed that these factors would soon make it all but impossible for the US to make hard choices at the WTO negotiating table.

The various WTO negotiating groups were sent back to work with the aim of preparing 'chair's texts' according to their respective mandates.

The result was another short burst of activity at the negotiating table. In March, for example, proponents of a high ambition outcome to the talks were prominent with two papers. First, the United States tabled its first and long awaited comprehensive proposal for new fisheries subsidies rules setting out a framework for new rules that included:[73]

- a broad ban, based on a pure top-down, negative list approach;
- a series of carefully articulated exceptions, building upon the approach previously proposed by New Zealand;
- an additional blanket exemption for 'small programmes' (which were not defined in this proposal);
- a definition of 'fishery subsidy' that fleshed out the emerging compromise on 'access-related' subsidies;
- an innovative 'amber light' allowing challenges to permitted subsidies that cause production distortions;
- an approach to S&DT and sustainability criteria that adopted the Argentine

approach (as revised in January 2007) as a starting point; and
- a requirement for notifications to include information relating to compliance with sustainability criteria.

This paper was politically important – because although the US had often played a less proactive role in leading the 'Friends of Fish' than, for example, New Zealand – it was broadly understood that the US was key in promoting new fisheries subsidies disciplines. As discussed above, it was the US that had spearheaded the formation of the 'Friends of Fish' coalition in the pre-Seattle days. Most importantly, in Seattle, Doha and Hong Kong, the willingness of the United States to keep fisheries subsidies on its short list of 'must have' results was the key to success when the final deals were cut among the leading players in the WTO's famous 'green room' sessions.[74] The March 2007 paper was an important signal that the US was willing to maintain its leadership.

As the spring progressed, the negotiations turned back to S&DT, with the chair again issuing a detailed 'non-paper', this time posing questions on the treatment of 'artisanal fishing' and on S&DT generally. While once more the level of technical detail in the chair's questions was not fully reciprocated in the answers provided by most delegations, papers submitted by the ACP and SVE groups, Brazil and Argentina responded to the chair's call and moved the talks to a new level of technical debate on the terms of S&DT.[75] In the same period, Japan (along with the Republic of Korea and Chinese Taipei) also tabled a paper reiterating and extending their resistance to several key elements of a 'high ambition' – opposing, for example, the use of sustainability criteria in the new rules and calling for a blanket exemption for subsidies to small scale fisheries.[76]

Taken all together, the formal and informal discussions from September 2006 to June 2007 brought the fisheries subsidies talks to a stronger posture in both political and technical terms. Although many questions remained unanswered, the chair had been given a great deal more to work with than had been available when the first chair's draft had been anticipated a year earlier.

In September 2007, UNEP and WWF released the final version of their 'Sustainability Criteria' paper, (UNEP-WWF, 2007b) which was received positively by many delegations. It suggested sustainability criteria for those subsidies that will remain outside the ban to reduce the risks of these subsidies. It proposes a series of basic tests to ensure that the fish stock is healthy, the fleet has room to grow and that there is adequate management. In each category, criteria are articulated at three different levels of environmental ambition, ranging from best practices at the national level to least ambitious requirements suitable for use in WTO rules (see Chapter 5, 'Sustainability Criteria for Fisheries Subsidies').

Elements of this proposal were reflected in the discussions at the NGR as well as in a number of country submissions. On 30 November 2007, the NGR chair issued a comprehensive text on 'rules' that raised eyebrows – along with hopes and concerns – among many delegations.[77] The text made international headlines mainly for its controversial treatment of 'anti-dumping' issues, where the chair had accepted a fundamental principle that was desired by the US but was vigorously opposed by nearly all other WTO delegations. But for 'Friends of Fish' and environmental stakeholders, the text was more remarkable for what it did on fisheries subsidies: the chair had unambiguously issued a highly ambitious draft.

Phase V: A Text on the Table Amidst Rising Politics and Delay

The chair's draft: 'All the elements of success'

The chair's draft of 30 November 2007 was the third major watershed moment in the fisheries subsidies negotiations: Doha had put the topic on the WTO agenda, Hong Kong had established a highly ambitious goal and the chair's draft put forward a concrete and comprehensive proposal for fulfilling these mandates.

Technically, the chair's draft (Annex 8), aimed for a highly ambitious result without giving a pure victory to the 'top down' desires of the 'Friends of Fish'. The proposed prohibition was built on a 'bottom up' positive list, but the list covered a broad set of subsidies including subsidies to:

- vessel construction, modification, renewal or repair;
- operating costs or inputs, such as fuel, bait, etc.;
- transfer of vessels (e.g., export of capacity);
- landing and 'in- or near-port' processing activities;
- port infrastructure 'exclusively or predominantly for activities related to marine wild capture fishing'; and
- income and price supports.

In addition, the draft proposed banning subsidies to any vessel engaged in IUU activities, as well as any subsidy 'conferred on any fishing vessel or fishing activity affecting fish stocks that are in an unequivocally overfished condition'. By covering not only directly capacity-enhancing subsidies but also effort-enhancing subsidies (especially operating costs) and by further including an explicit rule against subsidies on overfished stocks, the draft reflected the dual focus of the Hong Kong mandate on the twin problems of fleet overcapacity and excess fishing activity. The text also clearly excluded subsidies to aquaculture from the basic scope of the proposed new rules.

Alongside this broad list of banned subsidies, the chair's draft proposed a list of exceptions that captured most of the 'positive' subsidies that had been identified at the negotiating table, including subsidies for:

- aid for natural disaster relief;
- improvements for crew safety;

- improvements for more sustainable fishing techniques, environmental improvements, or compliance with management regulations;
- re-education of fishers towards alternative livelihoods or to allow early retirement or compensate for permanent cessation of fishing activities;
- capacity-reducing programmes such as vessel 'buy-backs'; and
- the allocation of property rights to establish individual or communal interests in fishing grounds or stocks.

The exceptions were couched in language to ensure the exempted subsidies did not result in increased fishing capacity and requiring the subsidizing government to maintain a fisheries management system meeting certain criteria. Moreover, the text proposed a general discipline that would allow WTO members to challenge any subsidy that causes 'depletion of or harm to, or creation of overcapacity in respect of' any stock in which the complaining member had an 'identifiable' interest under international law.

For developing countries, the chair's draft contained a lengthy article on 'special and differential treatment'. While it did not differentiate among developing countries (other than exempting 'least developed countries' from nearly all of the new disciplines[78]), it laid out a complex set off distinctions granting or withholding special treatment based on various combinations of the type of subsidy used, the scale of fishing involved and the location of the fishery within or outside of EEZ waters. The draft also subjected all of S&DT to the same fisheries management system requirement as applied to the general exceptions discussed above.

Beyond these core elements, the text tabled by the NGR chairman contained several other provisions that, although highly technical, were nonetheless important and groundbreaking. The text proposed, for example, that fisheries subsidies not duly notified under WTO transparency rules would be presumed prohibited until proven otherwise. Although this rule would be weaker than an absolute ban on un-notified subsidies (which some governments and stakeholders desired), it would be a significant departure from current WTO subsidy notification rules, which were essentially "toothless". The provisions of the chair's text would also require the notification of information relating to the management systems in place in subsidized fisheries and thus would tailor WTO notification rules to the particular needs of the fisheries subsidies context.

Another innovation was the chair's proposal to involve the FAO in a 'peer review' process that would play some vaguely defined role in judging (or helping the WTO judge) compliance with the fisheries management system requirement. Although the reference of factual questions to external authorities is not unknown within WTO rules, the involvement of an intergovernmental body whose mission and expertise were essentially environmental would stretch the envelope of routine WTO practice.

Taken all together, the chair's draft presented an innovative effort to achieve a robust 'win-win-win' outcome on fisheries subsidies. It was accordingly greeted with enthusiasm by environmental stakeholders and applauded by 'Friends of

Fish'. But it was also clear that the draft was far from a final text. Apart from the fact that delegations such as the US, Japan and the EU were still at odds over basic questions such as the scope of the desired ban, the draft itself contained some language that was more 'aspirational' than technically mature. There were, in short, both political deals yet to cut and technical details to hammer out. Still, as WWF put it the chair had proposed a text 'whose substance and architecture contain the necessary elements of success' (WWF, 2007a).

Technical doubts and political postures

A shift in the context

Beyond the 'Friends of Fish', the initial reaction to the chair's draft was mixed but basically encouraging. In formal terms, the text was widely embraced as a proper basis for continuing negotiations. This did not mean, of course, that all members accepted the provisions of the draft or even its overall high level of ambition. As discussed below, there was a lot to debate. But the draft was clearly accepted as a point of departure for further talks.

Importantly, the same could not be said for the chair's text on the volatile issue of 'anti-dumping', which accompanied the fisheries subsidies text as part of the chair's overall package on the NGR's negotiating topics. Anti-dumping was – and had long been – a hot potato within the multilateral trade system. At issue are domestic trade law procedures that, in theory, are used to combat unfair competitive practices. Many observers (and governments), however, feel that domestic anti-dumping laws are often subject to protectionist abuse. Efforts to use WTO rules to discipline domestic anti-dumping procedures command the full and passionate attention of major industries and powerful lawmakers in the US and other countries around the world and the issue ranks just a notch below agricultural subsidies and industrial tariffs as a potential 'round-breaker'. It is also an issue on which the United States stands in isolation against most of the rest of the world.[79]

The politics of anti-dumping matter to fisheries subsidies because the anti-dumping and fisheries subsidies texts were parts of a single report from the NGR chair, and because it is standard WTO negotiating practice to establish political links between disparate issues as a means of gaining negotiating leverage. The rhetoric of governments against the chair and against the US was very strong and it was couched in language rejecting the chair's work as a basis for negotiating. While polite statements accepting the fisheries subsidies text were being offered inside the negotiating room, strong statements condemning the proposed 'rules text' were dominating the public response and in the WTO corridors some delegates complained that on both fish and anti-dumping the chair seemed to favour the US.

Other major political factors within the Doha Round would soon start complicating the story. The chair's draft was released just a few weeks before Christmas 2007. When delegates returned from the winter break in mid January, the spectre of yet another 'don't let the Round die' ministerial meetings was already casting its shadow over Geneva. Given the usual pace of the WTO calendar, and the US elections coming up the following November, June 2008 seemed the deadline for modalities if George W. Bush was to sign the Doha deal in the final days of his presidency. And so, for the third time in as many years, the politics of the Round – and its strong north–south dialectic – dominated the late winter and spring in Geneva.

This time, however, the state of play on the chair's draft and the overall politics of the Round combined to change the fisheries subsidies negotiating dynamic in a significant way: the centrality of the S&DT issue turned suddenly from an asset to a potential liability for those seeking a final resolution.

One factor underlying this shift in context was that S&DT remained a difficult and unresolved issue in both the fisheries subsidies talks and in the Round as a whole. Especially as north–south politics heated up along with the crisis in the overall negotiations, issues surrounding S&DT within the fisheries subsidies talks were increasingly influenced by questions of systemic strategy and tactics. Moreover, the solution to the S&DT issue on fish would require two important innovations that were in direct tension with precisely such systemic considerations:

1 The 'no blank cheque' approach to S&DT, as promoted by almost all developing countries up to that point, meant accepting conditionality on S&DT. The unusual nature of developing country willingness to contemplate such conditionality had been explicitly noted by several developing country delegations in the course of the talks.[80]
2 The conditionality in question related directly to environmental performance in an area where many developing countries felt unable to meet such regulatory standards. This played on systemic fears that the Doha Round as a whole was heading towards unfairly burdening developing countries. It also raised potential echoes of developing country fears of 'green protectionism' from the early days of the trade and environment debate.

Compounding these challenges was the ambiguous and in some ways problematic treatment of S&DT within the chair's draft on fisheries subsidies. The text raised two sets of doubts in the minds of many developing country delegations:

1 How much S&DT would really be granted? Would the scope of S&DT really provide the 'policy space' many developing countries considered necessary on fisheries subsidies?
2 How burdensome would be the environmental conditionalities (and especially the fisheries management system requirements)? Would developing countries be expected to meet sustainability criteria that they felt were beyond their limited institutional and financial resources?

The first of these questions grew out of the chairman's multi-layered approach to setting the limits of S&DT. For example, the chair's text would allow developing countries to continue subsidizing vessel construction or the operating costs of fishing (e.g., fuel), but only if the subsidized vessels were fewer than ten metres long or would operate exclusively within the EEZs of the subsidizing country. The text was also ambiguous regarding the extent to which developing countries could subsidize value-added processing of fish products. Limits such as these cut directly at subsidies that some developing countries considered among their highest priorities.

Similarly, the second set of doubts grew directly out of the language of the chair's draft. The sustainability criteria relating to fisheries management were mainly set out in a single lengthy paragraph. The paragraph established a basic obligation on developing countries employing fisheries subsidies to 'operate a fisheries management system regulating marine wild capture fishing within its jurisdiction, designed to prevent overfishing'. The text then went on for more than 250 words to define the elements of such a system. Unfortunately, the text failed to distinguish clearly between specific practices that would become mandatory and those to which the text referred only as illustrations of how more general obligations might be implemented. For example, the text mentions electronic vessel monitoring systems, but leaves unclear whether these are required or are simply one optional means of implementing 'capacity and effort management measures'.

The concerns over the practical requirements of the proposed sustainability criteria were compounded by language in the chair's text that also required countries seeking to employ S&DT (or any of the general exceptions to the prohibition) to submit information about their management systems to 'the relevant body of the FAO' for a vaguely described 'peer review' process. The timing of this review, as well as nature of any evaluation or decisions to be made by 'the relevant body', were not clearly articulated in the chair's draft. As a result, a number of delegations (notably including Brazil) reacted sharply to the possibility that such a review process would amount to requiring a 'pre-approval' of any subsidy to be employed under the S&DT provisions. Fears of such a process strengthened the tendencies of many delegations to resist any formal role at all for the FAO in the administration of new WTO fisheries subsidies rules. In any case, if a role for the FAO was to be contemplated, it was quickly clear that acting as a 'pre-approval' screener was not an acceptable possibility.

Overall, the doubts raised by the chair's text on S&DT meant that the basic costs and benefits of the deal being offered to developing countries was fundamentally unclear. Despite all the progress that had been made within the NGR and parallel discussions on S&DT and sustainability criteria over the previous years, the hard answers were still missing.

New alliances, new actors

A tactical alliance between India and Indonesia, later to be joined by China, was emerging in April 2008. These large developing countries – all three of which had both deep political stakes in the north–south debate within the WTO and major fishing industries with expanding distant water subsectors – articulated a position on S&DT that split radically from the positions taken by Brazil,

Argentina and the many developing countries that had endorsed their basic approach. In essence, under the leadership of India, the three most populous developing countries rejected meaningful conditions on S&DT and sought what amounted to a blank cheque for their fisheries subsidies.[81] In the cases of India and Indonesia, this was not a wholesale departure from their past interventions. For China, it was a new and far more negative posture that took a number of delegations and stakeholders by surprise.

Moreover, as the summer ministerial meeting approached, India took the unprecedented step of seeking to include its demands on fisheries subsidies on the short list of issues to be discussed in the modalities negotiations. In other words, India struck a pose that sought some basic satisfaction of its demands for a blank cheque on fisheries subsidies as a condition for agreeing on modalities and taking the Round into its endgame. India's effort was rebuffed when none of the other major powers was ready to put fisheries subsidies on the ministerial agenda. But the level of 'linkage' and political priority that India gave to the fisheries subsidies issue was an important new development, even if some observers considered that the Indian and Chinese positions were heavily influenced by overarching tactical considerations not necessarily reflecting their ultimate interests in the fishery sector itself.

At the same time that the India–Indonesia–China proposal was being brought to the table, the NGR was continuing with meetings to review the chair's draft. These meetings were the occasion for a new spate of papers from various delegations, focused on different aspects of the chair's text. Some of these papers, such as a contribution by Norway in April 2008,[82] continued to develop the technical discussion on sustainability criteria, even as it positioned Norway to defend some of the subsidies it wished to maintain. Others, such as a paper from Canada seeking to re-introduce the concept of a *de minimis* exception,[83]

Figure 4.3 *Newspaper headlines after crash of Geneva Ministerial, July 2008*

seemed directed at laying down specific markers for negotiations over the scope of the ban. In general the papers tabled at this time were more reactive to the chair's text than proactive efforts at negotiating dialogue. Despite the formal acceptance by nearly all delegations of the chair's text as the basis for negotiations, the papers tabled in the spring of 2008 also revealed ongoing and substantial divisions over most of the key aspects of the proposed new rules.

During this period, UNEP was working with WWF and the US-based NGO Oceana to help delegations understand and improve the chair's draft and especially the approach to sustainability criteria that it contained. Starting at a jointly sponsored workshop in January 2008,[84] environmental stakeholders began to urge delegations to 'reduce the ambiguity without reducing the ambition' of the chair's text.

One other consequence of the chair's draft and the rising politics surrounding the fisheries subsidies talks was an expansion of the number of governments paying closer attention to the technical details of the issue, both through their Geneva delegates and through the increased involvement of their capital-based officials. To meet the expanding need to bring new actors up to speed on the key issues, UNEP published in May 2008 an introductory guidebook to the negotiations (UNEP, 2008b). Other environmental stakeholders similarly brought forward briefing papers on issues raised by the chair's draft or by the discussions within the NGR.[85]

Then, in July 2008, the entire Doha process reached a point of genuine crisis. Once more, WTO ministers gathered in Geneva and once more the meeting crashed amid acrimony along a north–south divide (with India and the US as the most overt antagonists). This time, however, the crash had a new and more definitive quality. With so many failed efforts behind them, and with the presidential election imminent in the US, governments knew there was no pretending that the process could just go on. Although there was no formal suspension of negotiations, there were explicit declarations that efforts to find a political solution to the modalities problem would be halted for the foreseeable future.

Another step forward? Or two steps back?

As 2008 drew to a close, the WTO was beginning to look for its new *modus vivendi* and Pascal Lamy was encouraging his negotiating group chairs to find ways to resume technical work in order to keep the Round moving forward as much as possible. In response to this, and in reaction to the variety of concerns raised about the chair's texts the previous spring, the NGR chair released a second set of 'consolidated texts' in December 2008. The main development in those drafts was the revision of the anti-dumping text, which added brackets to the language that had caused such outrage the previous year.

On fisheries subsidies the chair took a somewhat different approach. Rather than offer a revised negotiating text, the chair tabled a 'Roadmap for Discussions' that set out a detailed series of questions meant to guide discussions on the issues continuing to divide the negotiators. The chair commented that the continuing divisions 'go to the very concepts and structure of the rules' and even suggested that the dialogue needed to revert to the 'varied perceptions among participants as to the exact scope and meaning of the mandate'.[86] The roadmap went on to pose specific questions that ranged from fundamental (such as asking why fisheries subsidies should be banned or not banned) to highly technical. A series of NGR sessions was then scheduled at a pace that suggested a review of the roadmap would require the better part of a full year. At the time of writing, the roadmap remains the organizational focus of the NGR's work on fisheries subsidies.

Reactions to the roadmap, and interpretations of its meaning for the fisheries subsidies talks, were varied. Some delegations have seen it as a necessary return to basic principles. Others, along with some environmental stakeholders, were concerned that it appears to re-open questions about the mandate that had seemed already resolved. It was also unclear what the roadmap implied for the chair's draft of November 2007. The chair had repeatedly indicated that the 2007 draft remains the starting point for further negotiations. On the other hand, the roadmap itself raised a number of questions that appear to invite discussion as if from a clean slate. Meanwhile, among the uncertainties surrounding the process as of autumn 2009 has been the eventual timing of a revision to the chair's proposed fisheries subsidies text.[87]

Increasing and divergent engagement by developing countries

As the publication was being finalized the roadmap discussions had progressed again towards the issues surrounding S&DT, and developing country delegations were increasingly re-engaging in the debate – this time often with much greater involvement from their capital-based counterparts than before.

The interactions among developing country delegations on S&DT have also increasingly revealed differences of opinion about the goals and operation of S&DT for fisheries subsidies. The disagreement over S&DT for the high seas was becoming a point of sharp contention. Brazil, working closely with Mexico, soon emerged as the most active proponent of allowing S&DT on the high seas. Taking a strong opposite position were the delegations of Argentina, Chile, New Zealand and Norway, among others.[88] Later in 2009, potentially significant realignments were appearing plausible – most dramatically illustrated by a joint statement demanding S&DT on the high seas issued in September by Brazil, China, Ecuador and Mexico (later to be joined by Venezuela).[89] The appearance of Brazil and China on the same statement was particularly interesting, since Brazil had previously been such a strong proponent of sustainability criteria and China had relatively recently joined India and Indonesia in raising doubts about the acceptability of significant conditions of any kind on S&DT. Positions were clearly shifting, but whose and in which direction were not immediately clear. All of this activity signaled a clear rise in the sensitivity of delegations to their own national interests and perspectives. It also began to usher in a period in which a technical focus on sustainability and development was again the core of the debate.

The second half of 2009 saw growing interest in debating the juncture between S&DT and sustainability criteria. Accordingly, UNEP and other parallel process actors again dedicated resources to capacity building and facilitating technical dialogue on these key points. In April 2009, UNEP and WWF produced a joint briefing in Geneva aimed particularly at engaging newly arrived delegates. Then, in late July, UNEP joined forces with WWF, the Permanent Commission for the South Pacific and the Government of Ecuador to produce a major workshop in Guayaquil, Ecuador, on sustainability criteria for the benefit of capital-based officials across Latin America.

In September 2009, that work was extended to the Caribbean nations through UNEP's participation in a Geneva-based workshop hosted by the Group of Latin America and Caribbean Countries (GRULAC).

Conclusion

As noted at the beginning of this chapter, the history of the WTO fisheries subsidies negotiations remains incomplete. Considering the challenging innovations required to bring the talks to fruition, it can be said that much progress has been achieved towards a solution. But with more than a decade having passed since UNEP helped bring the problem of fisheries subsidies to worldwide attention, it may also be fair to say that a solution is long overdue.

The current hiatus in the Doha Round of negotiations has once again frustrated efforts to bring the fisheries subsidies talks to a close, but has also again created a context in which governments can take further steps towards refining and adopting appropriate sustainability criteria. Meanwhile, the crisis of fisheries depletion continues to threaten ecosystems and livelihoods around the world.

As this book went to press, the end of the fisheries subsidies story remained a question for fortune-tellers. At a minimum, however, the progress achieved at the multilateral level strongly suggests that the WTO can be made a real partner in the struggle to achieve sustainable patterns of production and consumption of the Earth's natural resources. UNEP will continue to work to help governments achieve that necessary goal.

> *WTO members are now negotiating to reform these subsidies programmes so that fishing becomes a sustainable industry and so that we can fully appreciate our oceans' bounty for generations to come. A deal in the WTO now, would mean richer oceans for the future generations.*
>
> Pascal Lamy, Director-General WTO,
> 8 June 2009, World Oceans Day

Endnotes

1 A timeline of the development of the fisheries subsidies issue can be found in Annex 1 of this book.
2 Note that almost all the studies and submissions referred to in this chapter can be found on the CD-ROM accompanying this book.
3 Can be found on accompanying CD-ROM.
4 See Statement of the Ninth APEC Ministerial Meeting Vancouver, Canada, 21–22 November 1997, Annex (Early Voluntary Sectoral Liberalization) (available at www.apec.org/apec/ministerial_statements/annual_ministerial/1997_9th_apec_ministerial/annex.html, last accessed 16/03/2010)
5 See, WTO Doc. WT/CTE/W/51, Environmental and trade benefits of removing subsidies in the fisheries sector – submission by the United States (19 May 1997); WTO Doc. WT/CTE/W/52, The Fisheries Sector – Submission by New Zealand, (21 May 1997).
6 United Nations General Assembly Resolution S/19-2, 28 June 1997 (UN Doc. No. A/RES/S-19/2).
7 See, e.g, Stone 1997, UNEP, 1998.
8 The General Agreement on Tariffs and Trade (GATT) was the predecessor of the WTO from 1947 to 1994. GATT was created and modified through a series of multilateral trade negotiating 'rounds' in the decades following World War II. While initially preoccupied with lowering tariffs and with preventing a return to the protectionism of the 1930s, GATT steadily expanded the scope of this role to impose disciplines on government actions that could erect 'non-tariff barriers' to trade, including 'trade-related environmental measures' such as the ban on 'dolphin unsafe' tuna that was at issue in the tuna–dolphin case.
9 The GATT Conference of Parties never formally adopted the GATT panel ruling against the United States, in light of a threat by the US to block its adoption through the GATT's consensus-based process. Under the more rigorous procedures of the WTO, the US would lack the ability to block dispute rulings adverse to its interests in this fashion.
10 See Trade and Environment, Decision of 14 April 1994, WTO Doc. No. MTN.TNC/45(MIN), meeting at ministerial level, Palais des Congrès, Marrakesh (Morocco), 12–15 April 1994, (06 May 1994), Annex II.
11 See, for example, White House press release 24 November 1999, 'The Clinton Administration Agenda for the Seattle WTO'.
12 WTO Doc. No. WT/GC/W/303, Preparations for the 1999 Ministerial Conference – Fisheries Subsidies – Communication from Australia, Iceland, New Zealand, Norway, Peru, Philippines and United States (6 August 1999). See also WT/GC/W/229, Preparations for the 1999 Ministerial Conference – Fisheries Subsidies – Communication from Iceland (6 July 1999); WT/GC/W/292, Preparations for the 1999 Ministerial Conference – Elimination of Trade Distorting and Environmentally Damaging Subsidies in the Fisheries – Communication from New Zealand (5 August 1999).
13 See WTO Doc. No. WT/CTE/W/99, 'Comments by the European Community on the Document of the Secretariat of the Committee on Trade and Environment (WT/CTE/W/80) on Subsidies and Aids Granted in the Fishing Industry' (6 November 1998); 'Preparations for the 1999 Ministerial Conference – Comments on the Fisheries Subsidies proposal submitted by Australia, Iceland, New Zealand,

Norway, Peru, Philippines and United States (WT/GC/W/303) – Communication from Japan', WTO Doc. No. JOB(99)/5367 (16 September 1999); WT/GC/W/221, 'Preparations for the 1999 Ministerial Conference – 'Negotiations on Forestry and Fishery Products – Communication from Japan' (28 June 1999).
14 WTO Doc. No. WT/CTE/W/99.
15 WTO Doc. No. WT/GC/W/221, 'Preparations for the 1999 Ministerial Conference – Negotiations on Forestry and Fishery Products – Communication from Japan' (28 June 1999).
16 WTO Doc. No. WT/GC/W/221.
17 See Japan's Basic Position on the Fisheries Subsidies Issue, WTO Doc. No. TN/RL/W/11 (2 July 2002).
18 Statement by H.E. Mr Alami Tazi, Minister of Commerce, Industry and Handicrafts, Morocco (opening plenary of Seattle Ministerial), WTO Doc. No. WT/MIN(99)/ST/29 (1 December 1999).
19 Mr Ibsen served as a senior official on Iceland's delegation to the Seattle ministerial and played a significant role in the preparation of Iceland's WTO proposal on fisheries subsidies.
20 WTO Doc. No. WT/MIN(01)/DEC/1, ministerial declaration adopted on 14 November 2001 (20 November 2001). NB: the reference to Paragraph 31 was unusual. That paragraph, entitled 'Trade and Environment', mandated negotiations on three technical areas unrelated to fish subs intended to enhance 'the mutual supportiveness of trade and environment'. A final and a separate clause of paragraph 31 stated simply, 'We note that fisheries subsidies form part of the negotiations provided for in paragraph 28.' This odd mutual cross-reference was clearly intended to indicate an environmental orientation to the fisheries subsidies talks, perhaps as a compromise in lieu of a more detailed environmental reference.
21 At various times, active members of the 'Friends of Fish' coalition have included Argentina, Australia, Chile, Ecuador, Iceland, New Zealand, Norway, the Philippines, Peru and the USA.
22 WTO Doc. No. TN/RL/W/3, 'The Doha mandate to address fisheries subsidies: Issues –submission from Australia, Chile, Ecuador, Iceland, New Zealand, Peru, Philippines and United States', (24 April, 2002).
23 These can be found on the accompanying CD-ROM.
24 WTO Doc. No. TN/RL/W/82, 'Submission of the European Communities to the Negotiating Group on Rules – Fisheries Subsidies' (23 April 2003).
25 WTO Doc. No. TN/RL/W/159, 'Fisheries Subsidies: Proposed Structure of the Discussion – Communication from Japan' (7 June 2004).
26 WTO Doc. No. TN/RL/W/136, 'Fisheries Subsidies' (14 July 2003).
27 See Chair's Summary of the UNEP Workshop on Fisheries Subsidies and Sustainable Fisheries Management, 26–27 April 2004, available at: www.unep.ch/etb/events/FishMeeting2004.php (last accessed 16/03/2010) and on the accompanying CD-ROM.
28 WTO Doc. No. TN/RL/W/166, 'Fisheries Subsidies – Communication from Argentina, Chile, Ecuador, New Zealand, Philippines and Peru' (2 November 2004).
29 See WTO Doc. No. TN/RL/GEN/41, 'Fisheries Subsidies: Programs for Decommissioning of Vessels and Licence Retirement – Communication from the United States' (13 May 2005). These concepts would later be reflected in papers by both the United States and New Zealand proposing legal language for new rules. See

WTO Doc. Nos. TN/RL/GEN/127 (United States, 24 April 2006) and TN/RL/GEN/141 (New Zealand, 6 June 2006).

30 WTO Doc. No. TN/RL/W/164, 'Proposal on Fisheries Subsidies – Paper by Japan' (27 September 2004).

31 The Pacific Islands Forum is an intergovernmental organization grouping 16 Pacific island states, including Cook Islands, the Federated States of Micronesia, Fiji, French Polynesia, Kiribati, the Marshall Islands, Nauru, New Caledonia, Niue, Palau, Papua New Guinea, Samoa, Solomon Islands, Tonga, Tuvalu and Vanuatu. With support from the Commonwealth Secretariat, the Forum's permanent mission to the WTO has played an active role representing the interests of Pacific island states in the fisheries subsidies talks. Although Australia and New Zealand are also formally members, the Forum has taken independent positions in the negotiations quite distinct from the 'Friends of Fish' *demandeur* group.

32 'Special and differential treatment' applies to numerous provisions throughout the WTO rule system that grant developing countries limited exceptions from certain WTO obligations.

33 The equitable claim to a share of fisheries resources is also extended by some developing countries seeking to subsidize expanded fishing beyond their EEZs, where competition for international stocks is particularly intense.

34 As outlined in Chapter 1, the data developed in the early efforts to catalogue and estimate fisheries subsidies suggested that as much as 90 per cent of the known subsidies were provided by only seven countries – Japan, the EC, the United States, Canada, Russia, the Republic of Korea and Chinese Taipei. More recent studies have suggested that developing country subsidies may be greater than generally assumed. Sumaila and Pauly (2006), for example, originally estimated that 48 per cent of all fisheries subsidies were granted by developing countries. The assumptions and extrapolations that underlay this conclusion have, however, been subject to continuing review and these original estimates may be subject to substantial downward revision in a revised version of the UBC paper intended for publication in 2010. Even if the original UBC conclusion is roughly correct, however, the fact would remain that richer developed countries would still far out-subsidize poorer countries on a 'per country' or 'per fleet' basis.

35 WTO Doc. No. TN/RL/W/166, 'Fisheries Subsidies – Communication from Argentina, Chile, Ecuador, New Zealand, Philippines, Peru' (2 November 2004).

36 See WTO Doc. No. TN/RL/W/154, 'Fisheries Subsidies: Overcapacity and Over-Exploitation – Communication from New Zealand' (26 April 2004), ¶ 18.

37 See, for example, materials from the UNEP Workshop on Fisheries Subsidies and Sustainable Fisheries Management, 26–27 April 2004, available at www.unep.ch/etb/events/FishMeeting2004.php, last accessed 16/03/2010. The prominence of developing country issues was also reflected in the workshop report, which is available as WTO Doc. No. TN/RL/W/161 (8 June 2004) – all contained in the accompanying CD-ROM.

38 See 'Artisanal Fishing: Promoting Poverty Reduction and Community Development Through New WTO Rules on Fisheries Subsidies, an Issue and Options Paper' (UNEP, 2005b); 'Reflecting Sustainable Development and Special and Differential Treatment for Developing Countries in the Context of New WTO Fisheries Subsidies Rules. An Issue and Options Paper' (UNEP 2005a); and 'Towards Sustainable Fisheries Access Agreements: Issues and Options at the World Trade Organization' (UNEP, 2008a). Excerpts can be found in Chapter 5.

39 See Chairs Summary of the Roundtable on Promoting Development and Sustainability in Fisheries Subsidies Disciplines, 30 June 2005, available at www.unep.ch/etb/events/2005rtGeneva.php (last accessed 16/03/2010) and on the accompanying CD-ROM.
40 WWF, for example, hosted a set of half-day 'technical roundtables' on developing country issues at the WTO in September 2004 and released a paper on developing country issues at a major conference on EU-ACP relations sponsored by the Commonwealth Secretariat in December 2004. (See, Schorr, 'Fishing Subsidies: Issues for ACP Countries' (WWF, 2004) available at: www.assets.panda.org/downloads/euacpschorrpresentation.doc, last accessed 16/03/2010). ICTSD held a workshop in Geneva in May 2005 dedicated largely to developing country issues at the intersection of fisheries and trade. (See 'Untangling Fisheries and Trade: Towards Priorities for Action', ICTSD, Geneva, Switzerland, 9–10 May 2005.)
41 The main Brazilian papers included WTO Docs. No. TN/RL/W/176 (31 March 2005), TN/RL/GEN/56 (4 July 2005) and TN/RL/GEN/79 (16 November 2005), along with substantial revisions to TN/RL/GEN/79 dated 21 February 2006, 21 April 2006, 2 June 2006 and 13 March 2007.
42 WTO Doc. No. TN/RL/W/196, 'Fisheries Subsidies – Paper from Brazil, Chile, Colombia, Ecuador, Iceland, New Zealand, Pakistan, Peru and the United States (22 November 2005).
43 UNEP also played a leading role helping to raise public expectations on fisheries subsidies for the ministerial meeting. For example, working jointly with WWF, UNEP co-authored a widely distributed op-ed calling for ministers not to lose the opportunity to act on fisheries subsidies. See UNEP-WWF (2005), 'Time to draw in the net on fishing subsidies', Opinion Editorial, Leape, J. and Toepfer, K., available at www.unep.ch/etb/events/Events2005/pdf/WWF-UNEP_opinion_editorial.pdf, last accessed 12/03/2010
44 WTO Doc. No. WT/MIN(05)/DEC, 'Doha Work Programme' – ministerial declaration adopted on 18 December 2005 (22 December 2005), Annex D, ¶ I.9.
45 C. Clover, 'EU split over return of fishing subsidies' in *The Telegraph* (22 May 2006) (www.telegraph.co.uk/news/worldnews/europe/1519089/EU-split-over-return-of-fishing-subsidies.html); see also 'Stop subsidising over-exploitation of fisheries' – NGOs call on governments to oppose the use of public funds for fleet modernization, even if this means rejecting the Council deal on the EFF in its entirety, joint public statement by WWF, Greenpeace, BirdLife International, Oceana and the Fisheries Secretariat (May 2006).
46 The most important of these was the adoption of the EFF regulation itself, Council Regulation (EC) No 1198/2006 of 27 July 2006 on the European Fisheries Fund, which was followed by a regulation allowing 'de minimis' exceptions to the EFF rules (Commission Regulation (EC) No 875/2007 of 24 July 2007); see also Communication from the Commission to the Council and the European Parliament on Improving the Economic Situation in the Fishing Industry, COM(2006) 103 final, 9 March 2006.
47 WTO Doc. No. TN/RL/GEN/134, 'Fisheries Subsidies – Submission of the European Communities' (24 April 2006).
48 WTO Doc. No. TN/RL/GEN/114, 'Fisheries Subsidies – Framework for Disciplines – Communication from Japan, the Republic of Korea and the Separate Customs Territory of Chinese Taipei, Penghu, Kinmen and Matsu' (21 April 2006).

49 WTO Doc. No. TN/RL/GEN/100, 'Fisheries Subsidies – Framework for Disciplines – Paper from New Zealand' (3 March 2006).
50 WTO Doc. No. TN/RL/GEN/127, 'Fisheries Subsidies – Communication from the United States' (24 April 2006).
51 Note that WWF had also proposed differentiation in the application of S&DT, but with a key difference: major fishing powers among developing countries would be excluded from S&DT, while sustainability criteria would continue to apply to all others. See WWF, 2004, § V.H.2.
52 Opposition to differentiation is a fundamental 'systemic' position of these large developing countries, who are adamant that developing countries (other than 'least developed countries') be treated as a single bloc throughout WTO rules.
53 See, for example, WTO Doc. No. TN/RL/W/159, 'Fisheries Subsidies: Proposed Structure of the Discussion – Communication from Japan' (7 June 2004), ¶¶ 13–15.
54 WTO Doc. No. TN/RL/GEN/100, 'Fisheries Subsidies Framework for Disciplines: Paper from New Zealand' (3 March 2006), proposed Art. 27bis.
55 This simple approach was known to rely on a sub-optimal metric. A preferred approach might have been to find a means of measuring the size of a country's fishing industry in relation to its available EEZ fisheries resources – a quantification of the 'room to grow' concept that often underlay developing country calls for S&DT. However, data on the value of EEZ fisheries and industry capitalization were not readily available for many countries. The 'share of global production' approach proposed by New Zealand was a crude but available alternative.
56 Raising the test for example to 1.0 per cent would leave countries such as China, Peru, Chile, Indonesia, India, Thailand, Republic of Korea, the Philippines, Viet Nam, Mexico, Malaysia, Chinese Taipei, Argentina, Myanmar and Morocco all still above the *de minimis* line.
57 WTO Doc. No. TN/RL/GEN/114, 'Fisheries Subsidies Framework for Disciplines – Communication from Japan, the Republic of Korea and the Separate Customs Territory of Taiwan, Penghu, Kinmen and Matsu' (21 April 2006), proposed Art. 27.16(b).
58 See, for example, 'Statement Delivered by Barbados on Behalf of the SVEs under Item 15.c of the Chair's List of Questions on Fisheries Subsidies' (24 September 2009).
59 Joint UNEP-ICTSD-WWF Briefing on the WTO Negotiations on Fisheries Subsidies: Issues and Options for Developing Countries, 27 April 2006, programme available at: www.unep.ch/etb/events/pdf/briefing%20agenda%20final27Apr06.pdf, last accessed 16/03/2010
60 See Chair's Summary of the Joint UNEP-ICTSD-WWF Workshop on Development and Sustainability in the WTO Fishery Subsidies Negotiations: Issues and Alternatives, 11 May 2006, available at: www.unep.ch/etb/events/2006ICTSD WWFMay11.php (last accessed 16/03/2010) and on the accompanying CD-ROM.
61 WTO Doc. No. TN/RL/W/203, 'Small Scale, Artisanal Fisheries – Submission by India' (6 March 2006).
62 See Brazilian papers citation, above. For Argentina's papers, see TN/RL/GEN/138 (1 June 2006 and its revision of 26 January 2007), along with TN/RL/W/211 (19 June 2007). Brazil and Argentina subsequently joined in several papers, including TN/RL/GEN/151 (17 September 2007) and its revision on 26 November 2007. Also TN/RL/GEN/79, TN/RL/GEN/79/Rev.1, TN/RL/GEN/79/Rev.2, TN/RL/GEN/79/Rev.3, TN/RL/GEN/79/Rev.4.

63 See materials of the Joint UNEP-ICTSD-WWF Workshop on Development and Sustainability in the WTO Fishery Subsidies Negotiations: Issues and Alternatives, 11 May 2006, available at: www.unep.ch/etb/events/2006ICTSDWWFMay11.php (last accessed 16/03/2010) and on the accompanying CD-ROM.
64 See ICTSD, 2006d.
65 See WTO Doc. No. TN/RL/GEN/79/Rev.2, 'Further Contribution to the Discussion on the Framework for Disciplines on Fisheries Subsidies – Paper from Brazil – Revision' (21 April 2006), proposed Arts. 1.4.1, 4.4, & 7.1(ii).
66 See WTO Doc. No. TN/RL/GEN/79/Rev.3, 'Possible Disciplines on Fisheries Subsidies – Paper from Brazil – Revision' (2 June 2006), proposed Arts. 1.3.1, 3.1(f).
67 The solution as eventually embodied in the chair's text drew on a combination of elements from the fourth revision of the Brazilian paper – TN/RL/GEN/79/Rev.4 (13 March 2007) – and a paper tabled almost simultaneously by the United States – TN/RL/GEN/145 (22 March 2007).
68 Brazil in some respects weakened its level of environmental ambition in successive drafts of its GEN/79 paper. For example, Brazil softened the requirements it proposed for the level of data to be required in subsidized fisheries and steadily reduced the role it was proposing for involving regional fisheries management organizations in setting a baseline for adequate management of international fisheries.
69 See, for example, WTO Doc. No. TN/RL/GEN/127, 'Fisheries Subsidies – Communication from the United States' (24 April 2006); TN/RL/GEN/134, 'Fisheries Subsidies – Submission of the European Communities' (24 April 2006); TN/RL/GEN/79/Rev.3, 'Possible Disciplines on Fisheries Subsidies – Paper from Brazil – Revision' (2 June 2006).
70 Chair's Summary of the UNEP-WWF Symposium on Disciplining Fisheries Subsidies: Incorporating Sustainability at the WTO & Beyond, 1–2 March 2007, available at: www.unep.ch/etb/events/2007fish_symposium.php and on the accompanying CD-ROM.
71 Launching event of the UNEP-WWF publication: *Sustainability Criteria for Fisheries Subsidies – Options for the WTO and Beyond*, 26 September 2007, www.unep.ch/etb/events/2007LaunchWWF_UNEPPubli.php, last accessed 16/03/2010.
72 See Lamy: 'We have resumed negotiations fully across the board' (WTO News Item 7 February 2007) (available at www.wto.org/english/news_e/news07_e/gc_dg_stat_7feb07_e.htm, last accessed 16/03/2010).
73 WTO Doc. No. TN/RL/GEN/145, 'Fisheries Subsidies: Proposed New Disciplines – Proposal from the United States' (22 March 2007).
74 The 'green room' is the informal name of the WTO Director General's private conference room within the GATT headquarters building. It has come to stand for meetings among selected groups of leading ministers during the most intense and high level (and usually 'eleventh hour') moments of negotiations.
75 See WTO Doc. No. TN/RL/W/209, 'Access Fees in Fisheries Subsidies Negotiations – Communication from the ACP Group' (5 June 2007); TN/RL/W/210, 'S&DT in the Fisheries Subsidies Negotiations: Views of the Small, Vulnerable Economies (SVEs) – Communication from Barbados, Cu[...]New Guinea and Solomon Islands' (6 June 2007) (and revisions dated 18 June and 22 June); TN/RL/W/211, 'Fisheries Subsidies: Continuation of Work on Special and Differential Treatment – Paper from Argentina' (19 June 2007), and TN/RL/W/212, 'Fisheries Subsidies: Fisheries Adverse Effects and S&D Treatment – Paper from Brazil' (29 June 2007).

76 WTO Doc. No. TN/RL/GEN/114/Rev.2, 'Fisheries Subsidies: Framework for Disciplines – Communication from Japan; the Republic of Korea and the Separate Customs Terri[...]u, Kinmen and Matsu – Revision' (5 June 2007).
77 WTO Doc. No. TN/RL/W/213, 'Draft Consolidated Chair Texts of the AD and SCM Agreements' (30 November 2007).
78 Even the most adamant opponents of 'differentiation' among developing countries have tended to accept special treatment for least developed countries, which is already a feature of WTO subsidy rules (see WTO SCM Agreement, Art. 27.3) and is common to several other elements of the WTO rule system (see, for example, WTO Agreement on Agriculture, Art. 16, WTO Agreement on Sanitary and Phytosanitary Standards, Art. 14; WTO Understanding on Rules and Procedures Governing the Settlement of Disputes, Art. 24).
79 The US is essentially alone in making use of a controversial valuation method – known as 'zeroing' – used in determining whether dumping has taken place. This valuation method is a centre-piece of US anti-dumping law, but has been increasingly called into question under emerging WTO jurisprudence. Among the top US priorities for the Doha Round (at least from the perspective of the US Congress) has been to clarify the WTO anti-dumping code to ensure zeroing is enshrined as legitimate, despite current case law trends.
80 See, for example, WTO Doc. No. TN/RL/W/210/Rev.2, 'S&DT in the Fisheries Subsidies Negotiations: Views of the Small, Vulnerable Economies (SVEs) – Communication from Antigua and [...]and Solomon Islands – Revision' (22 June 2007). This carefully worded paper notes the systemic position rejecting conditions on S&DT and does not formally concede the point in the context of fisheries subsidies. However, the text goes on at length to discuss specific concerns and preferences of SVEs with regard to how conditionality on S&DT could be viewed.
81 WTO Doc. No. TN/RL/GEN/155/Rev.1, 'Need for Effective Special and Differential Treatment for Developing Country Members in the Proposed Fisheries Subsidies Text –[...]Indonesia and China – Revision' (19 May 2008).
82 WTO Doc. No. TN/RL/W/231, 'Drafting Proposal on Issues Relating to Article V (Fisheries Management) of the Fisheries Subsidies Annex to the SCM Agreement [...]13 – Communication from Norway' (24 April 2008).
83 WTO Doc. No. TN/RL/GEN/156, 'Fisheries Subsidies – De Minimis Exemption – Communication from Canada' (2 May 2008).
84 See Workshop Report of the UNEP, 'WWF Technical and Informal Workshop on WTO Disciplines on Fisheries Subsidies: Elements of the Chair's Draft' (29 January 2008), available at: www.unep.ch/etb/events/2008FishSubWorkshop29Jan08.php (last accessed 16/03/2010) and on the accompanying CD-ROM.
85 See, for example, WWF (2008).
86 WTO Doc. No. TN/RL/W/236, 'New Draft Consolidated Chair Texts of the AD and SCM Agreements' (19 December 2008), Annex VIII: Fisheries Subsidies – Roadmap for Discussion.
87 As this chapter went to print, the chair had announced a plan to complete the roadmap discussions by the end of 2009 and then to open the table for new technical proposals as further inputs into an eventual second draft text.
88 WTO Doc. No. TN/RL/W/243, 'Fisheries Subsidies – Communication from Argentina, Australia, Chile, Colombia, the United States, New Zealand, Norway, Iceland, Peru and Pakistan (7 October 2009).

89 WTO Doc. No. TN/RL/W/241, 'Fisheries Subsidies – Communication from Brazil, China, Ecuador and Mexico' (28 September 2009) and Rev.1 (adding Venezuela) of 16 October 2009.

Chapter 5

Four Key Issues at the WTO

Introduction

The Ministerial Declaration adopted at the World Trade Organization's (WTO) Sixth Ministerial Conference in Hong Kong on 18 December 2005 explicitly calls on members to prohibit subsidies that contribute to overcapacity and overfishing. Yet reaching an agreement on how to discipline fisheries subsidies has been a long and sometimes difficult process. One of the main challenges has been the complexity of the technical questions involved. Since the beginning of the Doha Round, UNEP has sought to provide sound and expert analysis of these technical issues for negotiators, as well as for national policy-makers.

This chapter presents abbreviated versions of four papers published by UNEP between 2005 and 2008 covering four critical issues:

- the 'special and differential treatment' (S&DT) to be enjoyed by developing countries under new rules;
- the treatment of 'artisanal' fisheries;
- the treatment of subsidies related to fisheries access agreements; and
- the use of 'sustainability criteria' for fisheries subsidies.

Importantly, the four papers are unified by a strong focus on developing country concerns.[1] This focus reflects several factors. First, since it has been clear from the start of the talks that developing countries would be exempt from significant elements of any eventual fisheries subsidies ban, the ultimate environmental effectiveness of new WTO rules was known to depend in substantial part on the limits and conditions to be placed on the continuing right of developing countries to subsidize. Second, as discussed at length in Chapter 4, the political dynamics of the negotiations repeatedly made developing country issues central to the forward progress of the talks. Third – and resulting naturally from the first two factors – UNEP received many requests from member governments for analytic work on developing country issues and on the topics of these four papers in particular.

The first excerpt in this chapter is from the UNEP 2005 paper 'Reflecting Sustainable Development and Special and Differential Treatment for Developing Countries in the Context of New WTO Fisheries Subsidies Rules' authored by Vincente Paolo B. Yu III and Darlan Fonseca-Marti. The paper takes as its starting point the clear language of both the Doha and Hong Kong ministerial declarations requiring WTO fisheries subsidies negotiators to take into consideration the economic concerns of developing and least-developed countries. The paper underscores the real window of opportunity presented by fisheries subsidies reform to achieve sustainable development goals in the area of fish production and fish trade through S&DT that is truly 'sustainability-focused'.

The second excerpt focuses on a specific issue at the heart of the concerns of many developing countries with regard to new fisheries subsidies disciplines: the case of artisanal fishing. The artisanal fishing sector includes some of the poorest and most underdeveloped communities on Earth. But while governments universally recognize the developmental challenges facing many artisanal fishing communities, there has been less emphasis – and perhaps even less agreement – on the need for attention to the sustainability of artisanal fishing practices. 'Artisanal Fishing: Promoting Poverty Reduction and Community Development Through New WTO Rules on Fisheries Subsidies' (UNEP, 2005b), authored by David Schorr, emphasizes the irreducible link between the sustainable management of artisanal fisheries and the long-term economic and social health of the communities that depend on them. Some stakeholders suggest that artisanal fishing is too small in scale or otherwise underdeveloped to present any significant threat to fish resources. However, it is equally true that artisanal fisheries, in striving for modernization, are increasingly generating challenges to their own sustainability. Accordingly, the second paper argues that governments should not be prevented from investing in their underdeveloped artisanal fishing communities, but if long-term resource conservation is to be a priority, the potential sustainability impacts of subsidies to artisanal fisheries must not be ignored.

Another issue that has preoccupied a number of developing countries regards the treatment of subsidies that may result when governments act on behalf of their national fleets to procure access to fisheries within the Exclusive Economic Zones (EEZs) of other countries. As reflected in the third paper excerpted in this chapter, developing countries that are highly dependent on receiving access fees – such as small island states and some coastal African states – have been very worried that access fees might somehow be limited or discouraged under new WTO fisheries subsidies rules. Since the practice of selling such access is clearly enshrined as a right (or even, some might argue, an obligation) under the UN Convention on the Law of the Sea (UNCLOS), the impact of WTO rules on access agreements also raises significant questions of international law. On the other hand, 'flag nations' generally pass on the foreign EEZ fishing rights that they acquire to their industrial distant water fleets (DWFs) and often provide significant subsidies by doing so. Moreover, many access agreements have been criticized for their lack of transparency and often inequitable financial terms. Unfortunately, access agreements have also often been associated with unsustainable fishing practices by DWFs and with significant economic disadvantages to local artisanal fishing communities. The UNEP 2008 paper 'Towards Sustainable Fisheries Access Agreements: Issues and Options at the World Trade Organization', authored by Marcos Orellana, examines access arrangements with an eye to ensuring that WTO rules do not injure fee-reliant countries while also placing some disciplines on the subsidies that result from the further transfer of the access rights. It is hoped that such disciplines can also improve the transparency, equity and sustainability of access agreements.

The final paper presented in this chapter is the UNEP-WWF 2007 paper 'Sustainability Criteria for Fisheries Subsidies: Options for the WTO and Beyond', authored by David Schorr and John Caddy. While it is broadly agreed that fisheries subsidies should be disciplined, and some are likely to be prohibited, it is also clear that many significant categories of subsidies will probably continue to be permitted, especially to developing countries under the terms of S&DT. Since, as pointed out in Chapter 2, almost all classes of subsidies can have negative resource impacts in under-managed fisheries, many governments and stakeholders have called for new WTO rules to include some 'sustainability criteria' to be used as legal conditions on the use of permitted subsidies. However, in the special legal and institutional context of the WTO, such criteria must also be consistent with the WTO's limited mandate and capacities. The UNEP-WWF paper discusses sustainability criteria for fisheries subsidies, and proposes options for use at three levels of environmental ambition: 'minimum international requirements', 'minimum recommended conditions' and 'best practices'. The first of these, at the lowest level of environmental ambitions, is proposed as a starting place for further investigation and dialogue by WTO negotiators, while the higher levels of ambition are intended to guide national policy-makers in the domestic and regional use of subsidies.

Separately, the four papers excerpted here are sources of detailed technical advice. Read together, they reveal the depth of the linkages between environmental, trade and social issues that the fisheries subsidies debate has required and the remarkable degree to which questions of sustainability have penetrated into the WTO discussions of fisheries subsidies. The ensemble of the papers also illustrates the inter-related nature of many of the issues negotiators have confronted. The concern with 'artisanal fisheries', for example, recurs in three of the four papers, while the access-related subsidies paper deals with other communities at the most vulnerable end of the economic spectrum. Similarly, the need to balance rights and obligations is a constant theme, closely linked to the simultaneous goals of many developing countries to achieve rules that significantly restrain subsidies by richer countries (including in the context of growing 'south–south' competition) while also maintaining sufficient 'policy space' for their own subsidy programmes. The role of information about fisheries and the subsidies affecting them – and the costs of obtaining such information – also comes up repeatedly.

As a final note, the reader should bear in mind that each of the papers excerpted here was brought forward at a particular moment in the negotiations and each became the focus of at least one meeting sponsored by UNEP as part of the 'parallel process' that helped shape the WTO talks from the outset. (For a general history that discusses the parallel process and helps locate these papers within it, see Chapter 4.) All four papers have been shortened from their original versions, mainly to eliminate redundant background information and to reduce the number of technical citations. The full original versions of the papers are included in the accompanying CD-ROM.

The excerpts have not been updated for the purposes of this chapter and thus contain some remarks that reflect specific points in the negotiations. Nevertheless, the issues discussed have been central throughout the talks and reflect themes of basic importance to the dynamics of the fisheries subsidies issue. Indeed, the issues discussed here are also proving relevant beyond the WTO in the countries where governments have begun reviewing their own fisheries subsidies policies. The options and solutions discussed in this chapter are thus already proving relevant to fisheries subsidies reform activities at both international and domestic levels.

Special and Differential Treatment*

Sustainable development and developing countries

The achievement of sustainable development is a fundamental policy and institutional objective of the WTO. Explicit references to this objective can be found in both the WTO's constitutional legal instrument – the Marrakesh Agreement to Establish the World Trade Organization – and in other subsequent WTO legal instruments. The WTO Appellate Body has also stated that the explicit acknowledgment and recognition of the objective of sustainable development in the WTO Agreement's preamble showed that 'the signatories to that Agreement were, in 1994, fully aware of the importance and legitimacy of environmental protection as a goal of national and international policy' and that this preambular recognition 'informs not only the GATT 1994, but also the other covered agreements' of the WTO Agreement.[2]

Developing countries fully endorse the objective of sustainable development and the protection and preservation of the environment in a manner consistent with their development needs and concerns as stated in the preamble of the WTO Agreement. They recognize that environmental protection is an important policy objective within the concept of sustainable development, given that the environmental space within which the development process takes place is an indispensable prerequisite to the start and continuation of such process. Furthermore, developing countries have traditionally stressed that the economic and human development objectives within sustainable development are also equally as important in view of the massive social and cultural impacts that low levels of economic development could have on national political and social cohesion and hence on their national integrity.

When applied to the WTO context, the concept of sustainable development encompasses:

- a recognition that the different economic conditions of developing countries require S&D treatment (including sufficient policy space and flexibility) with respect to WTO rules and obligations; and
- support in the achievement of their sustainable development objectives through the expansion of market access opportunities for their exports, the

* Excerpted from: UNEP (2005a). The original report, commissioned by UNEP and co-authored by Vincente Yu and Darlan Fonseca-Marti from the South Centre, underwent a government review and was discussed at a UNEP workshop in June 2005 (Chairs Conclusions can be found on the accompanying CD-ROM). Since this discussion paper was finalized in preparation for the Hong Kong 6 Ministerial Conference in December 2005, it only takes into account the discussion and submissions before this date. Given the sensitivities of the issues raised, and the relatively early point in the negotiating process at which this paper originally appeared, only some among a range of possible S&DT options to be considered in the course of the negotiations were included. These are necessarily broad and general and were intended to stimulate the discussion rather than present the final solution. The report has been slightly adapted for the purposes of this book and some footnotes have been omitted from this excerpt. The full original version of this report is included in the accompanying CD-ROM.

provision of adequate technical and financial assistance so that their trade-related economic development policies and activities reflect environmental sustainability considerations (including transfers of environmental goods and technologies).

Role of fisheries in developing countries

Fishing is an economic activity of crucial importance that serves a wide range of purposes in developing countries. Not only are these countries responsible for about half of total world exports of fish, but beyond trade fisheries play a fundamental environmental and social role in such countries.

Fisheries as a source of livelihood

It is estimated that around 30 million people directly derive their income from fishing activities and it is further estimated that about 95 per cent of that employment is located in the developing world. The contribution of fisheries to employment, economic security, social integration and social advancement can therefore not be overstated.

For example, the Food and Agriculture Organization (FAO) estimates that each fisher creates occupation for another three additional workers. Processing activities; such as loining, canning, smoking, sun-drying and fermenting, and marketing activities, employ many families after the fish are landed. That corresponds to about 120 million people whose income derives fully or partly from fishing, whether marine, inland or aquacultured. It is further argued that these figures should be considered as a conservative lower ceiling because they may not capture seasonal workers and many workers for whom fishing or fish trade are a complementary, not principal, source of income.

Moreover, fisheries are crucial for gender relations. In fact, women play a pivotal role in the preparatory work, such as making and mending nets, as well as in processing activities. In artisanal or small scale fisheries, women help their husbands in the boats during difficult economic times, help unload the fish, sort it, clean it and process it. In western Africa, women are predominantly responsible for smoking (using, for instance, traditional *chorkor* ovens), salting and drying the fish. Women are also predominantly responsible for the link between production and consumption, because of their role in marketing the fish.[3]

In commercial fisheries, women are also intrinsically involved in the loining and canning processes and are largely employed in factories. In some developing countries, women have become important fish entrepreneurs, particularly in aquaculture, generating income not only for their household, but very often for the whole community. In any case, the monthly earnings obtained from fisheries trade is sufficient to pay for the school fees of children and other family expenditures or, at least, to contribute to the household income. Hence, fisheries are a fundamental contributor to social integration and advancement, particularly for women (Touray, 1996).

Fisheries are also a heavy employer in densely populated countries. In fact, 84 per cent of fishers and aquaculturists in the world are located in Asia. Besides, the pronounced reliance of many developing countries in fishery activities makes fish the main or even the only source of livelihood for the bulk of the local population. The contribution of fisheries to employment and social stability makes it a sector of extreme strategic importance in many countries.[4]

Finally, in addition to the contribution of fisheries to employment, gender and livelihood, it is also worthwhile mentioning the crucial importance of fisheries for subsistence and rural livelihood of the very poor in developing countries for whom fisheries may be the last ditch before hunger and misery. The value of subsistence fisheries is unlikely to be captured in economic terms.

Fisheries contribute to food security

The contribution of fish to domestic food security implies that the sustainability of and long-term access to these fish stocks is of great importance to developing countries. Fish production and trade also contribute largely to household income in developing countries, making the sector a fundamental contributor to food security in the way that the revenue generated from the sale of fish and fish products allows many families to purchase other items of food. In addition, fishers separate lower-value fish species from their daily catches for their own consumption. Finally, the produce of this economic activity, fish, is also a widely recognised highly nutritious source of animal protein, vitamins and minerals.

About 76 per cent of world fisheries production in 2002 was used for human consumption and fish is in fact the staple food in many areas of the world (FAO SOFIA, 2004). While the world average apparent consumption of fish in 2002 was estimated at 16.2kg per person (FAO SOFIA, 2004),[5] the yearly apparent consumption of fish is 187.3kg in the Maldives, 91.8kg in Palau, 75.5kg in Kiribati, 57.6kg in the Seychelles and 44.1kg in Gabon (FAO, 2004d).

Very pronounced geographical and regional disparities characterize fish consumption. As a general trend, as revenues rise the consumption of fish also increases. Therefore the highest average apparent consumption is found in industrialized countries (28.6kg) and the lowest averages are found in Africa and the Near East (4.1kg/year/per capita in East Africa). However, while fish represents only 7.7 per cent of the total protein intake in rich countries, it represents 50 per cent or more of animal protein in several small and island developing countries including Bangladesh, Cambodia, Congo, Gambia, Ghana, Equatorial Guinea, Indonesia, Sierra Leone and Sri Lanka.

The strategic importance of fish as a staple food can be observed, for instance, in densely populated countries such as China, where the growth of fish production has outpaced population growth during the 1987 to 2002 period. While the population increase in China was 1.1 per cent, food fish supply rose by 8.9 per cent. For governments, fish can be seen as a cheap source of protein that can enormously contribute, even if consumed in small amounts, to the fight against food insecurity and hunger.

Fisheries as a source of export earnings

Not only do fisheries make an enormous contribution to livelihood, employment and food security, but they are also a main source of export earnings and foreign

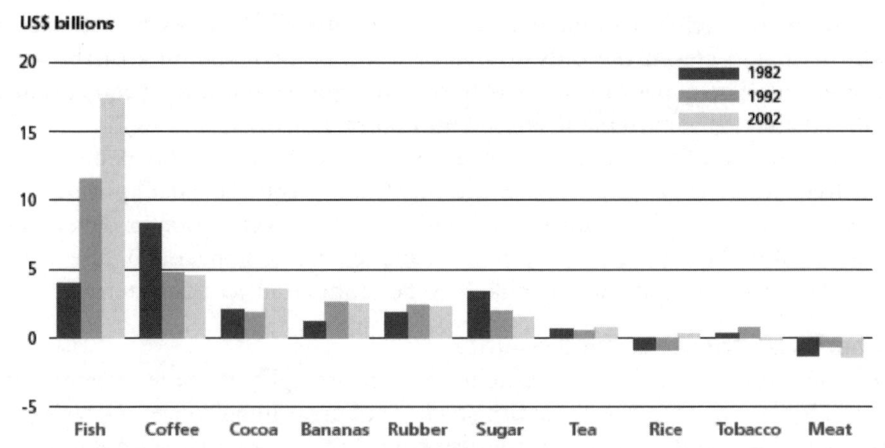

Figure 5.1 *Net Exports of Selected Agricultural Commodities in Developing Countries*

Source: FAO SOFIA (2004), Figure 31

exchange for a vast number of developing countries.

Although it is difficult to quantify the precise contribution of fisheries to national economies, there is clear evidence that earnings from exports are indeed very substantive. Developing countries are responsible for about half of the world production of fish and for 38 per cent of the production that enters international markets. They account for 49 per cent of fish exports by value and 55 per cent by volume. Low-Income Food-Deficit Countries (LIFDCs) alone account for 20 per cent of exports by value. Net receipts of fisheries foreign exchange (fish exports less fish imports) in developing countries are worth US$17.4 billion, or more than earnings from coffee, cocoa, bananas, rubber, sugar, tea, rice, tobacco and meat. Fish products are the single most valuable agricultural export[6] from developing countries as shown by Figure 5.1.

Apart from the weight and dynamism of the fisheries sector in the economy of developing countries as a whole, fish production and exports are often the major sources of export earnings. In fact, fish exports as a total of agricultural exports can be extremely high in several developing nations such as the Maldives (99.9 per cent), Seychelles (99 per cent), Angola, Tuvalu (96 per cent), Gabon (82 per cent), Bangladesh, Mauritania and Madagascar (over 70 per cent). Of course, for developing countries whose composition of exports is heavily dependent on only a small number of primary agricultural commodities, such high ratios can turn out to be a source of economic vulnerability. The ratio of fishery exports as a total of merchandise exports is above 20 per cent in several small island developing states as well as Bangladesh, Namibia, Senegal and Panama (FAO, 2004d).

The contribution of fish production to developing countries' GDP is estimated to be significant, despite the difficulties in calculating its exact value. The contribution is very high in small island developing states (33.56 per cent for the Maldives) and can be as high as 10 per cent in Cambodia (capture fisheries only) and 5.77 per cent in Laos (aquaculture only) (FAO, 2004c). Shrimp production employs about 100,000 workers in Madagascar and corresponds to 7 per cent of GDP (World Bank, 2003). Nevertheless, it is widely accepted that this contribution could be even higher since problems of data reliability difficulties in capturing the economic value of subsistence fishing produce inaccurate estimates.

To these earnings one must also add the substantial earnings derived from compensatory fees paid under access agreements (access fees) that many developing countries have concluded with Distant Water Fishing Nations (DWFN). Several nations in Africa and the Pacific region have settled such agreements mainly with the EU, USA, Japan, Taiwan and Korea. The aggregate value of such agreements concluded by the EU alone is estimated at around €170 million[7] for a single year and their weight within poor nations' overall government financing is often very high. The contribution of financial transfers under access agreements represented more than 60 per cent of overall government revenues in Guinea Bissau.

However, as Table 5.1 above indicates, there continue to be fewer developing countries than developed countries in the top 20 exporters of fish commodities. Of the top 20 exporters in 2002, only 8 were developing countries, accounting for 30.40 per cent of total world fish commodity exports (down from 31.25 per cent in 2001). 12 were developed countries accounting for 41.10 per cent (up from 40.45 per cent in 2001). Combined, these top 20 exporters accounted for 71.50 per cent of total world fish commodity exports in 2002.

Therefore, while developing countries are now increasingly becoming major exporters of fish products, in competition with both developed and other developing countries for access to global export market opportunities, developed countries continue to capture a dominant share of the global fish commodities export market.

Coastal areas, particularly in Small Island Developing States (SIDS) or in states with extensive marine boundaries, are usually very densely populated areas (true of Africa, Asia, Latin America and the Caribbean) often with fragile social and ecological environments. Protection and management of those areas are extremely important to ensure sustainable human and economic development.

In such areas, the cultural heritage associated with the sea and fisheries is extremely rich. Coastal ecosystems (for example, mangroves, reef atolls) are not only environmentally fragile but also fundamentally important beyond fisheries. For many poor developing countries, including Least Developed Countries (LDCs), 'sun and beach' and cultural tourism contain huge developmental potential.

In some developing countries, such as the Maldives, where tourism and fisheries are the two most important, and almost exclusive, economic activities, the protection of coastal ecosystems and adequate management of fish stocks are of vital importance.

However, many developing countries so far have lacked institutional, financial and human capacity to implement proper management and monitoring schemes of fisheries. Investment in technological improvement and innovation in traditional capture techniques and distribution can yield noticeable economic

Table 5.1 *Top 20 exporters of fish commodities in 2002 compared to 2001 (US$ 1000)*

Country	*2001*	*2002*	*%*
1. *China*	3,999,274	4,485,274	+12.2
2. *Thailand*	4,039,127	3,676,427	-9.0
3. Norway	3,363,955	3,569,243	+6.1
4. United States of America	3,316,056	3,260,168	-1.7
5. Canada	2,797,933	3,035,353	+8.5
6. Denmark	2,660,563	2,872,438	+8.0
7. *Vietnam*	1,781,385	2,029,800	+13.0
8. Spain	1,844,257	1,889,541	+2.5
9. *Chile*	1,939,295	1,869,123	-3.6
10. Netherlands	1,420,513	1,802,893	+26.9
11. *Taiwan Province of China*	1,816,865	1,663,821	-8.4
12. *Indonesia*	1,534,587	1,490,854	-2.8
13. Iceland	1,270,493	1,428,712	+12.5
14. *India*	1,238,363	1,411,721	+14.0
15. Russian Federation	1,528,022	1,399,369	-8.4
16. United Kingdom	1,306,042	1,353,123	+3.6
17. Germany	1,035,359	1,156,911	+11.7
18. France	1,018,843	1,088,572	+6.8
19. *Peru*	1,213,112	1,066,654	-12.1
20. Korea, Republic of	1,156,132	1,045,672	-9.6
Sub-total of Top 20 exporters	**40,280,176**	**41,595,669**	**+3.3**
Share of Sub-total of Top 20/World	71.7%	71.5%	
Sub-total of Top developing country exporters	**17,562,008**	**17,693,674**	**+0.75**
Share of Top developing country exporters/World	31.25%	30.40%	
Sub-total of Top developed country exporters	**22,718,168**	**23,901,995**	**+5.21**
Share of Top developed country exporters/World	40.45%	41.10%	
World total	**56,194,631**	**58,211,139**	**+3 .6**

Note: Italicized country names refer to developing countries.
Source: FAO (2004d)

and income gains without intensifying the current fishing effort. For instance, although artisanal fishing is known for minimal waste of fish, post-harvest losses can be further reduced by improving sanitary, processing, conservation and transport conditions.

Similarly, techniques such as the use of bottom trawls, that are not adapted to the tropical water seas of developing countries, can be discouraged. There is therefore a need to induce commercial fisheries to feel ownership and responsibility over fish stocks through management and conservation programmes. Developing country governments therefore have a central role to play in the expansion of their fishing activities by focusing not simply on growth of production, but also on implementing management policies to ensure the long-term sustainable exploitation of their stocks.

Prospective importance of fisheries

Despite the impressive figures about the dynamism of fisheries trade in developing countries, much of the potential contained in that sector has still to materialize. Fisheries have in fact great potential to lift a huge amount of people out of under-nourishment and poverty, and the prospects for expansion of production in many developing countries remain very large.

Recent growth of production and exports by many non-traditional developing country producers reveal that potential. Some of the fastest progressions in production (for both aquacultured and captured fish) between 2001 and 2002 were recorded in countries with limited tradition in fish exports, such as Iran, Laos, Brazil and Myanmar (FAO, 2004d).

The growth of fish exports from developing countries whose participation in international fish trade had been modest in the past indicates that the arrival of new entrants in commercial fishing activities could lead to growing competition for fish markets. More specifically, it could be a sign of growing competition among developing country producers for shares in high value markets (mostly, developed countries).

Fishing, particularly artisanal fishing, is an industry that requires relatively little investment and technologies, or investment and technologies that can be accessed relatively easily. Therefore, to the extent that fish stocks are sustainably managed and not overexploited, and to the extent that the kinds of fishing activities undertaken are both environmentally sustainable and economically beneficial, expanding fishing activities may be a means through which developing countries may diversify their economically productive sectors and thereby reduce their vulnerability resulting from dependency on only a few sectors or on primary commodities.

In addition, since fish is a highly perishable food product, it is predominantly traded in its processed form. This requires investment in infrastructure (refrigeration, smoking, canning and packaging), transport, market research and distribution. Such value-added activities may have very positive spill-over socio-economic effects. Moreover, cheaper operating and labour costs in developing countries may also constitute a significant drive for foreign direct investment and transfer of technology. Such is already a trend in some regions and certainly has the potential to extend to other developing countries.

The strategic future role of fisheries in the fight against poverty is unveiled by governmental and international efforts to mainstream fisheries in national developmental policies. International aid, environment and development agencies such as the World Bank, UNEP, FAO and the EU already cooperate with several poor developing countries to integrate fisheries in national development plans. Successes have been recorded for instance in the Maldives, Seychelles, Fiji and Saint Lucia (Thorpe et al, 2004).

Relevance of the fisheries subsidies negotiations for developing countries

The discussions above identify some of the reasons why fisheries can be so central to sustainable development. Consequently, international rules that impinge on the trade of fish products have potential important consequences for developing countries. There are two aspects of fish trade that are currently under negotiations of the Doha Work programme (DWP): market access and subsidies disciplines.

The first aspect, negotiated in the WTO Negotiating Group on Market Access (NGMA), concerns both the tariff treatment that is given to fish as well as non-tariff barriers (NTBs) that may operate as effective obstacles to the export of fish products. NTBs may include, among others, rules of origin, aspects of customs administration and abuse of anti-dumping, sanitary and phytosanitary (SPS) measures and technical barriers to trade (TBT). The negotiating mandate for market access is provided in Paragraph 16 of the Doha Ministerial Declaration (DMD).

The second negotiation that directly relates to fisheries is held in the WTO Negotiating Group on Rules (NGR) and concerns the adoption of improved disciplines to regulate subsidies that governments grant to the fisheries sector. The negotiating mandate to discuss subsidies is provided by Paragraph 28 of the DMD and includes an explicit reference to the need to regulate fisheries subsidies given the importance of that sector for developing countries. This section does not cover the market access aspect of the DWP and concentrates only on the latter negotiations, fisheries subsidies.

The fact that ministers have given particular consideration to fisheries within the negotiations for improved disciplines on subsidies opens a considerable window of opportunity to make a meaningful contribution to establishing sustainable trade in fish and fish products. There is a real opportunity to demonstrate that a 'win-win-win' outcome can be negotiated in the WTO, that is, the negotiated outcome could provide positive results for: (i) the development prospects of developing countries (through, for example, more operational S&DT in fisheries subsidies); (ii) for the fishing environment (for example, disciplines providing appropriate incentives for the reduction of fishing capacity and fishing effort and for the implementation of improved management of stocks) and (iii) trade (for example, increased shares by developing countries in global fish exports). Whether WTO members remain faithful to the developmental promises of the Doha Declaration, particularly in fisheries, will be crucial in the assessment of the benefits of the DWP for developing countries.

Developing countries' concerns can be incorporated in the fisheries subsidies negotiations in two ways. The first relates to the design of improved disciplines that promote more sustainable fisheries by effectively curbing effort and capacity-enhancing subsidies. This would hopefully create new market access opportunities for developing countries and would provide an incentive for more competitive industries in subsidizing developing countries. The second involves the incorporation in such disciplines of specific developmental needs in the form of operational special and differential treatment provisions so that, whatever new rules are agreed, they are not incompatible with the implementation of development policies.

Curbing trade distorting and capacity enhancing subsidies

Market distortions through the use of subsidies, and particularly complex subsidies, by some countries contribute to distortions in world fish prices and give an artificial competitiveness to subsidized fleets and multinational fish corporations. As a result, artificially competitive producers (mostly from developed countries) are those who benefit most from the levels of consumption and prices prevailing in high-value markets. These subsidies impact not only on capacity and fish stocks, but also distort market access opportunities of developing countries and have thereby a direct consequence for human development in those regions.

Moreover, subsidies have led to overinvestment and to a global, aggregate fleet capacity that is well beyond sustainable levels. Chapter 2 analysed the impact of fisheries subsidies under a variety of management and bio-economic conditions, and concludes that 'most subsidies have the potential to be harmful to fish stocks particularly in the absence of effective management...Subsidies that contribute directly to increased fishing capacity or effort are among the most harmful'. The idea that fisheries subsidies could lead to overcapacity and overexploitation seems to be widely accepted in the Negotiating Group on Rules and now needs to be translated into well crafted and tight disciplines. The Ministerial Declaration adopted at the Hong Kong WTO Ministerial in December 2005 for the first time explicitly links fisheries subsidies to overcapacity and over fishing and acknowledges the need for addressing this link.

Furthermore, complex mechanisms make certain fisheries subsidies harmful for fish stocks despite their innocuous or even positive design. For instance, it is known that poorly monitored and managed decommissioning programmes ('buy-backs') may lead to an increase in fishing capacity in domestic waters and the export of fishing capacity to foreign waters. Grants to support research and development of fishery technology may also result in increased fishing capacity. Marine insurance, management services and government-to-government access fees can provide significant benefits to the recipient fisheries industry. Because of their complexities, such instruments might have to be discussed in the negotiations so as to achieve rules that prevent 'box shifting'.[8] The Negotiating Group on Rules is increasingly aware of these complexities.[9]

In that context, the challenge facing negotiators is to establish improved disciplines on fisheries subsidies. The new disciplines should be efficient in curbing harmful subsidies (for example, capacity-enhancing programmes) and may also provide incentives for the establishment of stock management schemes.

Some WTO members have proposed a broad-based prohibition, elimination and reduction of fisheries subsidies, particularly the most capacity enhancing and trade distorting of them.[10] Other WTO Members agree that there is a need to ensure that the approach used for the identification of prohibited subsidies be effective in not creating loopholes in the final disciplines. For developing countries, whose financial, human and institutional capacity to engage in the dispute settlement mechanism is very limited, a broad-based prohibition of harmful subsidies would likely be the most efficient way of tackling trade distorting and capacity enhancing subsidies.

However, while developing countries, which are already actively engaged in international fish trade, would immediately benefit from a broad reduction of fisheries subsidies, new disciplines should still provide for flexibility in order to accommodate legitimate subsidy programmes that developing countries may need in order to pursue their environmental and developmental objectives.

For instance, programmes designed to monitor and manage resources, to retrain fishermen affected by the decommissioning of vessels or to fight against Illegal, Unregulated, or Unreported fishing (IUU) should ideally not be captured under a broad prohibition. Developed and developing countries alike would benefit from the recognition of the legitimacy of such subsidies, which could, for instance, be explicitly enumerated in a list of authorized subsidies.

Similarly, for many developing countries whose fisheries resources are largely under-exploited, subsidies may still have a role to play in supporting small-scale and artisanal fishing, creating fishing capacity where it currently does not exist and promoting local food security. Some developing countries have in fact suggested that new subsidies disciplines should be compatible with poor countries' development needs. This would, to some extent, include authorizing an increase in the fishing capacity of developing countries.[11] Such flexibilities could be provided for as special and differential treatment in favour of least developed and other developing countries.

It must be noted, however, that some developing countries, and particularly those that have a well-established fishing industry, might not necessarily favour the adoption of S&D provisions that allow other developing countries to expand their fishing capacity. The authorization of subsidies represents a distortion of trade conditions for the developing countries that are already engaged in international fish trade. Likewise, these developing countries that do not wish or cannot grant subsidies will face increasing competition from other developing countries for high value markets.

The complexity of the issues involved will require S&D provisions that are simple and operational, but which capture the divergent views of developed and developing countries.

Making sustainable development-based S&D operational in the fisheries subsidies negotiations

In discussing specific S&D provisions within the fisheries negotiations, it is worth undertaking a short review of S&D treatment in the WTO and the current Subsidies and Countervailing Measures (SCM) Agreement. This short review provides lessons about systemic failures regarding the operationalization of S&D. In fact, while the scope of S&D discussions in fisheries subsidies is and should remain specific to the fisheries context, drawing on past experience may offer an improved conceptual S&D framework, which can be used as a guideline for the crafting of new S&D provisions in the fisheries subsidies context.

Developing countries' experience with S&D in the WTO

S&D treatment is a fundamental and important part of the multilateral trading system and its legal framework under the WTO. This was explicitly reaffirmed by the WTO Ministerial Conference at Doha.[12] Its importance within the fisheries subsidies framework has been acknowledged in Paragraph 28 DMD and subsequently by several WTO members in their submissions to the negotiating group.

Therefore legally, S&D is not an exception to the application of multilateral trade rules provided for in the various WTO agreements. The fundamental premise of S&D is that countries continue to be at varying levels of economic development, with different economic needs, and should therefore have varying degrees of obligations commensurate to their levels of economic development. It is intended to achieve a key specific objective – that of providing a fair playing field for all WTO members in which the rules are adjusted to take different capacities and levels of development among the participants into account. The indiscriminate application of single rules to players with unequal abilities only tilts the playing field in favour of players who are more capable of playing – whether by virtue of economic or political strength – than others.

At its core, therefore, S&D is about creating a different set of multilateral trade rules crafted to meet and be commensurate with the needs of developing countries. These rules would be applicable to developing countries while they are still 'developing'. This different set of rules for developing countries could be about safeguarding their policy space and options to adopt and implement trade, economic and development policies.[13]

However, many developing countries have faced difficulties with respect to these S&D provisions. Some of them have pointed out that, under the WTO, the focus of S&D has shifted from addressing the issue of promoting economic development into assisting developing countries to implement their multilateral trade commitments more effectively (through mere grants of transition periods and technical assistance).[14] Others have also stressed that they 'could hardly benefit from the almost 145 S&D provisions (in the Uruguay Round Agreements) which mostly do not go beyond a best endeavour promise and therefore are not legally enforceable. Lack of any mechanism to ensure effective implementation of S&D provisions in the WTO has [also] been a major concern...'.[15]

These broad concerns of developing countries regarding the inefficacy of current S&D provisions in the WTO legal regime in supporting and promoting their development needs are at the core of the negotiating mandates on implementation-related issues and on S&D established by the WTO Ministerial Conference at Doha in 2001. However, these are now at a virtual standstill as members diverge on fundamental issues regarding the mandates.

Developing countries' experience with S&D in the SCM agreement

The implementation of the SCM Agreement seems to indicate that the S&D provisions contained therein have not provided the necessary flexibility for developing countries. This is shown in, for example, the general discussions relating to implementation-related issues and concerns on S&D.

Issues relating to the S&D provisions of the SCM agreement – Article 27 – figure prominently among the SCM agreement-related issues raised by developing country members in the WTO from before the 1999 Seattle ministerial conference to the present.[16] This is a sign that many developing countries have been dissatisfied with the implementation of S&D provisions contained in the agreement.

Some of the problems mentioned above stem from internal contradictions between Article 27 and the rest of the SCM. While this article states that subsidies may play an important role in developing countries' economic development, it also severely restricts the right of developing countries to use subsidies as a developmental policy instrument while further requiring the phasing out of many subsidies that developing countries are actually providing.

In addition, it has been argued that the nature of the SCM Agreement itself is such that S&D provisions can only have a limited scope. The definition of subsidies provided in Article 1 and the illustrative list under Annex I of the SCM Agreement may operate to create an imbalanced situation by forbidding subsidies that could be available to and accessible for poor and resource-constrained countries and authorizing those subsidies that can be provided only by richer, essentially developed, countries.

Furthermore, although S&D provisions under Article 27 generally contain the wording 'shall' and could thus be arguably construed as being mandatory in nature, the imposition of complex eligibility criteria for certain paragraphs (for instance, Art.27.4), the existence of exceptions where the provisions do not apply (for instance, Art.27.9) and the temporary nature of most of the flexibility allowed, may militate against making Article 27 fully effective, useful and operational for developing countries.

Among the improvements suggested by developing countries, proposals have been submitted with a view to:

- expand the range or extent of subsidies that may be provided by developing countries;
- expand the range or number of developing countries that may be allowed to provide subsidies;
- lessen the vulnerability of developing countries providing subsidies to WTO dispute settlement proceedings; and
- minimize the vulnerability of developing countries providing subsidies to the imposition of countervailing measures by other WTO Members.

In sum, many developing countries perceive the SCM Agreement and its S&D provisions to be inadequate in providing the necessary flexibility for policy instruments that could be used by resource-constrained developing countries to address their development needs. Moreover, the threat of a dispute within the context of the Dispute Settlement Understanding (DSU) lurks and has an inhibiting impact on developing countries intending to craft and provide new subsidies. Particularly in view of the importance of the fisheries sector to developing countries, the mandate to improve disciplines on fisheries subsidies provides a timely opportunity to address some of the asymmetries within the SCM Agreement through improved S&D provisions.

Possible elements for S&D treatment in the SCM agreement

In light of past experience with S&D and in light of the suggestions for improvement of the SCM submitted by developing countries, key elements for crafting more effective and operational S&D can be derived. The advantage of identifying key elements for S&D is that these elements may provide the skeleton that can then be fleshed out according to the specific concerns that need to be taken into account in a particular sector. They can also be incorporated in any of the approaches proposed to be used in the current fisheries subsidies negotiations ('top down' or 'bottom up'), while at the same time providing sufficient guidance as well as flexibility for negotiators to discuss and fine tune any of the operational details thereof.

The following, which are based on various developing country submissions to the WTO, might be seen as possible key elements for S&D in the SCM agreement:

- positive policy space and flexibility;
- positive impact in terms of increased market access opportunities
- mandatory applicability and enforceability;
- positive cooperation measures; and
- assessment-based and review-dependent implementation.

The maintenance and expansion of domestic policy space and flexibility is a core element of S&D. Operationally, this could be in the form of a lower level of obligations for developing countries, commensurate or adjusted to their current level of economic development, so as to afford them the necessary flexibility to pursue viable development-oriented policy options and encourage institutional and economic innovations capable of fostering industrialization, economic development and social advancement. This could also be reflected in a modest level of expectations with respect to their application and implementation of various multilateral trade obligations and commitments. In addition, longer and qualitatively better transition periods for the implementation of new commitments or obligations by linking the expiration of such periods to objective economic (for example, debt level, level of industrial development, human development index, etc.), social (for example, literacy and life expectancy) and environmental sustainability (for example, environmental and natural resource protection and conservation) criteria.

The utility of S&D provisions for developing countries depends largely on whether such provisions are mandatory in their application and enforceability. Hence, specific S&D provisions could be couched in mandatory language compliance with which can be enforced by developing countries, if needed and as appropriate, through the WTO's dispute settlement system (including notification requirements and the inclusion of these commitments in country schedules).

Positive cooperation among participants in the multilateral trading system would also be a key element in making S&D operational. This can take the form of, for example, the provision of additional, adequate and predictable financial and technical assistance and capacity-building support from developed to developing countries both in relation to the implementation of new rules or obligations and in ensuring that sustainable development benefits accrue from such implementation. The establishment and implementation of effective and operational positive technology transfer obligations from developed to developing countries could also be an important contribution in this area.

Finally, the different levels of economic development of developing countries need to be reflected in the multilateral trade regime as part of the S&D framework, by making the initial and continued implementation of new trade rules and obligations for developing countries dependent on positive evaluations from a pre-implementation assessment exercise and a periodic implementation review process.

For example, the initial application of new multilateral trade rules or obligations to developing countries could be conditioned on a positive evaluation. This evaluation could be based on a prior assessment of the development impact and implications of such new rules or obligations on developing countries (covering social, economic, environmental impacts and implementation costs). Such prior impact assessment could include looking at how such new rules or obligations facilitate the attainment of defined sustainable development targets. A prior evaluation of the implications of such new rules or obligations with respect to the implementation costs for developing countries is very important. Such prior evaluations could become the basis for determining the extent to which S&D provisions can be availed of and enforced by developing countries.

In addition, the legal and administrative requirements and procedures for making use of such S&D procedures need to take into account and reflect the fact that many developing countries are resource-constrained from the financial, human resource or technical perspective.

Furthermore, the continued application by developing countries of both S&D provisions and their general multilateral trade obligations could be subjected to a periodic implementation review mechanism that will look at the economic, social and environmental impact of such trade obligations at the national level with the aim of ensuring that such impacts contribute positively to the sustainable development prospects of the country concerned. Finally, a general review mechanism would be essential as a forum through which systemic implementation-related issues can be addressed and effectively resolved at the multilateral level.

Applying the suggested S&D elements to fisheries subsidies

Possible elements for fisheries subsidies-specific S&D

Applying the elements identified above to the fisheries subsidies negotiations, the primary objective of S&D treatment within new subsidies disciplines might therefore be to provide flexibility for developing countries to grant subsidies in the pursuit of their development priorities. Fisheries subsidies S&D treatment may then consist, among other things, of granting developing countries access to instruments that would otherwise be prohibited or actionable under the general disciplines agreed. S&D subsidies would be distinct and separate from other allowed subsidies, such as those contained in a possible 'Green Box'.

To safeguard the sustainability of global fisheries, however, such S&D should not just allow developing countries to provide any type or any amount of subsidies to their fisheries sector. Such a *carte blanche* approach may foster a 'race' to subsidization that could have detrimental effects on fish stocks. Any beneficial development effect that providing fisheries subsidies might have for a developing country would ultimately be negated if, in doing so, domestic fish stocks are depleted.

Moreover, it must also be noted that too wide an authorization of subsidization may also put at disadvantage those developing countries that either cannot (for example, because of their limited financial capacity) or do not wish to grant subsidies to their fishing industry. In that sense, WTO members designing new S&D provisions may wish to have regard to the differences between developing countries whose financial capacity and production potential differ enormously.

With this in mind, the fisheries subsidies S&D provision would then need to be focused on assisting developing countries to sustainably develop and manage their viable fish stocks in support of their development objectives, while at the same time ensuring that a harmful increase in capacity and the overexploitation of fish stocks as a result of subsidization programmes do not occur. For instance, it could be decided that programmes should be discontinued depending on certain economic and environmental conditions.[17]

Hence, in designing an appropriate S&D package in the context of the fisheries subsidies negotiations, and taking into consideration both the elements of the S&D conceptual framework discussed above as well as the general thrust of the proposals of developing countries with regard to the SCM Agreement in general, key elements of the fisheries subsidies-specific S&D package may include the following:

- *Market access*: the policy flexibility to adopt measures (including subsidies) designed to enable developing countries to take advantage of possible increased market access opportunities in other WTO members' markets while at the same time ensuring that their fisheries are managed in a sustainable manner. This would be particularly important for WTO members who have not yet developed their fishing industry to a level commensurate to their economic needs.
- *Sustainable development*: the necessary policy space to support fishing activities with a view to promoting sustainable development-oriented policy objectives, such as the alleviation of poverty in poor regions, the promotion of food security, sustainable utilization of resources, the organization of small-scale and artisanal fishers into cooperatives, improvement of transport and processing infrastructure, etc.
- *Conservation and management*: the policy flexibility and the resources needed to conserve, sustainably manage and develop the fisheries resources in their waters and on the high seas. This could include, for example,

measures that may limit access to specified fish stocks by fishers, impose certain landing, administrative, technical, or other requirements on fishers, measures against IUU, etc.; and

- *Technical and financial assistance*: the inclusion of positive measures that would require WTO members to provide, in a long-term, sustainable and adequate manner, technical cooperation and financial assistance to developing countries seeking to put in place within their waters effective and sustainable fisheries resource conservation and management regimes.

For the operationalization of these objectives, new WTO fisheries disciplines should therefore accommodate the policy instruments that are needed by, are useful for and are accessible to developing countries. The sustainability of the system would however require an arrangement that is fair for all developing countries, particularly those that will choose not to grant subsidies, and that provides an incentive for the improved management of stocks.

The instruments that would be allowed under the new S&D treatment could be drawn from a list and justified because of their design or public policy purpose. In that sense, S&D-authorized subsidies would be distinct and separate from subsidies authorized under a possible 'Green Box'. In fact, S&D is an instrument that covers the specific needs of developing countries (development), which provides a distinct basis for justification if compared to the 'Green Box' (for example, conservation, management, effort reduction). This reasoning would also justify the fact that only developing country members of the WTO would have access to the subsidies provided under S&D.[18]

Finally, S&D provisions should ideally be simple and transparent and they should also, as for the rest of the disciplines in fisheries subsidies, be compatible with the WTO trade mandate and expertise.[19]

The following is an illustrative, non-exhaustive list of the instruments that could be authorized under S&D treatment, compiled from negotiating proposals submitted by developing countries to the Negotiating Group on Rules as of July 2005:

1 subsidies for infrastructure development and construction, prevention and control of disease, scientific research and training and retraining of fishers skills;
2 subsidies or fiscal incentives for domestication and fisheries development;
3 support for the development of small-scale, artisanal fisheries sectors, provided that the fisheries resources accessible to small-scale, artisanal fishers are not threatened by the fishing activity;
4 payments received from other governments for access to the EEZ fisheries resources of the developing country; or to its quotas or any other quantitative limits established by a Regional Fisheries Management Organization;
5 development assistance to developing coastal states;
6 assistance to disadvantaged regions within the territory of a developing country pursuant to a general framework of regional development in the sense of Article 8.2 (b) of the SCM agreement;

7 emergency relief and adjustment to small-scale, artisanal fishers suffering significant loss of income as a result of reductions in fishing caused by conservation measures or unforeseeable natural disasters;
8 subsidies which increase fishing capacity or effort; and
9 fuel, bait or ice supplied for fishing activities.

Possible options to make fisheries subsidies S&D treatment operational

As pointed out above, the need for appropriate S&D treatment stems from the necessity that new disciplines on fisheries subsidies do not impair the current or future ability of developing countries to support their fisheries albeit in an environmentally and economically sustainable manner.

Taking into account the elements suggested above, some of the options available to incorporate S&D treatment in new disciplines for fisheries subsidies are spelled out below. Some of the advantages and disadvantages of each option are discussed. Different combinations of the options presented could also be envisaged since these options are neither mutually exclusive nor exhaustive.

Option 1: Definition of a 'subsidy'

One possible way would be to clarify, in the context of fisheries subsidies disciplines, exactly what kind of fisheries-related government transfers relevant for developing countries would fall under the existing definition of 'subsidy'.[20] For instance, government-to-government access fees paid under access agreements might not represent a subsidy if the agreed new definition of subsidy covers only government-to-industry direct transfers. Similarly, public investment in fishery-related infrastructure might not be seen as being a subsidy if a new definition covers only payments that benefit only an individual industry, and not all economic actors.

While this option may seem attractive because of its simplicity, such an approach may present several shortcomings in effectively disciplining subsidies that have negative effects over trade, fisheries resources and the environment. Firstly, this option is likely to considerably weaken the SCM agreement by introducing exceptions to the definition of subsidy. Secondly, it would leave many harmful subsidies totally unregulated (beyond the scope of the agreement) and could hence undermine the positive impact that new subsidies disciplines could have on the environment. Furthermore, there might be difficulties in clearly delineating which government transfers relevant *only* for developing countries would be covered by this approach, and which could give rise to implementation difficulties on the part of developing countries.

Option 2: 'Prior authorization' regime

Another option to operationalize S&D flexibilities in favour of developing countries could be through the establishment of a prior authorization requirement. Developing countries planning to implement a fisheries policy that includes subsidies would have to seek prior agreement of the WTO's SCM Committee before actually implementing the programmes. A set of minimum requirements concerning the information to be provided could be negotiated. The subsidies that may then require previous authorization could be drawn from an exhaustive list agreed in advance. For instance, one developing country could apply for a temporary authorization of a capacity-enhancing subsidy. Other WTO members could request information about the condition of the fishery and the management scheme maintained by the requesting member and could also enquire about the public policy objectives that prompt the design of a subsidy.

A variation of this system could include a list of subsidies for which there would only be enhanced notification requirements. Non-notified subsidies would be deemed to be harmful or prohibited. These subsidies would remain actionable under the new disciplines.[21]

While this option may present advantages for a global control over harmful subsidies, it would represent many challenges for the WTO and its members. Such a mechanism could risk forcing the SCM committee to undertake its own assessment of the quality and appropriateness of members' fisheries policy, an activity that might not necessarily be within its mandate or its technical competence. Moreover, the system may impose a significant administrative burden for the organization and its members, and the number of requests could become unmanageable.

Finally, such a mechanism may place too heavy a burden – both administrative and political – on developing countries, and particularly small developing countries. Moreover, the complexity of the system could operate as a dissuading factor, offsetting its potential benefits. Similarly, developing countries could also be disadvantaged in a dispute settlement scenario because of their limited human, institutional and financial resources.

Option 3: Positive list approach to subsidies that developing countries may apply

Other options to create built-in flexibilities in the new disciplines could consist in drafting and agreeing to a 'positive'[22] and exhaustive list of subsidies that developing countries would be authorized to apply. Such a list of subsidies could be directly linked to public policy objectives specific to developing countries, such as support for food security, subsistence and artisanal fishing and emergency actions.

This option presents the advantage of being easy to implement and manage. However, it places a considerable negotiating burden on developing countries. A related risk is that negotiations result in an overly restrictive list, which does not cover the full range of policy instruments that would be compatible with developing countries' specific needs.

Moreover, as regards the sustainability of the new disciplines, without further conditionality such a system could still lead to over-capacity and possible depletion of stocks because it is not contingent on an assessment of the impact of subsidies over fish stocks. Fisheries that are already overexploited could be further harmed by overinvestment, since the use of such flexibilities would be unlimited within the agreed list. Given the social, economic and environmental importance of fish in developing nations, such a system could jeopardize efforts to deliver more responsible global fish trade and production.

To prevent or address any undesirable impacts on the sustainability of fish stocks as a result of the provision of positive list subsidies, WTO members could consider: (i) the adoption of a maximum amount of subsidies over a given period of time; or (ii) the adoption of a *post-hoc* impact assessment mechanism.

In the first option, WTO members could agree on a quantitative limit (ceiling or '*de minimis*') for the use of the flexibilities, such as an overall, aggregate level of subsidization as a share of the total value of production over a period of time. For example, developing countries could be allowed to subsidize up to 30 per cent of their annual production of a specific fish product or all fish products.

It must be noted, however, that the limitations created by such an option are quantitative, not qualitative. In other words, this type of limitation is not contingent on the state of fish stocks. One option to improve the environmental sustainability aspect of the positive list approach would be to establish a mechanism for technical assessment of the impact of the provision of positive list fisheries subsidies on fishing capacity and effort, and their consequent impacts on fish stocks.

In the second option, WTO members could agree that a well-recognized international organization with technical competence in this area develops and implements an impact assessment mechanism to review periodically the environmental state of fish stocks and the level of fishing capacity and effort in countries providing positive list subsidies. In particular, the assessment could also look at the extent to which positive list subsidies have a positive or negative impact on the sustainability of fish stocks as a result of increases in fishing capacity and effort.

Ideally, the results of the assessment should be accepted by the WTO and not be open to challenge from WTO members in any WTO forum (including WTO dispute settlement), so as to avoid members stepping beyond the WTO trade mandate. Of course, the quality of this option would depend on the technical quality of the assessment undertaken and on WTO members' wide recognition of its validity.

A technical difficulty with such an assessment involves the establishment of a clear causal link between maintained subsidies, increases in fishing capacity and effort and the reduction of fish stocks. A related technical difficulty is a lack of data for certain regions and the regional differences in the quality of the data. The political difficulty is that such an assessment would require new institutional arrangements (see below).

Option 4: '*De minimis*' approach

WTO members may also consider the adoption of a '*de minimis*' level of aggregate amount of support within which developing countries can freely maintain fisheries subsidies programmes. In addition to fisheries subsidies that may fall inside any agreed-upon 'Green Box', developing countries would be free to provide fisheries subsidies as long as the aggregate amount of support granted does not exceed an agreed '*de minimis*' authorized level (ceiling). Fisheries subsidies authorized under the *de minimis* S&D provision could either be drawn from a positive list, from the whole universe of subsidies available outside of the 'Green Box', or from the whole universe of subsidies except for subsidies contained in a negative list.

The aggregate amount of support may be defined either as an absolute amount, for instance, US$100,000, or as a proportion of the total value of production, for instance, 20 per cent. The absolute amount approach would be necessarily arbitrary and may not capture the actual future needs of developing countries. The proportion approach imposes a greater restriction for developing countries whose total production is very low, while advantaging countries whose production is already larger.

Members would have to decide whether subsidies beyond the *de minimis* level would be prohibited or authorized under certain conditions (criteria approach).

Despite the simplicity of this approach, it may still lead to overinvestment and overcapacity and thus jeopardize the sustainability of fisheries, since access to S&D subsidies would not be contingent upon the existence of a management scheme or the actual state of fish stocks.

Finally, the efficacy of this approach is also greatly dependent on the availability of information about developing countries' overall production and about the amount of subsidies maintained in a given period. Only if this information was easily available would other members be able to monitor whether the *de minimis* level is being respected. Consequently, this option would require an enhanced and efficient notification mechanism. Similarly, this option would constitute a greater burden for all other members since its performance rests greatly on pro-active 'peer review' and pressure.

Option 5: Sustainable development-based S&D eligibility criteria approach

Finally, WTO members may consider crafting a list of criteria or conditions that would have to be met by developing countries to access S&D-related flexibilities. This mechanism would be compatible with aspects of the current ASCM S&D structure, which creates categories of developing countries depending on certain criteria (see Annex 11). However, because of the nature of fisheries, WTO members may consider extending the agreed set of criteria to cover both the environmental and trade aspects of fish production and trade.

The eligibility criteria would have to be designed with the objective of ensuring that: (i) as large a number of developing countries would be rendered eligible to apply S&D with respect to fisheries subsidies; and (ii) it reflects both economic development and environmental sustainability elements so as to ensure that S&D measures provide both environmental and developmental benefits.

In addition, the S&D measures that eligible developing countries could apply for could be those subsidies that, depending on the agreement of WTO members in the course of the negotiations: (i) are included in an exhaustive or indicative list of fisheries subsidies for use by developing countries which would be available and non-actionable, or (ii) in addition to subsidies generally available to all WTO members under any agreed 'Green Box', all other fisheries subsidies that would otherwise be prohibited or actionable.

The following discussion could be an example of how to make the sustainable development-based S&D eligibility criteria approach operational:

- *Only developing countries as S&D beneficiaries*: only 'developing countries' may avail of fisheries subsidies-related S&D provisions (in the case of positive measures relating to the provision of technical assistance and cooperation, only developing countries may be the beneficiaries)
- *Sustainable development-related eligibility criteria for S&D beneficiaries*: in addition, developing countries may benefit from fisheries subsidies S&D provisions only if they meet the following suggested set of cumulative criteria:
  1. The fisheries stocks in their internal waters, territorial sea or EEZ are not overexploited, depleted or recovering – in other words 'patently at risk'.[23] The determination of such a situation could be based on the result of an independent assessment by a well-recognized, competent international agency.
  2. They can present a national fisheries resource management regime[24] pertaining to fisheries stocks within their internal waters, territorial sea or EEZ that is being or will be implemented and is based on or conforms to the criteria set by international instruments, such as: (a) the FAO's Code of Conduct for Responsible Fisheries and its various International Plans of Action; or (b) standards or guidelines for fisheries management developed by relevant regional fisheries management organizations. This might have to be accompanied by information or other data laying out and evaluating or assessing the level and effectiveness of implementation of such a plan.[25]
  3. Their individual share of total global exports of fish commodities does not exceed a specified and agreed-upon share of such exports.

Firstly, regarding the assessment of the state of their fish stocks, WTO members may be confronted with the difficulty of identifying and mutually agreeing on an external and technically competent international organization that would be entrusted with the task of defining the operational parameters for and of undertaking such assessments. A mandate could be given to organizations, such as the FAO, UNEP and Regional Fisheries Management Organizations (RFMO) for the definition of a suitable monitoring mechanism.

Of course, such an arrangement would require strengthening and expanding the mandate and activities of existing organizations or creating innovative international agencies, which may prove to be a politically difficult task and in any case a task that goes beyond the current WTO negotiations.

These assessments of the environmental condition of fish stocks would have to be made frequently and repeatedly, so that at any point in time other WTO members can have access to an updated list of fisheries that may be overexploited. Moreover, the assessment and the maintenance of the list must be based on purely technical criteria to ensure that the findings are widely recognised by all members. However, it should be pointed out that such assessments may capture only the state of fisheries at any one point in time (a picture of the decline of fish stocks over time), but may not be able to capture or identify the causal relationship and the actual quantitative impact of the provision of fisheries subsidies to the environmental condition of the fish stocks assessed.

Naturally, the quality and smooth functioning of this option would very much depend on the quality of the assessments and their acceptance by the WTO membership. The quality of the assessments, in turn, relies largely on the availability of data on developing country fisheries, on specific species and on regions within a single country. The major obstacle to such assessments is that most regions of the world, in stark contrast with a few others, are not thoroughly monitored, or when they are, the quality of the data collected is very uneven. Overcoming this technical difficulty will be a determinant factor in evaluating the success of this arrangement.

Secondly, approaches to determine a fair share of world export markets could be:

1 *quantitatively*, for instance, in terms of a specific percentage that can be patterned: (a) for export competitiveness under Art. 27.6 of the SCM agreement (3.25 per cent of world trade of that product for two consecutive calendar years); or (b) the current percentage share of the developing country with the highest share of exports of fish commodities;[26] or
2 *qualitatively*, for example, the way in which GATT Article XVI:3 stipulates that export subsidies should be applied in a manner which results in a WTO member having more than an 'equitable share' of world export trade in that product.

This would ensure that fish trade, as well as the resources generated, is undertaken under a legal framework that allows the entry of new participants but imposes certain limits on existing ones, within the context of ensuring that fishing activities occur in an ecologically and economically sustainable way.

Moreover, adopting some limiting parameters may be desirable in order to minimize the risk that better-resourced developing countries may implement S&D flexibilities in a manner detrimental to smaller developing countries. Hence, when a developing country reaches a development level from which it is deemed to be in a position to adopt the full-fletched subsidies disciplines, it would have to phase out current subsidies and refrain from implementing additional schemes. This would further strengthen the principle that S&D treatment is about adjusting trade rules to the ability of different members to adopt and apply them.

However, it should be noted that this criteria approach may effectively create a new layer of procedural requirements that could eventually operate as a deterrent to the use of the S&D flexibility by developing countries. If compared to other types of authorized (for example, 'Green Box') subsidies, S&D subsidies would be of conditional access.

Nonetheless, subjecting access to fisheries subsidies S&D to certain eligibility criteria may be justified in view of the negotiating mandate under Paragraphs 28 and 31 DMD, which requires, among other things, that both developing countries' developmental concerns and environmental considerations be taken into account in the fisheries subsidies negotiations. By their very nature, the subsidies that may be permitted under S&D treatment, as compared to other allowed subsidies under the 'Green Box', could conceivably have a fishing capacity or effort-enhancing effect, and thus would more directly affect the state of fish stocks.

However, to reduce the burden of this approach for developing countries, WTO members may also consider: (i) drafting clear and simple requirements for the information to be provided pursuant to the criteria, so that developing countries, and particularly those with limited technical and institutional resources, will not find it difficult to provide such information; or (ii) exempting a limited amount of subsidies from the requirements through the adoption of a '*de minimis*' amount of support that would not be subject to the criteria and then applying the criteria approach to any fisheries subsidies that developing countries may wish to provide as S&D.

Options for criteria-based S&D treatment for fisheries subsidies

The specific kinds of subsidies that could be authorized for developing countries as S&D would depend on the exact list of subsidies, which would fall under the prohibited or actionable categories of subsidies. In fact, as explained above, the nature of authorized subsidies under the 'Green Box' and those authorized under S&D would be distinct. Independent of the specific contents of the various subsidies 'boxes' that might be agreed upon in the broader WTO rules negotiations, several options can be looked at in terms of identifying the kinds of fisheries subsidies that developing countries could provide as part of S&D treatment.

One option could be to allow all developing countries, or only those that meet the sustainable development-based eligibility criteria suggested above, to provide all fisheries subsidies that would otherwise be prohibited or actionable in addition

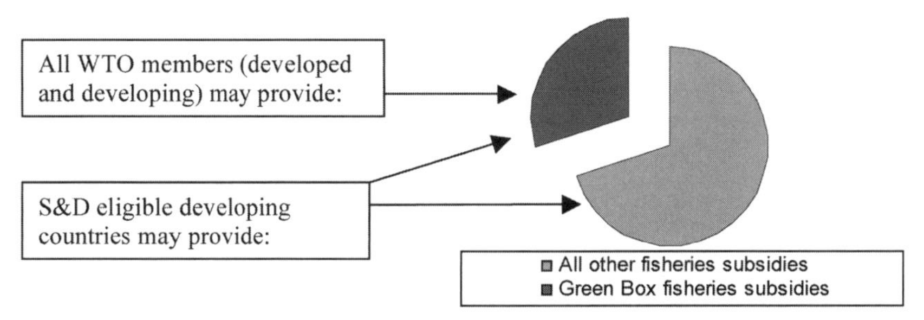

Figure 5.2 *Option 1 – S&D eligibility criteria without specific S&D subsidies list*

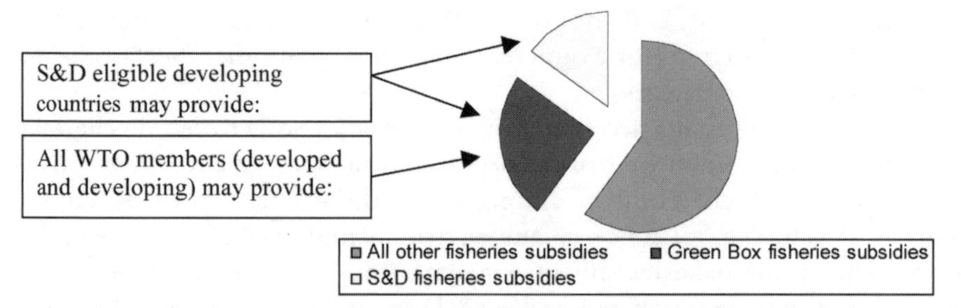

Figure 5.3 *Option 2 – S&D eligibility criteria plus specific S&D subsidies list*

to any agreed-upon subsidies that may be included in any 'Green Box'.

Another option could be designing an explicit 'S&D Box' that would list specific fisheries subsidies, not included in any 'Green Box', that developing countries meeting the suggested eligibility criteria, could provide.

Depending on what WTO members may agree on, these potential S&D subsidies could, among other things, include some or all of the subsidies that have been previously suggested by various developing countries in the context of the fisheries subsidies negotiations and others. Such subsidies could, qualitatively, be those that are effective and easily accessible to developing countries lacking adequate human, financial and institutional resources, and hence address the development needs of these countries. However, they should not result in IUU fishing, nor should they be allowed to enhance capacity beyond the scientifically determined sustainable level of exploitation.

Preserving the benefits of S&D for the beneficiaries

It might also be important to reflect on the specific relationship between the subsidies authorized under S&D provisions and: (i) on the one hand the dispute settlement mechanism, and (ii) on the other, the imposition of countervailing measures. Improved or clarified disciplines with respect to these two elements may ensure that whatever flexibilities flow from new subsidies disciplines, they are not impaired or offset by obligations contained in other provisions and agreements.

In the context of improved notification requirements, the current system of periodical review of subsidies programmes (both those inside and outside of S&D) needs to include their economic and environmental impacts.

Providing for positive measures on technical assistance and capacity building

In addition to the policy space and flexibility to adopt and implement subsidies programmes that would otherwise be prohibited, S&D could also include provisions that may require other WTO members, especially developed countries, to provide substantial, long-term and effective technical and financial assistance and capacity-building for the development and implementation of effective resource management systems, including methods and equipment for monitoring and surveillance methods. Such assistance and capacity-building support should be designed and implemented in consultation with the S&D beneficiary and could potentially have some correlation with the estimated amount, necessary to develop the beneficiary's domestic fishery industry in line with sustainable development considerations.

Transitioning out of fisheries subsidies S&D treatment

A fundamental element in ensuring the fairness of fisheries subsidies-specific S&D treatment is the establishment of a transition mechanism that would allow developing countries to gradually move out of S&D treatment towards assuming and implementing more stringent disciplines as and when they are able to do so. Not only is transitioning out of S&D important for the integrity of the S&D concept, but it is also fundamental for the fairness of the system, particularly for smaller developing countries.

In the case of the S&D eligibility criteria suggested above, when a developing country stops to fall under any one of the criteria, this could trigger an automatic implementation review procedure under the SCM committee that would allow the WTO member concerned to explain and address the reasons for non-compliance with the requisites and, should it wish to do so, bring itself back into compliance with such requisites.

Similarly, should a developing country member deem itself, or be considered by the SCM committee pursuant to the implementation review procedure, as no longer eligible for S&D, it could be provided with a certain time period to phase out and eliminate those subsidies programmes that it had provided as part of S&D.[27] Should a WTO member, formerly benefiting from S&D, be subsequently in compliance again with the S&D eligibility criteria above, other WTO members could agree that such a member should be able to once again benefit from the flexibility provided by S&D.[28]

With respect to the transitioning out of S&D and irrespective of the approach chosen for S&D treatment, an essential component of S&D-related policy space and flexibility is the provision of a qualitatively better transition period for developing countries, so as to enable them to adapt to new disciplines and obligations at a pace appropriate to their level of development.

Whereas existing SCM provisions offering transition periods stipulate a specific timeframe, a new S&D transition period could provide for sustainable development-based quantitative or qualitative parameters for determining exactly when the transition period ends. Such parameters may include, for example: (i) reaching an agreed-upon level of economic development, or (ii) reaching a state of fisheries resources beyond which further extraction would lead to overexploitation.

Conclusion

In conclusion, the creation of an S&D mechanism specifically designed for fisheries subsidies, with the objective of enhancing the flexibility and ability of developing countries to pursue their specific development policies in an economically-and ecologically-sustainable manner, could allow WTO members through the Negotiating Group on Rules to fulfil the interlinked Doha mandates to: (i) take into account the concerns and needs of developing and least-developed countries and the importance of the fisheries sector to them; and (ii) at the same time, enhance the mutual supportiveness of trade and environment. In fulfilling these mandates, the WTO may then move closer towards achieving its own institutional objective of promoting sustainable development.

This section has sought to highlight the twin key messages of ensuring policy space and flexibility for developing countries and that such space and flexibility are used in a manner designed to promote sustainable development. Given the rapid rate at which fish stocks are being depleted globally, and the great dependence of many developing countries on the continued existence of and their access to such fish stocks for their economic development and food security, WTO members will need to act to address both the causes and the symptoms of this crisis. One way of doing so is through designing an effective S&D mechanism for fisheries subsidies.

It should be noted that the suggestions for an S&D mechanism described in this section should simply be seen as among the range of possible options that could be considered in the course of the negotiations (WWF, 2004). These suggestions are necessarily broad and general, and will need further elaboration and operational clarification.

What could be stressed, in parting, is that the following key elements may need to be reflected in any fisheries subsidies-related S&D mechanism that may ultimately be designed:

- Policy flexibility in favour of developing countries to adopt measures (including subsidies) designed to enable their fisheries sectors to take advantage of increased market access opportunities while at the same time ensuring that their fisheries are healthy and managed in a sustainable manner.
- The necessary policy space to support fishing activities with a view to promoting sustainable development-oriented policy objectives.
- The policy flexibility and the resources needed to enable developing countries to conserve, sustainably manage and develop the fisheries resources in their waters (internal waters, territorial sea and EEZ) and on the high seas in pursuit of sustainable development-oriented policy objectives.
- The inclusion of positive international cooperation measures among WTO members for the provision of various forms of technical and financial assistance (including technology transfers) in a long-term, sustainable and adequate manner to developing countries seeking to put in place within their waters effective and sustainable fisheries resource conservation and management regimes.

The Special Case of Artisanal Fisheries*

Introduction

Since the earliest days of dialogue over fisheries subsidies at the WTO, it has been clear that 'artisanal fishing' presents a special case. Although never precisely defined, the term has been repeatedly used to identify a set of interests and people likely to need particular treatment under new WTO fishing subsidy disciplines.

There are good and fundamental reasons for this. In the midst of a global fisheries crisis caused mainly by large, high-tech, industrial fleets, 'artisanal fishing' refers broadly to small, underdeveloped and often severely impoverished fishing communities whose immediate survival depends on their ability to continue benefiting from local fisheries that in many cases are centuries old. These communities are highly significant in human, economic and environmental terms.

The artisanal fishing sector – regardless of any technical debate over its precise definition – provides direct employment to tens of millions of people and indirect employment to tens of millions more (many of them women involved in fish processing). Artisanal fishing comprises 90 per cent of all fishing jobs worldwide, approximately 45 per cent of the world's fisheries and nearly a quarter of the world catch. They provide critical income and edible protein to hundreds of millions across the globe. Moreover, artisanal fishers operate in some of the biologically richest and most sensitive waters on Earth, often in tropical coastal zones where interactions with coral reefs and land-based ecosystems introduce complex interdependencies.

The special concern of WTO delegations for artisanal fishing communities reflects a broadly shared desire to ensure that small, vulnerable and underdeveloped communities are not inadvertently harmed by new WTO rules that aim to eliminate unsustainable trade and production distortions in the fisheries sector. Subsidies to support such communities are in some crucial respects different in character from those granted to well-developed and globally competitive industries.

The urge to protect 'artisanal fishing' – and, in essence, to provide certain derogations from new fisheries subsidies disciplines for artisanal fishing – thus has a relatively clear basis. Less well understood, however, is how this urge can or should be translated into practice within new WTO fisheries subsidies rules.

---

* Excerpted from: UNEP (2005b). The original report was commissioned by UNEP to David Schorr as a consultant. The aim was to stimulate discussion on the treatment of artisanal fishing within efforts to clarify and improve WTO rules in a manner that contributes to sustainable development and the economic social and environmental health of world' fisheries. A draft version was circulated to all governments and reviewed at a UNEP sponsored workshop in June 2005 (chair's conclusions can be found on the accompanying CD-ROM). The report has been slightly adapted for the purposes of this book and some footnotes have been omitted from this excerpt. The full original version of this report is included in the accompanying CD-ROM.

As a contribution to the ongoing international dialogue, this section aims to elucidate some of the technical and political issues underlying that question. In particular, this section seeks to provide an analytical framework to facilitate discussion of two basic practical questions:

- What should be the scope of any special rules for subsidies to artisanal fisheries? In particular, what should be the definition of 'artisanal fishing' within the ASCM?
- What limits or disciplines should apply to subsidies to artisanal fisheries under new WTO rules? Are there substantive conditions that should be applied? Or procedural conditions?

Rather than proposing definitive answers to these questions, this section provides an analytical framework, and perhaps a few provocative words, in the hope of aiding discussion among governments and other stakeholders. The section does, however, take as its fundamental orientation the need to find an approach that maximizes incentives for truly sustainable development.[29]

Accordingly, this section is structured as follows: firstly it comments on the context in which this discussion is taking place, looking at both the overall challenges facing the artisanal fishing sector and the discussion of 'artisanal fishing' in the WTO so far. It then turns to the definitional debate that is preoccupying some delegations and stakeholders, exploring the difficulty of finding a ready-made definition of 'artisanal fishing' from the usage of the term outside the WTO context. The section then attempts to tease apart the 'why' and 'how' of subsidies to artisanal fisheries, reviewing basic policy objectives and the likely nature of subsidies to artisanal fisheries. Finally some of the practical issues surrounding new WTO rules on subsidies to artisanal fisheries are addressed in a general fashion, offering a few specific suggestions regarding both definitional and other issues.

Context

The sustainable development challenge

It is hard to imagine a human activity in which the twin imperatives of human development and environmental sustainability are more urgently united than in the case of artisanal fishing. The artisanal fishing sector includes some of the poorest and most underdeveloped communities on Earth, and it is little wonder that the development policies of many governments and intergovernmental organizations are focused on the artisanal fishing sector.[30]

Moreover, as in so many situations where underdevelopment is a predominant problem, the issue of artisanal fishing also raises significant questions of international equity. While it is generally accepted that 75 per cent of the world's commercial fisheries are either fished to the limits of their natural carrying capacity, or beyond (FAO SOFIA, 2004, p.32), and that aggregate global fishing capacity needs to be reduced (WWF, 1998b), the fact remains that many coastal developing countries have yet to enjoy the means to fully exploit the fisheries within their EEZs.

The developmental dimension of the artisanal fishing issue is thus fundamental. As the ministerial declaration adopted at the WTO's sixth ministerial conference in Hong Kong on 18 December 2005 for the first time explicitly called on members to prohibit those subsidies that lead to overcapacity and overfishing, a key goal within the current WTO negotiations must be to ensure that new fisheries subsidies rules do not prevent governments from investing in the improvement of their underdeveloped artisanal fishing communities, or from achieving equitable access to marine fisheries resources.

But while governments universally recognize the developmental challenges facing many artisanal fishing communities, there has been less emphasis – and perhaps even less agreement – on the need for attention to the sustainability of artisanal fishing practices. Some stakeholders in the debate, perhaps employing a degree of diplomatic license, have at times even gone so far as to suggest that artisanal fishing is too small in scale or otherwise underdeveloped to present any significant threat to the environment or to resource husbandry. Unfortunately, history has already shown that this is not the case.

Even if responsibility for the wholesale depletion of many of the world's major fisheries undoubtedly lies at the feet of the highly mechanized industrialized fleets, it is equally true that 'artisanal' fisheries around the world are increasingly facing challenges to their own sustainability. These challenges arise in a variety of circumstances. In some cases, the depletion of artisanal fisheries is again due – at least in part, and at times perhaps wholly – to the activities of industrial 'distant water fleets' (DWF) arriving to compete with traditional fishers.[31]

In a significant number of cases, however, challenges to the sustainability of artisanal fishing come directly from the actions of artisanal fishers themselves. This is true particularly where traditional patterns of fishing are undergoing change – which is probably the case in the majority of artisanal fisheries. In fact, the purely traditional fishery is fast becoming a thing of the past, as traditional fishers adopt new technologies, undertake new modes of social organization and aim at new markets for their fish. The motorization of vessels and modernization of fishing gear in artisanal fisheries is perhaps the most fundamental of these changes.[32]

These facts must, of course, be kept in perspective. Artisanal fisheries generally tend to be both 'cleaner' and more efficient than many industrial fisheries. Discarded by-catch, for example, is typically much lower (or even close to zero) in artisanal fisheries. Moreover, the kinds of changes just described are often the result of purposeful development policies. Cumulatively, however, they have been associated with a significant expansion in the capacity and fishing power of many artisanal fleets. In more than a few cases, these changes have led previously underutilized fisheries to the brink of overexploitation or beyond (FAO, 2003c). Unfortunately, in a small number of cases they have also been associated with the adoption of highly destructive fishing techniques, such as 'dynamite' fishing on tropical reefs (ADB, 1997, p.27).

Another factor affecting the sustainability of artisanal fisheries is the difficulty of establishing effective management regimes over them. Here again, it is worth confronting the notion that 'traditional' fisheries do not require formal management regimes: this hopeful idea is fast becoming an anachronism. The need to formalize and improve the management of artisanal fisheries is gaining worldwide acceptance and is the explicit policy goal of numerous governments.[33]

Unfortunately, many of the essential qualities of artisanal fisheries make them especially hard to manage.[34] Their small-scale and highly diffuse nature (thousands of small craft, often landing fish at scores of remote landing points), their often low level of technology (including the absence of communications or monitoring gear), the sometimes entrenched informality of their customary governance and even their often 'multispecies' character all compound the daunting difficulties faced by fisheries managers everywhere.

Finally, it is worth noting the evidence that inappropriate subsidy policies can pose a real threat to the health of artisanal fisheries. In the first place, fisheries subsidies in general have been shown more likely to do harm than good under most real world circumstances (see Chapter 2). Moreover, there is already clear evidence that improper subsidies can have negative consequences for artisanal fishing in particular. For example, a case study published by UNEP in 2002 described the negative economic and environmental impacts of subsidies to the artisanal fishery in Senegal (UNEP, 2002a). In other cases, evidence has been brought to light of subsidies inadvertently supporting IUU fishing by artisanal enterprises (Dallmeyer, 1989). The need to eliminate harmful fishery subsidies from the artisanal sector has been explicitly recognized by international development experts and governments alike (Christy, 1997). Of course, it is also true that appropriate subsidies to artisanal fisheries can have positive effects (FAO, 2004d).

In summary, the problem of sustainability facing artisanal fisheries is no less fundamental than the problem of development. It is clear that artisanal fishing communities cannot enjoy development if they run out of fish, and governments must now increasingly confront the need to reorient their policies accordingly. It follows that WTO fisheries subsidies rules need to take serious account of the sustainability dimension in their treatment of artisanal fishing, even as the weight of new disciplines is concentrated on curbing subsidies to the world's most rapacious industrial fleets.

The discussion of 'artisanal fishing' in the WTO so far

As noted above, the question of artisanal fishing has been on the table since the early phases of the WTO fisheries subsidies discussion. Apart from a recent attempt by Brazil[35] to develop possible legal language the disciplines, treatment of the issue has been relatively general and non-technical in character.

Undoubtedly, the WTO discussion so far does not reflect fairly the intricacy of interests or views held by governments in relation to artisanal fishing. Nevertheless, it may be useful to identify a few basic themes and issues that run through the existing papers:

- One theme common to a number of early submissions by the proponents of new rules (the '*demandeurs*') is that subsidies to artisanal fisheries are not the intended focus of new WTO fisheries subsidies disciplines.[36] The technical implications of these comments have yet to be fully clarified. The United States statement, for example, reportedly indicated that 'artisanal fisheries in developing countries...were unlikely to cause overcapacity and overfishing and were not an appropriate object of increased disciplines'.[37] A later submission by another *demandeur* suggested that subsidies to 'small-scale' fisheries should be left in an actionable 'amber' category, rather than subject to the prohibitions at the heart of proposed new rules.[38]

- Outside of the *demandeur* group, submissions have also tended to urge either the exclusion of subsidies to artisanal fisheries from new rules, or at least their special treatment. The strongest position offered to date, tabled by a group of small island developing states, called for 'measures undertaken by governments of small vulnerable coastal states to assist their artisanal fisheries sector' to be excluded from the definition of a fishing subsidy.[39] Japan and the Republic of Korea have similarly raised repeated concerns with the treatment of their own artisanal or small-scale fisheries.[40] The recent submission from Brazil constitutes the most concrete proposal so far, categorizing subsidies to small-scale and artisanal fishing into the green box, on condition that the fisheries are not 'patently at risk'.[41]

The widely (if vaguely) held desire to give some special consideration to subsidies to artisanal fisheries appears to be based on a loosely shared acceptance of two basic propositions:

1. that subsidies to artisanal fisheries are not likely to be harmful; and
2. that subsidies to artisanal fisheries may be important and necessary components of government policies aimed at poverty alleviation and development.

While there can be no doubt about the truth of the second of these propositions, the notion that subsidies to artisanal fisheries are simply harmless deserves close scrutiny and debate. Where the submissions discussed above have tended to take this view, they refer sometimes to the unlikelihood such subsidies will cause 'overcapacity or overfishing'[42] and at other times to the unlikelihood they will cause trade distortions.[43] These claims need to be combed apart and reviewed separately:

- The susceptibility of artisanal fisheries to overcapacity and overfishing – and the role poorly considered subsidies can play in causing these evils – has already been established, as discussed above. This is not to say that the risks of overfishing or the potential harm of subsidies in the artisanal context are the same as in larger industrialized fisheries. Certainly the scale of that harm in any given fishery will tend to be smaller in the artisanal case. This does not, however, mean that the harm is trivial, either from a local perspective or in terms of the overall impacts on the world's marine resources and ecosystems. It is important to recall that the artisanal sector represents a very significant proportion of total worldwide fishing. Indeed, several delegations have directly referred to the problem of achieving sustainability in artisanal fisheries and have clearly assumed that subsidies to artisanal fisheries should not be treated as inevitably safe.

- The argument that subsidies to artisanal fisheries will not generally lead to trade distortions raises an important point about the essential orientation of the fisheries subsidies negotiations: these talks are aimed not only at eliminating distortions caused by fisheries subsidies at the level of 'trade' (in other words, international sale) but also at the level of fisheries production, where the interaction between subsidies and sustainable fisheries management is vitally important. In other words, fisheries subsidies more directly distort the access of producers to resources than the access of exporters to markets – because, obviously, before you sell a fish you have to catch it. This focus on 'production distortions' has been repeatedly emphasized by governments and other stakeholders in the fisheries subsidies debate and is fully consistent with the mandate and competence of the WTO (WWF, 2004, pp.43–44). So once again the question returns to the issue of the impact of subsidies on the race for fish, where artisanal fisheries are undoubtedly implicated.
- The emphasis on eliminating distortions at the production level does not, however, fully answer questions that have been raised regarding the proper territorial reach of new WTO fishing subsidies rules. Some stakeholders have argued that the impacts of artisanal fishing are too localized within national jurisdictions to be of legitimate international concern.[44] In this view, artisanal fisheries are simply so small and 'coastal' that subsidies to them cannot have meaningful effects on international competition, even if the subsidies have negative impacts on sustainability. This argument, however, fails to recognize at least four ways in which artisanal fishing can (and increasingly does) have implications for international competition:

1. **Artisanal fishery products are increasingly oriented towards international markets.**[45] Indeed, several delegations have already highlighted the interest of their governments in using subsidies 'to raise income levels by expanding [artisanal fishing] into monetized activities for the domestic and speciality export market...'.[46] In other words, international competitive impacts are a direct objective of some subsidies to artisanal fishing. Some such supports would even likely qualify as 'export subsidies' under current WTO rules.
2. **Traditional nearshore artisanal fleets sometimes compete with foreign or export-oriented industrial fleets.**[47] Even where artisanal fishing is restricted to nearshore activities, conflicts between industrial and artisanal fleets can arise.
3. **Artisanal fishing activities are increasingly expanding to offshore fisheries where foreign or export-oriented fleets may be active.** The ambitions of governments and fishers alike for the development of artisanal fisheries often include the extension of fishing activities from traditional nearshore fishing grounds to more distant inshore or even offshore waters. The addition of motors, navigational aids and refrigeration has allowed many traditional

fleets to switch from day-long fishing trips to far more distant multi-day excursions. In this regard, it is interesting to note that in some cases artisanal fleets have proved more efficient and more successfully competitive than rival industrial fleets operating in the same waters (UNEP, 2002a).[48]

4 **Even a fishery that appears commercially isolated may be biologically linked to fisheries of international relevance.** Ocean ecosystems are complex and highly interdependent, often in ways that remain poorly understood. This interdependence can include biological connections between fisheries close to shore and those further away, even up to the high seas (Macy and Brodziak, 2001). Such links may be as simple as the migration of a given stock from nearshore spawning grounds to deep-sea homes. Or it could involve intricate relations among different species along a marine food chain. Or (as in the case of artisanal dynamite fishing on sensitive coral reefs), the larger ecosystem impacts may result from the degradation of habitats or the depletion of non-target species. In any case, biological interdependence in fisheries is tantamount to commercial interdependence in the race to catch fish and the assumption that any marine fishery is isolated is almost always dangerous and has often proved untrue.

In short, competitive interaction between artisanal and industrial fishing fleets is common and growing. Even artisanal fishing at its increasingly rare 'purely local' scale is a relevant factor in the international competition to catch and sell fish. As the UN has concluded: conflicts between artisanal and industrial fisheries for resources and on the market are increasingly frequent and may jeopardize development efforts.

Indeed, the international relevance (or potential relevance) of many artisanal fisheries directly underlies the strong interest of some governments in protecting their freedom to subsidize these fisheries. This international competitive relevance does not mean, of course, that subsidies to promote artisanal fisheries should necessarily be prohibited under new WTO rules – indeed, considerations of equity and development may strongly suggest the contrary. But it clearly means that subsidies to artisanal fishing cannot simply be excluded entirely from the scope of such rules because they are unlikely to have international competitive effects.

As discussed below, the WTO submissions to date also raise but do not resolve the issue of defining 'artisanal fishing', leaving a particular confusion around the relationship between the terms 'artisanal' and 'small-scale', and leaving open the delicate question as to whether special treatment for subsidies to 'artisanal fishing' should be limited to developing countries, or should extend to some fisheries in developed countries as well.

In short, as delegations well know, the discussion of 'artisanal fishing' within the negotiations to date has raised more questions than it has answered, leading several governments to call directly for more detailed examination of the relevant issues.

Relationship to the S&DT debate

Clearly, there is an important element of overlap between special rules for artisanal fishing and 'special and differential treatment' as introduced in the previous section. A few comments on the relationship between them may be in order:

- This section proposes an approach to artisanal fishing that focuses heavily on the goals of poverty alleviation, food security, and development – goals that also lie at the core of the broader S&DT question (see Chapter 5, 'Special and Differential Treatment'). To the extent that effective rules for S&DT are included in new WTO fishing subsidies disciplines, the need for specific treatment of artisanal fishing may be reduced.
- This section takes the view that special treatment of artisanal fishing within new WTO rules will be necessary and important, but also argues that such treatment is not without risks for delegations and stakeholders involved in the negotiations. In particular, it is critically important that the implications of special treatment for the sustainable management of fisheries be considered. Similar concerns clearly arise in the context of general approaches to S&DT on fishing subsidies, as reflected in the extensive treatment of the sustainability issue in Chapter 5, 'Special and Differential Treatment' and 'Sustainability Criteria for Fisheries Subsidies'.

The artisanal fishing and S&DT conversations also share a theoretical question raised by the focus of the fisheries subsidies issue on production distortions: to what extent should S&DT or provisions on artisanal fishing tolerate production distortions that would otherwise be discouraged by new fisheries subsidies disciplines?

- The ASCM as it presently operates (or purports to operate) creates more policy space for subsidies in developing countries than in developed countries. Underlying this normative architecture appear to be two fundamental tendencies that exist in some tension with one another. Firstly, at times the system seems to assume that subsidies in developing countries are less likely to cause trade distortions than developed country subsidies. Secondly, the system seems to tolerate some developing country subsidies even when they may lead to a certain degree of trade distortion. Some of the ongoing debate around the proper scope of S&DT for developing countries may revolve around the interplay between these assumptions and the degree to which one rather than the other should be given weight.
- But should, for example, the allowance for export subsidies by least developed countries serve as a model for tolerating 'capacity-enhancing' subsidies to artisanal fisheries? Here, it may be helpful to keep in mind two facts that distinguish the toleration of trade distortions from the toleration of production distortions in the fishery sector:
  1. In the case of fisheries, production distortions result not only in a redistribution of the benefits of economic activity, they may also reduce

overall activity by contributing to resource depletion. In other words, where trade distortions mainly effect how the pie is divided, production distortions can shrink the pie.

2 Under broadly subscribed international treaty commitments, such as the UN Convention on the Law of the Sea and the UN Code of Conduct for Responsible Fisheries, governments bear a responsibility for the conservation and husbandry of fisheries resources both on the high seas and within their EEZs and territorial waters.

Proper consideration of the distinction between a focus on trade distortions (as classically defined) and on production distortions is thus critical to achieving a true 'win-win-win' outcome for trade, environment and sustainable development in the context of fisheries subsidies, both as regards S&DT and as regards artisanal fishing.

From the foregoing, it is clear that the artisanal fishing and S&DT discussions cannot be carried on in isolation. Indeed, the overlap between them poses an important practical question: assuming new WTO rules eventually do include special treatment for artisanal fishing, should such treatment be granted in the context of S&DT, or separately?

The negotiating proposals tabled so far have varied in their answers to this question. When a group of eight small island developing states tabled the first clear demand for special treatment of artisanal fishing, they proposed amending ASCM Art. 1 to exclude subsidies to artisanal fishing from the ASCM's definition (TN/RL/W/136, p.3). Later, three of those delegations (joined by a fourth not involved in the first submission) tabled a proposal calling for special treatment of artisanal fishing subsidies as part of S&DT (TN/RL/GEN/57/Rev.1, p. 4 ¶ 16). (Chapter 5, 'Special and Differential Treatment' similarly includes artisanal fishing within the scope of the S&DT options.) Brazil, on the other hand, has proposed artisanal fishing subsidies be treated as 'non-actionable' under a revitalized green light, apparently leaving them outside the scope of S&DT provisions Brazil has simultaneously suggested (TN/RL/W/176; TN/RL/GEN/56; TN/RL/GEN/79).

Beneath this 'in or out of S&DT' issue are at least two specific policy questions:

1 Should special treatment for artisanal fisheries be limited to developing countries?
2 What (if any) conditions or disciplines should be imposed on subsidies to artisanal fishing? To what degree should such conditions and disciplines be the same as those associated with other classes of subsidies that might qualify for S&DT?

Ultimately, the answers to these questions – in other words, effective scope and conditionality of special rules for subsidies to artisanal fishing – matter more than the formalistic problem of 'in or out of S&DT'.

As evident in the discussion below, this section tends to support the restriction of special treatment of artisanal fishing to developing countries and argues in favour of disciplining special treatment for artisanal fishing in ways that would probably need to apply to subsidies falling under general S&DT rules as well (for example, the conditions imposed to prevent subsidies from contributing to unsustainable production distortions).

The definitional debate

The definitional debate in perspective

Given the increasingly visible likelihood that some special treatment of 'artisanal fishing' will be necessary within new fisheries subsidies rules, the problem of defining the term has come to the fore. In fact, as this section was being drafted several other efforts to discuss or propose definitions of 'artisanal fishing' were underway.[49]

The definitional question is obviously critical, since a technical definition would provide one principal means of delineating the potential scope of any special treatment granted to 'artisanal fishing' under new rules. In particular, the suggestion by some governments of a broad carve-out for subsidies to artisanal fisheries implies that the definition of 'artisanal fishing' would be a critical substantive provision.

As noted above, the question of defining 'artisanal fishing' has been given some direct attention within the WTO conversation so far. Indeed, a few rudimentary definitions have even been offered. The United States at one point, in an oral answer to an inquiry by Japan, stated: 'As regards the term "artisanal fisheries", it referred to small-scale fisheries that employed labour intensive harvesting, processing and distribution technologies to exploit marine and inland fishery resources. Such fisheries typically targeted local rather than export markets' (TN/RL/M/8, p.9, ¶ 39).

One of the sponsors of the small island developing states paper also offered an oral definition: 'On artisanal fisheries, [the sponsor] defined it to be the small-scale fisheries which are local in nature'.[50]

In an early CTE submission, Japan seems to have taken a similar approach, parenthetically defining 'artisanal fisheries' as 'small-scale coastal fisheries', (WT/CTE/W/173, p.3, ¶ 14).

The most fully elaborated definition so far was offered by Brazil in a submission to the informal negotiating process in July 2005, in which it proposed artisanal fishing be defined as follows:

> (a)...fisheries activities performed at an in-shore basis with non-automatic net-retriever devices;
>
> (b)...activities carried out on an individual basis (including, but not necessarily, the family members);
>
> (c)...the basic scope of the activities encompasses both family livelihood and a small profit trade; and there is no employer-employee

relationship on the activities carried out. (TN/RL/GEN/56, p.2, fn. 5., TN/RL/GEN/79, p.3)

One confusing element quickly evident in this early definitional dialogue is the juxtaposition of the terms 'artisanal' and 'small-scale'. With the exception of the Brazilian approach, the definitions noted above all appear to treat 'artisanal' as a subset of 'small-scale'. Other submissions to the negotiations have varied in the degree to which they distinguish between these terms – at times seeming to use them interchangeably and at times distinguishing them. Only the Brazilian paper has attempted clearly separate definitions, proposing that 'small-scale' be taken to mean:

Activities carried out by vessels with total length not exceeding 24 meters and with a total catch not over 250 tons per year. In addition:

(a) if the fishery is under the management of a Regional Fisheries Management Organization (RFMO):

(a.1) and if a country limit is set to a specific specie, the total catch of the country small-scale fleet for that specific specie shall not exceed 10 per cent of the limit set to the country for that specific specie by that RFMO; or

(a.2) and if no country limit is set to a specific specie, the total catch of the country small-scale fleet for the specific species that have no individual limits shall not exceed 5 per cent of the limit set to the country by that RFMO for those specific species that have no individual limits; or

(a.3) and if a global limit is set to a specific specie, the total catch of the country small-scale fleet for that specific specie shall not exceed 0.5 per cent of the global limit set to that specie by that RFMO; or

(b) if the fishery is not under the management of a RFMO, the annual increase of the volume catch by the country small-scale fleet for that specific specie shall not exceed 3 per cent of the most recent volume catch data reported to a competent international organization (TN/RL/GEN/56, p.2, fn. 4; TN/RL/GEN/79).[51]

In any case, as several delegations have noted, 'there is no agreed definition of small-scale fisheries and each country has its own criteria based on the circumstances surrounding its fishery sector' (TN/RL/W/172, p.4, ¶ 17). The same could surely be said of the term 'artisanal fisheries' as well.

In fact, as discussed below, the definitional issue cannot really be resolved in isolation from other questions about the purpose and operation of special WTO rules for subsidies.

In this regard, it is important to keep the entire discussion of special provisions for artisanal fisheries in perspective. The goal of such provisions should be to preserve certain rights of governments to subsidize artisanal fisheries *for reasons relating to their artisanal nature*, understanding that governments may also wish to provide subsidies to artisanal fisheries for reasons *not* related to their artisanal nature. Indeed, a given subsidy may be:

- available only to artisanal fisheries;
- available to artisanal and non-artisanal fisheries, but on a preferential basis to artisanal fisheries; or
- equally available to artisanal fisheries and others.

For example, a government policy (or a WTO rule, for that matter), could make subsidies to onboard life-saving equipment uniquely available to artisanal fishers, preferentially available to artisanal fishers, or equally available to both artisanal and non-artisanal fishers. This section is concerned only with the WTO's treatment of the first two categories. If particular kinds of subsidies in the third category would require special treatment under new WTO disciplines (whether S&DT or provisions relating to specific classes of subsidies), these should be treated separately rather than 'shoehorned' into provisions on artisanal fishing.

Obstacles to a universal definition of 'artisanal fishing'

Any effort to craft a definition of 'artisanal fishing' must obviously take account of how the term is commonly used by fisheries scientists and policy-makers. Indeed, the term frequently appears both in formal governmental documents and in the substantial technical literature devoted to fishing and fisheries. A limited sampling of definitions drawn from formal and technical sources is set forth in Annex 12 of this book.

As illustrated below, even a brief review of primary and secondary sources quickly reveals that no single prevailing definition of 'artisanal fishing' exists. Indeed, the term is so variously used that several authorities have concluded that it is impossible to find a definition that is both precise and generally applicable.[52] Moreover, as already seen, this polyphony is compounded by the frequent use of related terms, such as 'small-scale' and 'traditional'.

There are, of course, a number of elements that recur within the most common definitions and usages. An analysis of these elements may simultaneously illustrate the difficulty of discovering a single universal definition and help make it easier to comb out the specific interests and policy objectives at stake in the ongoing WTO debate. The most common 'definitional elements' contributing to definitions of 'artisanal fishing' can be divided into four broad categories,[53] as follows.

Table 5.2 helps illustrate the difficulty in arriving at a single universal definition of 'artisanal fishing' in two ways:

1. Firstly, existing definitions mix and match elements across categories in different ways. Regulatory definitions, for example, tend to focus on physical attributes of vessels or the specific fisheries targeted, while scientific definitions commonly look to social structure and economic condition.
2. Secondly, and most compelling, existing definitions vary widely in the specific characteristics they associate with the various definitional elements. In fact, for each of the 'typical examples' identified in Table 5.2, it is possible

Table 5.2 *Elements of a Definition of 'Artisanal Fishing'*

Category	*Definitional Elements*	*Typical examples*
Physical Attributes	Vessel type	Canoe, dory
	Vessel size	Short (e.g., < 10m); light
	Vessel motor	Un-motorized or small engine
Pattern of Fishing	Fishing gear/technique	Manual or small nets; passive; low tech
	Location of land base	Rural
	Location of fishery	Inshore
	Target type	Multi-species
Social Structure	Of fishery	Traditional (clan or community)
	Of fishing enterprise	Family crew, owner on board
Economic Condition	Market orientation	Direct consumption or local market
	Income level	Subsistence or very poor

to find counterexamples. Thus, for instance, among fisheries labelled 'artisanal' it is possible to find cases where the vessel is a trawler, or is quite large, or runs a powerful engine, or uses advanced technology, or is based in a city, or involves an off-shore fishery, or is targeted on a single pelagic species, or is organized around formal corporations with non-family crew, or is oriented towards export, or is returning a middle-class income to the fishers. Indeed, at least one case exists in which reference has been made to an 'industrial artisanal fleet,' (Fundacion Patagonia Natural, 2005).

As one expert has put it with gentle understatement: 'There is thus no elegant definition' of artisanal fishing (FAO, 2003c, p.52).[54]

For those engaged in the WTO fisheries subsidies negotiations, the implications are clear: no definition of 'artisanal fishing' is possible that is simultaneously universal, precise and coherent. If the WTO were to adopt a definition covering every fishery currently considered 'artisanal' by some relevant authority, the result would be so broad that widely disparate fisheries would fall within its scope. Such a definition would open a large and vague loophole in any new fisheries subsidies disciplines, with unpredictable consequences.

Accordingly, a definition of 'artisanal fishing' will need to be tailored to the WTO context. But how, and on what basis? These questions cannot be fully answered without examining the goals of the proposed WTO provisions on subsidies to artisanal fisheries and the kinds of subsidies that may be involved.

Underlying objectives and practices

Why are governments so interested in the artisanal fishing issue within the context of subsidies and the WTO? In particular, two questions may help inform a more focused and coherent discussion:

- What are the policy objectives that appear to underlie interest in 'artisanal fishing' in the WTO context?
- What are the kinds of subsidies governments use today or appear likely to use in the future with regard to 'artisanal fishing'?

Why subsidize artisanal fishing?

It is difficult at present to know in detail what may motivate governments to provide subsidies to 'artisanal fishing'. Although there is a large technical literature on artisanal fishing generally, relatively little of it discusses subsidies policy. And within the bounds of the WTO debate so far, only a few governments have made references to the interests underlying their demands. Nevertheless, even if documentary support for any analysis may be thin, this section will suggest five basic motives that, to one degree or another, appear to underlie the 'artisanal fishing' debate:

1. **Poverty alleviation:** as noted throughout the technical literature, artisanal fishing communities (and particularly those associated with 'subsistence' economic lifestyles) are often extremely poor. In many cases, artisanal fishing provides a source of 'employment of last resort' when land-based economic activities (especially rural agriculture) suffer downturns. Indeed, this 'safety net' role of artisanal fisheries has in some cases been associated with overexploitation of stocks when subsistence fishing activities have increased dramatically to absorb the newly unemployed.

   Poverty alleviation, including employment stability, is thus one obvious policy goal in the artisanal fishing sector (ADB, 1997, pp.37–40). For the purposes of this section, 'poverty alleviation' is narrowly defined to mean the removal or prevention of extreme deprivation, or (in positive terms) the granting of assistance to meet basic human needs, including the need to engage safely and honourably in productive economic activity.
2. **Food security:** as described in the previous section, fisheries are without a doubt a critical source of food protein for human nutrition. Artisanal, 'subsistence' and 'small-scale' fisheries make an important contribution to this food supply, and the FAO has noted the need for increasing this contribution to food security through purposeful government action, including subsidies.
3. **Community/social development:** closely related to poverty alleviation (but not identical) is the goal of community and social development, defined in this section to mean the evolution or transformation of the socio-economic circumstances of a community or of a productive sector through technological improvements, increased education, multiplication of personal and economic options and general improvement of the standard of living.
4. **Improving sustainable fisheries management:** as noted throughout this section – and indeed throughout the international dialogue over fisheries subsidies generally – the most critical challenge facing the fisheries sector is the sustainability challenge. Without improved husbandry of fisheries

resources, there will be little chance of reversing the trend towards increased overexploitation that has plagued fisheries worldwide for the past several decades.[55] As noted above, this sustainability challenge confronts artisanal and industrial fisheries alike. While inappropriate subsidies are clearly a contributing factor underlying the current crisis of fisheries depletion, it is also true that properly designed subsidies can play a positive role in promoting the transition to sustainability.

5. **Preservation of fishing cultures and lifestyles:** some governments appear interested in subsidizing 'artisanal fishing' as a means to maintain particular cultures and lifestyles, especially in relatively rural areas. This goal does not necessarily call for poverty alleviation, food security, development or improved fisheries management, but merely for the maintenance of desired types of community in the face of market pressures that would otherwise injure or even doom them. This motive is more likely typical of developed than developing countries.

Two comments on the foregoing objectives:

1. Firstly, outside of the subsidies context, there may be other significant objectives underlying a government's policies towards 'artisanal fishing'. For example, governments may be concerned with 'artisanal fishing' from a regulatory perspective, using the term to distinguish some fishing fleets under its jurisdiction from others, whether for the allocation of fishing rights or other regulations.
2. Secondly, the goal of community/social development stands out as unique in at least one important respect. Unlike the other objectives, 'development' tends to imply the pursuit of substantial change in the patterns or intensity of fishing. Put another way, subsidies to promote the 'development' of artisanal fishers means interventions aimed at altering some essential aspects of a fishery, whether by changing the extent or manner of fishing, the social or economic organization of fishing activity or a combination of these. In many cases, such a policy will be aimed at expanding fishing, or at making fishing more efficient, or at creating improved processing and transportation. As noted at the outset, this may be consistent both with rational social policy and with establishing a more equitable distribution of access to the Earth's common natural resources. Nevertheless, it also obviously implies the need for attention to fisheries management and conservation of the aquatic ecosystem.

The preceding analysis suggests two intermediate conclusions:

1. Firstly, it seems relatively non-controversial to suggest that subsidies to the artisanal fishing sector aimed at promoting poverty alleviation, food security, development of underdeveloped communities and improved fisheries management should be allowed under new WTO rules. It is a more detailed

question – and one that delegations and stakeholders may wish to discuss explicitly – whether the preservation of fishing cultures and lifestyles should likewise be adopted as a legitimate goal for the relaxation of any new fisheries subsidies disciplines. This question also appears closely related to the question of whether special provisions for artisanal fishing should be applicable universally or only to developing countries.

2 Secondly, the 'sustainability challenge' remains front and centre. It will not be possible to use definitional language alone to ensure that allowances for subsidies to artisanal fisheries promote sustainable development. Some of the principal underlying motives for such subsidies – including job creation and community/social development – directly raise issues of sustainability and fisheries management.

What kinds of subsidies are involved?

Obviously, another key to understanding the scope of the current debate would be a review of existing trends in the subsidization of artisanal fishing. Unfortunately, this effort must be based, at least for the moment, on some degree of speculation, for two reasons. Firstly, the subsidization of artisanal fisheries has been the subject of only a few published studies, and even the general question whether artisanal fishing is commonly subsidized appears to be the subject of differing views.[56] Secondly, many stakeholders in the current negotiations appear to be more concerned with securing the future 'policy space' for subsidies to artisanal fisheries than with protecting current subsidy practices.

Nevertheless, a brief review of the existing literature suggests that among the kinds of subsidies most likely to be associated with artisanal fishing are:[57]

- vessel/gear modernization (including motorization);
- landing and processing infrastructure (port facilities, refrigeration, roads/transport);
- export;
- fuel;
- other inputs (for example, ice);
- training;
- capital (cheap money).

Subsidies to artisanal fisheries, where they exist, also appear to take on a wide variety of forms, including direct cash transfers, subsidized loans and tax deferrals, among others.

In short, governments (at least in the aggregate) are likely to be interested in using a broad range of subsidies for their artisanal fishing sectors, including some that appear likely to fall within the scope of general fisheries subsidies disciplines now under negotiation. It is particularly important, from the sustainability perspective, to note that some of these subsidies would be directly 'capacity-enhancing' or 'effort-enhancing' – among the class of subsidies that have often been identified as the most problematic.

Practical implications for WTO rules

Following on the analysis of this section, what are the implications for a further discussion of the treatment of subsidies to artisanal fisheries under new WTO rules? Does the foregoing discussion provide any basis for crafting a WTO definition of 'artisanal fishing'? What does it suggest about other provisions that may be necessary to achieve balanced and effective rules?

The definitional question, revisited

In light of the 'objectives and practices' discussed above, one approach to resolving the definitional problem identified would be to agree that the principal aim of special treatment for subsidies to artisanal fisheries is to promote poverty alleviation, food security and development. Such an approach would lead towards a definition that targets situations where poverty or underdevelopment are prevalent, and might, for example:

- emphasize poverty and/or subsistence economic patterns;
- emphasize very small vessels and/or vessels with small or no engines;
- emphasize low levels of technological development and high labour-intensitivity;
- emphasize fishing very close to shore.

In addition, such a definition could either be limited to developing countries, or impose different tests and higher burdens on developed countries seeking to apply new rules to their own 'artisanal fishing' sector.[58]

From the perspective of sustainability, such an approach would have some clear advantages. First, the focus on poverty alleviation itself helps create conditions for improved fisheries management. As has been noted by the Asian Development Bank: 'The widespread poverty in artisanal fishing communities contributes to resource degradation; hence, poverty reduction needs to be a priority concern' (ADB, 1997, p.37).

Secondly, while very small-scale, low-technology fisheries can be subject to overexploitation or other forms of environmental degradation, especially where destructive fishing practices are involved, it is probable that there is less risk of overexploitation with fishing of this kind than where more powerful fleets are involved. In other words, to the extent 'artisanal fishing' means fishing at the lower end of the technology and developmental scales, subsidization (and particularly subsidization aimed at basic poverty alleviation) would seem to carry relatively light risks for sustainable fisheries management.

However, the parameters for a definition of 'artisanal fishing' proposed above are not intended to comprise only the poorest and least developed of fisheries. Moreover, the analytic separation offered here between 'poverty alleviation' and 'development' is obviously somewhat artificial – and in any event subsidies directed at achieving significant development must clearly fall within the scope of any special treatment for artisanal fishing under new WTO rules.

In other words, even if limited to the purposes of poverty alleviation and development, special rules for subsidies to artisanal fishing are likely to cover a relatively broad range of government programmes, including programmes that, if improperly designed, could contribute to trade or production distortions and the accompanying overexploitation of resources. It would seem, then, that even with a carefully circumscribed definition, subsidies to artisanal fisheries will require some WTO discipline and that a definition of 'artisanal fishing' should not be seen as the basis for a simple carve-out from new fishing subsidies rules.

Other elements of a WTO approach

Assuming, then, that subsidies to artisanal fisheries are to be subject to some disciplines, there are at least three basic (and familiar) ways in which such disciplines might be applied:

1. Distinctions could be made among the kinds of subsidies falling within the scope of the artisanal fishing provisions, with different levels of discipline applicable to each.
2. Substantive conditions, particularly related to the 'fisheries management' context, could be applied.
3. Procedural requirements could be imposed to encourage transparency and appropriate implementation.

Each of these will be considered briefly in turn.

Distinguishing different types of fisheries subsidies

As noted before, governments are likely to be interested in preserving their right to apply a broad range of subsidy types to their artisanal fishing sectors. Just as in the case of fisheries subsidies generally, some of these are likely to carry higher risks of production (as well as trade) distortions than others. For example, there is little question that capacity- or effort-enhancing subsidies raise more direct risks of distorting fisheries production than subsidies to land-based infrastructure.

Moreover, it may be possible to classify subsidies to artisanal fisheries according to a combination of their type and their underlying policy objective. Particularly recalling the analytical distinction offered above between the objectives of 'poverty alleviation' and 'development', a very rough correlation of risk to subsidy class might take the form of Table 5.3.

Obviously, even if the risk identifications in Table 5.3 were accepted as accurate, this table would not provide a basis for simply including or excluding particular classes of subsidy programmes from any provisions adopted for subsidies to artisanal fisheries. In fact, distinguishing programmes aimed at poverty alleviation from those aimed at development may be both theoretically and practically difficult. Still, a discussion of the various levels of risk associated with different categories within Table 5.3 may at least help distinguish the easier from the more difficult cases.

Table 5.3 *Possible Risks of Distortion to Production or Trade*

	Poverty alleviation		*Community/ Social development*		*Cultural conservation*	
	Risk of Production Distortion	*Risk of Trade Distortion*	*Risk of Production Distortion*	*Risk of Trade Distortion*	*Risk of Production Distortion*	*Risk of Trade Distortion*
Vessel/gear modernization	–	–	High	Med	Med	Med
Land-based infrastructure	Low	Low	Med	Med	Low	Med
Export	Med	Low	High	High	Med	High
Fuel	Low	Low	High	Med	Med	Med
Other inputs	Low	Low	High	Med	Low	Med
Training	Low	Low	Med	Low	Low	Low
Capital	Med	Low	Med	Med	Low	Med

Notes: (i) this table is not based on any specific empirical analysis, and must be considered a rough suggestion for the purposes of stimulating dialogue; (ii) a 'high' risk of trade or production distortion is not necessarily the same as a high risk of a resulting harm; (3) however, regarding the possible wisdom of less toleration of production distortions than trade distortions in the fisheries sector.

SUBSTANTIVE CONDITIONS ON SPECIALLY PERMITTED SUBSIDIES

The recurring difficulty is how to ensure that subsidies to artisanal fisheries do not contribute to production distortions and the resulting exhaustion of fisheries. One general direction towards a solution that has been proposed is to allow special treatment for subsidies to artisanal fishing only under certain conditions, such as in fisheries that are not considered 'patently at risk'. This concept, and others that might be grouped under the general heading of the 'fisheries management context', is discussed later in this chapter. At this point, some of the specific issues will be pointed out that might arise in an effort to attach such conditionality to subsidies to artisanal fisheries.

With regard to both the biological and the industrial condition of a fishery, 'context sensitive' disciplines will likely need to refer to facts or judgements about things like the status of a fish stock or the effective capacity of a fishing fleet (see Chapter 5, 'Sustainability Criteria for Fisheries Subsidies'). But in artisanal fisheries, data limits can often be very severe due to the diffuse and sometimes remote nature of the fishery and to the lack of resources for investigation. Thus, even the most basic need of modern fisheries management – to have some idea of 'what's out there' – can present a formidable problem.

With regard to the regulatory condition of a fishery, 'context sensitive' disciplines will probably need to refer to facts about the regulatory infrastructure of a fishery (such as, for example, whether it is purely 'open access' or subject to management). Here, the artisanal nature of the fishery – particularly if defined as suggested above – often implies a formal management system that is either rudimentary or absent (although in many cases there may be a traditional community-based management system). There is today a growing trend towards the active formal management of artisanal fisheries, but the social, political and financial challenges remain significant.

Any rule system seeking to impose 'fisheries management context' conditions on subsidies to artisanal fisheries must take account of these realities. But how? On the one hand, the problems mentioned above are not only real, they may be the focus of policies that go hand in hand with some of the subsidies for which governments are interested in preserving policy space. On the other hand, the acceleration of fishing capacity or effort in fisheries characterized by lack of data and thin regulation can be especially dangerous. And so the dog chases its tail.

The fundamental problem is one of sequencing. From a strictly environmental perspective, the 'right' answer is relatively clear: governments should give priority to investments in data and management and should avoid capacity- or effort-enhancing subsidies to artisanal fisheries until a sufficiently robust 'fisheries management context' has been established. Unfortunately, this simple advice may be impractical where the demands of poverty alleviation and development are severe. One potential solution, therefore, would be to relax but not altogether relinquish the sequential imperative. Two examples of such an approach might be:

1 to require, for certain high risk classes of subsidies to artisanal fisheries, parallel and possibly equal investments in fisheries management infrastructure;
2 to allow high risk subsidies to artisanal fisheries in the absence of adequate management infrastructure, but only for a limited period of time.

Note, however, that provisions such as these should not be taken to extremes. Where a fishery is purely open access, or where even the most basic assessment of stocks or fleet capacities have not been performed, investment in capacity- or effort-enhancing subsidies seems a folly.

The foregoing discussion illuminates one other noteworthy aspect of the issue. The need for technology transfer and development assistance to developing countries in the artisanal fishing sector is increasingly recognized as a high international priority. Whether the dynamics of the current fisheries subsidies negotiations at the WTO provide an opportunity to encourage or secure increases in such assistance is a political question beyond the scope of this section, but certainly not beyond the scope of the discussion this section hopes to encourage.

Procedural limits and conditions

A third approach to providing special treatment for subsidies to artisanal fisheries, or to disciplining them, would naturally be the application of typical ASCM 'procedural' devices, such as:

- Notification requirements: it is standard practice within the WTO system to require advanced notification of measures which are permitted by exceptional rules or other derogations from core WTO obligations.[59] While notification requirements are often considered onerous, particularly by developing country governments, they are an essential component of functioning rules. Chapter 5, 'Special and Differential Treatment' for example, considers advance notification requirements a part of a robust approach to S&DT.

Subsidies to artisanal fishing benefiting from special treatment should be similarly conditioned upon advanced notification. Moreover, in the fisheries context, effective disciplines will likely require information about the conditions of production, including some rudimentary information about the 'fisheries management context' in which fishing subsidies are to be applied (WWF, 2004).

- Distribution of the burdens of proof: the effectiveness of rules and the costs of their implementation often depend on which party to a potential dispute must carry the burden to prove a claim. In the case of disciplines on subsidies to artisanal fishing, these burdens should be allocated to maximize the ability of developing countries to pursue poverty alleviation and development, without losing sight of the need to prevent unsustainable production distortions. Thus, for example, if new WTO rules were to allow capacity-enhancing subsidies to artisanal fisheries, but subjected such subsidies to stricter disciplines than non-capacity-enhancing subsidies, the burden to prove the capacity-enhancing character of a given measure might rest with the party challenging the subsidy. At the same time, however, if such a measure were allowed only in fisheries not considered 'patently at risk', the burden to demonstrate the elements of 'not patently at risk' (such as, for example, the existence of a capacity management plan) might rest on the government granting the subsidy.[60]

Conclusions and general recommendations

The discussion above presumes that subsidies to artisanal fisheries deserve and require special treatment under any new WTO fisheries subsidies disciplines. However, the section emphasizes the irreducible link between the sustainable management of artisanal fisheries and the long-term economic and social health of the communities that depend on them. Further, the section adopts the broad presumption – based on such empirical evidence as exists – that many classes of fisheries subsidies, and particularly those likely to increase fishing capacity or effort, can contribute significantly to the risk of overexploitation in artisanal fisheries.

In this context, the section draws the following basic conclusions:

1. A universally applicable definition of 'artisanal fishing' cannot be established solely on the basis of how the term is applied in practice outside the WTO context.
2. It is desirable to find a relatively narrow definition of 'artisanal fishing' for application in the WTO context, in order to avoid opening unintended loopholes in any new fisheries subsidies disciplines.

3. In crafting an appropriate definition of 'artisanal fishing', close attention should be paid to the interests and policy objectives underlying the concerns of governments with this issue in the particular context of the WTO.
4. Among the most compelling objectives for attention to artisanal fishing are poverty alleviation, food security and the development of underdeveloped communities and fisheries. However, the 'development' objective inherently includes transformations of fishing practices, and often implies the intensification of fishing capacity and effort. Thus, to the limited extent that poverty alleviation programmes can be distinguished from fisheries development programmes, associated subsidies may merit different treatment under new WTO rules.
5. An emphasis on poverty alleviation and development suggests substantial overlap between the 'artisanal fishing' and 'S&DT' issues. To the extent that governments adopt robust S&DT for developing country fishing subsidies, the need for specific provisions for artisanal fishing may be reduced.
6. The potential application of subsidies to the cultural preservation of artisanal fishing communities (regardless of income or developmental levels) requires further discussion, and relates in part to the question whether WTO rules should recognize 'artisanal fisheries' in developed countries.
7. Even accepting a more limited definition of artisanal fishing, it appears likely that governments will seek to protect their ability to apply a broad range of types of subsidies in their artisanal fishing sectors. Nevertheless, a discussion of the various types of subsidies that may be applied is necessary to illuminate the particular challenges and risks they raise within the artisanal fishing sector. It may also be useful to classify subsidies to artisanal fisheries in accordance with both their type and their underlying objectives.
8. It is unlikely that even an aggressive and successful effort to limit the definition of 'artisanal fishing' and to distinguish those subsidies most likely to be harmful in the artisanal fishing context will solve the basic tension between the need for developmental subsidies and the dangers associated with the intensification of fishing. Thus, it will probably be necessary to apply disciplines to some permitted classes of subsidies to artisanal fisheries (particularly those classes covering subsidies likely to be effort-or capacity-enhancing).
9. The approach of disciplining subsidies through conditionality related to the fisheries management context – already proposed by some delegations outside of the artisanal fishing discussion – merits serious exploration, bearing in mind: the limits imposed by the proper scope of the WTO's institutional competence and the need to relax (but not abandon) the 'sequencing imperative' that counsels in favour of establishing a robust fisheries management context prior to the purposeful expansion of artisanal fishing capacity or effort.
10. Effective rules on subsidies to artisanal fisheries will probably require nuanced use of procedural provisions already familiar to the ASCM, including notification requirements, distributions of burden of proof and adjustments of available remedies.

The Special Case of Access Agreements*

Introduction

The treatment of subsidies related to 'access agreements' has emerged as a sensitive topic within the current WTO fisheries subsidies negotiations. While a consensus appears to have emerged on the need to discipline fisheries subsidies that contribute to over-capacity and over-harvesting, in light of their negative impacts on international trade, the marine environment and sustainable development more generally, subsidized fishing enabled by access arrangements has emerged as a controversial issue. Delegations from countries highly dependent on access fees have been especially uneasy, fearing that new WTO rules could reduce the north–south monetary transfers these fees represent. Other countries have pointed to the lack of transparency surrounding these agreements and to their negative impacts both on sustainability and on international markets. The discussion on this topic has been further compounded by difficulties of terminology, which have made it difficult to clearly identify what is the subsidy element, if any, involved in access arrangements.

The access arrangements at issue here generally involve government-to-government payments in return for foreign access to developing countries' EEZs.[61] Such access arrangements constitute significant sources of income for some developing countries, in particular SIDS, and thus may be important to meeting legitimate development needs. In light of their importance for the budget of certain coastal, developing countries, SIDS and other countries have proposed excluding access arrangements from any definition of fishing subsidy.

At the same time, fisheries access arrangements now form the main supply for fishery species such as tuna, some demersal fishes and molluscs to the EU and Japan, which are major DWFNs. Fisheries access payments, subsidies on fishing vessels, financial credits and compensation on joint ventures with third countries form significant fisheries budgets in these DWFNs.

Environment and development implications of fisheries access arrangements

It is clear that developing countries do not always get the best end of the access arrangement bargains (ICTSD, 2006d). The terms of the arrangements often leave the host country with only a fraction of the actual resource value, and more than a few access arrangements have led to the depletion of host country stocks.

Case studies in Senegal and Argentina have shown how distant water fishing

* Excerpted from: UNEP (2008a). UNEP had commissioned the original paper to Marcos Orellana from the Centre for International Environmental Law (CIEL) based on a request by governments to provide analysis on how the issue might be handled in the WTO context. The original paper has benefited from review and discussion at a UNEP-ICTSD-WWF workshop in May 2006 (see accompanying CD-ROM for chair's conclusions of this workshop). The report has been slightly adapted for the purposes of this book and some footnotes have been omitted from this excerpt. The full original version of this report is included in the accompanying CD-ROM.

nations, facing overexploitation and fisheries collapse in their own waters, have transferred their problem of overcapacity to distant waters via bilateral access agreements (UNEP, 2004c).[62] Research on former Euro-African fishing agreements reveals how distant water fleets significantly contributed to overfishing and declining yields in African waters (WWF, 1997; UNEP, 2006a). Moreover, as the Senegalese example illustrates, vessels often discard as 'by-catch' fish that are not of the required species or size agreed in the arrangements in order to maximize the value of their output. Furthermore, in the absence of proper means and equipment for monitoring fishing activities, fishing by foreign fleets in Senegal's EEZ takes place virtually without control on the part of the Senegalese authorities (UNEP, 2004c). Such practices compound the problems raised by the fact that access agreements are only very rarely accompanied by thorough stock assessments – hence, frequently neither the surplus of a fishery is determined before concluding arrangements (ADE-PWC-EPU 2002),[63] nor are precise provisions consistently included related to effort or catch limits (Clark, 2006; Sporrong et al, 2002).

Declining fish stocks, and related impacts on the marine ecosystem, entail severe social and economic consequences for the local fishing population and the development of the island or coastal state. Senegalese agreements have attracted particular attention not only because they comprise high-volume fish catches, but also since they involve species that are endangered or used locally, in other words, that are strategic from the point of view of food security. Frequent conflicts between distant fishing fleets and local small-scale fleets relate to competing for the same stocks, gear conflicts when vessels occupy the same fishing grounds and destruction of locally-important habitats such as reefs and seagrass beds (ICTSD, 2006d). In addition, access arrangements have often been criticized as unfair given that they are very rarely based on resource rent principles – access agreements in the southwest Indian Ocean and also the western and Pacific Ocean, for example, are estimated to account for not more than 5–10 per cent of the value of the catch.

The fact that distant water fleets are often highly subsidized exacerbates the impacts depicted above. By lowering the production costs of fishing units, fishing agreements may encourage foreign vessels to fish beyond the economic optimum compatible with sustainable resource management, discourage the exit of fishing vessels from troubled fishing industries and encourage overfishing as mentioned above (WWF, 1997). To the extent that distant water fleets receive subsidised access when their governments acquire access rights for them, the subsidized access itself can further distort competitive relationships on the international level.

On the positive side, and in addition to the important contribution of access payments to SIDS economies mentioned above, access arrangements have the potential to help integrate developing-country fishing or fish-processing industries into the global economy.[64] Likewise, if properly designed and implemented, they can help promote conservation and sustainable fisheries management (WWF, 2004, pp.53ff).

International debate on fisheries access agreements and UNEP involvement

The mixed social, economic and ecological consequences of fisheries access arrangements have been explored by a multitude of actors on the national and international level, such as the Coalition for Fair Fisheries Agreements (CFFA) (Gorez, 2005), WWF,[65] Enda Diapol,[66] the International Centre for Trade and Sustainable Development (ICTSD, 2006d), the OECD[67] and the EU (Bartels et al, 2007) itself. Coming from different perspectives, ranging from an environmental to a policy coherence dimension, these actors have denounced the negative impacts of inappropriately designed fisheries access agreements and have called for international action to make them supporting instead of undermining sustainable development.

As mentioned above, subsidies related to access arrangements are often part of a broader package of subsidies having negative impacts on fishing stocks and associated ecosystems. Fees paid by European ship owners covered by the agreements, for example, represent only about 10 per cent of the price of access paid by the European Commission, which itself represents only a fraction of the actual value of the target fisheries resources (UNEP, 2004c). Some of the complex environmental and development impacts of such subsidies have been covered by UNEP case studies (UNEP, 2004a). The question of how the specific characteristics of a fishery (for example, its level of exploitation and its management regime) may affect those impacts has been reviewed in Chapter 2.

The legal and policy implications of the issues raised above have been discussed in several workshops organized by UNEP. A June 2005 UNEP roundtable on fisheries subsidies resulted in a request by governments for UNEP to study the question of access arrangements in the broader political economy framework, with a view to providing analysis on how the issue might be handled in the WTO context.[68]

On the basis of some preliminary results, the May 2006 discussion (UNEP-ICTSD-WWF Workshop) on access arrangements helped clarify some key points.[69] Firstly, there appeared to be agreement that new WTO rules should not treat government-to-government access fee payments as 'subsidies' flowing between distant water fleet nations and host EEZ nations. Secondly, participants from all perspectives appeared united in the view that new WTO rules should not impede or discourage the access payments on which many small vulnerable economies depend. Thirdly, participants also generally agreed that the lack of transparency of current access arrangements poses significant problems and that there might be a role for the WTO in this regard. However, some differing views existed with regards to measuring the subsidy element of access agreements.[70]

In December 2006 UNEP, jointly with WWF, organized a discussion that showed growing consensus on the fact that there is no subsidy in the access arrangement itself, but rather there is *only* a subsidy under ASCM Article 1 when a DWFN fails to receive sufficient payment from its distant-water fishing fleet in exchange for onward transfer of rights to fish in foreign EEZs. The legal analysis that supports this conclusion is presented below. It starts from the assumption that *any subsidy that might be found within access arrangements can only arise between the DWFN and its own domestic fleet*, on whose behalf the DWFN secured access to foreign fishing grounds.

Since December 2006, WTO negotiations have slowly resumed, including debates in the Rules Negotiating Committee on fisheries subsidies reform. In this context, a number of proposals addressing access agreements have been submitted and discussed. These submissions reflect the evolution of the debate by clearly distinguishing access agreements, the payments pursuant to such arrangement and the further transfer of rights to the distant water fleet. Moreover, several country positions adhere to the idea of conditionality by suggesting environmental and transparency requirements – which reveals a growing attention to significant sustainability concerns related to fisheries access agreements.

Issue, scope and terminology

This section builds on the above-mentioned developments. It analyses the legal framework governing access arrangements and explores options for improved disciplines on fishing subsidies. In that regard, this section takes the view that if negotiators can agree on appropriate means to include subsidies associated with fishing access arrangements within new WTO rules, it could contribute to reforms that help: (a) create a fair trading field for competing DWFN fleets, as well as for the local fleets of host nations operating in their own EEZ's; (b) enhance sustainable development of SIDS and coastal developing countries; (c) establish greater international transparency in cases of subsidized access, and (d) avoid over-exploitation of fisheries resources and associated ecosystems.

More particularly, this section will analyse the legal issues relevant to the treatment of access agreements under the disciplines of the WTO Agreement on Subsidies and Countervailing Measures (ASCM), with a view to clarifying how subsidized fishing enabled by access arrangements falls under the subsidies definition of the ASCM.

In approaching this task, the section will take as a given the reader's familiarity with the ASCM and will only refer to its architecture when the analysis so requires. The ASCM may be found in Annex 10 of this book.

Further, at least three preliminary distinctions between '*government payments*', '*access to fisheries*' and '*moneys collected*' are relevant to this topic, namely:

- *payments* from a DWFN to a coastal State to secure access to its EEZ fisheries;
- *access* to foreign EEZ fisheries granted to a distant-water fleet pursuant to government-to-government access arrangement;
- *remuneration* collected by DWFN from its distant-water fleet in exchange for the transfer of access rights to fish in a foreign EEZ.

Lastly, as to the issue of terminology, which has been the source of considerable confusion and controversy, this section will use the terms access arrangements or access agreements interchangeably.

Access arrangements and the Law of the Sea

In order to properly contextualize the Doha Negotiations on fishing subsidies, particularly as they relate to access arrangements, references to the origins of the exclusive economic zone (EEZ) highlight the importance of access arrangements in fostering development and food security. In addition, the UN Convention on the Law of the Sea (UNCLOS) and the FAO Code of Conduct provide relevant legal context to access arrangements.

Access arrangements and the UN Convention on the Law of the Sea

Under the UNCLOS, coastal states retain sovereign rights over the natural resources found up to 200 nautical miles from their coasts. The UNCLOS qualifies this sovereign right in important ways. For example, the UNCLOS provides that where the coastal state does not have the capacity to harvest the entire allowable catch in its EEZ, it shall, through agreements or other arrangements give other States access to the surplus of the allowable catch. Again here, the objectives of food security and development are apparent in the attempt to ensure the optimum utilization of fisheries.

This obligation to grant access to its EEZ via access arrangements in turn calls for several important considerations. Firstly, the coastal state determines the allowable catch and this determination is excluded from compulsory dispute settlement; so in practice the obligation to grant access remains at the discretion of the coastal State (UNCLOS, Articles 61&297(3)). Secondly, in granting access, preference should be accorded to developing land-locked and geographically disadvantaged states, taking into account economic considerations and nutritional needs (UNCLOS, Articles 62, 69 & 70). Thirdly, EEZ laws and regulations shall also apply to nationals of countries granted access, including those relating to among others: vessel position reports, species, seasons, gear, etc. and requirements for local landings (UNCLOS, Article 62).

To recap, customary law confers a sovereign right on the Coastal State over its EEZ and under the LOS Convention the Coastal State is under an obligation to grant access to its EEZ via access arrangements where it does not have the capacity to fish the allowable catch. This framework under the Law of the Sea governing access arrangements to foreign EEZs would stand in direct conflict with the WTO if it were to be concluded that access arrangements, per se, constituted a violation of WTO law. While instances of conflicts of norms may be found in international law, the situation in respect of access arrangements does not seem to present such conflict among the WTO and the LOS Convention, as examined further below. In other words, the WTO and the LOS Convention can be applied concurrently. Before analysing WTO law and the SCM Agreement in particular, however, a final element of context is relevant to situate access arrangements.

Access arrangements and the FAO Code of Conduct

In addition to the LOS Convention, the Code of Conduct for Responsible Fisheries adopted under the auspices of the FAO contains several provisions relevant to access arrangements. Generally, the Code places strong emphasis on enhancing the ability of coastal States in developing their own fisheries (FAO, 1995a, Article 5.2), as well as on securing a livelihood for subsistence, artisanal and small-scale fishers (FAO, 1995a, Article 6.18). More particularly, the Code of Conduct provides that states should not condition access to markets on access to resources (FAO, 1995a, Article 11.2.7). The Code also states that this principle does not preclude fishing agreements between states which include provisions referring to access to resources, trade and access to markets, transfer of technology, scientific research, training and other relevant elements. However, it links the 'right to fish' to the obligation to do so in a responsible manner so as to ensure effective conservation and management of the living aquatic resources (FAO, 1995a, Article 6) and contains detailed provisions related to management measures for the long-term conservation and sustainable use of fisheries resources (FAO, 1995a, Article 7).

Having established the legal framework that enables and governs EEZ access agreements, it is now necessary to examine how WTO Law and particularly the ASCM relate to access arrangements.

WTO law and access agreements

The problem of how access agreements relate to the ASCM raises complex legal issues. These issues have been compounded both by terminology difficulties, as well as by the types of questions asked. Indeed, while early discussions focused on whether access agreements themselves constituted a subsidy, more recent comments have sought to identify certain elements or practices enabled by access agreements that could constitute a subsidy.

In approaching the legal issues involved in the examination of access agreements under the ASCM, it is important to recall the basic components of these agreements. The starting point is: a government-to-government agreement that establishes the right of access – under agreed conditions – for the distant water fleet of one government to EEZ fisheries of the other. According to the particular *structure* of a specific agreement, this right of access might be exchanged for a fee, for 'free' (in other words, without explicit *quid pro quo* on the face of the agreement), or in exchange for non-pecuniary rights, which might consist of the reciprocal access to fisheries. Then, variations in the *modes of implementation* could include, among others:

- the level of fees paid by one of the governments;
- the form in which fees are calculated and paid, for example, lump sum or catch contingent;
- the degree to which the government paying the fees recovers them from its industry;

- the kind of conditions and regulations introduced into the agreement, including, for example, vessel monitoring systems, local landings, quotas, seasons, technology transfer, etc.; and
- the level of transparency in negotiations and reporting.

These modes of implementation are relevant to the analysis of how the ASCM applies or relates to access arrangements because they will determine whether a government grants a financial contribution to its industry that confers a benefit, in other words a subsidy.

Definition of subsidy under the ASCM

The starting point in the analysis of access arrangements under the SCM Agreement is, of course, the definition of 'subsidy' (ASCM Article 1, see Annex 10). For a subsidy to be deemed to exist within the scope of the ASCM two elements need to be present. Firstly, there must be a financial contribution by a government, or by a private body 'entrusted' or 'directed' by the government. Secondly, a benefit must thereby be conferred (ASCM, Article 1.1) (WTO Appellate Body Report, 'US – Softwood Lumber IV', para. 51). When these two elements are found in a governmental measure, a subsidy exists under the ASCM.

WTO jurisprudence has clarified the meaning of these terms to a large extent. Still, several areas remain subject to interpretation and thus are open for discussion. For example, while the Appellate Body observed that Article 1.1(a)(1) sets out a 'wide range of transactions' that fall within the meaning of a financial contribution (WTO Appellate Body Report, US – Softwood Lumber IV, para. 52.), the particular scope of application of these 'transactions' involves a degree of uncertainty. With regard to the second element of a subsidy, the term 'benefit' is not defined at all in the text of the ASCM (Panel Report, EC – Countervailing Measures on DRAM Chips, para 7.173). Due to the scarcity of textual guidance in the definition of a subsidy, the analysis that follows emphasizes jurisprudential developments.

The following sections focus on the two elements of the definition of subsidy in the ASCM (financial contribution and benefit), with a view to identifying how the ASCM, as it currently stands, applies to subsidized fishing under access arrangements. Recalling the terminological distinction highlighted above, this analysis will not concern the *payments* from a DWFN to a coastal state, but only refer to the onward transfer of access rights and the corresponding *remuneration* collected by the DWFN from its distant-water fleet.

Interpretation of 'financial contribution'

Out of the two elements of a subsidy, financial contribution presents less of a definitional problem. As noted by the Panel in 'EC – Countervailing Measures on DRAM Chips', 'Article 1.1(a)(1) subparagraphs (i) to (iii) set forth three situations that are considered to constitute such a financial contribution by the government', while subparagraph (iv) 'adds that a financial contribution may also be considered to have been provided by the government, in cases where the government has entrusted or directed a private body to provide one of the types of financial contributions' (Panel Report, EC – Countervailing Measures on DRAM Chips, para 7.48).

Stated differently, the overall parameters of what constitutes a financial contribution are well delineated in the ASCM. In that regard, the Appellate Body observed that:

> ...a financial contribution may be made through a direct transfer of funds by a government, or the foregoing of government revenue that is otherwise due...in addition to such monetary contributions, a contribution having financial value can also be made *in kind* through governments providing goods or services, or through government purchases. (Appellate Body Report, 'US – Countervailing Duty Investigation on DRAMs', para 52) [Author's emphasis]

The emphasis added in the quote above clarifies that financial contribution need not take the form of money, but can also occur through in kind contributions, such as the provision of goods. In this ambit, one issue that immediately surfaces in respect of access arrangements concerns the provision of rights to goods, such as the right to fish. This issue is addressed further below.

For the purposes of examining access arrangements under the terms outlined by the ASCM, as interpreted by the Appellate Body, the following questions appear most relevant.

IS THERE A FINANCIAL CONTRIBUTION BY THE GOVERNMENT OR A PUBLIC BODY WITHIN THE TERRITORY OF A MEMBER?

According to ASCM Article 1, 'there is a financial contribution by a government or any public body within the territory of a member (referred to in this agreement as 'government')'. Thus the question whether there is a 'government' that provides a financial contribution.

The answer to this question will operate as a threshold and determine whether the examples of financial contributions listed in the ASCM apply to access arrangements. It appears that in the context of a government-to-government access agreement which secures access to fish in the coastal State EEZ, the DWFN is a 'government' for the purposes of the ASCM. Further, the access rights so obtained by the government are then transferred by the government to its fishing industry. Consequently, subsidized fishing under access arrangements falls under the scope of ASCM Article 1.

It has been argued, however, that no government-to-government transfers occur 'within the territory of a Member', as access arrangements occur between governments and that consequently these agreements are beyond the scope of the ASCM. This interpretation appears to read into Article 1 a territorial limitation of the ASCM. In that sense, this interpretation confuses the territorial application of the ASCM, on the one hand, with the definition of public body, on the other.

On account of the text in its context, and in light of the object and purpose of the ASCM, the better interpretation reads the phrase 'within the territory of a Member' as a way to distinguish and qualify the 'public body' that immediately antecedes the phrase. In other words, a public body within the territory of a member that provides a financial contribution will be subject to the ASCM. Consequently, the place where the financial contribution takes place is not relevant. This interpretation is also consonant with the object and purpose of the ASCM, which, among other things, attempts to reduce distortions in the conditions of competition in international trade.

Is there a government practice involving a direct transfer of funds?

According to ASCM Article 1, there is a financial contribution where a government practice involves a direct transfer of funds (for example, grants, loans and equity infusion), potential direct transfers of funds or liabilities (for example, loan guarantees); thus the question. This question received some attention by commentators before the Appellate Body decision in the *Softwood Lumber* case, examined further below.

At first sight the wording of Article 1.1(a)(1)(i) does not explicitly cover government-to-government access payments, but it does not exclude a coverage either.

It has been observed by Porter that access payments are not a 'direct' transfer of funds from government to industry, but rather a transfer of funds from government-to-government (WWF, 1998b, p. 63). This view is also shared by Stone, who observed that such payments would be indirect and thus outside the scope of the ASCM (Stone, 1997).

On the other hand, it has been argued that payments are a direct transfer of funds from the government that confers a benefit. Schorr has explored this vein, noting that the term 'direct' does not require that the subsidy flow 'directly to the subsidized party' (WWF, 2004). Schorr bases his argument on the fact that for a subsidy to exist under Article 1 there must not only be a 'financial contribution' (Art.1.1(a)) but also a 'benefit' conferred (Art. 1.1(b)). These requirements are set out in separate paragraphs of Article 1. This separate treatment implies that it is only the benefit and not necessarily the financial contribution that must run to the subsidized party. According to this reasoning the word 'direct' is only meant to define one form of a financial contribution (the actual transfer of money) in contrast to other forms listed in Article 1.1(a)(1) (for example, contribution through an intermediary funding mechanism named in Article 1.1(a)(1)(iv)). Schorr's argument is therefore that not only 'direct' but also 'indirect' transfers of benefits are covered by the definition of Art. 1.1(a)(1)(i).

As noted by Chang, another argument that supports the view that a financial contribution may accrue indirectly through a transfer of funds is that Article 1 of the ASCM does not require that the recipient and the beneficiary of a 'financial contribution' be identical (Chang, 2003). Chang further underlines that indirect subsidies are not a unique feature of the fisheries sector but are just as likely to occur in other manufacturing sectors as well, citing the WTO panel's decision in the 'United States Lead'.[71]

These 'indirect transfer' analyses, however, falter in the case of access agreements insofar as they focus on government-to-government payments instead of on the relationship between a distant water nation and its industry. A different and ultimately more satisfactory approach arises under ASCM Art. 1.1.(a)(1)(iii). This alternative approach – which has been explored in detail by the Appellate Body's 'Softwood Lumber' decision (WTO, n.d.) and might be more relevant to the issue of access arrangements in the fisheries subsidies context – will be examined next.

IS THERE A PROVISION OF GOODS OR SERVICES BY THE GOVERNMENT?

According to the ASCM, there is a financial contribution where the government provides goods or services. This question is probably the most directly applicable to the access rights acquired by the distant water fishing nation and the transfer of these fishing rights to its distant water fleet. By entering into access arrangements, do governments provide goods or services to its industry? Or rather, by transferring the fishing rights obtained by virtue of access agreements to its industry, do governments provide goods or services to its industry?

On the one hand, it could be argued that access rights (licenses, permits) are not a provision of goods given their regulatory character. Just like other laws or regulations permitting economic activities, such licenses do not constitute a provision of goods. On the other hand, it could be argued that access rights to fish – and particularly of rights to fish *in a foreign EEZ* – do constitute a 'provision of goods' under the ASCM. In that vein, the purpose of Article 1 is to reach any government measure that provides capital, operating resources or other services to specific industries on terms better than could be obtained on the open market.

This issue, whether rights to fish constitute a provision of goods, is illuminated by the 'US – Softwood Lumber IV' case, as explained next. The discussion is divided into two parts: firstly the question of what constitutes a 'good' and secondly the question of what does it mean to 'provide' goods.

Interpretation of 'goods' in Article 1.1(a)(1)(iii)

In 'US – Softwood Lumber IV' Canada appealed the final determination of the US Department of Commerce that 'Canadian provincial governments made a financial contribution because, through stumpage arrangements, those governments *provide goods* to timber harvesters,' (Appellate Body Report, 'US – Softwood Lumber IV', para 46). The Panel had ruled that 'providing standing timber to the timber harvesters through the stumpage programmes' fell within the scope of Article 1.1(a)(1)(iii), as providing the access rights to standing timber could be classified as providing goods (WTO, n.d., Appellate Body Report, 'US – Softwood Lumber IV', para 47).

Appealing this decision, Canada argued, 'standing timber, that is, trees attached to the land and therefore incapable of being traded as such, are not "goods"' (WTO, n.d., para. 48). Canada's argument was based primarily on the contention that the term 'goods' only encompasses 'tradable items with an actual or potential tariff classification' (WTO, n.d., para. 54.).

The argument that only 'harvested timber' is traded, and not 'standing timber' – and therefore, because the latter is not a 'tradable item,' it is not a 'good' – was rejected by the Appellate Body. In rejecting this argument, the Appellate Body decided that stumpage contracts that provide access to an area of land implicitly provide the individual trees for purposes of harvesting (WTO, n.d., para 66). For this reason, the Appellate Body found no reason to exclude standing timber from the scope of 'goods' based on the notion that stumpage agreements do not explicitly provide the individual harvested trees. Also, in considering the ordinary meaning of the term 'goods' in Article 1.1(a)(1)(iii), the Appellate Body found that it does not 'exclude tangible items of property, like trees, that are severable from land' (WTO, n.d., para 59).

The argument that 'goods' must have a 'potential or actual tariff classification' was also rejected, as it implies that the term 'goods' in the SCM Agreement have the same definition as 'products' in the GATT 1994 (WTO, n.d., para 61). In the Appellate Body's view, the scope of the meaning of 'goods' should not be limited by the definition of the term 'products' in Article II of the GATT 1994. Such an interpretation would 'undermine the object and purpose of the SCM Agreement, which is to strengthen and improve GATT disciplines' (WTO, n.d., para 64). Accordingly, the Appellate Body decided "Goods' in Article 1.1(a)(1)(iii) of the SCM Agreement and 'products' in Article II of the GATT 1994 are different words that need not necessarily bear the same meanings in the different contexts in which they are used'(WTO, n.d., para 63).

In sum, the Appellate Body found that,

> nothing in the text of Article 1.1(a)(1)(iii), its context, or the object and purpose of the SCM Agreement, leads us to the view that tangible items – such as standing, unfelled trees – that are not both tradable as such and subject to tariff classification, should be excluded...from the coverage of the term 'goods' as it appears in that Article (WTO, n.d., para 67).

This decision is relevant to the question of subsidized fisheries under access agreements in light of the clear parallels between trees and fish. Both trees and fish are tangible goods that can be harvested from the land or the sea. Both trees and fish are generally fungible goods, except in rare circumstances. Both trees and fish are harvested by virtue of regulatory permits that usually specify conditions of harvest, such as quotas and location. Further, both trees and fish become the object of possession and exclusive appropriation when harvested. Thus, if trees standing in provincial lands are 'goods', it would be extremely odd if fish swimming in the EEZ were not.

In sum, trees under stumpage agreements as well as fish under access agreements are 'goods' under Article 1 of the SCM Agreement. Thus, if they are provided by the government, there will be a 'financial contribution'. This leads to the question of what does it mean to 'provide' goods, addressed next.

Interpretation of 'provides' in Article 1.1(a)(1)(iii)

The second issue regarding financial contribution raised in the 'US – Softwood Lumber IV' dispute is the scope of the term 'provides' in Article 1.1(a)(1)(iii). Canada argued, 'stumpage arrangements do not "provide" standing timber…all that is provided by these arrangements is an intangible right to harvest' (WTO, n.d.,para 68). That is, the access agreement only 'makes available' standing timber. Canada's contention was that 'makes available' is not the same as 'provides.' The Panel and Appellate Body were not impressed by these arguments.

In 'US – Softwood Lumber IV', the Appellate Body observed that:

> …the Panel found that stumpage arrangements give tenure holders a right to enter onto government lands, cut standing timber, and enjoy exclusive rights over the timber that is harvested. Like the Panel, we conclude that such arrangements represent a situation in which provincial governments provide standing timber…By granting a right to harvest, the provincial governments put particular strands of timber at the disposal of timber harvesters and allow those enterprises, exclusively, to make use of the resources' (WTO, n.d.,para 75).

Moreover, as the Appellate Body observed, 'the evidence suggests that making available timber is the *raison d'être* of the stumpage arrangements' (WTO, n.d., para 75). Accordingly, the Appellate Body concluded that 'by granting a right to harvest standing timber, governments provide that standing timber to timber harvesters' (WTO, n.d., para 75).

For this reason, the Appellate Body upheld the findings of the US Department of Commerce, that providing standing timber through stumpage programmes is the same as providing a good, and therefore falls within the meaning of providing a financial contribution in Article 1.1(a)(1)(iii) (WTO, n.d., para 76).

Again here, the similarities between stumpage arrangements and access arrangements are ostensible. To paraphrase, making available fish is the *raison d'être* of the access arrangements. And by granting a right to harvest fish in the foreign EEZ, DWFNs provide that fish to fishers.

In light of these similarities, the conclusion is warranted that the transfer of foreign EEZ access rights by the DWFN to its fleet constitutes a financial contribution. Still, the fact that there is a financial contribution does not mean that there is a subsidy, as the second element of the definition of subsidy, in other words, a benefit, also needs to be satisfied.

Interpretation of 'Benefit'

The second element that must be satisfied for a subsidy to be deemed to exist is the conferral of a 'benefit'. However, the ASCM does not define what is meant by the term 'benefit'. In that regard, the determination of the definitional scope of the term is an issue where WTO jurisprudence is particularly helpful. Another issue relevant to the determination of whether a benefit exists is the appropriate benchmark that should be used to calculate the numerical value of a benefit. These two issues are examined in light of the WTO jurisprudence, where available.

The most useful source for establishing what is meant by the term 'benefit' is the Appellate Body Report in 'Canada – Aircraft'. The issue in this case was whether the Panel had erred in its interpretation of 'benefit.' The decision process followed by the Panel, as quoted by the Appellate Body, set out that:

> [dq]...the ordinary meaning of 'benefit' clearly encompasses some form of advantage...In order to determine whether a financial contribution confers a 'benefit', i.e., an advantage, it is necessary to determine whether the financial contribution places the *recipient* in a *more advantageous position than would have been the case but for the financial contribution*. In our view, the only logical basis for determining the position the recipient would have been in absent the financial contribution is the *market*. Accordingly, a financial contribution will only confer a 'benefit', i.e., an advantage, if it is *provided on terms that are more advantageous than those that would have been available to the recipient on the market*' (Panel Report quoted in Appellate Body Report, 'Canada – Aircraft', para 149 [author's emphasis].

The Appellate Body, in rejecting Canada's appeal, upheld the Panel's interpretation of the term 'benefit' (para 161). The decision is based on several considerations. First, it argued, the term 'benefit' implies that there must be a recipient, and this 'provides textual support for the view that the focus of the inquiry under Article 1.1(b) *should be on the recipient* and not on the granting authority' (para 154) [author's emphasis]. This is backed up by the ordinary meaning of the word 'confer' in Article 1.1(b), which 'calls for an inquiry into *what was conferred on the recipient*' (para 154).

Article 14 provides contextual support for this interpretation. The 'explicit textual reference to Article 1.1' in Article 14 indicates that the two Articles are using the term 'benefit' in the same way (para 155). Therefore, 'the reference to 'benefit *to the recipient*' in Article 14 also implies that the word 'benefit,' *as used in Article 1.1*, is concerned with the 'benefit *to the recipient* and not with the 'cost to government' (para 155).

In this discussion, three elements may be distinguished:

1 whether the recipient is left in a more advantageous position than would have been the case but for the financial contribution;
2 whether the recipient is left better off, regardless of any cost to the government;

3 whether the terms of the financial contribution are more advantageous than those that would have been available on the market.

Although these three elements are intertwined, they are addressed separately in turn.

Whether the recipient is left in a more advantageous position than would have been the case but for the financial contribution

The Panel in 'EC – Countervailing Measures on DRAM Chips' equated the ordinary meaning of 'benefit' to 'that of an "advantage", something which leaves the recipient better off' (Panel Report, 'EC – Countervailing Measures on DRAM Chips', para 7.173).

On the one hand, it has been argued that access arrangements improve the competitive position of an industry that acquires access to a resource that it otherwise would not have had. Stated differently, if it were not for the financial contribution, certain fishing industries might not be able to capitalize on their investments and thus would be forced off the market. On the other hand, it has also been argued that such access in itself does not make the recipient better off if other conditions are present, such as recovery of adequate fees by the government, etc. Under this light, the access itself does not appear determinative of whether the recipient is left in a more advantageous position, but rather an inquiry is due to the *conditions associated to the financial contribution*, in particular whether the industry has been charged *adequate fees* for the goods it has been granted.

Whether the recipient is left better off, regardless of any cost to the government

One implication of the Panel's findings in 'Canada – Aircraft' is that, in interpreting the term 'benefit,' no consideration needs to be given to the 'cost to government,' (WTO, n.d., para 150). That is, when seeking to establish if a benefit has been conferred by a financial contribution, what needs to be taken into account is the relative position of the recipient and not the government.[72] This is particularly important for access agreements, as consequently the analytical focus is not on the cost to the government, in other words, the level of access fees that the DWFN paid to the coastal State, but on the value of the goods or services that the fishing industry received from the DWFN.

Further, in the context of subsidized fishing under access arrangements, the *recipient* appears to be the distant water fleet that gains access to an EEZ fishing right. Whether such a recipient is left better off as a result of the transfer of such access rights must be determined by *reference to the market*, as explored next.

Whether the terms of the financial contribution are more advantageous than those that would have been available on the market

In accordance with the Appellate Body, the word 'benefit' in Article 1.1(b) implies some kind of comparison. When this requirement is assessed in the contextual light of the ASCM , the Appellate Body decided that the marketplace is the 'appropriate basis for comparison,' (WTO, n.d., para 158). The Panel in

'EC – Countervailing Measures on DRAM Chips' agreed with the Appellate Body decision in the 'Canada – Aircraft' case that the appropriate benchmark for determining whether the recipient has received a benefit is the market, based on a contextual reading of Article 14. By implication, while general criteria can be identified regarding the existence of a benefit, ultimately the presence of such benefit will require a case-by-case analysis.

Thus, a benefit is conferred 'if the recipient has received a financial contribution on terms more favorable than those available to the recipient in the market,' (Ibid., para 159). This reference to the 'market' raises several questions in regards to access agreements.

In the 'Softwood Lumber' case, the Appellate Body concluded that 'a benefit is conferred when a government provides goods to a recipient and, in return, receives insufficient payment or compensation for those goods' (Ibid., para 159). When this reasoning is applied to access rights to foreign EEZ fish, the issue immediately turns on whether the recipient fishing industry has paid an *adequate price* to its government in exchange for the access rights. Where industry has received access rights for free, there will be a strong case that a benefit has been conferred. But when industry has received access rights in exchange for some amount of payment, the question then becomes how to determine the adequacy of remuneration.

In accordance with Article 14 of the ASCM, 'the adequacy of remuneration shall be determined in relation to prevailing market conditions for the good or service in question in the country of provision or purchase (including price, quality, availability, marketability, transportation and other conditions of purchase or sale).'

The starting point in the analysis is by reference to the prices of the goods in relation to the prevailing market conditions in the country of provision. As anticipated, the reference to the 'market' raises several questions. Firstly, is it a market of access rights or of fish? As the financial contribution refers to the provision of goods, the analysis could thus focus on which *goods* are provided. In that sense, it appears that it is not just fish generally, but live fish in a foreign EEZ. If this is right, then there may not always be a market for such swimming fish in the water. And secondly, is it a market in the DWFN, in the coastal state, or a world market? As the financial contribution refers to the provision of goods, the country that *provides* the goods is the DWFN. However, the fish are located in a different jurisdiction, which raises difficult interpretative problems.

These questions have not been addressed in dispute settlement and so there is little guidance from the Appellate Body. In any event, while the answer to these questions will be key to determining whether in a specific case a subsidy has been granted, for the purposes of this analysis, we do not need to arrive at definitive answers to these questions, because what is clear is that there will be a 'benefit' when the fishing industry has not paid an adequate price to its government in exchange for fishing rights.

As much as we do not need to answer the questions outlined above, we do need to demonstrate nevertheless that the questions are *answerable*. Stated differently, is it possible to reach a determination of 'benefit' under the current rules of the ASCM? This question surfaces in a submission by the African, Caribbean and Pacific Group of States (ACP) to the Negotiating Group on Rules, which states, 'Since the fishery access payments made are usually the result of a series of bilateral negotiations with the DWFNs, there appears to be no workable "market" benchmark against which one can examine whether the recipient is better off than it would otherwise have been' (TN/RL/W/209 (2007) para 9).

In this regard, there appears to be some indicia in WTO jurisprudence which could guide the analysis relating to the question of whether a workable 'market' benchmark can be found or constructed. In this vein, in the 'US – Softwood Lumber' case the Appellate Body addressed the issue whether 'an investigating authority may use a benchmark, under Article 14(d) of the ASCM, other than private prices in the country of provision' (WTO, n.d., para 82). The Appellate Body found that:

> Members are obliged, under Article 14(d), to abide by the guideline for determining whether a government has provided goods for less than adequate remuneration. However, contrary to the views of the Panel, that guideline does not require the use of private prices in the market of the country of provision in every situation. Rather, that guideline requires that the method selected for calculating the benefit must relate to, or be connected with, the prevailing market conditions in the country of provision, and must reflect price, quality, availability, marketability, transportation and other conditions of purchase or sale, as required by Article 14(d),' (WTO, n.d., para 96).

Stated differently, prices in the market of the country of provision are the primary, but not the exclusive, benchmark for calculating a benefit (WTO, n.d., para 96). The question that then surfaces is: when is it permissible to consider a benchmark other than private prices in the country of provision, for the purposes of calculating a benefit? In this regard, the Appellate Body observed that, 'an investigating authority may use a benchmark other than private prices of goods in question in the country of provision, when it has established that those prices are distorted, because of the predominant role of the government in the market as a provider of the same or similar goods' (WTO, n.d., para 103).

The question of alternative benchmarks has also received some attention by the Appellate Body, which noted that, 'alternative methods for determining the adequacy of remuneration could include proxies that take into account prices for similar goods quoted on world markets or proxies constructed on the basis of production costs' (WTO, n.d., para 106). The Appellate Body, however, observed that in the particular case it did not need to determine the consistency of any method with the ASCM, as such evaluation will be determined by the way that any such method is applied in a particular case.

Consequently, it is submitted that it is possible to establish a workable market benchmark to determine whether a benefit has been conferred on the recipient of a financial contribution. This benchmark may need to be constructed on the basis of production costs or may take into account world markets. Either way, a benefit will be conferred to a distant water fishing industry when it fails to pay adequate remuneration for the rights to fish in a foreign EEZ. Clearly, the measure of 'adequate remuneration' is not the amount paid from DWFN government to EEZ government for access rights. In the light of the figures mentioned above,[73] it is more likely that the 'adequate remuneration' – since it is supposed to reflect the actual value of those access rights – substantially exceeds the amount paid by the DWFN government under the access agreement.

Conclusion on access agreements under the ASCM

This section analysed the two elements of the definition of subsidy under the ASCM Agreement (as summarized in Table 5.4 below). It found that a financial contribution exists where a DWFN provides its fleet with access rights to fish in a foreign EEZ. It also found that such financial contribution confers a benefit where the DWFN fails to receive sufficient payment in exchange for the right to fish that it provides to its distant-water fishing fleet. This chapter also empha-

Table 5.4 *Summary of Legal Analysis: The onward transfer of rights in the context of fisheries access agreements falls under the ASCM definition of subsidies of Art. 1 ASCM.*

Subsidy Definition under ASCM Article 1	*Analysis with regards to fisheries access agreements*	*Result*
"Financial contribution" Art. 1.1. (a)(1)		✔
...by the government or a public body?	DWFN is a "government" for the purposes of the ASCM.	✔
...within the territory of a member?	This refers to the "public body" – the place where financial contribution takes place is irrelevant.	✔
Government practice involving a direct transfer of funds? Art. 1.1(a)(1)(i)	Access agreements include a direct transfer of funds from government to government – the recipient and the beneficiary of the "financial contribution" do not have to be identical.	?
Provision of goods or services by the government? Art. 1.1.(a)(1)(iii)		✔
...good?	Given the clear parallels between trees and fish, the Appellate Body's interpretation of trees being a "good" under the ASCM in US - Softwood Lumber IV can be transferred to fish under access agreements.	✔
...provides?	Given the similarities between stumpage arrangements and access arrangements, the affirmative answer of the Appellate Body with regards to "provision" in US – Softwood Lumber IV can be transferred to the case of access agreements.	✔
"Benefit" conferred Art. 1.1.(b)	A benefit is conferred when the fishing industry has not paid an adequate price to its government in exchange of fishing rights. It is possible to establish a workable market benchmark to determine if this is the case, but this has to be done on a case-by-case basis.	✔

sized that only a case-by-case analysis can show whether these two elements are present in any particular situation.

In sum, access agreements themselves do not breach any rules of the ASCM, but certain fisheries enabled by access agreements may fall within its disciplines.

In light of this conclusion, there are several options available to WTO members to improve existing rules. Before exploring some of these options, the next section will summarize the proposals that have directly addressed access payments in the Negotiating Group on Rules.

WTO submissions addressing access arrangements

Several WTO members have submitted proposals to the Negotiating Group on Rules that address access payments directly. Table 5.5 summarizes, in the simplest terms, the various positions submitted up to August 2007 when the original paper was drafted. These submissions reflect different views on the role and legal status of access arrangements – with none of them proposing to include access agreements under new fisheries subsidies disciplines. They vary in (i) totally exempting access agreements from new disciplines and (ii) conditioning the exemption of access agreements upon the non-existence of a subsidy element, upon environmental and/or transparency criteria.

While early submissions referred to access payments or fees, more recent submissions more clearly distinguish between the access arrangements, the payments pursuant to such arrangements and the further transfer of rights to the distant water fleet. On the basis of the analysis conducted in this section, it is submitted that this distinction is key to improving subsidies disciplines under the ASCM. To this end, several options are available for WTO members.

Options for improving the ASCM

WTO negotiations on fishing subsidies offer the possibility of improving disciplines to ensure that access agreements contribute to the development of coastal and other states, to removing trade distortions in international fish markets and to the sustainable harvest of fish stocks. Several options are available to WTO members in approaching these negotiations. These options range from:

- inaction;
- improving the definition of subsidies;
- clarifying potential remedies;
- introducing an exception for developing countries that meet certain criteria, and
- strengthening transparency requirements.

These options are explored below. Concrete textual suggestions presented under these options vary from rather stand-alone elements only on access-related subsidies[74] to passages where access-related subsidies are embedded into a more general fisheries subsidies language. Nonetheless, all options have obviously to

Table 5.5 *Key Elements of Country Submissions*

WTO member(s)	*Key elements of position towards access agreements*	*Conditions for exemption*
Small & Vulnerable Economies (SVE)[76]	Propose to exclude access fees in fisheries access agreements from subsidies disciplines on account of special and differential treatment. However, are generally willing to examine possible disciplines which seek to minimize environmental and ecological damage so long as they are mutually supportive of the developmental priorities of SVE and other similarly situated developing countries.	None.
New Zealand[77]	Proposes to allow access payments but subject them to strict transparency provisions.	Transparency provisions.
Brazil[78]	Considers that a fishery subsidy shall be deemed to exist if a benefit is conferred in the onward transfer of access rights from the paying government, and proposes to prohibit such fishery subsidy. In addition, Brazil subjects access payments and transfer of access rights to strict transparency requirements.	Access agreements do not include *subsidy element*; Transparency provisions.
Japan, Korea and Taiwan[79]	Propose to include access payments in a green box (non-actionable), provided that they comply with transparency and environmental criteria.	Transparency and environmental criteria.
Norway[80]	Is not proposing to include access fees in the discipline; however, Norway is willing to consider suggestions that make it necessary for the fishing industry of developed members to reimburse their governments for the financing of such access agreements.	Potentially: DWFN government is *reimbursed* by its fishing industry for financing of access agreements (= no subsidy element).
Argentina[81]	Distinguishes between payments pursuant to government-to-government agreements (outside of the scope of the ASCM) and the transfer of access rights by a government to specific enterprises if not done in exchange for a fair trade price (covered by the ASCM).	Transfer of access rights by a government to specific enterprises is done in exchange for a *fair trade price* (= no subsidies element).
The ACP Group[82]	Notes the general agreement amongst the WTO membership that government-to-government payments are not subsidies. The Group also argues that any secondary transfer of rights should be non-prohibited and non-actionable, on account of the difficulties in identifying a workable 'market' benchmark against which the existence of a 'benefit' could be determined	None.
United States[83]	Proposes to include the onward transfer of access rights to a member's fleet within the definition of subsidies, but to exclude such transfer from the prohibition if in compliance with substantive economic, transparency, and environmental requirements.	Fleet pays compensation to its government comparable to the *cost it would otherwise have to pay for access to the fisheries resources* (= no subsidies element); Transparency and environmental requirements.
Indonesia[84]	Proposes to include the onward transfer of access rights to a member's fleet within the disciplines, but to exclude such transfer from the prohibition provided that a benefit is not conferred by the onward transfer of such rights to the Member's fishing fleet and that agreements are in compliance with environmental and notification requirements.	Member's fleet pays *compensation comparable to the value of the access of the resource* (= no subsidies element); Environmental and notification requirements.

be considered in the context of the overall reform – single useful elements may thus be adapted accordingly and flow into broad proposals.

Maintaining the current situation

The first option is, naturally, inaction; that is, maintaining the current situation with respect to subsidized fishing under access agreements.[75] This option is not without implications, however, given that access agreements and fishing subsidies are closely linked. At one level, maintaining the current situation could represent a lost opportunity to introduce effective disciplines and thus achieve the objectives articulated in the Doha Mandate. At another level, countries that suffer injury from artificially low prices or barriers to market access that result from subsidized fisheries under access arrangements might explore dispute settlement.

In light of the conclusions reached above, any challenge to subsidized fishing under access agreements will be stronger where the foreign fishing industry does not pay or pays a minimal amount in exchange for the access rights. In this regard, the theory presented by some countries that access arrangements constitute an advantage that provides a benefit, on account of their use as tools to access resources otherwise off-limits, might be asserted in a confrontational, legal context. Still, any such challenge will face the difficulty identified above of determining the appropriate benchmark to demonstrate that a benefit has been conferred to the fishing industry.

Yet at another level, maintaining the current situation at the ASCM could displace the fisheries subsidies discussion regarding access agreements in different forums.

Improving the definition of access-related subsidies in the ASCM

Improving the definition of subsidies in the ASCM by explicitly referencing transfers of access rights acquired by virtue of access arrangements would provide for a comprehensive coverage of fishing subsidies and related practices in the improved disciplines. This improved definition could address both the 'financial contribution' element and the 'benefit' element of the subsidies definition. In this regard, it may be more important to clarify the 'benefit' element, as the 'financial contribution' element has been clarified by the Appellate Body. The analysis above shows that this clarification would not change existing law but would simply make explicit what appears already to be implicit in it, reducing potential disputes over the treatment of access agreements as part of potential new fisheries subsidies disciplines.

If this avenue of clarification is pursued, an important element to be considered is an exception for developing countries that could safeguard the income of Small Island Developing Countries (SIDS). In this vein, it could be considered whether this exception should be subject to environmental, economic and transparency criteria.

OPTION 1: CLARIFYING SUBSIDY ELEMENT

The definition of subsidy could be improved to include particular language clarifying the specific subsidy element involved in access arrangements:

> Article 1
> Definition of a **Fisheries** Subsidy
>
> 1.1 For the purpose of this Agreement, a **fisheries** subsidy shall be deemed to exist if:
>
> (a)(1) there is a financial contribution by a government or any public body within the territory of a Member (referred to in this Agreement as 'government'), i.e. where:
>
> (v) **a government provides to its nationals direct or indirect access to fish under the jurisdiction of third states.**
>
> and
>
> (b) a benefit is thereby conferred, **i.e. where:**
>
> (i) **the government fails to recover from its nationals the value of the access provided in (v) above.**

COMMENTARY ON OPTION 1

Option 1 addresses the relationship between access agreements and the ASCM by focusing on the two definitional elements of a subsidy.

Firstly, the financial contribution element explicitly covers the situation where a government provides to its nationals direct or indirect access to fish under the jurisdiction of a third state. The reference to 'direct' or 'indirect' is meant to encompass a broader range of situations. It must be noted that this clarification will clearly cover foreign EEZs. Additionally, reference to 'indirect' may also encompass a situation where a party to a Regional Fisheries Management Organization is selling its fishing quota to another party.

Secondly, the benefit element explicitly covers the situation where a government fails to recover the value of such access from its fleet.

The merits of Option #1 include: no tension between the Law of the Sea, which encourages and in certain circumstances requires access agreements, and the ASCM, which would not cover access agreements. In addition, this option addresses the concerns of small developing countries regarding access agreements, as these arrangements would not be deemed illegal.

The demerits of Option 1 centre on its workability in situations where the value of the access is hard to determine, for instance as a result of market distortions or lack of transparency. In such situations, techniques designed to determine the value of access to fish resources are further explored in Option 2 below.

OPTION 2: TECHNIQUES TO DETERMINE THE VALUE OF ACCESS

In addition to improving the definitional elements of a subsidy, the rules could explicitly address the difficulties in determining the value of access to fish resources. Under this option, the definition of a fisheries subsidy could read as follows:

> Article 1
> Definition of a **Fisheries** Subsidy
>
> 1.1 For the purpose of this Agreement, a **fisheries** subsidy shall be deemed to exist if:
>
> (a)(1) there is a financial contribution by a government or any public body within the territory of a Member (referred to in this Agreement as 'government'), i.e. where:
>
> (v) **a government provides to its nationals direct or indirect access to fish resources under the jurisdiction of third states,**
>
> and
>
> (c) a benefit is thereby conferred, **i.e. where:**
>
> (i) **the government fails to recover from its nationals the value of access to fish resources granted in (v) above (referred to in this Agreement as 'fisheries subsidy').** *Where it is difficult to determine the value of access to fish resources as a result of market distortions, lack of transparency, or any other reason, a benefit will be deemed to exist where a prevailing market price reveals a failure to recover the value of access to fish resources. Alternative techniques to determining a prevailing market price include, but are not restricted to, the construction of a price on the basis of costs of production or the consideration of world market prices.*

COMMENTARY ON OPTION 2

Option 2 addresses the difficulty in establishing the value of access to fish resources in situations where markets are distorted, lack adequate transparency or are otherwise incapable of providing a workable benchmark for comparison to determine whether a benefit has been conferred. In such situations, the definition utilizes a constructed market price to determine whether the amounts recovered confer a benefit to the distant water fishing fleet.

The advantages of Option 2 include its flexibility, as the phrase 'prevailing market price' is sufficiently broad to encompass various techniques. Some techniques were explored in the 'Softwood Lumber' case, where the Appellate Body noted that the ASCM did not require the use of any particular one. For purposes of guidance and clarity, certain examples of alternative benchmarks constructed by reference to costs of production or world market prices are included.

The drawbacks of Option 2 include its ambiguity, as the 'prevailing market price' could be different depending on the technique employed. While this issue may lead to a degree of uncertainty, in case of a dispute, in light of the 'Softwood Lumber' case referred to above, the WTO is equipped to address and determine a 'prevailing market price'.

Improving the remedies associated with covered subsidies

If the improved rules are to cover the transfer of access rights for insufficient price in the definition of a fisheries subsidy, then the first question that arises is whether it should be prohibited. A second level of analysis is whether there should be any exception to the prohibition, and if so, if such exception should be subject to conditions. This section addresses the prohibition discussion.

Option 1: Prohibited subsidies

> Article 3
> Prohibition
>
> 3.1 Except as provided in the Agreement on Agriculture, the following subsidies, within the meaning of Article 1, shall be prohibited:
>
> (a) subsidies contingent, in law or in fact, whether solely or as one of several other conditions, upon export performance, including those illustrated in Annex I;
>
> (b) subsidies contingent, whether solely or as one of several other conditions, upon the use of domestic over imported goods.
>
> 3.2 **Fisheries subsidies.**
>
> 3.3 A Member shall neither grant nor maintain subsidies referred to in **this Article** ~~paragraph 1~~.

Commentary on Option 1

Option 1 provides that a fisheries subsidy cannot be granted or maintained, including the subsidies arising from the onward transfer of foreign access rights (assuming that this has been clarified in Article 1 as suggested in (b)). In the light of current negotiations as well as the submissions presented above, it is clear that this 'simple blanket ban' approach is not a realistic option for reformed subsidies disciplines, neither for subsidies in general nor for access-related subsidies. For the sake of the logical sequence of this section it is nevertheless presented at this point – more sub-options will follow below.

The merits of Option 1 include its clear, bright line and associated remedies that contribute both to reducing market distortions as well as to securing sustainable fish stocks.

In the strict ambit of trade, it would not be necessary for a claimant to establish adverse effects in order to bring an action against the subsidy, because such effects would be presumed to result from the prohibited subsidy.

In the ambit of sustainability, Option 1 has the advantage of introducing greater protection to the fish stocks, which may suffer from over-exploitation if the offending subsidy is not removed. That is, remedies other than the removal of the subsidy do not necessarily ensure the sustainability of the fisheries resources. In addition, the removal of the prohibited subsidy is in the interest of all WTO members and does not concern the economic interests of any one member alone.

Another merit of this option is its workability, as it is easier to show the existence of a prohibited subsidy than it is to show adverse effects in the marketplace. Given its greater workability, this option is better suited to inducing the removal of such access-related subsidies.

The disadvantage of Option 1 is that if a country fails to remove the prohibited subsidy, countermeasures (in other words, the denial of concessions) are then the only remedy available. Still, this is not different from other prohibited subsidies and according to the ASCM Article 4, an accelerated remedies process is available in such instances.

Establishing exceptions to the prohibition and conditioning the transfers of access rights to environmental and economic criteria

As mentioned above, negotiators might want to provide for exceptions to the prohibition of subsidies, potentially accompanied by certain conditions. The following options address these elements.

Option 1: Prohibited subsidies with exception

Article 3
Prohibition

3.1 Except as provided in **Article 4 bis and in** the Agreement on Agriculture, the following subsidies shall be prohibited:

(a) subsidies contingent, in law or in fact, whether solely or as one of several other conditions, upon export performance, including those illustrated in Annex I;

(b) subsidies contingent, whether solely or as one of several other conditions, upon the use of domestic over imported goods.

3.2 Fisheries subsidies except as provided in Article 4 bis.

3.3 A Member shall neither grant nor maintain subsidies referred to in **this Article** ~~paragraph 1~~.

Commentary on Option 1

Option 1 provides an exception (labelled Article 4 bis) to the prohibition to the granting or maintenance of an access-related fisheries subsidy, subject to certain requirements. Such requirements in effect condition that transfer of access rights to certain criteria, examined further below (Option 2). If this exception subject to conditions were included in the rules, the environmental and economic criteria would both secure an income for SIDS as well as secure the transition toward the sustainable management of fish stocks.[85]

The merits of this option include its emphasis on securing the income that SIDS obtain from access arrangements. Subject to certain requirements, the state securing access to foreign waters for its fleet by way of an access arrangement may not need to recover the full value of the access it has procured and transferred to the fishing industry.

A variant of this exception option could relate to a sunset clause, whereby the exception would lapse after a specified period of time (see Option 3). This sunset clause could be designed to enable coastal states to acquire the capacity necessary to benefit from their natural resources. The effect of the sunset clause is that the foreign fleet would be required to pay in full the value of the access to fish resources that it has obtained from its government.

The demerit of this option includes the fact DWFNs could still provide access to fish to its fleet for free or in terms that provide a benefit. As has been documented, such subsidies both distort international trade and create pressures leading to stock depletion. This latter element could be addressed by strict requirements in the conditions established in the exception, explored in turn.

An exception for subsidized fishing under access arrangements that meet certain criteria could strengthen the contribution of improved disciplines on fishing subsidies to sustainable development, in accordance with the WTO mandate. In addition, this exception secures an important source of revenue for SIDS while introducing important environmental and social elements.

Option 2: An Exception with conditions

Article 4 bis
Exception to the Prohibition in Article 3

4.1 bis Access-related fisheries subsidies that comply with all requirements set out below are exempt from the prohibition in Article 3.

(a) **Members shall notify all the terms and conditions of access arrangements that they sign, accede or ratify, including their financial terms;**

(b) **Members granting access to the waters under their jurisdiction shall adopt and enforce laws and regulations necessary to ensure the sustainable exploitation of fish stocks, including requirements for effective reporting of catches and vessel position, in accordance with applicable international law;**

(c) **Members shall ensure that rules of origin relating to access agreement do not constitute a market access barrier to the fish products of the coastal State.**

(d) **Members shall ensure that access arrangements contemplate effective programs for capacity-building, technology transfer, and other tools for local development.**

COMMENTARY ON OPTION 2

An exception does not necessarily mean *carte blanche*. In order to secure special and differentiated treatment for SIDS and other coastal developing states, the options for an exception could consider certain cumulative criteria.[86] These criteria could emphasize the need to transition towards sustainable fish stocks management. In addition to the sustainability dimension, the options could consider how these criteria enable a better functioning of the disciplines discussed in the options above, particularly with respect to transparency and fees. Further, the criteria could address issues of local development and technology transfer.

In that light, the criteria in the carve-out in Option 2 address several issues, including:

- *Process*: transparency in negotiations and disclosure.
- *Sustainability*: in accordance with the UNCLOS, EEZ fishing and related access arrangements should be subject to various laws and regulations designed to ensure the sustainable exploitation of the stocks, including, among other things: environmental management, level of catch, control measures and multilateral negotiations for highly migratory species.[87]
- *Trade*: Access arrangements could address the protectionist elements written into rules of origin, which operate as market barriers to fish products from developing countries.
- *Development*: capacity-building initiatives associated to access arrangements are key to securing the developmental benefits of fisheries for coastal states. In this vein, local landings and technology transfers (for example, to meet sanitary and phytosanitary SPS requirements in foreign markets) would enable greater value-added, creation of jobs and better export capabilities.

While some of these criteria may be controversial, they reflect the opportunities opened by improved disciplines on fisheries subsidies.

OPTION 3: AN EXCEPTION SUBJECT TO 'SUNSET'

Article 4 bis
Exception to the Prohibition in Article 3

4.1 bis Access-related fisheries subsidies that comply with all requirements set out below are exempt from the prohibition in Article 3.

(a) **Members shall notify all the terms and conditions of access arrangements that they sign, accede or ratify, including their financial terms;**

(b) **Members granting access to the waters under their jurisdiction shall adopt and enforce laws and regulations necessary to ensure the sustainable exploitation of fish stocks, including requirements for effective reporting of catches and vessel position;**

(c) **Members shall ensure that rules of origin relating to access agreement do not constitute a market access barrier to the fish products of the coastal State.**

(d) **Members shall ensure that access arrangements contemplate effective programs for capacity-building, technology transfer, and other tools for local development.**

4.2. bis <u>This Article will remain in force twenty years following the entry into force of this Agreement</u>.

COMMENTARY ON OPTION 3

As noted above, a carve-out could be set to expire after a given period of time. Such 'sunset' provision may be justified in order to enable developing coastal States and SIDS to develop their own capacity to exploit their sovereign rights over marine living resources in their EEZs. Moreover, a sunset provision would arguably move the world to more ecologically responsible fisheries without overcapacity/overfishing from DWFNs by reverting, at the time of sunset, to a presumption that an access-related subsidy was ecologically harmful and trade-distorting. As noted above, the effect of this sunset provision would be that the DWFN would, at the time of the sunset, be required to recover from its fleet sufficient remuneration in exchange for the transfer of EEZ access rights. At no time, either before or after the sunset, would access agreements themselves be prohibited.

Improving transparency

The need for transparency in the operation of access agreements cuts across a number of areas. For example, transparency is relevant to strengthening the bargaining position of coastal states, as well as for obtaining adequate reporting on fish stocks. Transparency may also aid in the adequate determination of the value of the fish resources in question.

Still, at a conceptual level, the key challenge appears not simply to require transparency, but to attach particular consequences to practices devoid of transparency, including with respect to subsidized fishing under access arrangements. In that regard, it may be that the use of presumptions of illegality that cannot be rebutted could induce countries to ensure transparency – although this would be likely seen by some governments as an extreme remedy and is accordingly not reflected in the following hypothetical text.

Article 25
Transparency

25.1 bis Access-related fisheries subsidies are subject to the following transparency requirements:

(a) Members shall notify all the terms of government-to-government access arrangements that they sign, accede or ratify, including their financial terms;

(b) Members shall notify all the terms and conditions of fishing licenses and permits that it grants to foreign vessels;

(c) Members granting access to its EEZ to foreign vessels shall establish effective monitoring schemes regarding the biological status of the species subject to harvest, and shall continuously publish the data collected in the monitoring schemes;

(d) Members granting access to its EEZ to foreign vessels shall periodically notify the level of fishing capacity that operates in its EEZ.

COMMENTARY ON TRANSPARENCY PROVISIONS

Transparency requirements included above encompass not only government-to-government access arrangements, but also situations where a member grants access to its EEZ through private deals. Disclosure of information regarding these practices is important for introducing greater transparency to the relevant markets in the coastal states.

Transparency requirements included above also refer to certain environmental issues that are key to ensuring that a fishery does not become over-harvested or depleted. In particular, the coastal state that grants access to its EEZ to foreign vessels is required to establish effective monitoring schemes regarding the biological status of the targeted fish stocks and associated populations and to publish the data produced by the monitoring schemes. This data should be published continuously, as soon as the schemes produce the data, so that at any given time accurate information exists regarding the biological status of the fisheries.

Finally, as a means to prevent over-capacity and over-exploitation, every member is required to notify the level of fishing capacity that is authorized to be fished in its EEZ.

Conclusion

This section has sought to clarify how WTO subsidies rules relate to fisheries access agreements concluded between two or several governments. The results and related suggestions were intended to flow into ongoing WTO fisheries subsidies negotiations where the treatment of access agreement has emerged as a divisive topic amongst negotiating parties. This is mainly due to significant foreign currency flows that these access agreements represent for many small coastal and island developing countries – provoking the fear that these financial flows might cease once they are subjected to WTO disciplines.

However, the detailed legal analysis of the WTO Agreement on Subsidies and Countervailing Measures and relevant jurisprudence provided by this section leads to the conclusion that access agreements themselves do not breach any rules of the ASCM. Only where the Distant Water Fishing Nation is not sufficiently reimbursed by its fleet for the provision of access rights, the corresponding financial element of access agreements may be covered by ASCM disciplines. The amount of this 'subsidy element', arising between a Distant Water Fishing Nation and its own fleet, would have to be determined on a case-by-case basis.

The analysis of the current legal situation concerning access agreements is followed by textual suggestions for integrating access-related subsidies into potential new fisheries subsidies disciplines. These options have been elaborated while negotiations in the WTO were ongoing and an initial draft text has been proposed by the Chair of the Rules Negotiating Group.

The contribution of WTO rules towards more sustainable access regimes will depend on the practical application of the norms. Most importantly, the question of how to determine the value of access rights has to be explored in a detailed way, since this would constitute the basis for calculating the existence and amount of subsidization. Moreover, all criteria for access related subsidies would have to be designed in a way to incentivize and ensure technical assistance for developing countries to sustainably manage their EEZs.

This said, disciplining access-related subsidies via WTO fisheries subsidies rules, as explored in this section, might have several potential consequences. The most direct one would be the revision of current access regimes, in a way (i) to ensure full reimbursement of access payments by distant water fleets to their government, (ii) to guarantee that levels of access payments reflect the value of the access rights and (iii) to design, in collaboration between DWFN and host countries, access arrangements that contain adequate provisions for ensuring sustainable fisheries management of the EEZ, including technical assistance.

However, even with increased sustainability of government-to-government agreements, there are broader questions related to access fishing that are beyond the scope of this chapter but need to be addressed. Increasingly, other forms of arrangements are used to regulate the access to foreign resources, such as joint ventures or private-to-government agreements – that is, industry associations or individual companies negotiating access to foreign EEZ fisheries without any cover agreement concluded by their government. It has been noted as well that the departure of fleets under governmental agreements might not necessarily lead to a decrease in fishing intensity, but might entail a change to a flag of convenience and/or an increase in private fishing agreements. It has been argued that these government-to-private sector arrangements provide weaker governance regimes and are often more opaque and less beneficial for EEZ countries than governmental agreements.

In this regard, a growing number of private agreements would still raise concerns of sustainability and fairness, even if the access-related subsidy element of government-to-government agreements had vanished. Indeed, private agreements do not contain a subsidy in the meaning of Article 1 ASCM. However, it has been discussed that private agreements should nevertheless be covered by specific management and transparency requirements.

Such an approach would acknowledge that most of the sustainability concerns related to fishing under access regimes are not limited to government-to-government agreements.

Still, the need remains to identify other fora, complementary to the WTO, to address these concerns in a consistent manner. A first step might take the form of a high-level engagement of coastal states and DWFNs. Even if the latter are not always directly involved via negotiating access rights for their fleets, they provide the legal framework for the activities of their distant water fleets and have some instruments at their disposal to regulate their behaviour in distant waters – as in the case of IUU fishing. A potential future international initiative to promote more sustainable access regimes might include the development of binding transparency requirements for access-related financial and information flows, as well as concrete sustainability criteria for access fishing, related to the state of the stocks, to fishing capacity and to the management system of the EEZ. This could be accompanied by provisions to encourage on-shore investments by Distant Water Fishing Nations and companies to the benefit of the local fishing community and its fisheries management structures.

These considerations, intended to put the access-related elements of WTO fisheries subsidies negotiations into a broader context, should however not question the significance of including corresponding provisions into a revised ASCM. With a clarification of 'access-related' subsidies as part of reformed fisheries subsidies disciplines, potential litigation at a later stage could be avoided. Moreover, the establishment of criteria on access-related subsidies would ensure that government-to-government access agreements, as a key element of international fisheries policy, do not compromise but contribute to the sustainable development of small island and coastal developing countries.

Sustainability Criteria for Fisheries Subsidies*

Introduction

Ten years after UNEP and WWF co-sponsored the first international symposium on the relationship between subsidies and overfishing, governments around the world are increasingly engaged in efforts to eliminate subsidies that contribute to excess fishing capacity and deplete fisheries resources. In the WTO, governments are moving towards adopting binding new limits on fisheries subsidies, including an outright ban on certain classes of them. Beyond the WTO domestic policy-makers in many countries have begun reviewing and reforming their own fisheries subsidies policies.

Perhaps the most fundamental question facing government reconsidering fisheries subsidies is the scope of the WTO ban they have agreed to adopt – whether, for example, the ban will cover subsidies to land-based fisheries infrastructure or post-harvest processing and marketing and how far the ban should be relaxed in accordance with the principal of 'special and differential treatment' for developing countries.

This section addresses a separate but related question: how should governments deal with subsidies that fall outside the scope of a new ban in order to ensure that they avoid contributing to overcapacity and overfishing? This question presents itself not only to WTO negotiators, who must decide on the legal conditions and limits that will apply to those fisheries subsidies that remain permitted. In domestic and regional fora outside the WTO, governments will need to confront the inherent risks posed by the continued use of fisheries subsidies and will need to be proactive in pursuing policies that minimize or eliminate those risks to the greatest extent possible.

It is broadly agreed that fisheries subsidies are least dangerous where fisheries are underexploited, undercapitalized and well-managed. But it is a more complicated matter to translate these general principles into practical guidelines or binding rules. As elaborated below, workable criteria for the use of fisheries subsidies need to be sufficiently specific to guide policy-makers, while also being widely acceptable and applicable to a variety of circumstances. In the special context of the WTO, criteria must also be consistent with the WTO's institutional mandate and capacities.

---

* Excerpted from: UNEP/WWF (2007b). The original report was jointly commissioned by UNEP and WWF to David Schorr and John Caddy, a former senior fisheries management expert at the FAO. A preliminary draft was discussed at a UNEP-WWF Symposium on Disciplining Fisheries Subsidies: 'Incorporating Sustainability at the WTO and Beyond' in March 2007, with more than 120 symposium participants from national governments, IGOS, NGOs, regional fisheries management organization and academic institutions (chair's conclusions can be found on the accompanying CD-ROM). The role of the FAO and WTO in reviewing, presenting and clarifying issues needs to be particularly acknowledged. The original report has been slightly adapted for the purposes of this book and some footnotes have been omitted from this excerpt. The full original version of this report is included in the accompanying CD-ROM.

The aim of this section is to assist governments in the identification of criteria for the use of fisheries subsidies, with the dual ambition of helping WTO negotiators craft new international law and providing domestic governments with useful advice as they pursue responsible fisheries subsidies policies. These two ambitions obviously overlap, but are also somewhat distinct. WTO rules cannot embody robust policy advice for fisheries managers, but can only set a few simplified (but important) legal constraints on the 'policy space' governments enjoy for fisheries subsidies.

It should be noted that this section is not intended to promote the use of fisheries subsidies. Many economists and fisheries experts would argue that fisheries subsidies are rarely, if ever, a rational policy alternative. The conditions discussed in this section should be viewed as necessary – but not necessarily sufficient – for reducing the risk of fisheries subsidies to tolerable levels. Moreover, this section assumes that many if not most of the most dangerous capacity-and effort-enhancing fisheries subsidies will be subject to the terms of a binding WTO prohibition consistent with the decision taken by WTO members at the 2005 WTO ministerial conference in Hong Kong.[88]

Management and subsidies: Linked issues

Amidst the failure of many governments to manage their fisheries responsibly, harmful subsidies are a secondary but still significant contributing factor. Yet inadequate management and harmful subsidies have an intertwined relationship: bad management compounds the dangers of subsidies and inappropriate subsidies contribute to bad management (Beddington et al, 2007, p.1714). In the early days of debate, some governments adopted half of this equation to argue against WTO negotiations on fisheries subsidies. The problem, they said, was not subsidies but inadequate management – and that with proper management subsidies would only alter profits, not deplete resources.[89]

This argument was ultimately unpersuasive to the WTO membership as a whole and the consensus has now been clearly stated – including by heads of state at the Johannesburg World Summit on Sustainable Development (United Nations, 2002, 31(f).): that management alone cannot solve the fisheries subsidies problem.

But it is equally true that the subsidies problem cannot be fixed without attention to management. Unless all subsidies to the fisheries sector are to be broadly and absolutely disallowed – a solution that a few stakeholders seem to consider wise and none seem to consider plausible – it will remain a relevant fact that the more poorly managed a fishery, the more likely subsidies will drive resource depletion. This negative correlation has been examined in Chapter 2 of this book.

Choosing good indicators

Meta-criteria

In drafting sustainability criteria for fisheries subsidies it is important to clarify the objectives. This section seeks criteria with the following characteristics:

- *Concrete Specificity*: criteria should be sufficiently concrete and specific to allow unambiguous judgments about whether the conditions they describe have been fulfilled; they must refer to specific facts, not general characteristics or circumstances.
- *Predictive Power*: criteria should be strong predictors of positive or negative conditions relevant to judgments about the likelihood capacity- or effort-enhancing subsidies will contribute to overcapacity or overfishing (recalling again that no global criteria can reduce this risk to zero).
- *Acceptability*: criteria should be based on broadly accepted principles rooted in prevailing international norms.
- *Plausibility*: criteria should require behaviour that is plausible for all stakeholders.
- *Consistency*: criteria should be at least roughly consistent in practice (for example, criteria for judging acceptable stock health should not depend on assessment practices far exceeding those required for a minimally acceptable management infrastructure).
- *Institutional Appropriateness*: criteria should require data and judgments that are appropriate to the institutional context in which the criteria will be applied.

Criteria at three different levels of ambition

The 'meta-criteria' set forth above clearly have different implications (and different weights) in different contexts. The need for 'institutional appropriateness', for example, imposes much stricter limits on criteria adopted into WTO rules than on criteria for national application; criteria for use by regional fisheries management bodies might lie somewhere in the middle of this spectrum.

In light of the different needs associated with different contexts, it may be useful to distinguish three different levels of ambition at which criteria for fisheries subsidies could be articulated:

1. At the top end of the spectrum, it is possible to describe 'best practices'. These would set out something approaching the ideal conditions governments would achieve in order to minimize to the greatest possible extent the dangers inherent in capacity- or effort-enhancing subsidies. Accordingly, such criteria would have the flexibility needed to take full account of differences in ecological and economic circumstances, governmental capacities, and expert opinions.[90]
2. At a more universal level, but better suited to policy guidelines than to binding international rules, might be 'minimum recommended conditions' for fisheries subsidies. These would aim to give relatively standardized but

still flexible criteria that governments could be encouraged to apply in their domestic subsidies policies. Criteria such as these might also be appropriate in the context of a voluntary international normative instrument governing fisheries subsidies, were one ever to be developed.

3 At the most basic level, governments might agree to 'minimum international requirements' for fisheries subsidies, such as could be incorporated into binding international rules – which, for the purposes of this section, principally means the ASCM. Such criteria would give special weight to the institutional context in which they operate.

The special case of artisanal fisheries

As has been described in the previous section on artisanal fishing, many governments and other stakeholders feel that 'small-scale' or 'artisanal' fisheries present a special case for subsidies policy. Some governments have proposed to exempt small-scale fisheries (or even simply small vessels) from new disciplines entirely. Others are seeking exemptions only for a narrowly defined class of 'artisanal' fisheries. This section will not enter too deeply into this debate. For the sake of a thorough analysis it will be assumed that subsidies to small-scale and artisanal fisheries could fall within the scope of new WTO fisheries subsidies disciplines, but that many of them – at least in the case of developing countries – will fall beyond the terms of a new WTO ban.

Technical discussion

Most analyses suggest that subsidies are especially likely to contribute to overcapacity or overfishing unless:

- the affected stocks are well below sustainable levels of exploitation;
- the affected fleets are well below sustainable levels of capacity; and
- the affected fisheries are subject to effective management.

Many would argue that even where these conditions prevail, subsidies can pose significant dangers. At a minimum, it is clear that the absence of any one of these biological, industrial or regulatory characteristics puts a fishery at significant risk from capacity- or effort-enhancing subsidies. Accordingly, good policy (and effective WTO rules) will often depend on the application of sustainability criteria in each of these dimensions.

Stock-related criteria

The Necessity – but insufficiency – of stock-related criteria

The major instruments establishing international norms for responsible fisheries require governments to assess their fish stocks and maintain them at levels consistent with long-term sustainability. Moreover, the ultimate goal of efforts to reform fisheries subsidies is to prevent them from contributing to the depletion of fish stocks. Clearly, stock-related considerations must be included among the minimum criteria for fisheries subsidies.

Nevertheless, of the three sets of criteria for fisheries subsidies to be discussed, those related to stock conditions are in some sense the weakest guarantors of sustainability. Unlike low fleet capacity or effective management, the presence of a robust stock is not in itself evidence that a fishery faces a reduced threat of depletion. There is ample evidence of how quickly an abundant stock can become overexploited if it is subject to the pressures of overcapacity and inadequate management. In short, evidence of stock health should not be considered an argument in favour of fisheries subsidies, but evidence of depletion (or the absence of evidence of good stock health) should be considered an important, if not decisive, factor weighing against the use of effort-or capacity-enhancing subsidies.

The question of access to data

Stock-related criteria obviously will require (or at least encourage) governments to obtain certain kinds of information about their fish stocks. As stated above, this is consistent with prevailing international norms. Yet assessment practices vary in both kind and intensity, with consequent variety in the stock-related information available in specific cases. They also vary widely in cost – a fact often considered important by governments with limited budgetary and human resources.

Since stock-related criteria depend on data-based conclusions, they will reflect implicit judgments about the assessment practices prerequisite to the use of fisheries subsidies. There is thus a direct link between the choice of stock-related criteria and the assessment component of management-related criteria. As will be described below, it is not only consistent with international norms to require governments to undertake basic stock assessments as a prerequisite to subsidies, it also seems reasonable and plausible.

Indeed, according to the FAO, 80 per cent of reported global marine catches come from stocks for which '[s]tock assessment information allowing some estimate of the state of exploitation is available' (FAO, 2005b, p.6).[91] This information is most frequently collected by governments or government institutes, sometimes with the involvement of non-governmental actors, such as academic institutes or private consultants. At the supra-national level, some RFMOs sponsor fisheries assessments, as do other inter-governmental bodies dedicated exclusively to developing fisheries-related data and/or advice. In short, while there may be legitimate questions regarding the sufficiency of some of this information, the data needed for at least rudimentary stock assessment are available in the majority of commercial marine fisheries, and should be considered potentially available in any fishery where subsidies are intended.

'Optimal' exploitation levels, MSY and 'reference points'

Capacity- or effort-enhancing subsidies obviously should not be employed to expand or maintain fishing beyond (or perhaps even approaching) optimal levels of resource exploitation. But how should such 'optimal levels' be defined? A leading candidate for this benchmark is the concept of 'maximum sustainable

Box 5.1 *FAO Terminology for Describing Stock Conditions*

Underexploited = Undeveloped or new fishery. Believed to have a significant potential for expansion in total production.

Moderately Exploited = Exploited with a low level of fishing effort. Believed to have some limited potential for expansion in total production.

Fully Exploited = Operating at or close to an optimal yield level, with no expected room for further expansion.

Overexploited = Exploited above a level which is believed to be sustainable in the long term, with no potential room for further expansion and a higher risk of stock depletion/collapse.

Depleted = Catches are well below historical levels, irrespective of the amount of fishing effort exerted.

Recovering = Catches are again increasing after having been depleted or after a collapse from a previous high.

Source: FAO 2005b, p.213

yield' (MSY). Both the UNCLOS and the UN Code of Conduct for Responsible Fisheries identify MSY as the overarching objective of fisheries management, as do a number of other international instruments. According to the FAO, MSY can be defined as:

> The highest theoretical equilibrium yield that can be continuously taken (on average) from a stock under existing (average) environmental conditions without affecting significantly the reproduction process. (FAO Fishery Glossary, FAO, 2005a)[92]

As suggested in the FAO's terminology for describing stock conditions (Box 5.1) MSY is also an element (albeit not the sole basis) of the vocabulary used by the FAO in reporting on the status of world fisheries.[93]

Proposals at the WTO negotiating table have also included MSY as a measure of stock health in setting conditions on fisheries subsidies. However, it is important to note that the application of MSY remains subject to significant international technical debate, in several respects (FAO, 2002c).

Firstly, where multi-species fisheries are involved, simple reliance on MSY is often impractical. While a theoretical MSY limit may exist for each target species in a multi-species fishery, it is not always possible simply to set overall fishing limits for the fishery by merging in some way the MSY estimates for individual stocks. While MSY may be a useful concept for discussing the health of various elements of a multi-species fishery, it may not be readily available as a benchmark for controlling output from the fisheries in question (although in such cases the MSY limit of the most heavily exploited or most vulnerable stock may be seen as the relevant benchmark).

Secondly – and more fundamentally – even where MSY can be directly applied to a fishery it has been increasingly viewed as a risky upper limit rather than as the best target for sustainable fishing yields. Advancing science has done little to reduce the inherent uncertainties in calculating MSY (due to both exogenous factors, such as natural stock cycles, and endogenous factors, such as the imprecision of assessment techniques). In fact, as knowledge of fisheries biology has grown, so has an appreciation of the complex linkages among species (and between species and habitats) in marine ecosystems. The emerging trend towards 'ecosystem-based management' has raised new challenges for managers, while highlighting the need for strong precaution when establishing target catch levels.[94] A similar need to look beyond simple reliance on MSY can result from the need for 'integrated coastal zone management', in which fisheries must be viewed as part of a larger marine and land-based complex of resources, economic opportunities, and environmental threats.

There may also be solid economic arguments for fishing below the effort corresponding to MSY.[95] In many cases, rents associated with a fishery are maximized prior to reaching MSY, at a point called 'maximum economic yield' (MEY). Maximizing the profits of a particular fleet (such as a sport fishing industry, or an export-oriented industry) may sometimes imply 'optimal' fishing levels that are even lower still (for example, in order to preserve the number of highly valued larger individuals within the target stock) (Caddy and Mahon, 1995).[96]

Both the need for precaution and the variety of economically optimal biomass levels have given rise to the terminological distinction between 'limit reference points' (LRPs) – which describe the outer limit of optimal yields – and 'target reference points' (TRPs) – which set the actual target level desired in a fishery.[97] As discussed below, robust management systems may depend on the formal use of both types of reference points, in addition to 'threshold reference points' (ThRPs) that provide an additional layer of precaution and regulatory control. It should also be recalled that, in today's world of widespread fisheries depletion, actual TRPs and ThRPs often must be set in the context of strategies for stock recovery – in other words, far below the levels possible in healthy fisheries. In such cases, the reference points are those needed to allow stocks to be rebuilt towards the 'optimal' TRPs based on MSY.

Of the various limit reference points discussed above, only MSY (suitably adjusted by precaution) most directly relates to the goal of maintaining the long-term biological health of target stocks. Given the objective of avoiding subsidized overfishing and depletion, this would seem the right starting point for stock-related fisheries subsidies criteria. But if stock-related criteria are to have the strong predictive power required by the meta-criteria, a precautionary approach to MSY will be necessary. In effect, this means treating MSY as a 'limit reference point' rather than a 'target' reference point (FAO, 2002c). In other words, capacity- or effort-enhancing subsidies should be disallowed or strongly discouraged wherever stocks are not at a given target level above the biomass at MSY equilibrium.

Quantitative versus qualitative indicators

An important question in crafting sustainability criteria for fisheries subsidies is the extent to which those criteria can be quantitative as opposed to qualitative. Where they are appropriate quantitative approaches to fisheries subsidies, quantitative criteria may offer the benefits of precision and clarity. But while quantitative benchmarks may make criteria less ambiguous, they generally involve an unavoidable degree of arbitrariness in the specific numbers they contain.

In the context of rules systems such as at the WTO – where ambiguity can lead to unpredictable outcomes in the course of future disputes – a certain degree of arbitrary quantitative line-drawing may be necessary and desirable. Indeed, WTO disciplines in the ASCM and elsewhere clearly reflect this. For example, the now-lapsed provisions of ASCM Art. 8 (Non-Actionable Subsidies) contained quantitative tests to allow subsidies for assisting 'disadvantaged regions' within a member's territory. In defining the level of economic development to be considered 'disadvantaged', Art. 8.2(b)(iii) set out mandatory criteria as follows:

> (iii) the criteria shall include a measurement of economic development which shall be based on at least one of the following factors:
>
> – one of either income per capita or household income per capita, or GDP per capita, which must not be above 85 per cent of the average for the territory concerned;
>
> – unemployment rate, which must be at least 110 per cent of the average for the territory concerned;
>
> as measured over a three year period; such measurement, however, may be a composite one and may include other factors.

Obviously, the numerical terms for GDP and unemployment in this provision are somewhat arbitrary. There could be regions that reasonable observers would agree are 'disadvantaged' but that do not fit this precise quantitative mould. Conversely, there may be regions whose 'disadvantaged' status would be highly debatable but where these statistical tests could easily be met. Still, the drafters of this provision preferred to draw a clear if somewhat arbitrary line rather than leave the definition of 'disadvantaged' to the interpretation of WTO dispute panellists. A similar exercise in line-drawing is found in the exemption from subsidy disciplines granted by the WTO Agreement on Agriculture to payments for relief from natural disasters.[98]

It follows that quantitative benchmarks could be used for fisheries subsidies criteria, where appropriate – including in the context of WTO rules. However, in the case of stock-related criteria the trade-offs between clarity and arbitrariness may be especially discouraging, particularly if the criteria seek to establish static cut-off points based on biomass or catch levels. For example, a quantitative stock-related criterion might look something like the following:

> Subsidies should be discouraged/disallowed unless there is room to expand catches by at least xx per cent before sustainable catch levels are reached/exceeded;

or

> Subsidies should be discouraged/disallowed unless current stock biomass is greater than xx per cent of the biomass at MSY equilibrium.

The numerical terms in these possible criteria would need to be set to ensure an adequate degree of precaution in setting benchmarks of stock health, as discussed above. But, even leaving aside possible debate over how much precaution is necessary, numbers such as these may be unacceptably arbitrary. The margin of safety applicable in any particular case could be heavily affected by the reproductive biology of the stock, the type of fishing, the impacts on the marine habitat and the existing capacity level of the fleet, among other highly variable factors.

In short, static quantitative benchmarks may have little application in stock-related criteria (they may make more sense in the context of capacity-related criteria). But this does not mean that quantitative terms in stock-related criteria could never be used. On the contrary, trends in certain bio-economic indicators can provide significant input into qualitative judgments about stock health. Quantitative terms referring to these trends would be much less arbitrary, and thus should be considered plausible elements of stock-related criteria. Accordingly, they are included in specific criteria proposed below.

Before turning to those specific options, however, it is worth asking what kinds of qualitative benchmarks of stock health might be used where quantitative approaches are inappropriate. Here, the FAO's standard vocabulary for describing stock status provides what may be the best answer. As is apparent in Box 5.1, the FAO terms are largely focused on the degree to which a stock can withstand expanded production (in other words, increased catches) – precisely the question at issue when capacity- or effort-enhancing subsidies are contemplated. Given the care with which those terms have been crafted, and their widespread acceptance in the arena of international fisheries policy, they seem well fitted for use in the context of fisheries subsidies criteria. Accordingly, FAO vocabulary could be used as the basis of stock-related criteria, including in the context of 'minimum international requirements' for use in the WTO.

Options for stock-related criteria

The analysis so far has sought to establish that stock-related criteria should:

- be data-based, understanding that the nature of the data required has implications for (or, perhaps, is determined by) the nature of the stock assessments that may be required by management-related criteria;

- depend on the availability of reliable data about catch or landing levels, and in most cases levels of fishing effort, in fisheries where subsidies are being considered;
- be based on MSY, applied using a precautionary approach that accounts fully for both scientific uncertainty in establishing biologically optimal fishing levels and the uncertainties inherent in predicting the impacts of capacity- or effort-enhancing subsidies;
- avoid setting arbitrary static quantitative benchmarks, while making judicious use of quantitative terms related to trends in key bio-economic indicators; and
- draw on the FAO stock status vocabulary where qualitative descriptions of stock status are employed.

Based on the foregoing, and on the considerations, it is possible to propose some specific stock-related criteria, ranging from WTO-appropriate 'minimum international requirements' to 'best practices' for implementation by national governments (see summary table, Annex 13).

At the most basic level, stock-related criteria suitable for use as 'minimum international requirements' could require that a target stock be declared 'underexploited' (or the equivalent) on the basis of a science-based stock assessment using reliable catch or catch-plus-effort data.[99] This would give stock status reports provided by the FAO a presumption of validity. In other words, where such reports indicate that a stock is anything other than underexploited, the burden of proof would lie on governments to show that stocks targeted for subsidized fishing are significantly below MSY biomass, and are able to withstand long-term increases in fishing pressure.

At a higher level of rigor, 'minimum recommended conditions' could encourage governments to use scientific survey assessment techniques, where appropriate. In addition, stock-related criteria could discourage capacity- or effort-enhancing subsidies where trends in key bio-economic indicators provide indirect evidence of stock depletion. Sample 'red flag' benchmarks of this kind could include:

- landings (over the past three years) are less than (50 per cent) of the average that applied for the best three years on record;
- landings have declined by more than (X per cent) over the last Y years;
- average catch rates for a standard commercial vessel category have declined by more than (10 per cent) over the last (5) years;
- average catch rates for a standard research vessel (or a standard chartered commercial vessel) over a fixed series of stations have declined by more than (20 per cent) over the last (5) years;
- prices for the product have grown by more than (20 per cent) over the last (5) years (allowing for inflation). The price on international markets or the price on domestic markets has risen by more than (20 per cent) over the last (5) years.

While indirect indicators such as these may not be decisive reasons to reject fisheries subsidies, they are serious warning signs to be considered by governments in their domestic policy-making processes. Thus, the absence of these red flags could be considered among the 'minimum recommended conditions' for subsidies use.

Finally, 'best practices' in the application of stock-related criteria by domestic governments could combine the rigors of quantitative benchmarks with the flexibility of fishery-by-fishery regulation, and would do so in the context of a regulatory regime meeting 'best practices' standards such as the management-related criteria discussed below. Thus – in addition to using the criteria above – best practices would encourage governments to develop quantitative benchmarks for each fishery proposed to be subsidized, employing both 'fishery dependent' data (in other words, data obtained in the course of commercial fishing) and data collected through regular scientific surveys. The criteria would be set in conjunction with formally adopted and precautionary threshold reference points for controlling fishing effort, output or capacity. For example, if a threshold reference point had been established in a given fishery requiring limits on fishing if a target stock falls below 110 per cent of biomass at MSY equilibrium, then an additional and more precautionary threshold reference point (for example, 115 per cent of equilibrium) could be established below which capacity- or effort-enhancing subsidies could be disallowed by domestic regulations.

Capacity-related criteria

As detailed below, international norms of responsible fishing require governments to assess and control the capacity of their fishing fleets. Moreover, overcapacity is often a critical link between subsidies and overfishing, as recognized in the explicit mention of overcapacity in the mandate issued by trade ministers at the sixth WTO ministerial meeting in Hong Kong. Thus, there are strong grounds to emphasize capacity-related issues in minimum criteria for fisheries subsidies. But there are also certain technical difficulties that relate to the definition of 'capacity' and to the variety of methods used to assess it.[100] An authoritative discussion of the issues surrounding the definition and measurement of capacity can be found in a 2004 FAO publication by J.M. Ward et al (FAO, 2004b), which provides the principal basis for the following synopsis.

The Definition of Capacity

After significant review and debate, the FAO recently concluded that fishing capacity can best be defined as:

> [T]he amount of fish (or fishing effort) that can be produced over a period of time (e.g. a year or a fishing season) by a vessel or a fleet if fully utilized and for a given resource condition. Full utilization in this context means normal but unrestricted use, rather than some physical or engineering maximum. (FAO, 2000a, §1.1)

The FAO's definition combines two basic approaches to defining fishing capacity: 'input-based' and 'output-based'. Input-based measures of capacity look at the factors of production used to harvest fish, such as the number of vessels active in a fishery or the level of effort they apply (days at sea, number of traps deployed, etc.). Output-based measures describe capacity in terms of potential levels of production – in other words, in quantities of fish. While input-based measures are often found in the vocabulary of fisheries regulators, output-based measures may make more intuitive sense to the layperson. In talking about the capacity of an automobile factory, for instance, it would be more common to speak of the number of cars it can produce per day, rather than the number of conveyor belts or factory workers employed in the production.

Ward et al review various uses of input-based and output-based definitions, and conclude that the approaches are 'not necessarily incompatible' and possibly complementary (Id., § 1.1(J.M. Ward et al (FAO, 2004b)). In practice, the definition adopted by the FAO – and the advice promulgated by Ward et al and other FAO publications – suggests that neither approach can really be excluded. Similarly, the definition of capacity adopted in the context of WTO rules would be strongest if it adopted this dual-basis approach.

It should be noted that the definition of capacity accepted by the FAO is fundamentally different from simplistic definitions of capacity that are sometimes used in discussions of fisheries policy. It is not uncommon for fishing capacity to be equated with one or more rudimentary characteristics of a fishing fleet, such as the number of vessels of a given size or engine power. These crude physical attributes cannot provide genuine measures of fleet capacity for two reasons. Firstly, in many cases such physical attributes are not reliable indicators of fishing power. For example, vessels of similar size may deploy very different levels of fishing power, depending on the gear they use. Similarly, engine size may be highly relevant in some cases (for example, trawl fisheries) and less so in others.

Secondly, these crude physical measures look only at the input side of the capacity equation and so cannot support a complete definition of fishing capacity. As noted below, however, even simple physical inventories of fleets can play a meaningful role in the management of fishing capacity.

CAPACITY, 'OVERFISHING', AND THE NEED FOR STRONG PRECAUTION

Capacity is a critical variable in any fishery and the most important link between subsidies and overfishing. Even an underexploited fishery is in significant danger if it is the target of overcapacity fleets.

Where subsidies are involved, a fishery that is approaching full capacity is fraught with risk due to the following factors:

- the difficulty of knowing reliably and precisely how much capacity is appropriate or how much capacity is actually in a fishery;
- the consistent trend in fisheries towards technological advances in effective capacity, often without much visible change in the configuration or number of licenses of a fleet;

- the frequent practice of replacing old licensed vessels with vessels of higher efficiency;
- the difficulties of effectively controlling capacity growth in many fisheries and especially in fisheries where illegal fishing is a significant factor;
- the often significant and unpredictable impact of exogenous causes of fish mortality, including both natural ecosystem cycles and anthropogenic threats such as pollution and climate change; and
- the lifespan of fishing vessels that, with regular refitting, may continue to operate for up to 40 years or more, such that the subsidy decision has long-term consequences (Caddy, 1994).

What all this implies is that capacity- or effort-enhancing subsidies to any fishery that is not substantially under-capacity are inherently very risky. Indeed, even at less than full capacity levels, regulatory management of capacity through licensing or other mechanisms is often necessary just to compensate for the tendency of capacity to rise inexorably. All this suggests that capacity-related minimum criteria need to be strongly precautionary if they are to include a margin of safety sufficient to offset the factors listed above. Where such margins are not present, it may even be safer to use subsidies to reduce capacity rather than to increase it.

Assessing capacity: An unmet international commitment

Very few fisheries appear to have been the object of formal capacity assessments, and in this sense the practice of capacity assessment lags far behind the practice of stock assessment. This circumstance persists despite repeated commitments made by governments in both binding and voluntary international instruments. The UNCLOS, for example, requires coastal states to determine the capacity of their fleets to harvest their EEZ fisheries (UNCLOS Article 62.2). The first objective that the Code of Conduct identifies for the sustainable management of fisheries is to avoid excess fishing capacity (FAO, 1995a, Art. 7.2 .2 (a)). The FAO, for its part, has called the regular assessment of capacity 'essential' (FAO, 2004b, § 1.4.).

Most specifically, in 1999 FAO members adopted an International Plan of Action for the Management of Fishing Capacity (IPOA Capacity) that calls on governments to undertake a series of steps to assess their major national fishing fleets by the end of 2000 and to develop preliminary capacity management plans by 2002. Full completion of the assessment, diagnosis and management planning of fleet capacity was to have been completed by the end of 2005. Similarly, the IPOA calls on regional fisheries organizations to undertake similar steps for the assessment and management of capacity within their zones of responsibility, on the same timetable as national governments.

To date, only a single 'national plan of action' has been submitted to the FAO and posted on its website.[101] This low level of formal compliance with the IPOA Capacity may in part reflect rational priority-setting on the part of overworked and under-funded national fisheries administrations. Moreover, it may be that attention to capacity management is increasing more than is reflected in formal capacity assessments or management plans. Nevertheless, where subsidies come into play, there are strong and obvious arguments for raising both the priority and formality accorded to capacity assessment and management.

VARIOUS APPROACHES TO CAPACITY ASSESSMENT

Capacity assessments can be grouped into three basic types that correspond roughly to three levels of data availability. At the lowest level, 'crude fleet inventories' can provide basic information about the number, size and physical characteristics (for example, engine power, hold volume, etc.) of the vessels active in a fishery. As noted above, such inventories do not constitute genuine assessments of capacity since they include no correlation to outputs (for example, average catch per vessel), and because they may not correctly capture important determinants of fishing power (for example, type of gear employed). Nevertheless, as long as they are regularly updated, crude fleet inventories can be used to establish important trends in gross levels of fleet capacity, and may be used in combination with other bio-economic indicators to provide rough estimates of current capacity levels.

At the other end of the scale are various kinds of 'direct capacity assessments' which are based on scientific survey methods and involve often elaborate techniques to describe the relationship between characteristics of fishing inputs and the resulting level of outputs (catches). Impossibly simplified, a quantitative assessment depends on deriving a formula that says something like:

1 vessel = 10,000 tonnes of fish per year

Unfortunately, however, the relationship between fishing inputs and outputs is rarely so simple. There are multiple relevant inputs (vessel size, vessel shape, engine power, gear type, quantity of gear, fishing technique, etc. – as well as the varying abundance of the target stock), each of which can have very different relevance and weight in different fisheries. Moreover, fishing firms can vary substantially in their efficiency and random factors (for example, weather) may need to be considered.

These facts result in both theoretical and practical complications. Theoretically, the equations used to quantify capacity can be enormously complex, while in practice it is impossible to define input-output relationships (such as the relative importance of vessel size or engine power) without extensive empirical observations. A quantitative capacity assessment, then, is a theoretically complex and data-intense business. It is no wonder that, despite their obvious value, only a relatively few have been undertaken.

The third category of capacity assessments – occupying a middle level between crude fleet inventories and direct quantitative assessments – are 'indirect capacity assessments'. Although based on scientific methods, these assessments do not depend on measuring and correlating specific mixes of inputs with output levels. Instead, they look at basic bio-economic conditions in a fishery in order to reach gross judgments about whether a state of overcapacity exists. According to the FAO treatise, indirect indicators that suggest overcapacity include:

- stock is depleted;
- catches exceed their target reference points;
- quotas are used up prior to the end of the fishing season, or the effective fishing season has been progressively reduced from year to year;
- there is a trend towards unused (or 'latent') fishing permits (such that vessels with permits remain tied up at dockside for a significant proportion of the total fishing days); or
- there is a declining catch or value per unit effort or expenditure.

It should also be noted that much of the data used for indirect assessments is identical to that used for indirect assessments of the biological health of a fishery – in other words, data about levels of fishing effort and catches.

Although indirect assessments cannot establish whether or to what degree a fishery is under the optimal capacity level, they can be useful tools for identifying fisheries that are already 'overcapacity'. Moreover, some indirect indicators can support quantitative estimates of overcapacity (for example, the ratio of unused or latent fishing permits to total permits).

Options for capacity-related criteria

Capacity-related criteria are more critical to the proper regulation of fisheries subsidies than stock-related criteria, but may also raise greater challenges. The fundamental dilemma is this: the responsible use of fisheries subsidies requires reliable capacity assessment, yet capacity assessment is technically difficult and remains the exception rather than the rule in fisheries management practice. From this flow three important implications for capacity-related criteria:

1. they will be more likely than stock-related criteria to require governments to improve current management practices as a prerequisite to employing fisheries subsidies;
2. they need to be especially precautionary; and
3. they can only be effective in conjunction with strong management-related criteria aimed at ensuring good capacity assessment and management practices.

With these considerations in mind, proposals for capacity-related criteria can once again be offered at the three successive levels of ambition.

Beginning this time at the top, it is relatively clear what 'best practices' in capacity-related criteria entail. Before granting capacity- or effort-enhancing subsidies (to the extent they remain permitted by new WTO rules), governments should be encouraged to conduct thorough quantitative capacity assessments based on direct scientific observations of fleet characteristics, fishing practices and stock conditions. Armed with those assessments, governments would proceed with subsidies only if the total current capacity in a target fishery is far below the full capacity that can be supported by the biological productivity of the resource. While judgments will vary from fishery to fishery, even in accordance with the kind of subsidies contemplated, the margin of safety needs to be generous. A good rule of thumb would be to avoid subsidies unless total current capacity in a fishery is 50 per cent or less of the capacity needed to take the MSY by full time operation. Moreover, subsidies should be discouraged where capacity has been growing in their absence, except possibly in new or very underdeveloped fisheries where historical growth rates have been positive but very low, such that the 50 per cent threshold would not be breached during the economic life of the subsidies.

Given the importance of capacity-related criteria to responsible subsidies policies, the 'minimum recommended conditions' for subsidies would depart only slightly from the foregoing. In other words, where capacity- or effort-enhancing subsidies are employed, responsible fisheries policy may require something very close to best practices in capacity assessment and management. Minimum recommended conditions would thus require governments to arrive at science-based quantitative estimates of actual and optimal fleet capacity levels, using scientific survey techniques to the greatest extent possible.

The greatest difficulty comes in identifying 'minimum international requirements' of the type that could be incorporated into WTO rules. Here, the basic question is whether new international rules should require science-based quantitative demonstrations of under-capacity or whether it would be enough to rely on qualitative assessments that suggest the absence of overcapacity. It could be argued that, in conjunction with robust stock-related and management-related criteria, a qualitative demonstration of 'not overcapacity' status would suffice.

There are, however, good grounds to reject this view. As noted above, even strong stock-related criteria are insufficient guarantors against overfishing. And, as evident in the discussion to follow, it seems unlikely that WTO rules can ensure that all subsidized fisheries are well-managed. Since both stock-related and management-related criteria will have inherent weaknesses, there is good reason to view strong capacity-related criteria as an indispensable element of effective new WTO rules. Moreover, considering the links between subsidies, overcapacity and overfishing – and recalling the particularly undeveloped state of capacity management practices worldwide – it can be argued that improvements in capacity assessment and management are among the highest priority goals WTO fisheries subsidies disciplines should aim to achieve.

Accordingly, 'minimum international requirements' could require total capacity in a target fishery to be quantified and to be far below full capacity, so that subsidized fishing entails little or no risk of causing overcapacity in the foreseeable future. To be meaningful, the quantitative requirement would demand a 'science-based' assessment, although it might be possible to indicate (through interpretive footnotes or otherwise) that calculations could be based on a combination of crude fleet inventories and indirect bio-economic indicators rather than on specialized data collected through scientific surveys. Given the need for especially strong precaution, the level of total capacity above which subsidies should be avoided or forbidden should be relatively low – for example, in the order of 50 per cent.

A far weaker – and possibly insufficient – alternative to a rule including a quantitative threshold would be to adopt a highly precautionary verbal formula requiring a declaration of undercapacity supplemented by a requirement that governments track a basket of indirect indicators in order to establish that the 'red flags' of an overcapacity fishery are not present. Such indirect 'red flag' indicators could include (FAO, 2004b):

- the active fleet capacity in the target fishery is more than 20 per cent higher than active capacity was during the last of the best three years' landings on record;
- quotas are used up prior to the end of the fishing season;
- the effective duration of the fishing season has declined from year to year;
- there are unused (or 'latent') fishing permits;
- catch per unit effort is declining, and/or unit value of the resource (or, where fishing rights are marketable commodities, their value) is rising.

This alternative approach would effectively require little more than indirect evidence of the absence of overcapacity. If this weaker alternative were to be adopted, it could be reinforced through an additional criterion requiring quantitative inventories of fleets (for example, number, type, size and power of vessels) to be annually updated, along with strict limits on capacity growth rates established by a licensing body.

Management-related criteria

The foregoing sections have discussed the biological and industrial conditions necessary to help reduce the risks inherent in fisheries subsidies. Unfortunately, however, even underexploited and undercapacity fisheries can be subject to overfishing and rapid depletion in the absence of effective management. Fisheries subsidies thus cannot be responsibly employed without attention to the regulatory condition of target fisheries. But management-related criteria for fisheries subsidies are very different from those that focus on biomass or fleet capacity. 'Good management' cannot be quantified, and saying what it is in any given case can raise substantial controversy.

Unlike with stock health and fleet capacity, there is no international obligation on governments to assess the quality of their fisheries management regimes. There is not even an organized process – such as exists at the WTO with regard to national trade policies – for the periodic international review of national fisheries policies, although it is worth noting current international efforts to develop criteria for reviewing the performance of RFMOs (IISD, 2007). In fact, at the moment there appears to be only one globally-focused institution that regularly attempts formal evaluations of fisheries management regimes – the private certification and labelling regime known as the Marine Stewardship Council.[102]

This does not mean, of course, that the tools for assessing management are undeveloped or that assessments of management regimes do not take place. The FAO among others has invested substantial energy into developing instruments for assessing the adequacy of management, of which two relevant examples are a 1996 checklist for management issues associated with the implementation of the Code of Conduct (Caddy, 1996)[103] and guidelines adopted in 2005 for the eco-labelling of fish products (FAO, 2005c). Moreover, the FAO and many other institutions are regularly engaged in studying and writing about the effectiveness of specific management regimes all around the world (FAO, 1999a).

But the Marine Stewardship Council (MSC) and FAO guidelines clearly illustrate that assessing good management cannot easily be reduced to a few basic indicators or rules of thumb. Rather, to discover whether a management regime is fully meeting the tests of sustainability (or of 'responsible' fishing) requires a detailed, case-by-case investigation. This does not pose any analytical impediment to describing the 'best management practices' governments should employ in subsidized fisheries – they are nothing less than the best practices that can be recommended in the absence of subsidies.

But the problem is quite different when it comes to proposing globally applicable minimum preconditions (whether recommended or required) for fisheries subsidies. The practical reality is that simple tests of 'good management' meeting all of the meta-criteria set out above probably cannot be found.

It follows that management-related criteria for fisheries subsidies (other than at the level of best practices) must aim at something less than identifying 'well managed' fisheries. Instead, the following discussion seeks to elaborate criteria for judging whether the minimum elements of good management are in place – or, in the context of WTO rules, whether the most obvious management failures are being avoided.

The basic elements of responsible management

At the outset, management-related criteria should ensure that subsidized fisheries are subject to all the basic necessary elements of a management regime recognized by the Code of Conduct and other international fisheries norms. While the application of those norms can be complex in practice, the rudiments of good management that they require are simple and few. They include:[104]

- science-based assessments (counting and functional analysis) of fish stocks and fishing fleets;
- appropriate regulatory limits on fishing and fishing capacity (sometimes referred to as 'controls');
- surveillance and enforcement of those limits.

Some observers might argue that this list should be expanded to require other elements, such as 'ecosystem-based management'[105] or the involvement of all stakeholders in a participatory management regime (FAO, 2002d, p. 8).[106] There is little doubt that elements such as these would be widely accepted as 'best practices'. Other observers, tending in a different direction, might argue that the three aforementioned elements are inapplicable or inappropriate in some small scale developing country fisheries – a significant question that is given separate treatment in the discussion of 'artisanal' fisheries, below. But these issues do not significantly detract from the global consensus – reflected in multiple international norms and national management systems – that the three elements outlined above form the backbone of responsible fisheries management.

To be meaningful, management-related criteria for fisheries subsidies need to address each of the three basic elements just outlined. The pressing question is how to do this while balancing specificity and rigor against the need for criteria that can be broadly and fairly applied.

Assessment

The assessment of fish stocks is a necessary step towards several important aspects of fisheries management. As stated in a comprehensive treatise recently published by the FAO:

> Quantitative data are required under the precautionary approach, to evaluate the performance of the fishery in meeting its selected goals and objectives, and to enable managers to make rational decisions 'based on the best scientific evidence available' (FAO, 2006b, p.51).

The quote within this quote serves as a reminder that stock assessment is also a basic requirement of UNCLOS, the Code of Conduct, and other international instruments (FAO, 2007).

Still, stock assessments are often difficult, expensive and imprecise undertakings. In order to know how much assessment can be reasonably required by practical criteria for fisheries subsidies (in accordance with the 'meta-criteria'), a rudimentary review of stock assessment techniques is in order.

For the purpose of this analysis, it may be useful to classify stock assessment methods into five broad categories according to the techniques they employ and the kind of data they require.[107] They are listed in Table 5.6, in declining order of data-intensiveness (and thus also of cost).

The variety of assessment practices illustrated in Table 5.6 derives in part from the diversity of fishery ecologies and theories of population dynamics and in part from the choices of governments regarding the resources they dedicate to assessment activities. In other words, the variety relates both to the kinds of assessments and to the degree of assessment rigor that may be applied in given circumstances. These might be respectively called 'technical' and 'political' sources of variety.[108]

The most powerful assessment practices combine some scientific surveys on target stocks with at least one of the next two techniques ('analytic methods' and 'biomass dynamic modeling'). But many governments lack the financial and/or human capacities necessary to conduct scientific surveys on all of their fisheries.

In fact, on a global basis, regular scientific surveying remains the exception rather than the rule. This suggests that such surveys cannot be reasonably required by 'minimum international requirements', but could be encouraged by criteria at the two higher levels of ambition.

At the other end of the scale of quantitative rigor, assessments that depend only on economic analyses cannot provide more than general or even supplementary indications of likely stock conditions or trends. Except where economic information points strongly to an advanced state of depletion, these assessments alone would usually be considered to lack strong predictive power. Rapid informal assessments, for their part, are generally considered a sufficient basis for policy only in artisanal fisheries in impoverished developing country settings, where data needed for more quantitative approaches are simply unavailable.

Table 5.6 *Various Approaches to Stock Assessment*

Name	*Method*	*Data*	*Result*
Scientific Surveys	Primary research, usually by specially equipped scientific vessels, sometimes over multiple years. May include 'scientific fishing'[109] as well as technologies such as acoustic surveys and satellite tagging. May also include dissection of fish to study diets and reproductive biology.	'Fishery-independent' data of many types, collected by or under the control of fisheries scientists.	Direct calculations of biomass and target biomass levels.
Analytical Methods	Analysis of size, age and species composition of catches via biological theories or assumptions about growth, reproduction, and predator/prey characteristics of target species.	'Fishery-dependent' data (in other words, from commercial fishing) on size, age, and species composition of catches (data must be taken at sea, since discards must be included).	Indirect estimates of biomass and/or trends in stock conditions.
Biomass Dynamic Modelling	Analysis of fishing effort and/or catch trends via sophisticated statistical models (also based on biological theories or assumptions) that relate these trends to stock conditions.	'Fishery-dependent' catch and/or effort data (ideally collected at sea, since reliability is lower when data taken at the landing place only)	Indirect estimates of biomass and/or trends in stock conditions.
Economic Analyses	Analysis of costs, revenues and product prices.	Economic data at various levels of detail (industry-wide to enterprise-specific).	Qualitative estimates of trends in stock conditions and economic performance.
Informal Rapid Assessments	Local surveys, interviews and observations coupled with expert knowledge of the dynamics and history of specific fisheries. Analysis may be partly or mainly non-quantitative.	Various; often anecdotal; does not include comprehensive data on catches or effort.	Informed expert opinion about basic stock conditions and trends in resources and earnings.

The remaining two assessment types set out in Table 5.6 – analytical methods and biomass dynamic modelling – provide the bulk of quantitative assessments carried out today. Moreover, they depend on data much of which are considered basic in the documentation of fishing activities. Biomass dynamic modelling in particular, with its heavy dependence on catch and effort data, seems eminently 'plausible', as it involves data whose collection are essentially mandated by prevailing international norms.

The foregoing analysis suggests that stock-related minimum criteria for fisheries subsidies (leaving aside the artisanal case) could be based on basic catch and effort data and on the analytical methods or biomass dynamic modelling that employ such data. In some cases, it may even be acceptable to base assessments purely on catch data. However, assessments that rely exclusively on catch data can be misleading, particularly where data are available for only a few years. Especially in a new or developing fishery, catches may rise consistently until well past the point of MSY. The 'minimum' nature of stock-related criteria that rely on such assessment methods should not be overlooked. As developed below, such criteria should be accompanied by other requirements, such as mandatory transparency in assessment practices.

Controls

The establishment of controls on fishing is important and requires some further elaboration. There is substantial technical discussion and debate over various approaches to setting limits on fishing activities. It is not necessary for the purposes of establishing feasible management-related criteria to enter far into these technical issues, since the criteria sought here are specifically intended to avoid them. What is critical is that minimum criteria for fisheries subsidies would explicitly require controls on fishing to be in place in any fishery where subsidies are intended or applied. Criteria could also require (or encourage) these controls to have certain basic characteristics to promote their effectiveness and their consistency with international norms. For example, criteria could require controls on fishing activities to:

- be contained in legally binding legislative or regulatory provisions;
- be part of formal management plans designed to achieve the long-term sustainability of the target fisheries (with or without explicit mention of maintaining stocks at or above MSY equilibrium and/or the need for precaution) (Note: Particular consideration should be given to requiring controls to include a formal capacity management plan consistent with the FAO International Plan of Action for the Management of Fishing Capacity);
- include clear reference points (see above) articulating targets and limits for levels of capacity, and biomass and/or catches; and
- include regulations mandating specific actions to be taken when target reference points (or 'threshold reference points') are exceeded.[110]

Surveillance and enforcement

Criteria related to surveillance and enforcement of controls are also of obvious importance; good management depends in significant part on effective enforcement. Moreover, the Code of Conduct urges, and the Compliance Agreement obliges, governments to undertake effective enforcement of their fisheries management laws (FAO, 1995a, Art. 8.2.7; FAO, 1995b, Art. III.8).[111] To the extent that fisheries subsidies criteria are intended simply as a guide to domestic policy-makers, there seems no particular obstacle to criteria that look directly at the effectiveness of enforcement in target fisheries.

But where criteria are to be adopted into WTO rules – which have sharper teeth than the Compliance Agreement – governments may find it more difficult to accept criteria that subject their enforcement practices to international quality control. The problem is especially pointed when it comes to the actions of developing countries, many of which face real limits on the resources available to police their fisheries effectively, and may confront the impacts of foreign vessels active in their EEZs or operating just outside of it.

Still, trade provisions requiring the enforcement of domestic laws are not without precedent. The WTO's Agreement on Trade-Related Aspects of Intellectual Property Rights (TRIPS), for example, not only requires enforcement procedures to be included in domestic Intellectual Property Rights (IPR) laws, but also that these provisions permit effective action against IPR infringements (WTO Agreement on trade-related aspects of intellectual property rights (TRIPS), Art. 41:1). The TRIPS text goes on to set out in significant detail some key operating elements of the required enforcement procedures (WTO Agreement on trade-related aspects of intellectual property rights (TRIPS), Arts. 42–49). Also of interest are the provisions of several US bilateral and regional trade agreements that require parties to enforce their environmental laws effectively (side agreement to the North American Free Trade Agreement, Art. 5:1). While these precedents are interesting, they may also be controversial and are perhaps unlikely to serve as models for binding fisheries subsidies criteria.

Nevertheless, given the importance of enforcement to achieving responsibly managed fisheries, evaluating effective enforcement cannot be left out of 'minimum international requirements' for fisheries subsidies. Nor would it seem sufficient for minimum criteria to stop at merely requiring enforcement provisions to exist on section. Instead, a verbal formula could be adopted that requires enforcement efforts to be reasonable, and to be at least sufficiently effective to prevent a significant pattern of illegal fishing. For some countries, even this relaxed standard might require significant additional resources to be directed at enforcement efforts. Where capacity- or effort-enhancing subsidies are on offer, however, this is perhaps not an entirely unreasonable requirement.

Monitoring, control and surveillance (MCS) infrastructure

Beyond criteria that require management regimes to include the three elements outlined in the previous section, governments should consider adopting criteria to look to the basic administrative apparatus necessary to implement them. Such basic administrative elements of management are sometimes called the Monitoring, Control and Surveillance (MCS) infrastructure for a fishery and they are increasingly the focus of international efforts aimed at improving and establishing cooperative links among them.

An excellent example is the maintenance of public vessel registry information. Various international instruments – including UNCLOS (Art. 94.2(a)),[112] the Code of Conduct (FAO, 1995a, Art. 8.2.1), the IPOA-Capacity (IPOA Capacity 17), and the UN Compliance Agreement (FAO, 1995b, Art. IV)[113] – establish a clear norm requiring every government to maintain a registry (or 'record') of vessels authorized to fish under their flag and to cooperate in the sharing and harmonization of registry information. Consistent with these norms, governments have already undertaken significant cooperative efforts to establish international vessel registries. In accordance with the Compliance Agreement, the FAO maintains the High Seas Vessel Authorization Records (HSVAR) database (www.fao.org/fishery/collection/compliance-agreement/en, last accessed 16/03/2010), while the EU administers the international EQUASIS system as part of its vessel safety programme (www.equasis.org, last accessed 16/03/2010). And currently, efforts are accelerating towards the establishment of a Comprehensive Record of Fishing Vessels (or 'Global Record') to be administered by the FAO.[114]

The Compliance Agreement and HSVAR are of particular interest, since they establish specific information requirements that governments could use as elements of management-related fisheries subsidies criteria.[115]

The prominence of these efforts, and the essential role of vessel registries in tracking both capacity and subsidies, strongly suggests that registration should be required for all vessels active in a subsidized fishery. In addition, criteria could require such national registries to participate in or cooperate fully with any applicable regional or global registry systems.

A second basic element of MCS infrastructure – and a corollary to vessel registration – are fishing license regimes that require all vessels active in a fishery to be formally authorized to fish and that public records of those authorizations be maintained (FAO, 1995a, Arts. 8.1.1 & 8.1.2). Here again, international norms already establish that fishing should be subject to mandatory licensing (FAO, 1995a, Arts. 8.2.2). Catch documentation schemes are similarly fundamental to responsible fisheries management. As discussed above, maintaining catch records is the minimum level of data necessary for all but informal methods of stock assessment. Catch documentation schemes have not yet been subject to the same degree of international harmonization and cooperation as vessel registries, but relevant efforts are underway, including a new international consortium for the sharing of fisheries data (the Fishery Resources Monitoring System, or 'FIRMS')[116] and the FAO Strategy for Improving Information on Status and Trends of Capture Fisheries (FAO, 2003a). Here again, full participation in these efforts could be an important management-related criterion. Similarly, catch documentation systems should fully satisfy the requirements of any applicable RFMO or other international cooperative instrument (for example, the catch documentation programme for toothfish set up by the Commission for the Conservation of Antarctic Marine Living Resources (CCAMLR).

Finally, consideration should be given to the extent to which on-board observers could be treated as a required element of catch documentation schemes. In the context of WTO rules, this may be a step too far. But for 'minimum recommended conditions' or 'best practices', the presence of observers could be an important element. Such basic elements of MCS are clearly very important, and their presence or absence is easily ascertained.

Rapid evaluation of management regimes

The criteria discussed in the previous two sections would help ensure that the rudiments of an adequate management regime are in place in a fishery where subsidies are to be utilized. Although significant, this would hardly amount to a guarantee that subsidized fisheries are well managed. Unfortunately, as noted above, criteria sufficient to judge when a fishery is well managed are likely to require case-by-case application of detailed indicators, such as those typical of the MSC or of the FAO's checklist for implementation of the Code of Conduct. Such criteria are not necessarily suitable for use in the globally applicable criteria which are the main focus of this section.

But this does not mean that tools for the evaluation of overall fishery conditions are irrelevant to discussion of fisheries subsidies criteria. On the contrary, since responsible fishing policies generally require that capacity-and effort-enhancing fisheries subsidies be avoided in any fishery that is not truly well managed, such tools are ultimately necessary.

In an effort to steer a course between the inadequacy of rudimentary criteria and the unacceptable complexity of fully mature instruments for assessing fisheries management, it is possible to articulate a set of rudimentary benchmarks to serve as a guide to policy-makers seeking a rapid evaluation of the efficacy of their management regime. Such benchmarks could look to bio-economic trends to establish whether some basic warning signs of overfishing[117] or depletion are present. The data required for these benchmarks would be of a kind that are – or should be – readily available in commercial fisheries where subsidies are contemplated. A proposed set of such benchmark indicators is set forth in Annex 14 of this book. Taken together, benchmarks such as those presented in Annex 14 can serve as a preliminary scorecard for evaluating the health of a fishery and the strength of its management system. The further elaboration of such an approach could produce a valuable tool for governments in their domestic policies and in establishing benchmarks for comparing management conditions on an international basis. It remains an open question whether it is presently possible to integrate criteria even at this level of intermediate detail directly into binding international rules. One option might be to use them as an illustrative guide wherever rules require information about overall compliance with international norms of responsible fishing.

International fisheries

Note that the foregoing discussion refers mainly to domestic fisheries management regimes. But the analysis applies equally, if not more strongly, to fisheries involving migratory, straddling, or high seas stocks (in other words, any fishery not contained entirely within the territorial waters or EEZ of a single nation). In the case of such international fisheries, responsible management requires international cooperation, without which the criteria outlined above cannot be met satisfactorily. Thus, wherever subsidies affect international fisheries, criteria should require the existence of a binding international management regime and the criteria articulated here should apply to that regime. In some respects, the criteria for international fisheries would need to be strengthened. For example, MCS on the high seas poses significant logistical hurdles, some of which might be seen to require the use of satellite tracking systems and/or on-board observers for effective surveillance.

Options for management-related criteria

As evident in the foregoing analysis, the complexities of fisheries management make the identification of management-related criteria for fisheries subsidies a stiff challenge. At the level of 'best practices', criteria for using fisheries subsidies would require management regimes to pass the most rigorous and detailed tests available, such as those set forth in MSC guidelines, FAO eco-labelling guidelines and the FAO checklist for implementing the Code of Conduct.

But where simpler and more broadly applicable criteria are required, the discussion above suggests a three-tiered approach that includes:

1. setting out the basic regulatory elements of a management regime (assessment, control and enforcement);
2. requiring management regimes to include certain key elements of MCS administrative infrastructure; and
3. employing a set of simplified benchmarks to help inform qualitative judgments about the basic health of the fishery and its management regime.

Some options within these steps have already been discussed above and will only be collected here into the three classes of indicators set forth above (this time working from the lowest to the highest level of ambition).

With regard to 'minimum international requirements' (suitable for use in WTO rules) governments could consider adopting criteria that require:

- science-based stock assessments using reliable catch data or catch–plus-effort data (in other words, assessments that employ 'analytical methods' or 'biomass dynamic modelling') with data collected for at least three years prior to subsidization and ongoing annual assessments during the life of the subsidies;
- science-based capacity assessments that result in quantitative estimates of total fleet capacity (active and latent) in target fisheries, including assess-

ments of trends in capacity for at least three years prior to subsidization, and ongoing annual assessments during the life of the subsidies (but noting that these assessments can be based on indirect indicators rather than fully-fledged direct scientific surveys);

- adoption of a formal management plan for each target fishery, including a capacity management plan consistent with the FAO International Plan of Action for the Management of Fishing Capacity;
- legally binding precautionary target and limit reference points for both stocks and capacity based on science-based assessments, taking MSY equilibrium as the outer limit of acceptable limit reference points for stock biomass;
- pre-determined mandatory regulatory responses to be taken in the event target reference points are breached;
- mandatory registration of all vessels active in the target fishery in a public registry that includes all 'mandatory' information required by the HSVAR database, and provision of all requisite information to any applicable international registry system;
- mandatory licensing of all vessels in the target fishery, detailing their authorization to fish and maintenance of license information in a public license registry;
- mandatory reporting of catches or landings by all vessels active in the target fishery;
- enforcement provisions and procedures sufficient to permit reasonably effective action against illegal fishing activities in the target fishery and to prevent significant patterns of illegal fishing therein;
- enforcement provisions to include mandatory withdrawal and repayment of subsidies received by any vessel found to have engaged in illegal fishing activities; and
- use of rapid evaluation benchmarks (see Annex 14) to provide additional information during implementation of rules.

As noted in the next section, in the context of possible WTO rules, the foregoing criteria could be combined with a broad obligation to maintain a management system consistent with the Code of Conduct and related norms. The advantages and disadvantages of such an approach are discussed at that point.

With regard to 'minimum recommended conditions' for fisheries subsidies – that is, criteria of the kind that could plausibly be used by national governments or adopted into voluntary or 'soft law' international standards of subsidy practice – the criteria just listed could be supplemented and strengthened by adopting criteria that, in addition to the required conditions, encourage:

- stock assessments to be based on catch data (not only landing data), and to be supplemented wherever possible by scientific surveys of target fisheries and by investigations into ecosystem or coastal zone considerations, including changes in trophic levels of catches and marine environmental productivity;

- capacity assessments based on fleet inventories and indirect methods to be supplemented by scientific surveys and direct capacity assessment techniques;
- management plans to include ecosystem-based management and, where appropriate, coordination with integrated coastal zone management plans;
- vessel registry information to include all 'optional' information sought by the HSVAR database;
- mandatory reporting of catches (including all discards), to be verified by at least partial onboard observer coverage in target fisheries;
- enforcement procedures to include a public record of enforcement actions;
- structured and regular use of rapid evaluation benchmarks (see Annex 14) as part of ongoing evaluation of fishery.

Finally, with regard to management-related criteria establishing 'best practices' for fisheries subsidies, only a few specific ideas will be presented here. In general, 'best practices' would require governments to implement best management practices across the board in any subsidized fishery. In essence, such fisheries would meet the highest standards for management practices (such as would characterize fisheries warranting certification and ecolabelling).[118]

Nevertheless, building on the criteria elements discussed above, it may be useful to suggest that best practices criteria would additionally require:

- scientific surveys and stock assessments to be conducted on all subsidized fisheries;
- legally binding reference points to include 'threshold' reference points triggering restrictions on subsidies;
- vessel registry information to include all 'additional' information sought by the HSVAR database;
- full on-board observer coverage of all vessels active in target fisheries; and
- enforcement procedures to include independent public review of enforcement actions and effectiveness thereof.

The 'simple reference to international norms' approach

As an alternative to spelling out specific criteria such as those just described, some governments have proposed a simple and broad approach that would rely on a single 'mega-criterion' requiring subsidizing countries to have in place a fisheries management system in line with the FAO Code of Conduct (see Chapter 1).

This 'simple reference' approach has inherent strengths. The Code of Conduct provides a substantial and growing body of international norms for responsible fishing. The Code itself enjoys a breadth of support and an absence of dissent that is rare even for a 'voluntary' agreement. And many of its core elements are replicated in binding international instruments, including the Law of the Sea, the UN Stocks Convention and the UN Compliance Agreement, to name a few. Using the Code in establishing minimum criteria – at the WTO or elsewhere – would thus appear to fulfil many of the 'meta-criteria'. Depending on how a blanket reference to the Code of Conduct is interpreted, it could also set a fairly high bar for the management-related prerequisites to the use of permitted fisheries subsidies.

But therein lies the main problem with this approach – it fails to provide specific and concrete guidance and would thus require substantial interpretation in the course of rule implementation. While the core elements of the Code may be easy to identify, they are not simple to apply.

It might be suggested that this interpretive challenge is manageable. Proponents argue that dispute panels need only consider the narrow circumstances of the case before them, and thus would have an easier time implementing a brought reference to the Code than negotiators would have in seeking to craft more specific minimum criteria. The analysis set forth in the preceding section, however, has hoped to remove some of the difficulty in identifying appropriate criteria for use in 'minimum international requirements' fisheries subsidies.

The best approach may simply be to combine all of the above. A rule requiring compliance with the Code of Conduct could be much more manageable if it were accompanied by the other specific management-related criteria proposed in the previous section. Certainly, such a rule could add significant strength to those detailed criteria.

Criteria for 'artisanal' fisheries

As noted earlier, the question has been repeatedly debated whether 'small-scale' or 'artisanal' fisheries should receive special treatment under new WTO rules. For the purposes of this section, the specific question is whether sustainability criteria for subsidies affecting these fisheries should depart from those that apply more generally.

The question is a significant one, because even if the social and developmental needs of artisanal communities require special consideration, artisanal fisheries cannot be considered immune to overfishing and depletion, or to the potential harms of inappropriate fisheries subsidies. Indeed, responsible management is especially important where subsidized fisheries development is intended to alter fishing patterns or transform a fishery's traditional economics. Even where subsidies to artisanal communities may be perceived as necessary to offset competition from subsidized foreign fleets, it is important to ensure that subsidy policies are not used in ways that cause inadvertent environmental or economic harms.

In order to determine what, if any, special accommodations should be made in the minimum criteria for subsidies to artisanal fisheries, it may help to review the basic arguments offered in favour of special treatment. These arguments generally fall into three broad classes:

1 those arguing that competing national policy goals make proper fisheries management prohibitively expensive for some developing countries;

2 those arguing that social policy goals for artisanal fishing communities may be more important than achieving 'optimal' fishing practices; and
3 those arguing that the techniques of modern, science-based fisheries management are inappropriate for many artisanal fisheries.

The first argument accepts (or at least does not reject) both the goal of MSY and the suitability of science-based command-and-control management techniques, but argues that achieving these goals is prohibitively expensive in some developing country contexts. The second argument does not necessarily reject 'modern' management techniques, but rejects the need to fish within the limits of MSY (for example, where maximizing employment in a given fishery takes priority over maximizing catch levels or even over maximizing total catch value). The third argument accepts the goal of fishing within the limits of MSY, but rejects science-based command and control management techniques as the best means to achieve that goal in the context of artisanal fisheries.

All three arguments are clearly rooted in the real-world experiences of many developing countries. But where fisheries policy includes granting subsidies, it is not certain that they present persuasive reasons for altering sustainability criteria. The first argument, to start with, seems hard to apply where governments have the resources to subsidize increased fishing. If public funds exist to support more fishing pressure, they would seem to exist to support improved data collection and management in parallel.

The second argument – in favour of maximizing employment rather than the biomass of fish or the aggregate income of fleets – seems more compelling,[119] but still ultimately fails where subsidies are available. Whatever the rationality of purposefully fishing beyond MSY to protect livelihoods,[120] it seems an unnecessary policy in the context of fisheries subsidies. Subsidizing people to fish past MSY would be a 'lose-lose-lose' scenario in which subsidies promote overfishing while failing to maximize total industry revenue. If subsidies are available, a far more rational policy would be to grant income supports to people who restrain their fishing effort or leave the industry, thus guaranteeing livelihoods while allowing greater total production of both protein and profits.

The third argument – which accepts the basic goals of responsible fishing but questions the applicability of data-intensive, centralized management techniques – is the most consistent with prevailing international fisheries norms and obligations. It simply insists that the social realities of artisanal communities require flexibility in how responsible fishing practices are defined and implemented – a claim with strong apparent merits.

This 'right goals, wrong techniques' argument also has direct implications for the definition of 'artisanal' fishing. As discussed in the previous section on artisanal fishing, 'artisanal' has been used with a wide and inconsistent variety of meanings in the academic language of fisheries science, in the formal instruments of fisheries law and in the course of the WTO fisheries subsidies talks. This section will follow the previous section by adopting the following functional definition focused on precisely those economic, social and physical characteristics that make artisanal fisheries poor candidates for data-intensive, command and control management regimes:

> 'Artisanal fisheries' are fisheries in developing countries consisting of a large number of small, owner-operated vessels using low-tech fishing gear (such as manual net retrieval) in nearby inshore fisheries, whose products are destined for consumption by the fishers' own households or for sale in highly localized markets, and whose poverty, geographic location, traditional social organization, diffuse patterns of fishing and landing, and disconnection from centralized markets make them particularly difficult to manage through data-intensive, command and control techniques. (UNEP-WWF, 2007b)

Fisheries having these characteristics do require special treatment under sustainability criteria for fisheries subsidies. Criteria that require formal stock and capacity assessments or key elements of an MCS system may simply run counter to artisanal realities. The organization of management in these communities needs to be adapted to their social and economic circumstances with more flexibility than a standard checklist of assessment, regulatory and MCS criteria allows.

Thus, minimum criteria for subsidies to artisanal fisheries need to reflect their condition. This may require a deeper analysis of two aspects:

1. What are the precise constraints that limit application of the standard criteria discussed above? Which elements of fisheries administration are most difficult in a given context, and which most plausible?
2. Are the constraints in question to be considered permanent or temporary (and, if the latter, how rapidly subject to change)? What is the vision for how the fishery will be organized when it achieves a better developed condition, and when is that likely to occur?

These questions in turn imply at least two possible responses to the challenge of attaching minimum criteria to subsidies to artisanal fisheries: adaptation of the criteria to artisanal contexts and providing for phase-in periods for their application.

These two elements of flexibility for artisanal fisheries could be built into minimum criteria, and into WTO rules, in ways that provide flexibility while still giving priority to public investments in good fisheries management before subsidizing increases in fishing capacity or effort. For example, assuming adoption of a careful and narrow definition of 'artisanal fishery':

- criteria for stock and capacity assessments could be relaxed for a limited number of years to allow the use of non-quantitative informal methods, so long as such informal assessment processes (and their results) are transparent to the public and based as closely on scientific methods as conditions allow;[121]

- if the stock-or capacity-related criteria adopted for non-artisanal fisheries include quantitative benchmarks (such as 'fleet capacity less than 50 per cent of capacity needed to harvest at long-term MSY'), these could be relaxed without being discarded by requiring informal assessments to establish explicitly that such conditions likely prevail;
- criteria related to vessel registries, licensing and catch documentation could likewise be delayed or relaxed, or even put aside indefinitely if certain other conditions are fulfilled, such as maintaining the local, in-shore character of the fleet, or requiring phase-in of catch documentation to the extent that the fishery moves towards an export orientation.

Ideas such as these obviously require further development. It should be noted that a number of them may entail substantial risks. For example, some experts argue forcefully in favour of quantitative stock assessment methods (FAO, 2006a, p.51). The fundamental concept, however, would be that where social conditions genuinely make 'modern' approaches to fisheries management inapplicable, the rules would allow appropriate adjustments to – but not the elimination of – minimum criteria.

Conclusion

This section has sought to ascertain whether, from the complex science and policy of fisheries management, a set of relatively simple criteria can be distilled which governments can apply to reduce the risks inherent in the use of fisheries subsidies. Most immediately, the goal has been to contribute ideas of utility to WTO negotiators as they craft appropriate limits and conditions on fisheries subsidies that fall beyond the scope of a proposed new WTO ban.

Accordingly, this section has laid out biological, industrial and regulatory criteria for policies designed to reduce the risks associated with the use of fisheries subsidies. In each area, criteria have been articulated at three levels of environmental ambition: 'minimum international requirements', 'minimum recommended conditions' and 'best practices'. Specific proposals for criteria in each category have been set out above, and are summarized in Annex 13.

It may be interesting to note some general implications of the proposals made with regard to the WTO in particular. The 'minimum international requirements' proposed would impose new constraints on the freedom of WTO members to subsidize their fisheries. But these constraints would vary significantly from category to category in the degree to which they imply forward-looking policy reforms.

For example, the stock-related 'minimum international requirements' (and the assessment practices associated with them) mainly require governments to do what many of them are already doing – and what all are obliged to do by well-established (if imperfectly implemented) international norms. There is little real innovation required by these criteria, even if they might impose some real limits on the number of fisheries eligible for subsidies.

The capacity-related criteria, in contrast, would require many elements of 'best practices' in capacity assessment and management to be treated as minimum international requirements. Here, before employing capacity- or effort-enhancing subsidies, governments often would have to undertake significant policy reforms. While all of these reforms are clearly promoted – and arguably even required – by existing international instruments, they are still far from standard practice today. But it may be particularly appropriate to require tangible improvements in capacity management in the context of rules aimed directly at preventing subsidized overcapacity.

Finally, the management-related 'minimum international requirements' reflect a difficult compromise. On the one hand, they aim only to identify those fisheries in which the most rudimentary tests of adequate management have been met. On the other hand, by highlighting key aspects of fisheries regulation, and then focusing on a number of very specific and concrete elements of management infrastructure, they could result in significant improvements in the administration of fisheries where subsidies are to be used.

Regardless of these differences, all of the criteria discussed above have the potential to improve existing fisheries subsidies policies and practices, even if they are not always sufficient to eliminate the risks posed by fisheries subsidies altogether.

This section has sought to demonstrate that sustainability criteria for fisheries subsidies can be plausible, solidly rooted in accepted international norms and practices and tailored for use by national governments as well as in the WTO. It is hoped that the options and technical suggestions outlined here will serve as a starting place for further investigation and dialogue.

Endnotes

1 Even the fourth paper – on 'sustainability criteria' generally – has a significant emphasis on developing country perspectives, although the paper as a whole relates to questions affecting both developed and developing countries.
2 WTO Appellate Body WT/DS58/AB/R, 12 October 1998, para. 129.
3 In Western Africa and Asia, 80 per cent of seafood is marketed by women (FAO, available at www.fao.org/FOCUS/E/fisheries/women.htm, last accessed 16/03/2010).
4 About 3.3 million people work in capture fisheries in the Philippines, 10.6 million are directly engaged in fishing and fish farming in India and fisheries is the most important or second most important industry in a very large number of countries of the Pacific and Indian Ocean, the Caribbean and in certain African regions. See FAO, 2004b.
5 This average includes China, which is responsible for 33 per cent of world production of fish. The figure excluding China is 13.2kg per person per year.
6 It is worth mentioning that fish products are not defined as agricultural products in the WTO and are not subject to the specific disciplines of the WTO Agreement on Agriculture (see Annex 1 of the Agreement for a list of agricultural products).

7 The EU has bilateral fishing agreements in force with 17 countries: Angola, Cape-Verde, Comoros, Côte d'Ivoire, Gabon, Greenland, Guinea, Guinea-Bissau, Kiribati, Madagascar, Mauritania, Micronesia, Morocco, Mozambique, São Tomé and Principe, Seychelles and Solomon Islands. European Commission 'Bilateral fisheries partnership agreements between the EC and third countries', available at www.ec.europa.eu/fisheries/cfp/external_relations/bilateral_agreements_en.htm, last accessed 16/03/2010

8 This term refers to the agricultural negotiations where subsidies have been placed into categories, or 'boxes'. In order to avoid a subsidy falling under a more regulated or prohibited category, subsidizing WTO members (mostly developed countries) design, frame and adapt their programmes in such a way that they fall under authorised categories of subsidies. Through that practice developed countries avoid reducing their overall levels of subsidization.

9 See WTO Doc. TN/RL/GEN/36, Fisheries Subsidies to Management Services, Paper from New Zealand, 23 March 2005 and WTO Doc. TN/RL/GEN/41, Programmes for Decommissioning of Vessels and Licence Retirement, Communication from the USA, 13 May 2005.

10 The group is not a formal grouping of WTO members and its composition may vary. It usually includes Argentina, Australia, Chile, Ecuador, Iceland, New Zealand, Peru, Philippines and the United States.

11 See, for instance, WTO, 'Brazil – Contribution to the discussion on the framework for disciplines on fisheries subsidies', TN/RL/GEN/56, 4 July 2005, at 4.

12 WTO, 'Ministerial Conference – 2001 Doha Ministerial Declaration', WT/MIN(01)/DEC/1, 20 November 2001, paragraph 44 stating that S&D is an integral part of the WTO agreements and Paragraph 50 which stresses the principle of S&D for developing countries and LDCs in the context of the Doha-mandated negotiations.

13 The WTO Secretariat classified the S&D provisions contained in various WTO agreements into six types: '(i) provisions aimed at increasing the trade opportunities of developing country Members; (ii) provisions under which WTO Members should safeguard the interests of developing country Members; (iii) flexibility of commitments, of action, and use of policy instruments; (iv) transitional time periods; (v) technical assistance; and (vi) provisions relating to least-developed country Members (WTO, 'Secretariat – Implementation of Special and Differential Treatment Provisions in WTO Agreement and Decisions', WT/COMTD/W/77/Rev.1, 21 September 2001, para. 3).

14 WTO, 'Cuba, Dominican Republic, Honduras, India, Indonesia, Kenya, Malaysia, Pakistan, Sri Lanka, Tanzania, Uganda, and Zimbabwe – Proposal for a Framework Agreement on Special and Differential Treatment', WT/GC/W/442, 19 September 2001, paras 7 and 9.

15 Ibid., para 9.

16 These issues are reflected in the 'Compilation of Outstanding Implementation Issues Raised by Members' (Job(01)/152/Rev.1, 27 October 2001), the 2001 'Doha Ministerial Decision on Implementation-Related Issues and Concerns' (WT/MIN(01)/17, 20 November 2001, paras 10.1 to 10.6) and the 'General Council Chairman's proposal with respect to Agreement-specific S&D proposals' (Job3404, 5 May 2003, pp.30–32).

17 For example, as the development situation of the S&D beneficiary improves or the environmental condition of the fish stock worsens, a progressive tightening of

fisheries subsidies disciplines could take place (leading towards the full application of the general disciplines).

18 'Developing countries' are those WTO Members to whom Article 27 of the SCM agreement is currently applicable.

19 The subtle boundary between the WTO's mandate on trade issues and the much larger environmental issues has been referred to by WWF as 'the thin green line'.

20 The term 'subsidy' is explicitly defined in Article 1.1 of the SCM agreement as involving a financial contribution by a government or any public body with the territory of a Member or any form of income or price support that confers a benefit to the recipient. The UNEP study contained in Chapter 2 suggests that there are eight basic kinds of fisheries subsidies. These include: (i) subsidies to fishing infrastructure; (ii) management services; (iii) subsidies to securing fishing access; (iv) subsidies to decommissioning of vessels; (v) subsidies to capital costs associated with fishing activities; (vi) subsidies to variable costs associated with fishing activities; (vii) income supports and (viii) price supports.

21 Criteria to dispute such subsidies could be based on current SCM language of 'serious prejudice' for instance.

22 The term 'positive list' is used in this paper in the sense that anything inside the list is allowed and everything else is not allowed. The opposite phrase – 'negative list' – would then refer to a situation in which anything inside the list is not allowed and everything else is allowed.

23 In this connection, Brazil has suggested that a fishery could be considered 'patently at risk' if its status of exploitation is 'not known or uncertain', 'overexploited', 'depleted' or 'recovering' according to the FAO or by a competent regional or international authority having jurisdiction over the fishery. WTO Doc. TN/RL/GEN/79, *Further Contribution to the Discussion on the Framework for Disciplines on Fisheries Subsidies* – Paper from Brazil, 16 November 2005

24 Ideally, this could be designed as an 'effective management' regime – in other words, one that 'combines scientifically-based catch and effort controls, adequate monitoring and surveillance measures and socio-economic incentives for sustainable fishing' (see Chapter 2).

25 While this may not currently exist, a mandate could be provided to international organizations such as the FAO or UNEP to develop mechanisms, in consultation with regional fisheries organizations and their member states' fisheries agencies, for such evaluation and data collection.

26 Based on FAO statistics for global fish product exports covering the years 2001 to 2003, this share corresponds to 8.29 per cent and is held by China (see FAO, 2003f), Table A-3, 'International trade in fishery commodities by principal importers and exporters', available at www.fao.org/fi/statist/statist.asp, last accessed 16/03/2010

27 Members can negotiate this timeframe. It could be, for instance, five years for developing countries and ten years for LDCs and small vulnerable coastal states.

28 This would be an arrangement similar to that decided upon by the WTO ministerial conference at Doha with respect to developing countries whose GNP per capita falls back below US$1000. See WTO, 'Ministerial Conference – 2001 Ministerial Decision on Implementation-Related Issues and Concerns', WT/MIN(01)/17, 20 November 2001, para 10.4.

29 More particularly, this paper takes as its frame of reference UNEP's institutional mission, the mandate of ¶¶ 28 & 31 of the WTO Doha Declaration, and the call for the elimination of harmful fishing subsidies issued by the 2002 World Summit On

Sustainable Development in Johannesburg (WSSD Plan of Implementation, ¶ 31(f)).

30 See, for example, ADB, 1997, p.35 (the ADB's intervention in the artisanal fishing sector 'provides great opportunity for addressing the crosscutting concerns of poverty reduction and environmental protection').

31 The technical literature on artisanal fishing is replete with references to the problems caused by competition between off-shore fleets and artisanal fishers. See, for example, SFLP Dakar Declaration 2001. See also UN Code of Conduct for Responsible Fisheries, § 6.18 (referring to the need to grant 'preferential access' to artisanal fishers in inshore waters).

32 See, for example, FAO (2003c), p.47 ('With the widespread adoption of motorization, small-scale fisheries have grown significantly over the past two decades. The rapid expansion of artisanal fishing capacity under open access regimes has begun to exert overfishing pressures on coastal fisheries resources, especially in Asia and Africa. There are increasing conflicts between different gear groups as a result of increased mobility of fishing vessels, capacity expansion and overfishing pressures.'). Regarding challenges to the sustainability of artisanal fisheries generally, see ADB, 1997, pp.37–40; van Bogaert, 2003; FAO, 2002a.

33 For a sample of the literature on the challenges to managing artisanal fisheries, see FAO, 2003c; FAO 2002a; FAO 1993c.

34 See generally 'Governance of Small-scale Fisheries' in *United Nations Atlas of the Oceans* (referring to the 'severe constraints faced by artisanal fisheries in terms of management').

35 WTO Doc. TN/RL/GEN/79, *Further Contribution to the Discussion on the Framework for Disciplines on Fisheries Subsidies* – Paper from Brazil, 16 November 2005 (Note: the Brazil text was tabled during the final stages of authoring this paper and after the completion of peer review. Accordingly, Brazil's text is not analysed here.)

36 See, for example, WTO Doc. TN/RL/M/2, Summary Report of the Meeting Held on 6 and 8 May 2002 – Note by the Secretariat, 11 June 2002, p.3, ¶ 16; WTO Doc. TN/RL/W/77, Possible Approaches to Improved Disciplines on Fisheries Subsidies – Communication from the United States, 19 March 2003 , ¶ 3 & fn. 1.

37 WTO Doc. TN/RL/M/7, Summary Report of the Meeting Held on 19-21 March 2003 – Note by the Secretariat, 11 April 2003, p.6 , ¶ 25 (referring to comments by the United States). Whether this accurately reflects the full US position remains to be seen.

38 WTO Doc. TN/RL/W/115, *Possible Approaches to Improved Disciplines on Fisheries Subsidies* – Communication from Chile, 10 June 2003, pp.2–3, ¶ 1.5.

39 WTO Doc. TN/RL/W/136, Fisheries Subsidies, Antigua and Barbuda et al, 14 July 2003, p.3.

40 WTO Doc. TN/RL/W/160, Questions and Comments from Korea on New Zealand's Communication on Fisheries Subsidies (TN/RL/W/154), 8 June 2004 (WTO 2004f), pp.3–4, 6, 11; WTO Doc. TN/RL/W/172, *Contribution to the Discussion on the Framework for the Disciplines on the Fisheries Subsidies* – Communication from Japan; the Republic of Korea; and the Separate Customs Territory of Taiwan, Penghu, Kinmen and Matsu, 22 February 2005, pp.4, 15–16.

41 WTO Doc. TN/RL/GEN/79, *Further Contribution to the Discussion on the Framework for Disciplines on Fisheries Subsidies* – Paper from Brazil, 16 November 2005, pp.2–3.

42 See, for example, WTO Doc. TN/RL/M/7, Summary Report of the Meeting Held on 19-21 March 2003 – Note by the Secretariat, 11 April 2003 p.6, ¶ 25 (comments by the United States).
43 See for example WTO Doc. TN/RL/W/115, *Possible Approaches to Improved Disciplines on Fisheries Subsidies* – Communication from Chile, 10 June 2003, pp.2–3, ¶ 1.5;) WTO Doc. TN/RL/M/10, Summary Report of the Meeting Held on 18 – 19 June 2003 – Note by the Secretariat, 17 July 2003 , pp.8–9, ¶ 27 (comments by Chile).
44 This argument was made forcefully by some participants at the June 2005 UNEP expert workshop convened in Geneva to review early drafts of this and the S&DT paper contained in Chapter 5, 'Special and Different Treatment'. The argument does not appear to have been included in formal WTO submissions to date.
45 In Mauritania, the artisanal sector increasingly exports fish of high commercial value to international markets (UNEP, 2006a).
46 WTO, 2003b, p.3.
47 For purposes of this paper, the terms 'nearshore' 'inshore' and 'offshore' are used as defined by the FAO: nearshore is 'shallow waters at a small distance from the shore', inshore is 'waters of the shallower part of the continental shelf' and offshore is 'waters located well beyond the shores (beyond the edge of the nearshore or inshore waters)...part of the oceanic environment' (FAO, 2005a).
48 In Senegal, subsidies to encourage development of an industrial fleet failed because the less-subsidized artisanal fleet proved more competitive.
49 For example, see the Commonwealth Secretariat's compilation of definitions in use for the terms 'artisanal' fishing and fisheries, 'small-scale' fishing and fisheries and 'subsistence' fishing and fisheries, WTO Doc. TN/RL/W/197, Definitions Related to Artisanal, Small-Scale and Subsistence Fishing – Note by the Secretariat, 24 November 2005
50 WTO Doc. TN/RL/M/11, Summary Report of the Meeting Held on 21 – 22 July 2003 – Note by the Secretariat, 8 September 2003 , p.6, 27; WTO Doc. TN/RL/GEN/57/Rev.1, WTO Fisheries Subsidies Disciplines – Architecture on Fisheries Subsidies Disciplines, Paper from Fiji; Jamaica; Papua New Guinea; and the Solomon Islands, Revision, 4 August 2005, p.4, ¶ 16(ii). The meeting summary does not identify the delegation making these remarks, other than as one of the sponsors of the paper under discussion. Consistent with this approach, a later submission by some small island developing states also refers consistently to 'artisanal or small-scale fisheries'.
51 Note that in its more recent submission, Brazil no longer uses a total catch of 250 tonnes per year as criteria for artisanal fishing.
52 See, for example, *United Nations Atlas of the Oceans*, 'Governance of Small Scale Fisheries' ('Small-scale fisheries, often also referred to as artisanal fisheries, are difficult to define unambiguously, as the term tends to apply to different circumstances in different countries'); FAO, 2005a (entry for 'artisanal fishing'): 'In practice, definition [of 'artisanal fisheries'] varies between countries...'; FAO, 2003c, p.52: 'The definition of what constitutes traditional, artisanal or small-scale could be any one or a combination of [a wide variety of] characteristics'; FAO, 2001b, pp.4–5.
53 These examples are not universal. See § 2.10(ii).
54 Mathew presents this conclusion at the end of a compelling narrative illustration of the variety and inconsistency in the uses of the term 'artisanal fishing'.
55 The FAO has noted 'a consistent downward trend in the proportions of stocks offering potential for expansion' along with 'an increasing trend in the proportion of

overexploited and depleted stocks' dating from the 1970s and continuing through the most recent assessments. FAO SOFIA, 2004, p.32.

56 It is common to hear government officials and experts suggest that subsidies to artisanal fisheries are virtually non-existent, a view supported by FAO, 2003c, p.50. Certainly, however, this has not been true historically and it is clear that there are current cases in which subsidies do flow to artisanal fisheries. See, for example, www.cas-geneve.ch/index.php, last accessed 16/03/2010; Christy, 1997; UNEP, 2002a.

57 The following list is compiled from sources such as Blasé, 1982; Christy, 1997; UNEP, 2002a.

58 The approach suggested here excludes attention to the 'cultural preservation' objective within any special rules for 'artisanal fishing'. That is not, however, meant to imply that subsidies to maintain 'traditional' fishing communities should be considered illegitimate under new WTO rules. But where such communities are not mired in poverty or underdevelopment, and particularly where they are in developed countries, it may be best to ensure that subsidies to them are consistent with the general fishing subsidies disciplines now under negotiation.

59 For a catalogue of WTO notification requirements, see 'Updating of the Listing of Notification Obligations and the Compliance Therewith as Set Out in Annex III of the Report of the Working Group on Notification Obligations and Procedures', G/L/223/Rev.12 (Council for Trade in Goods, 3 March 2005) and G/L/223/Rev.12/Corr.1 (29 March 2005).

60 This example is offered as an illustration. Note the potential relevance of the 'sequencing' issues discussed in above.

61 Certain government-to-government access arrangements do not exchange access for payments, but may establish reciprocal EEZ access or other arrangements. Other types of access agreements – that are not subject to potential new WTO subsidies disciplines – are concluded between government and the private sector (industry associations or single companies).

62 The EU's second generation agreements with Argentina, establishing joint-enterprise companies operating with EU vessels, have proved especially disastrous for Argentine hake fishery.

63 This report shows that even where scientific analysis comes to necessary conclusions, this advice is often ignored (for example, octopus in EU-Mauritania 2001–2006).

64 Michaud (2003) illustrates, for example, how industrial tuna fishing and the tuna canning factory have become indispensable pillars of the Seychelles economy, in great part due to the access agreements which allow distant water fleets from the EU, Japan and Taiwan to fish for tuna in Seychelles waters.

65 A stream of work by the WWF started in 1998 with the publication *The Footprint of Distant Water Fleets on World Fisheries*, WWF International, 1998.

66 Enda Diapol (Senegal) has undertaken several activities in the context of programmes related to fish (Programme of Fisheries, Trade and Environment in West Africa (PCEAO), Network on Fishing Policies in West Africa (REPAO)).

67 See, for example, OECD, 2004; Clark, 2006.

68 See chair's summary of the Roundtable on Promoting Development and Sustainability in Fisheries Subsidies Disciplines, 30 June 2005, available at: www.unep.ch/etb/events/2005rtGeneva.php (last accessed 16/03/2010) and on the accompanying CD-ROM.

69 See chair's summary of the Joint UNEP-ICTSD-WWF Workshop on Development and Sustainability in the WTO Fishery Subsidies Negotiations: Issues and Alternatives, 11 May 2006, available at: www.unep.ch/etb/events/2006ICTSDWWFMay11.php (last accessed 16/03/2010) and on the accompanying CD-ROM.

70 Some participants argued that a subsidy exists to the extent the access fees are not repaid to the DWFN government by its industry. Others referred to the difference between the commercial value of the access enjoyed by the private fleet and the amount it paid to its government in return for the securing of that access.

71 Panel Report, 'United States – Imposition of Countervailing Duties on Certain Hot-Rolled Lead and Bismuth Carbon Steel Products Originating in the UK', WT/DS138/R, adopted on 6 July 1998, as upheld by the Appellate Body Report (WT/DS138/AB/R, 7 June 2000), Appellate Body Report, Canada-Aircraft, Canada – Measures Affecting the Export of Civilian Aircraft – AB-1999-2 – Report of the Appellate Body, WT/DS70/AB/R, 2 August 1999, para 154.

72 This interpretation is based on a contextual reading of the ASCM with particular regard given to Article 14. At the same time, Annex IV was found to be irrelevant to the context of 'benefit'. See Appellate Body Report, 'Canada – Aircraft', para 150.

73 An additional example to those mentioned above is the case of Guinea–Bissau, where the EU compensation and license payments by the EU vessel owners in 1996 were equal to 10.5 per cent of the estimated value of resources taken by EU vessels from the Guinea-Bissau coastal waters; see Kaczynski and Fluharty (2002), pp.75–93.

74 'Access-related subsidies' are subsidies that arise out of the relationships surrounding the procurement or transfer of foreign access rights, and that are – or under new rules would become – cognizable by the ASCM. Consistent with the discussion in Chapter 4, this oblique term is preferable to the commonly used term 'access subsidies' in order to clarify that the granting of access by a host country is not itself a subsidy, but that other elements of access relationships, such as the onward transfer of access rights, may be.

75 This option is still seen in the context of a fisheries subsidies reform, which in this case would not mention fisheries access agreements.

76 WTO (2003b); WTO Doc. TN/RL/GEN/57/Rev.2, *WTO Fisheries Subsidies Disciplines – Architecture on Fisheries Subsidies Disciplines*, Paper from Antigua and Barbuda; Barbados; Dominican Republic; Fiji; Grenada; Guyana; Jamaica; Papua New Guinea; St. Kitts and Nevis; St. Lucia; Solomon Islands; and Trinidad and Tobago, Revision, 13 September 2005; WTO Doc. TN/RL/W/210/Rev.2, S&DT in the *Fisheries Subsidies Negotiations: Views of the Small, Vulnerable Economies (SVEs)*, Communication from Antigua and Barbuda, Barbados, Cuba, Dominican Republic, El Salvador, Fiji, Guyana, Honduras, Jamaica, Mauritius, Nicaragua, Papua New Guinea, and Solomon Islands, Revision, 22 June 2007

77 WTO Doc. TN/RL/GEN/100, *Fisheries Subsidies Framework for Disciplines* – Paper from New Zealand, 3 March 2006; WTO Doc. TN/RL/GEN/141, *Fisheries Subsidies – Exhaustive List of Non-Prohibited Fisheries Subsidies* – Paper from New Zealand, 6 June 2006

78 WTO Doc. TN/RL/GEN/79/Rev.4, *Possible Disciplines on Fisheries Subsidies* – Paper from Brazil – Revision, 13 March 2007

79 WTO Doc. TN/RL/GEN/114, *Fisheries Subsidies Framework for Disciplines*, Communication from Japan; the Republic of Korea; and the Separate Customs

Territory of Taiwan, Penghu, Kinmen and Matsu, 21 April 2006; WTO Doc. TN/RL/GEN/114 Rev.2, *Fisheries Subsidies: Framework for Disciplines*, Communication from Japan; the Republic of Korea; and the Separate Customs Territory of Taiwan, Penghu, Kinmen and Matsu, Revision, 5 June 2007

80 WTO Doc. TN/RL/GEN/144, *Fisheries Subsidies* – Proposal by Norway, 26 January 2007

81 WTO Doc. TN/RL/GEN/138/Rev.1, *Fisheries Subsidies: Special and Differential Treatment* – Paper from Argentina – Revision, 26 January 2007

82 WTO Doc. TN/RL/W/209, *Access Fees in Fisheries Subsidies Negotiations* – Communication from the ACP Group, 5 June 2007

83 WTO Doc. TN/RL/GEN/145, *Fisheries Subsidies: Proposed New Disciplines* – Proposal from the United States, 22 March 2007

84 WTO Doc. TN/RL/GEN/150, *Fisheries Subsidies: Proposed New Disciplines* – Proposal from the Republic of Indonesia, 22 July 2007

85 It should be noted that the exemption and corresponding conditions examined here and in the following sections only refer to access-related fisheries subsidies. Another key issue in current WTO fisheries subsidies negotiations is how to deal with exemptions for fisheries subsidies in general (for example, when exemptions are justified in the context of Special and Differential Treatment provisions; for this discussion, please see Chapter 5, 'Special and Differential Treatment').

86 See Chapter 5, 'Special and Differential Treatment' for a more detailed analysis on S&DT.

87 See Chapter 5, 'Sustainability Criteria for Fisheries Subsidies' for a more detailed analysis.

88 See Ministerial Declaration adopted on 18 December 2005, WT/MIN(05)/DEC (22 December 2005), Annex D, 9.

89 See, for example, WTO Doc. TN/RL/W/11, Japan's Basic Position on the Fisheries Subsidies Issue, 2 July 2002

90 It is beyond the ability of this chapter paper to propose 'best practices' in fisheries policy. Nevertheless, the analysis below hopes to offer at least a preliminary identification of some criteria. For want of a better term, these will be presented as 'best practices', with all the suitable caveats implied.

91 While this 80 per cent figure probably provides a good indication of the extent of assessments conducted on stocks whose harvests enter commerce, it is almost certainly an overstatement of the percentage of fisheries subject to assessment globally. The 2005 FAO review covered 584 'stock or species groups being monitored on which at least general catch trends are reported'. Estimates of the total number of marine stocks being fished worldwide range into the thousands – transboundary stocks alone may number 1000–1500. (Caddy, 1997). Many of these unmonitored fisheries are small and localized. Their actual catch levels are difficult to ascertain.

92 See www.fao.org.fi/glossary,

93 The FAO categories do not, however, depend simply and directly on MSY. In other words, 'fully exploited' is not equivalent to 'fished at MSY'. Rather, analysts may consider other factors, including precaution, in assigning exploitation status to stocks.

94 A corollary to ecosystem-based approaches – the marine trophic index (MTI) – can simultaneously reveal and help address the limits of single-species stock assessment. Ecosystem approaches often focus on predator-prey relations across the food web.

Since historically commercial fisheries have focused on large, high value species, changes in the 'trophic level' of fisheries landing may be important indicators of ecosystem (and stock) health. The MTI has been developed to help describe these phenomena. See Watson et al, 2004.

95 In more accurate technical terms, 'below MSY' should be taken to mean fishing so that the biomass of the target stock is greater than at 'MSY equilibrium'. This distinction is significant because catch yields may be 'less than MSY' either where fishing is restrained by economics or regulation or where the target stock is overfished (i.e., has already declined to a biomass below MSY equilibrium).

96 See also FAO, 2006b, pp.24–25. There may also be socio-economic arguments (in other words, to maximize employment rather than total rent) in favour of purposely fishing a stock down to less than biomass at MSY equilibrium. This argument is discussed in the context of artisanal fisheries in Chapter 5, 'The Special Case of Artisanal Fisheries'.

97 For technical definitions of these terms, see the FAO's online fisheries glossary, available at www.fao.org/fi/glossary/, last accessed 16/03/2010; for an introductory discussion of reference points, see Caddy and Mahon 1995; see also FAO, 2002c, pp.99–106.

98 Article 8(a) of Annex 2 to the agreement includes in its definition of 'natural disaster' the requirement that an affected agricultural industry suffer 'a production loss which exceeds 30 per cent of the average of production in the preceding three- year period or a three-year average based on the preceding five-year period, excluding the highest and the lowest entry.'

99 While some consideration might be given to using 'moderately exploited' as the appropriate benchmark, the FAO definition of this term (see Box 5.1) suggests this would be an insufficiently precautionary benchmark.

100 These considerations are also highly relevant to aspects of the WTO debate over the scope of an eventual prohibition on certain fisheries subsidies, where various notions of 'capacity' are very much at issue. See WWF: Schorr 2006.

101 See NPOA-Capacity of the United States, available at www.fao.org, last accessed 16/03/2010

102 For information on the MSC generally, see www.msc.org, last accessed 16/03/2010. The MSC 'Principles and Criteria' set out a general framework for the evaluation of fisheries. These are then filled out in significant detail on a case-by-case basis during the certification process. See, for example, MSC Assessment Report: The United States Bering Sea and Aleutian Islands Pollock Fishery, 15 February 2005 (available at www.msc.org/assets/docs/AK_Pollock/BSAI_Pollock_Final_Rpt_15Feb05.pdf, last accessed 16/03/2010).

103 Hereafter also called the 'FAO Checklist'.

104 The FAO has proposed a 'working definition' of fisheries management that encapsulates these three basic elements: '*The integrated process of information gathering, analysis, planning, consultation, decision-making, allocation of resources and formulation and implementation, with enforcement as necessary, of regulations or rules which govern fisheries activities in order to ensure the continued productivity of the resources and accomplishment of other fisheries objectives.*' FAO 1997, p.7 [author's emphasis]. See also FAO, 2002b, p.3; FAO, 2002d, p.8. For other synoptic overviews of the components of responsible management, see, for example, FAO, 1999a. See generally Caddy, 1996.

105 Note that ecosystem considerations can be treated as possible sub-elements of stock assessment or of the regulatory establishment of fishing limits. For example, ecosystem considerations may have an impact on stock assessments by identifying inter-species relationships that alter predictions about the reproductive potential of a stock. Or they may affect the establishment of fishing limits where it is necessary to maintain a balance between populations of stocks connected by a predator-prey relationship.

106 Many descriptions of good management include the consideration of socio-economic factors/goals. Id.; FAO 1997, § 1.5; FAO, 2002d, esp. pp.14–16.

107 Various more technically sophisticated approaches to classifying stock assessment methods can be found in FAO, 2006b, pp.43ff and the treatises cited therein; see also FAO, 2002c, pp.106–110.

108 A candid review of assessment practices today would suggest that governmental choices about how much assessment 'rigor' to support are not always consistent with even a lenient interpretation of the norms of responsible fishing. Thus, to some extent, the variety of assessment practices is a symptom of the undermanagement that underlies the global fisheries crisis.

109 Scientific fishing in the most rigorous sense implies fishing with a predetermined spatial and temporal pattern combined with a careful identification and measurement of catches. Often, the gear used for scientific fishing may be different from that used for commercial fishing, for example, through the use of nets with smaller holes purposefully intended to collect juvenile fish so that the age structure of a stock can be directly studied. Strictly scientific fishing is not generally combined with commercial operations. However, commercial vessels are sometimes said to engage in 'scientific' fishing when their activities are subject to temporal and spatial limits, the catch carefully documented, and the data made available for use in a pre-designed study.

110 This seemingly formalistic requirement is receiving increased attention among fisheries management specialists, who point out that breaching target reference points often means a fishery is in trouble – with associated economic and political stressors that can delay or prevent appropriate regulatory responses (such as halting or restricting fishing). Accordingly, legally predetermined actions triggered by breaching targets can be important to effective management. See, for example, Beddington et al, 2007.

111 These provisions contain nearly identical language. The question of effectiveness is the object of some drafting finesse: the term 'effective' is used to describe the sanctions associated with the enforcement measures rather than with the measures themselves.

112 The UNCLOS obligation refers to all vessels of a flag state and not just fishing vessels.

113 The Compliance Agreement applies only to vessels engaged in fishing on the high seas.

114 A key step towards implementation of this system was taken in March 2007 when the FAO's political level Committee on Fisheries accepted a report recommending the establishment of the Global Record, and agreed to convene an expert consultation to begin hammering out the technical details (FAO, 2007a, p.70.) See the related background report, FAO 2007b; see also FAO 2007c, p.20.

115 HSVAR, for example, distinguishes between 'mandatory', 'optional', and 'additional' information to be provided by participating governments. The use of these different categories in specific management-related criteria is discussed below.

116 Launched in 2004, FIRMS is an important and ambitious effort to centralize national, regional, and international management-relevant data on fisheries on a global basis. The effort is based on a partnership among international organizations (including RFMOs) and national governments, with the FAO serving as secretariat.

117 Note that 'overfishing' – which is a matter of effort – can occur on a stock that is not yet 'overfished'.

118 A possible conundrum: the presence of subsidies may be viewed as a contra-indication against certification. See, for example, MSC, 2002, Principle 3, 6; see also FAO, 1999a; UNEP, 2009.

119 The argument also has some grounding in the Code of Conduct itself, which refers in Art. 7.2.1 to the objective of achieving MSY 'as qualified by relevant environmental and economic factors, including the special requirements of developing countries'. This language has already received some attention at the WTO negotiating table (see WTO Doc. TN/RL/W/176, Contribution to the Discussion on the Framework for Disciplines on Fisheries Subsidies – Paper from Brazil (31 March 2005), fn. 2 and accompanying text).

120 This is, at best, a risky argument. As noted above, the current trend is towards an increasingly precautionary approach to using MSY as a target reference point. Moreover, even the proponents of this argument agree that fishing intentionally beyond MSY should never be carried to the point of causing irreversible stock depletion. Accordingly, good fisheries management would continue to depend on at least rudimentary stock assessment and mechanisms for controlling capacity and/or effort.

121 Methodologies for the assessment of artisanal fisheries, including so-called 'rapid rural assessments', have been a topic of substantial technical discussion. See, for example, FAO, 2005d; FAO, 2005f (esp. Part 5); FAO, 2001d. These methodologies may depend on informal survey techniques (where 'survey' refers to investigations on land based on interviews with participants in the fishery). Although they may be performed with rigor, the limited and often anecdotal nature of the data on which they depend makes such approaches tantamount to 'the subjective assessment of individuals who are in a position to provide an informed judgment' (FAO, 2004f, § 3.3.2). One review of assessment techniques applicable to tropical fisheries, where small scale operation is often the rule, concludes that true 'stock assessments' cannot depend on participant surveys alone, but must include some data on catch or catch and effort (FAO, 1998e, p.348.) Still, while the informal and subjective character of these participant survey assessments makes them less than ideal management tools, they can be of critical importance to the sustainable development of artisanal fisheries. See FAO, 1999b, pp.51–52.

Chapter 6

Conclusion and Way Forward

The materials presented in this volume – reflecting over a decade of research, analysis and experience – reveal both the importance and the maturity of fisheries subsidies as a cutting-edge international issue. Although improper subsidies are just one of several factors driving the crisis of overfishing, their very real impacts and the urgent need for their reform have been thoroughly documented and widely recognized. Moreover, the political and technical impediments to reform have by now been catalogued and explored in substantial detail. In short, the fisheries subsidies debate has reached a point where both the nature of the problem and the path towards effective solutions are reasonably clear.

The preceding chapters, along with the other materials collected in the accompanying CD-ROM, provide a rich basis for understanding the details of the fisheries subsidies issue. This book can thus serve as a technical reference on a broad range of topics underlying the debate. But beyond its value as a reference and as a sampling of primary sources, this book provides an opportunity to look at the fisheries subsidies issue in a synthetic fashion, to draw a few fundamental lessons and to indicate a way forward. Accordingly, this final chapter identifies seven overarching challenges in need of urgent attention by governments and stakeholders along the path towards sustainability and fisheries subsidies reform.

Challenge 1: Improving Fisheries Management and Implementing Sustainability Criteria

The link between the threats posed by fisheries subsidies and the inadequacy of fisheries management has been a focal element of the debate from the outset. In the earliest years of the debate, governments and stakeholders who wished to resist progress towards international disciplines on fisheries subsidies frequently argued that management, not subsidies, was the real problem. Then, with the publication of UNEP's seminal 'matrix' analysis, contained in Chapter 2, a more technical discussion was opened on the link between management contexts and subsidy impacts. The paper demonstrated that under almost all real-world management conditions, most subsidies pose a significant threat to fisheries resources. Accordingly, there is an urgent need for improved management systems and for implementation of the international declarations and codes of conduct developed over the past ten years (see Chapter 1).

However, improvements in management alone will not be able to solve the subsidies problem in the near future. Since a complete elimination of subsidies seems unlikely and politically unfeasible, a solution to the fisheries subsidies problem will require both improving fisheries management and changing subsidies practices. The concept of 'sustainability criteria' for fisheries subsidies establishes a crucial link between these complementary paths. As discussed in detail in Chapter 5, sustainability criteria for fisheries subsidies consist of a set of relatively simplified conditions that should be met before the use of capacity- or effort-enhancing subsidies can reasonably be contemplated. These conditions – which relate to the biological status of fish stocks, to the level of capacity of fishing fleets and to the adequacy of fisheries management regimes – can serve as a basic guide to government action at various levels. They are intended, at a minimum, to ensure that subsidies are not employed where stocks are already overfished, where fleets are above sustainable levels of capacity, or where management regimes are plainly inadequate. The further development, dissemination and implementation of such criteria will be one of the most important tasks ahead if harmful fisheries subsidies are to be eliminated.

As discussed in Chapter 5, the implementation of sustainability criteria will mean different things at national and international levels. Within the WTO, an agreement on criteria that are plausible and universal will require identifying a few basic and general requirements. At the national level, sustainability criteria can and should present a much higher bar to the use of those capacity- and effort-enhancing subsidies that will remain permitted under eventual WTO rules. Either way, a commitment to sustainability criteria is tantamount to a choice to sequence investments in the fisheries sector, giving priority to improving management before promoting fishing.

But the implementation of sustainability criteria will not simply eliminate the risk that subsidies contribute to overcapacity and overfishing. As noted in Chapter 2, even well-developed command and control regulation cannot effectively eliminate the risks posed by subsidies in any fishery that remains 'open access'. Both basic economic theory and practical experience indicate the same bottom line: almost all kinds of fisheries subsidies can promote overfishing in any regime where economic incentives reward the race for fish.

To rationalize the economics of fishing, therefore, it will be necessary to improve traditional fisheries management techniques, to eliminate harmful subsidies and to create economic incentives that promote sustainability. In recent years, a number of tools have emerged to create such incentives, two leading examples of which are 'rights-based management' and eco-labelling.

Rights-based fisheries management (RBM) employs legal instruments or customary practices to create economically valuable private interests in fisheries resources. RBM mechanisms can range from 'individual tradable quotas' that give specific enterprises a direct and quantified share of allowable catches (such as the system successfully adopted in Norway, as discussed in Chapter 3) to traditional structures for communal ownership of local fishing rights. In whatever form, successful RBM mechanisms ensure that the owners of fishing rights will receive a meaningful share of the long-term rewards for good resource husbandry. RBM thus aims at having an effect precisely opposite to that of inappropriate subsidies: whereas subsidies can insulate fishers against the short term consequences of depletion, RBM internalizes the costs of depletion, while also internalizing the eventual benefits of stewardship. Some experts would argue that rights-based management is the only means of creating long-term self interest in fisheries conservation. Certainly, as reflected in the leading work of Elinor Ostrom, the 2009 Nobel Laureate for Economic Governance, it is clear that RBM need not mean the individual privatization of fisheries, but can be deeply rooted in community-based management where the people share rights and obligations associated with fisheries resources.

A second, and perhaps more private-sector focused, mechanism for creating economic incentives that favour sustainable practices is eco-labelling. Independent certification of sustainable practices, combined with effective labelling and purchasing by both institutional and individual consumers, is already starting to provide real incentives for responsible fishing. Such different types of incentives are just beginning to grow and need to receive further support from governments and other stakeholders to be further developed in line with sustainability criteria.

Challenge 2: Pursuing Equity and Sustainable Growth for Developing Countries

The stakes in the fisheries subsidies debate are particularly high for developing countries – a fact that has been repeatedly emphasized by governments and stakeholders from the outset. As outlined in Chapter 1, the fishery sector is a matter of fundamental importance for food security and livelihoods in many developing countries. The depletion of fisheries is thus a significant threat to development. Moreover, as shown in Chapter 3, developing countries that cannot afford massive subsidies, for example, Ecuador, have been put at a significant disadvantage in the international competition for fish resources and for export markets. Other case studies, such as Senegal, have illustrated the human and economic costs when developing country subsidies go awry in the context of export-driven development policies.

It is thus clear that developing countries have a lot at stake in the effort to eliminate subsidies that drive overfishing. But many developing countries have also felt that their essential interests could be threatened by the wrong kind of new WTO rules. As discussed in Chapter 4, the importance and sensitivity of developing country concerns has made discussion of 'special and differential treatment' (S&DT) a centerpiece of the WTO negotiations such that the issue has received a degree of priority that is unusual in the context of most WTO negotiating processes.

The same combination of political relevance and substantive complexity has made developing country concerns a dominant focus of the work brought forward by UNEP as it has helped foment the 'parallel process' of discussions outside the formal negotiating rooms. This is reflected, for example, in Chapter 5, 'Special and Differential Treatment'. The artisanal fishing and access agreement issues discussed in Chapters 5 further illuminate some of the key issues for many of the worlds' economically most vulnerable communities and countries. And the discussion of sustainability criteria takes careful account of the challenges such criteria may pose to developing countries and artisanal communities.

Fundamental and longstanding issues of international equity underlie the S&DT debate. Durable and effective solutions to the fisheries subsidies problem will only be possible if they help end historic imbalances while paying full attention to the needs of developing countries for long-term growth, employment, food security and poverty alleviation.

But progress towards solutions on fisheries subsidies will also require governments to move past simplistic notions of monolithic interests in either developing or developed country 'blocs'. In the area of fisheries subsidies, there has been a remarkable level of common interest between various developing and developed countries on both sides of the debate. One of the keys to the effectiveness of the 'Friends of Fish', especially in the early days, has been the participation in the coalition of both developing and developed countries. Similarly, north–south lines have been blurred where some developing countries have joined those industrialized countries that have tended to oppose new rules on fisheries subsidies.

Indeed, beyond the traditional 'north–south' divide there have emerged significant and often obvious differences of interest among developing countries themselves. As discussed in Chapter 4, developing countries have been divided both by their very different economic circumstances and by differences of policy or ambition with regard to the development of their national fishing industries. The future of fishing, and of trade in fish products, will thus increasingly be influenced by so-called 'south–south' competition.

In practical terms, then, there is simultaneously a need for increased sensitivity to the shared needs of developing countries and a need for realism and candor with regard to how those needs are met. As the debate moves forward, this is just one more area where a solution to the fisheries subsidies problem will require moving beyond 'business as usual'.

Challenge 3: Strengthening Capacities for National and Regional Action

Although the majority of governmental energy dedicated to the fisheries subsidies problem has been focused on work in global fora such as the WTO, FAO,

OECD and UNEP, effective solutions will ultimately depend on the reform of domestic policies as well as on the efforts of regional and intergovernmental organizations (IGOs) where the management of international fisheries is at stake. After all, fisheries subsidies are administered by national and sub-national governments and fisheries management remains a matter mainly controlled at these administrative levels. Accordingly, a critical challenge going forward will be to ensure the capacity of domestic and regional governmental entities to implement sustainability criteria for fisheries subsidies and to administer subsidies in a manner that is fully integrated with resource management.

The nature of this challenge should not be underestimated. International rules can establish basic norms and minimum levels of acceptable practice. But, as demonstrated in Chapter 2, it requires best practices to eliminate the significant threats posed by fisheries subsidies to effective resource husbandry. The sustainability criteria developed by UNEP and WWF recognize that WTO fisheries subsidies rules can only establish the 'least common denominator' of fisheries management behaviours across the global community of nations. More ambitious criteria are needed for national governments acting in their individual sovereign capacities.

Moreover, as clearly shown in the case studies of Chapter 3 and in the detailed treatment of issues such as artisanal fishing in Chapter 5, governments will continue to face complex policy choices in any situation where fisheries subsidies continue to be used. Important elements of fisheries policy must always remain fisheries-specific. The proper integration of the economic, social and environmental dimensions of fisheries policies (including subsidies) often requires not only inter-agency coordination, but close cooperation between local communities and national authorities.

Similarly, increased attention to subsidies by regional governmental bodies will be needed. Even if strong WTO fisheries subsidies disciplines are ultimately adopted, global rules will have only limited effectiveness if not complemented by action on subsidies within the context of regional fisheries management organizations (RFMOs). The economic priorities of RFMOs must turn from a preoccupation with balancing competitive interests towards the administration of systems that create economic incentives for sustainability. Today, weak RFMOs are asked to control the powerful pressures of counterproductive economic signals and the resulting politics of short-term acquisitiveness. Instead, RFMOs will need to be strengthened in order to be able to be active partners in the implementation of global norms regarding the use of fisheries subsidies.

Meeting these challenges will require both political will and strengthened administrative capacities, especially in many developing countries. The resources needed to build such capacity can, perhaps, come partly from redirecting the billions spent on irrational subsidies today. But further attention will need to be given to budgetary priorities, to the possible application of user fees, and – in the case of developing countries – to increased flows of technical and financial development assistance.

Challenge 4: Improving Policy Integration and Coherence

As the materials presented in this book reveal, the fisheries subsidies issue is multi-faceted and fundamentally interdisciplinary in nature. This interdisciplinary character has been at once a source of political energy and a challenging difficulty. The raw politics of globalization helped make fisheries subsidies the first major test of the WTO's ability to deliver a genuine 'win-win-win' outcome for a critical natural resource sector. But moving towards that outcome has at times required managing the collision of fundamentally different cultures within the governmental and non-governmental worlds of trade, development and environmental policy-making.

Fortunately, as described in Chapters 1 and 4, the fisheries subsidies debate has already advanced significantly towards policy-making integration and coherence. Ten years ago, it would not have been possible to hear interdisciplinary discussions of the kind now taking place within WTO negotiating rooms. When the WTO was being born in the early 1990s, 'environment' was a topic that many preferred to keep quite apart from 'trade', and when the two fields undeniably overlapped, the problem was seen more as a matter of managing conflict than of seeking synergies. Today, in the context of the fisheries subsidies talks, trade negotiators are deeply engaged in discussing the direct relationship between proper environmental management and the establishment of a level commercial playing field.

Similarly, environmental stakeholders have undergone their own maturation of vocabulary and institutional orientation with regard to the international trade system. Whereas early environmental discussions of the fisheries subsidies problem were focused almost entirely on the resource impacts and the links between subsidies and fleet overcapacity, today's contribution by environmental stakeholders clearly demonstrates a good understanding of the WTO system. An example is the detailed attention paid to WTO law and the demands of the WTO's institutional context in the materials presented in Chapter 5.

An equally important dimension must be the integration of development priorities into both trade and environmental policy-making, especially where the most vulnerable developing countries are concerned. The fisheries subsidies discussion, although certainly less than perfect, has set a good example by dedicating serious diplomatic and 'parallel process' resources to issues at the heart of development. One example of how progress can be achieved in this regard is the development of the discussion of 'access related subsidies' addressed in Chapter 5. Here, the ultimate direction of the talks was towards a balanced outcome, increasing transparency and equity in access agreements, confronting the need for environmental responsibility, but taking full account of the vital role played by access payments in the economies of many developing countries.

As efforts to reform fisheries subsidies continue, governments and the individuals who constitute them will need to maintain and expand the kind of open-mindedness and balance through which the advances made within the

WTO fisheries subsidies talks have been achieved. This will be necessary not only as negotiators finish the job at the WTO, but perhaps even more so as national and local authorities proceed with domestic reforms.

Challenge 5: Increasing Inter-institutional Innovation Internationally

The necessity for integrated policy-making implies the need for an interdisciplinary policy-making process. The progress made on fisheries subsidies so far, especially in the WTO, has depended heavily on the ability of actors within the process to reach beyond institutional boundaries. How this was accomplished holds important lessons for extending and reinforcing interdisciplinary processes in the future.

One key lesson in this regard comes from the significant role played in the WTO debate by the 'parallel process' convened by UNEP and leading NGOs such as WWF. A careful comparison of the papers emerging from the parallel process and the vocabulary of the formal WTO debate clearly reveals the degree to which concepts and possible solutions first explored outside the negotiating arena have become possible elements of the final technical outcome.

The parallel process had two important characteristics. Firstly, it was convened by institutions not directly involved in the formal WTO negotiations and was conducted in an informal and 'off the record' environment. Secondly, it was organized with the explicit goal of promoting a balanced interdisciplinary conversation. UNEP is, and has acted as, an intergovernmental body. This has given it a useful power to convene and the ability to provide governments with some assurance of political neutrality in UNEP events, notwithstanding UNEP's clear environmental mission and orientation. The general acceptance by governments of UNEP's role in this process was perhaps facilitated by the fact that UNEP had no direct institutional interest in the outcome of the fisheries subsidies debate. Thus, while both the FAO and the WTO have themselves hosted consultations across agency disciplines, UNEP has perhaps enjoyed a palpable and useful additional measure of freedom in facilitating dialogue.

Beyond the informal parallel process, coherent fisheries subsidies policies will also require effective and formal integration among inter-governmental institutions. The interdisciplinary nature of the fisheries subsidies issue puts it beyond the scope of any single intergovernmental institution. Just as the parallel process has depended on bringing trade, fisheries and environment officials together informally, international legal solutions to the fisheries subsidies problem will require formal links between trade and fisheries authorities. As noted in Chapters 4 and 5, a measure of dialogue has already begun between the WTO and the FAO. Eventually, properly integrated policies – as well as effective WTO fisheries subsidies rules – will depend on formalizing a degree of shared authority over various aspects of the issue. Mechanisms for such collaboration, as outlined in Chapter 5, must be considered, including, for example, the possibility of a formalized memorandum of understanding under which the FAO is made a partner to the WTO in producing judgments of facts and policy advice from the perspective of fisheries authorities.

Challenge 6: Improve Stakeholder Involvement and Coordination in Domestic Processes

While international inter-institutional innovation is clearly relevant for subsidies reform, the challenge of inter-institutional innovation at the national level is also important and may in some ways be even more severe than that faced at the WTO negotiating table and between international organizations. Even in countries where discussion of fisheries subsidies is most advanced, there remains a strong need for better coordination among economic and environmental policy-making agencies. In fact, the innovative culture shift required for effective reforms is further advanced within the WTO than within the practice of many national governments. The frequent workshops and symposia sponsored or co-sponsored by UNEP were often the only venues in which trade, environment and fisheries officials were able to hold policy discussions together. The usefulness of providing an opportunity to interact with counterparts from other ministries within their own governments and with civil society from their own countries were too often noted by the participants – providing evidence for the lack of effective communication channels at capitals.

One advantage of discussing fisheries subsidies in the WTO has been precisely the ability of that forum to require governments to develop positions representing the interests of multiple domestic agencies. Where national governments are left to themselves, it generally requires specific legal structures to establish such interagency coordination. In the realm of fisheries policy, few such domestic legal structures currently exist. Particularly, globalization and the demands for moving towards a 'Green Economy' require increased coherence among economic, environmental and social policies. To achieve this goal, mechanisms for effective inter-ministerial cooperation will also need to grow.

Similarly, there is an even greater need to ensure stakeholder involvement at the domestic level than internationally. In some countries, legal and administrative mechanisms exist to require and facilitate stakeholder participation in policy-making. In the area of fisheries subsidies, as with fisheries regulation generally, these mechanisms need to be in place and effective.

Challenge 7: Improving Transparency and Accountability in Subsidies Programmes

An important factor impeding fisheries subsidies reform is the continuing lack of public information about the subsidies programmes governments have in place. This lack of information is not only complicating efforts to negotiate new WTO rules, but also contributes to the harm fisheries subsidies can cause. Without transparency, accountability is difficult or impossible. And in the highly competitive realm of fisheries, where subsidies are in effect a tool of resource allocation, lack of accountability is a step on the path to overcapacity and stock depletion. If political decisions are being made on behalf of private interests, open information about subsidies is necessary for the creation of a level commercial playing field.

Transparency is also a prerequisite to effective international disciplines on fisheries subsidies. The importance of transparency to functioning WTO rules is nearly axiomatic. But, as many governments have recognized in their formal proposals at the fisheries subsidies negotiating table, current WTO transparency rules are not adequate to meet these needs. As noted in many studies, in fact, the great majority of fisheries subsidies have historically not been properly reported to the WTO, as already reported by WWF in 2001. While there may have been some improvement in practices, lack of transparency is still the rule rather than the exception.

In short, real subsidies reform will clearly need more transparent national practices and more effective international transparency rules. These should become immediate priorities.

The Bottom Line: The Oceans Can't Wait

A fact clearly evident across the chapters of this book is that the stakes involved in the fisheries subsidies issue are very high for both marine ecosystems, human communities and the economy. The scale and urgency of the worldwide fisheries crisis is obvious and has motivated governments, civil society organizations and businesses to pursue solutions in a wide variety of national, regional and global fora.

Concluding the fisheries subsidies negotiations at the WTO must remain a high priority. Unfortunately, the pace and ultimate fate of the Doha Round will be determined by factors far beyond the fisheries subsidies issue. It will take political will to restart and conclude the agriculture and industrial tariffs negotiations. And many other significant political and technical hurdles remain to be overcome.

But progress on fisheries subsidies since 2001 has been substantial, with much of the hardest technical work already done. This can already set a precedent illustrating governments' willingness to really align trade, environment and development goals. If the WTO can manage to agree on new fisheries subsidies rules incorporating clear sustainability criteria, this can certainly be done for other sectors such as energy and agriculture. The key now is for governments to continue the talks in the innovative spirit that has been evident so far and to prevent the fisheries subsidies issue from becoming a political hostage to other aspects of the overall Doha Round.

The debate portrayed across the pages of this book suggest that a dozen years can be both a short time and also far too long. The earnest goodwill of many participants in the fisheries subsidies debate has produced remarkable progress, even as institutional and political realities have combined to delay widespread changes in tangible obligations towards meeting the challenges outlined above.

But, as government leaders have frequently reminded us when they have stood before microphones to explain their determination to end subsidized overfishing, 'the clock is ticking' and 'the fish cannot wait'. This is not mere rhetoric, but a dangerous fact.

Annex 1
Timeline – The Emergence of an International Process on Fisheries Subsidies

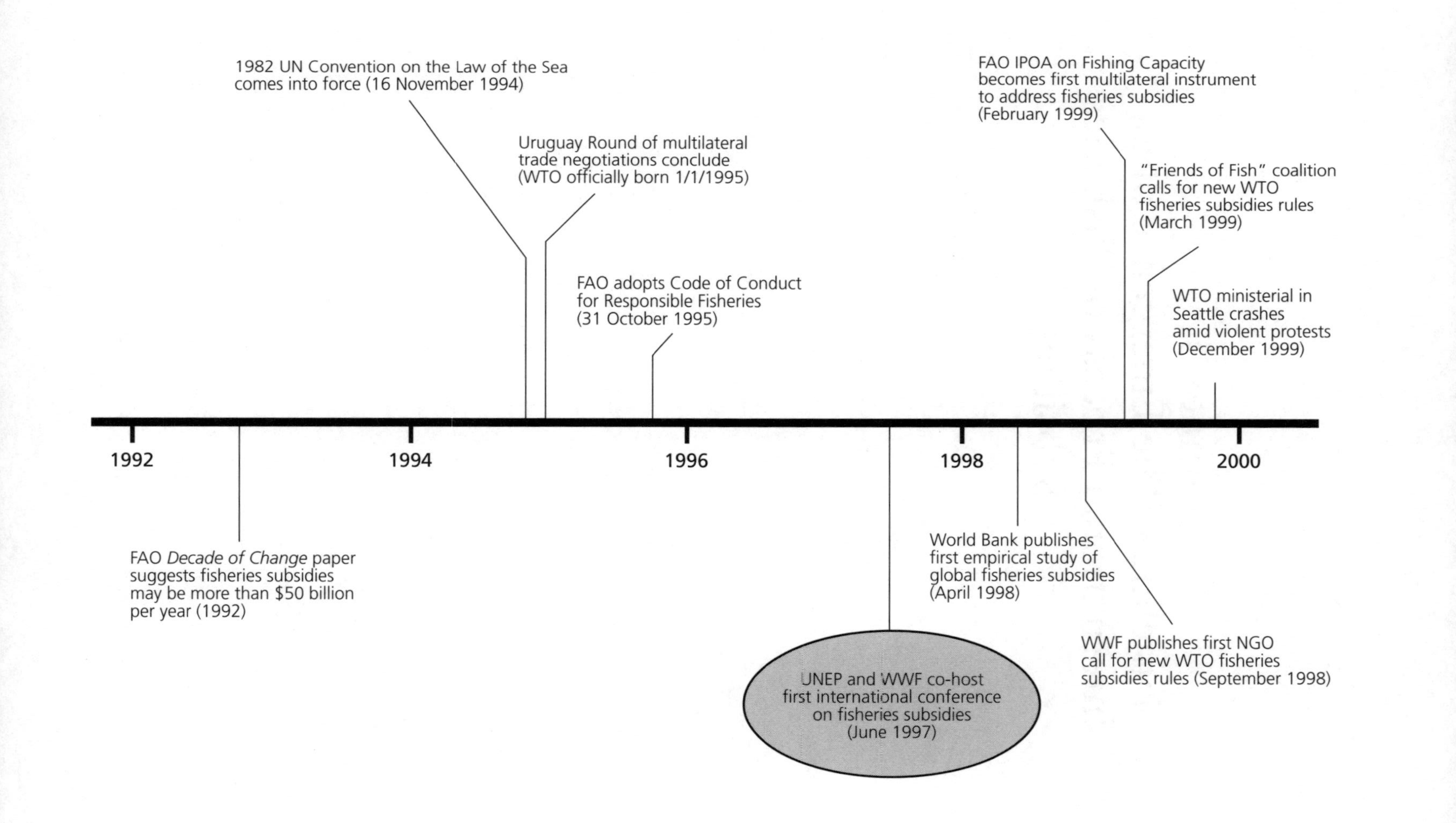
1982 UN Convention on the Law of the Sea comes into force (16 November 1994)
Uruguay Round of multilateral trade negotiations conclude (WTO officially born 1/1/1995)
FAO adopts Code of Conduct for Responsible Fisheries (31 October 1995)
FAO IPOA on Fishing Capacity becomes first multilateral instrument to address fisheries subsidies (February 1999)
"Friends of Fish" coalition calls for new WTO fisheries subsidies rules (March 1999)
WTO ministerial in Seattle crashes amid violent protests (December 1999)
1992
1994
1996
1998
2000
FAO Decade of Change paper suggests fisheries subsidies may be more than $50 billion per year (1992)
UNEP and WWF co-host first international conference on fisheries subsidies (June 1997)
World Bank publishes first empirical study of global fisheries subsidies (April 1998)
WWF publishes first NGO call for new WTO fisheries subsidies rules (September 1998)

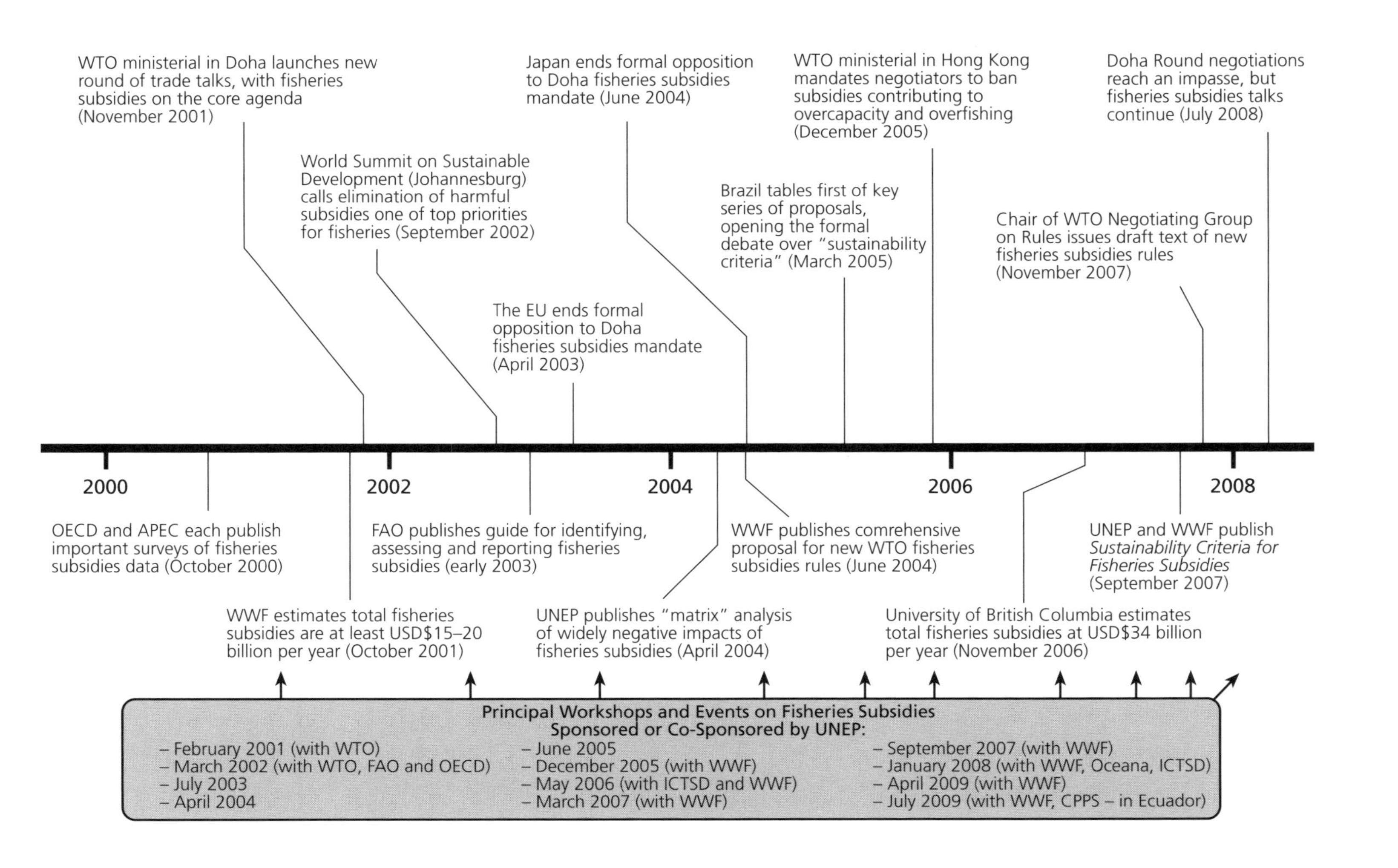
WTO ministerial in Doha launches new round of trade talks, with fisheries subsidies on the core agenda (November 2001)
World Summit on Sustainable Development (Johannesburg) calls elimination of harmful subsidies one of top priorities for fisheries (September 2002)
The EU ends formal opposition to Doha fisheries subsidies mandate (April 2003)
Japan ends formal opposition to Doha fisheries subsidies mandate (June 2004)
Brazil tables first of key series of proposals, opening the formal debate over "sustainability criteria" (March 2005)
WTO ministerial in Hong Kong mandates negotiators to ban subsidies contributing to overcapacity and overfishing (December 2005)
Chair of WTO Negotiating Group on Rules issues draft text of new fisheries subsidies rules (November 2007)
Doha Round negotiations reach an impasse, but fisheries subsidies talks continue (July 2008)
2000
2002
2004
2006
2008
OECD and APEC each publish important surveys of fisheries subsidies data (October 2000)
FAO publishes guide for identifying, assessing and reporting fisheries subsidies (early 2003)
WWF publishes comrehensive proposal for new WTO fisheries subsidies rules (June 2004)
UNEP and WWF publish *Sustainability Criteria for Fisheries Subsidies* (September 2007)
WWF estimates total fisheries subsidies are at least USD$15–20 billion per year (October 2001)
UNEP publishes "matrix" analysis of widely negative impacts of fisheries subsidies (April 2004)
University of British Columbia estimates total fisheries subsidies at USD$34 billion per year (November 2006)
Principal Workshops and Events on Fisheries Subsidies Sponsored or Co-Sponsored by UNEP:
– February 2001 (with WTO)
– March 2002 (with WTO, FAO and OECD)
– July 2003
– April 2004
– June 2005
– December 2005 (with WWF)
– May 2006 (with ICTSD and WWF)
– March 2007 (with WWF)
– September 2007 (with WWF)
– January 2008 (with WWF, Oceana, ICTSD)
– April 2009 (with WWF)
– July 2009 (with WWF, CPPS – in Ecuador)

Annex 2
Summary of International Fisheries Related Agreements

1946 **International Convention for the Regulation of Whaling:** this Convention established the International Whaling Commission (IWC), a system of international regulation for the whale fisheries to ensure proper and effective conservation and development of whale stocks.

1971 **Ramsar Convention on Wetlands of International Importance Especially as Waterfowl Habitat:** signed in Ramsar, Iran, this treaty provides the framework for national action and international co-operation for the conservation and wise use of wetlands and their resources.

1982 **United Nations Convention on the Law of the Sea (UNCLOS):** this fundamental convention provides the basic legal framework for international regulation of the ocean and the exploitation of its resources.

1993 **FAO Agreement to Promote Compliance with International Conservation and Management Measures by Fishing Vessels on the High Seas:** this Agreement extends and defines certain of the UNCLOS obligations with regard to regulation of fishing vessels by their flag states.

1995 **Jakarta Mandate of the Convention on Biological Diversity:** this mandate sets out general guidelines for applying the Convention on Biological Diversity (CBD) to economic activities in marine and coastal areas such as mariculture and fisheries. While these guidelines are not binding, they offer common principles for the design of marine protected and conservation areas, for aspects of aquaculture and for the relationship of coast dwellers and resource users with concepts of biodiversity protection and use.

1995 **UN Agreement on the Conservation and Management of Straddling and Highly Migratory Stocks:** this Agreement calls on coastal states and states fishing on the high seas to pursue co-operation in relation to straddling and highly migratory stocks either directly or through the creation of appropriate sub-regional or regional organizations or arrangements.

1995 **FAO Code of Conduct for Responsible Fisheries:** this code establishes principles and standards for the elaboration and implementation of national policies for responsible conservation of fisheries resources and fisheries management and development.

1999 **FAO International Plan of Action (IPOA) on Management of Fishing Capacity:** the Code of Conduct for Responsible Fisheries provides that states should take measures to prevent or eliminate excess fishing capacity and should ensure that levels of fishing effort are commensurate with sustainable use of fishery resources. The IPOA is a voluntary FAO instrument that applies to all states and entities and to all fishers.

1999 **FAO International Plan of Actions on the Incidental Catch of Seabirds in Longline Fisheries:** the objective of this plan of action is to reduce the incidental catch of seabirds in longline fisheries where this occurs.

1999 **FAO International Plan of Action on Shark Fisheries:** the objective of this plan of action is to ensure the conservation and sustainable management of sharks.

2001 **FAO International Plan of Action to Prevent, Deter and Eliminate Illegal, Unreported and Unregulated Fishing (IPOA-IUU):** this plan of action is a voluntary instrument that applies to all states and entities and to all fishers. It sets objective principles and measures to prevent, deter and eliminate IUU fishing. These measures focus on all state responsibilities, flag state responsibilities, coastal state measures, port state measures, internationally agreed market-related measures, research and regional fisheries management organizations.

Annex 3
Regional Fisheries Bodies

Management Bodies (RFMOs)

CCAMLR	Commission for the Conservation of Antarctic Marine Living Species
CCBSP	Convention on the Conservation and Management of the Pollock Resources in the Central Bering Sea
CCSBT	Commission for the Conservation of Southern Bluefin Tuna
GFCM	General Fisheries Commission for the Mediterranean
IATTC	Inter-American Tropical Tuna Commission
IBSFC*	International Baltic Sea Fishery Commission
ICCAT	International Commission for the Conservation of Atlantic Tuna
IOTC	Indian Ocean Tuna Commission
IPHC	International Pacific Halibut Commission
IWC	International Whaling Commission
LVFO	Lake Victoria Fisheries Organization
NAFO	Northwest Atlantic Fisheries Organization
NASCO	North Atlantic Salmon Conservation Organization
NEAFC	North-East Atlantic Fisheries Commission
NPAFC	North Pacific Anadromous Fish Commission
PSC	Pacific Salmon Commission
RECOFI	Regional Commission for Fisheries
SEAFO	South East Atlantic Fisheries Organization
SIOFA	South Indian Ocean Fisheries Agreement
SPRFMO	South Pacific Regional Fisheries Management Organization
WCPFC	Commission for the Conservation and Management of Highly Migratory Fish Stocks

Advisory Bodies

APFIC	Asia-Pacific Fishery Commission
BOBP-IGO	Bay of Bengal Programme
CARPAS**	Regional Fisheries Advisory Committee for the Southwest Atlantic
CECAF	Fishery Committee for Eastern Central Atlantic
CIFAA	Committee for Inland Fisheries of Africa
COMHAFAT	Atlantic Africa Fisheries Conference
COPESCAL	Commission for Inland Fisheries of Latin America
COREP	Comite Regional des peches du Golfe de Guinee
CPPS	Permanent South Pacific Commission
CTMFM	Joint Technical Commission for the Argentina/Uruguay Maritime Front
EIFAC	European Inland Advisory Fisheries Commission
FCWC	Fishery Committee of the West Central Gulf of Guinea
FFA	South Pacific Forum Fisheries Agency
ICES	International Council for the Exploration of the Sea
MRC	Mekong River Commission
NAMMCO	North Atlantic Marine Mammal Commission
OLDEPESCA	Latin American Organisation for the Development of Fisheries
PICES	North Pacific Marine Science Organization
RECOFI	Regional Commission for Fisheries (Indian Ocean)
SEAFDEC	South East Asian Fisheries Development Center
SPC	Secretariat of the Pacific Community
SRFC (CSRP)	Sub-regional fisheries commission (Eastern Central Atlantic Ocean)
SWIOFC	South West Indian Ocean Fisheries Commission
WECAFC	Western Central Atlantic Fishery Commission

Other Fisheries Related Institutions and Networks

ACFR	Advisory Committee on Fishery Research
CWP	Coordinating Working Party on Fishery Statistics
FIRMS	Fishery Resources Monitoring System
GESAMP	Group of Experts on Scientific Aspects of Marine Environmental Protection
ICES	International Council for the Exploration of the Sea
NACA	Network of Aquaculture Centres in the Asia-Pacific Region

Endnotes

* Ceased to exist on 1 January 2007.

** has no regulatory powers and has been inactive since 1974.

Annex 4
Regional Seas Programmes and Conventions

UNEP Administered Programmes	
Wider Caribbean	Convention for the Protection and Development of the Marine Environment of the Wider Caribbean Region
East Asian Seas	n/a
Eastern Africa	Convention for the Protection, Management and Development of the Marine and Coastal Environment of the Eastern African Region
Mediterranean	The Convention for the Protection of the Marine Environment and the Coastal Region of the Mediterranean (Barcelona Convention)
Northwest Pacific	Action Plan for the Protection, Management and Development of the Marine and Coastal Environment of the North-West Pacific Region
Western Africa	Abidjan Convention for Co-operation in the protection and Development of the Marine and Coastal Environment of the West and Central African Region
Non-UNEP Administered Programmes	
Black Sea	The Convention on the Protection of the Black Sea Against Pollution
Northeast Pacific	The Convention for Cooperation in the Protection and Sustainable Development of the Marine and Coastal Environment of the Northeast Pacific
Red Sea and Gulf of Aden	Regional Convention for the Conservation of the Red Sea and Gulf of Aden Environment
ROPME Sea Area	Kuwait Regional Convention for Cooperation on the Protection of the Marine Environment from Pollution
South Asian Seas	South Asian Seas Action Plan (SASAP)
Southeast Pacific	Convention for the Protection of the Marine Environment and Coastal Areas of the South-East Pacific
Pacific	Convention for the Protection of Natural Resources and Environment of the South Pacific Region

Independent Programmes

Arctic Region	n/a
Antarctic	The Antarctic Treaty, Protocol on Environmental Protection to the Antarctic Treaty; The Convention on the Conservation of Antarctic Marine Living Resources
Baltic Sea	The Baltic Sea States signed the Convention on the Protection of the Marine Environment of the Baltic Sea Area, The Baltic Sea Joint Comprehensive Environmental Action Programme Year Adopted: 1992 (UNEP, 2001)
Caspian Sea	Framework Convention for the Protection of the Marine Environment of the Caspian Sea; Caspian Environment Programme
Northeast Atlantic	The Convention for the Protection of the Marine Environment of the Northeast Atlantic

Annex 5
2001 Doha Ministerial Declaration

WORLD TRADE ORGANIZATION

WT/MIN(01)/DEC/1
20 November 2001
(01-5859)

MINISTERIAL CONFERENCE
Fourth Session
Doha, 9–14 November 2001

MINISTERIAL DECLARATION

Adopted on 14 November 2001

1. The multilateral trading system embodied in the World Trade Organization has contributed significantly to economic growth, development and employment throughout the past fifty years. We are determined, particularly in the light of the global economic slowdown, to maintain the process of reform and liberalization of trade policies, thus ensuring that the system plays its full part in promoting recovery, growth and development. We therefore strongly reaffirm the principles and objectives set out in the Marrakesh Agreement Establishing the World Trade Organization, and pledge to reject the use of protectionism.

2. International trade can play a major role in the promotion of economic development and the alleviation of poverty. We recognize the need for all our peoples to benefit from the increased opportunities and welfare gains that the multilateral trading system generates. The majority of WTO Members are developing countries. We seek to place their needs and interests at the heart of the Work Programme adopted in this Declaration. Recalling the Preamble to the

Marrakesh Agreement, we shall continue to make positive efforts designed to ensure that developing countries, and especially the least-developed among them, secure a share in the growth of world trade commensurate with the needs of their economic development. In this context, enhanced market access, balanced rules, and well targeted, sustainably financed technical assistance and capacity-building programmes have important roles to play.

3. We recognize the particular vulnerability of the least-developed countries and the special structural difficulties they face in the global economy. We are committed to addressing the marginalization of least-developed countries in international trade and to improving their effective participation in the multilateral trading system. We recall the commitments made by Ministers at our meetings in Marrakesh, Singapore and Geneva, and by the international community at the Third UN Conference on Least-Developed Countries in Brussels, to help least-developed countries secure beneficial and meaningful integration into the multilateral trading system and the global economy. We are determined that the WTO will play its part in building effectively on these commitments under the Work Programme we are establishing.

4. We stress our commitment to the WTO as the unique forum for global trade rule-making and liberalization, while also recognizing that regional trade agreements can play an important role in promoting the liberalization and expansion of trade and in fostering development.

5. We are aware that the challenges Members face in a rapidly changing international environment cannot be addressed through measures taken in the trade field alone. We shall continue to work with the Bretton Woods institutions for greater coherence in global economic policy-making.

6. We strongly reaffirm our commitment to the objective of sustainable development, as stated in the Preamble to the Marrakesh Agreement. We are convinced that the aims of upholding and safeguarding an open and non-discriminatory multilateral trading system, and acting for the protection of the environment and the promotion of sustainable development can and must be mutually supportive. We take note of the efforts by Members to conduct national environmental assessments of trade policies on a voluntary basis. We recognize that under WTO rules no country should be prevented from taking measures for the protection of human, animal or plant life or health, or of the environment at the levels it considers appropriate, subject to the requirement that they are not applied in a manner which would constitute a means of arbitrary or unjustifiable discrimination between countries where the same conditions prevail, or a disguised restriction on international trade, and are otherwise in accordance with the provisions of the WTO Agreements. We welcome the WTO's continued cooperation with UNEP and other inter-governmental environmental organizations. We encourage efforts to promote cooperation between the WTO and relevant international environmental and developmental organizations, especially in the lead-up to the World Summit on

Sustainable Development to be held in Johannesburg, South Africa, in September 2002.

7. We reaffirm the right of Members under the General Agreement on Trade in Services to regulate, and to introduce new regulations on, the supply of services.

8. We reaffirm our declaration made at the Singapore Ministerial Conference regarding internationally recognized core labour standards. We take note of work under way in the International Labour Organization (ILO) on the social dimension of globalization.

9. We note with particular satisfaction that this Conference has completed the WTO accession procedures for China and Chinese Taipei. We also welcome the accession as new Members, since our last Session, of Albania, Croatia, Georgia, Jordan, Lithuania, Moldova and Oman, and note the extensive market-access commitments already made by these countries on accession. These accessions will greatly strengthen the multilateral trading system, as will those of the 28 countries now negotiating their accession. We therefore attach great importance to concluding accession proceedings as quickly as possible. In particular, we are committed to accelerating the accession of least-developed countries.

10. Recognizing the challenges posed by an expanding WTO membership, we confirm our collective responsibility to ensure internal transparency and the effective participation of all Members. While emphasizing the intergovernmental character of the organization, we are committed to making the WTO's operations more transparent, including through more effective and prompt dissemination of information, and to improve dialogue with the public. We shall therefore at the national and multilateral levels continue to promote a better public understanding of the WTO and to communicate the benefits of a liberal, rules-based multilateral trading system.

11. In view of these considerations, we hereby agree to undertake the broad and balanced Work Programme set out below. This incorporates both an expanded negotiating agenda and other important decisions and activities necessary to address the challenges facing the multilateral trading system.

WORK PROGRAMME

Implementation-Related Issues and Concerns

12. We attach the utmost importance to the implementation-related issues and concerns raised by Members and are determined to find appropriate solutions to them. In this connection, and having regard to the General Council Decisions of 3 May and 15 December 2000, we further adopt the Decision on Implementation-Related Issues and Concerns in document WT/MIN(01)/17 to address a number of implementation problems faced by Members. We agree that

negotiations on outstanding implementation issues shall be an integral part of the Work Programme we are establishing, and that agreements reached at an early stage in these negotiations shall be treated in accordance with the provisions of paragraph 47 below. In this regard, we shall proceed as follows: (a) where we provide a specific negotiating mandate in this Declaration, the relevant implementation issues shall be addressed under that mandate; (b) the other outstanding implementation issues shall be addressed as a matter of priority by the relevant WTO bodies, which shall report to the Trade Negotiations Committee, established under paragraph 46 below, by the end of 2002 for appropriate action.

Agriculture

13. We recognize the work already undertaken in the negotiations initiated in early 2000 under Article 20 of the Agreement on Agriculture, including the large number of negotiating proposals submitted on behalf of a total of 121 Members. We recall the long-term objective referred to in the Agreement to establish a fair and market-oriented trading system through a programme of fundamental reform encompassing strengthened rules and specific commitments on support and protection in order to correct and prevent restrictions and distortions in world agricultural markets. We reconfirm our commitment to this programme. Building on the work carried out to date and without prejudging the outcome of the negotiations we commit ourselves to comprehensive negotiations aimed at: substantial improvements in market access; reductions of, with a view to phasing out, all forms of export subsidies; and substantial reductions in trade-distorting domestic support. We agree that special and differential treatment for developing countries shall be an integral part of all elements of the negotiations and shall be embodied in the Schedules of concessions and commitments and as appropriate in the rules and disciplines to be negotiated, so as to be operationally effective and to enable developing countries to effectively take account of their development needs, including food security and rural development. We take note of the non-trade concerns reflected in the negotiating proposals submitted by Members and confirm that non-trade concerns will be taken into account in the negotiations as provided for in the Agreement on Agriculture.

14. Modalities for the further commitments, including provisions for special and differential treatment, shall be established no later than 31 March 2003. Participants shall submit their comprehensive draft Schedules based on these modalities no later than the date of the Fifth Session of the Ministerial Conference. The negotiations, including with respect to rules and disciplines and related legal texts, shall be concluded as part and at the date of conclusion of the negotiating agenda as a whole.

Services

15. The negotiations on trade in services shall be conducted with a view to promoting the economic growth of all trading partners and the development of

developing and least-developed countries. We recognize the work already undertaken in the negotiations, initiated in January 2000 under Article XIX of the General Agreement on Trade in Services, and the large number of proposals submitted by Members on a wide range of sectors and several horizontal issues, as well as on movement of natural persons. We reaffirm the Guidelines and Procedures for the Negotiations adopted by the Council for Trade in Services on 28 March 2001 as the basis for continuing the negotiations, with a view to achieving the objectives of the General Agreement on Trade in Services, as stipulated in the Preamble, Article IV and Article XIX of that Agreement. Participants shall submit initial requests for specific commitments by 30 June 2002 and initial offers by 31 March 2003.

Market Access for Non-agricultural Products

16. We agree to negotiations which shall aim, by modalities to be agreed, to reduce or as appropriate eliminate tariffs, including the reduction or elimination of tariff peaks, high tariffs, and tariff escalation, as well as non-tariff barriers, in particular on products of export interest to developing countries. Product coverage shall be comprehensive and without a priori exclusions. The negotiations shall take fully into account the special needs and interests of developing and least-developed country participants, including through less than full reciprocity in reduction commitments, in accordance with the relevant provisions of Article XXVIII bis of GATT 1994 and the provisions cited in paragraph 50 below. To this end, the modalities to be agreed will include appropriate studies and capacity-building measures to assist least-developed countries to participate effectively in the negotiations.

Trade-Related Aspects of Intellectual Property Rights

17. We stress the importance we attach to implementation and interpretation of the Agreement on Trade-Related Aspects of Intellectual Property Rights (TRIPS Agreement) in a manner supportive of public health, by promoting both access to existing medicines and research and development into new medicines and, in this connection, are adopting a separate Declaration.

18. With a view to completing the work started in the Council for Trade-Related Aspects of Intellectual Property Rights (Council for TRIPS) on the implementation of Article 23.4, we agree to negotiate the establishment of a multilateral system of notification and registration of geographical indications for wines and spirits by the Fifth Session of the Ministerial Conference. We note that issues related to the extension of the protection of geographical indications provided for in Article 23 to products other than wines and spirits will be addressed in the Council for TRIPS pursuant to paragraph 12 of this Declaration.

19. We instruct the Council for TRIPS, in pursuing its work programme including under the review of Article 27.3(b), the review of the implementation of the TRIPS Agreement under Article 71.1 and the work foreseen pursuant to

paragraph 12 of this Declaration, to examine, inter alia, the relationship between the TRIPS Agreement and the Convention on Biological Diversity, the protection of traditional knowledge and folklore, and other relevant new developments raised by Members pursuant to Article 71.1. In undertaking this work, the TRIPS Council shall be guided by the objectives and principles set out in Articles 7 and 8 of the TRIPS Agreement and shall take fully into account the development dimension.

Relationship between Trade and Investment

20. Recognizing the case for a multilateral framework to secure transparent, stable and predictable conditions for long-term cross-border investment, particularly foreign direct investment, that will contribute to the expansion of trade, and the need for enhanced technical assistance and capacity-building in this area as referred to in paragraph 21, we agree that negotiations will take place after the Fifth Session of the Ministerial Conference on the basis of a decision to be taken, by explicit consensus, at that Session on modalities of negotiations.

21. We recognize the needs of developing and least-developed countries for enhanced support for technical assistance and capacity building in this area, including policy analysis and development so that they may better evaluate the implications of closer multilateral cooperation for their development policies and objectives, and human and institutional development. To this end, we shall work in cooperation with other relevant intergovernmental organisations, including UNCTAD, and through appropriate regional and bilateral channels, to provide strengthened and adequately resourced assistance to respond to these needs.

22. In the period until the Fifth Session, further work in the Working Group on the Relationship Between Trade and Investment will focus on the clarification of: scope and definition; transparency; non-discrimination; modalities for pre-establishment commitments based on a GATS-type, positive list approach; development provisions; exceptions and balance-of-payments safeguards; consultation and the settlement of disputes between Members. Any framework should reflect in a balanced manner the interests of home and host countries, and take due account of the development policies and objectives of host governments as well as their right to regulate in the public interest. The special development, trade and financial needs of developing and least-developed countries should be taken into account as an integral part of any framework, which should enable Members to undertake obligations and commitments commensurate with their individual needs and circumstances. Due regard should be paid to other relevant WTO provisions. Account should be taken, as appropriate, of existing bilateral and regional arrangements on investment.

Interaction between Trade and Competition Policy

23. Recognizing the case for a multilateral framework to enhance the contribution of competition policy to international trade and development, and the need for enhanced technical assistance and capacity-building in this area as referred to

in paragraph 24, we agree that negotiations will take place after the Fifth Session of the Ministerial Conference on the basis of a decision to be taken, by explicit consensus, at that Session on modalities of negotiations.

24. We recognize the needs of developing and least-developed countries for enhanced support for technical assistance and capacity building in this area, including policy analysis and development so that they may better evaluate the implications of closer multilateral cooperation for their development policies and objectives, and human and institutional development. To this end, we shall work in cooperation with other relevant intergovernmental organisations, including UNCTAD, and through appropriate regional and bilateral channels, to provide strengthened and adequately resourced assistance to respond to these needs.

25. In the period until the Fifth Session, further work in the Working Group on the Interaction between Trade and Competition Policy will focus on the clarification of: core principles, including transparency, non-discrimination and procedural fairness, and provisions on hardcore cartels; modalities for voluntary cooperation; and support for progressive reinforcement of competition institutions in developing countries through capacity building. Full account shall be taken of the needs of developing and least-developed country participants and appropriate flexibility provided to address them.

Transparency in Government Procurement

26. Recognizing the case for a multilateral agreement on transparency in government procurement and the need for enhanced technical assistance and capacity building in this area, we agree that negotiations will take place after the Fifth Session of the Ministerial Conference on the basis of a decision to be taken, by explicit consensus, at that Session on modalities of negotiations. These negotiations will build on the progress made in the Working Group on Transparency in Government Procurement by that time and take into account participants' development priorities, especially those of least-developed country participants. Negotiations shall be limited to the transparency aspects and therefore will not restrict the scope for countries to give preferences to domestic supplies and suppliers. We commit ourselves to ensuring adequate technical assistance and support for capacity building both during the negotiations and after their conclusion.

Trade Facilitation

27. Recognizing the case for further expediting the movement, release and clearance of goods, including goods in transit, and the need for enhanced technical assistance and capacity building in this area, we agree that negotiations will take place after the Fifth Session of the Ministerial Conference on the basis of a decision to be taken, by explicit consensus, at that Session on modalities of negotiations. In the period until the Fifth Session, the Council for Trade in Goods shall review and as appropriate, clarify and improve relevant aspects of Articles V, VIII and X of the GATT 1994 and identify the trade facilitation needs and

priorities of Members, in particular developing and least-developed countries. We commit ourselves to ensuring adequate technical assistance and support for capacity building in this area.

WTO Rules

28. In the light of experience and of the increasing application of these instruments by Members, we agree to negotiations aimed at clarifying and improving disciplines under the Agreements on Implementation of Article VI of the GATT 1994 and on Subsidies and Countervailing Measures, while preserving the basic concepts, principles and effectiveness of these Agreements and their instruments and objectives, and taking into account the needs of developing and least-developed participants. In the initial phase of the negotiations, participants will indicate the provisions, including disciplines on trade distorting practices, that they seek to clarify and improve in the subsequent phase. In the context of these negotiations, participants shall also aim to clarify and improve WTO disciplines on fisheries subsidies, taking into account the importance of this sector to developing countries. We note that fisheries subsidies are also referred to in paragraph 31.

29. We also agree to negotiations aimed at clarifying and improving disciplines and procedures under the existing WTO provisions applying to regional trade agreements. The negotiations shall take into account the developmental aspects of regional trade agreements.

Dispute Settlement Understanding

30. We agree to negotiations on improvements and clarifications of the Dispute Settlement Understanding. The negotiations should be based on the work done thus far as well as any additional proposals by Members, and aim to agree on improvements and clarifications not later than May 2003, at which time we will take steps to ensure that the results enter into force as soon as possible thereafter.

Trade and Environment

31. With a view to enhancing the mutual supportiveness of trade and environment, we agree to negotiations, without prejudging their outcome, on:

(i) the relationship between existing WTO rules and specific trade obligations set out in multilateral environmental agreements (MEAs). The negotiations shall be limited in scope to the applicability of such existing WTO rules as among parties to the MEA in question. The negotiations shall not prejudice the WTO rights of any Member that is not a party to the MEA in question;

(ii) procedures for regular information exchange between MEA Secretariats and the relevant WTO committees, and the criteria for the granting of observer status;

(iii) the reduction or, as appropriate, elimination of tariff and non-tariff barriers to environmental goods and services.

We note that fisheries subsidies form part of the negotiations provided for in paragraph 28.

32. We instruct the Committee on Trade and Environment, in pursuing work on all items on its agenda within its current terms of reference, to give particular attention to:

(i) the effect of environmental measures on market access, especially in relation to developing countries, in particular the least-developed among them, and those situations in which the elimination or reduction of trade restrictions and distortions would benefit trade, the environment and development;

(ii) the relevant provisions of the Agreement on Trade-Related Aspects of Intellectual Property Rights; and

(iii) labelling requirements for environmental purposes.

Work on these issues should include the identification of any need to clarify relevant WTO rules. The Committee shall report to the Fifth Session of the Ministerial Conference, and make recommendations, where appropriate, with respect to future action, including the desirability of negotiations. The outcome of this work as well as the negotiations carried out under paragraph 31(i) and (ii) shall be compatible with the open and non-discriminatory nature of the multilateral trading system, shall not add to or diminish the rights and obligations of Members under existing WTO agreements, in particular the Agreement on the Application of Sanitary and Phytosanitary Measures, nor alter the balance of these rights and obligations, and will take into account the needs of developing and least-developed countries.

33. We recognize the importance of technical assistance and capacity building in the field of trade and environment to developing countries, in particular the least-developed among them. We also encourage that expertise and experience be shared with Members wishing to perform environmental reviews at the national level. A report shall be prepared on these activities for the Fifth Session.

Electronic Commerce

34. We take note of the work which has been done in the General Council and other relevant bodies since the Ministerial Declaration of 20 May 1998 and agree to continue the Work Programme on Electronic Commerce. The work to date demonstrates that electronic commerce creates new challenges and opportunities for trade for Members at all stages of development, and we recognize the importance of creating and maintaining an environment which is favourable to the future development of electronic commerce. We instruct the General Council to consider the most appropriate institutional arrangements for handling the Work Programme, and to report on further progress to the Fifth Session of the Ministerial Conference. We declare that Members will maintain their current practice of not imposing customs duties on electronic transmissions until the Fifth Session.

Small Economies

35. We agree to a work programme, under the auspices of the General Council, to examine issues relating to the trade of small economies. The objective of this work is to frame responses to the trade-related issues identified for the fuller integration of small, vulnerable economies into the multilateral trading system, and not to create a sub-category of WTO Members. The General Council shall review the work programme and make recommendations for action to the Fifth Session of the Ministerial Conference.

Trade, Debt and Finance

36. We agree to an examination, in a Working Group under the auspices of the General Council, of the relationship between trade, debt and finance, and of any possible recommendations on steps that might be taken within the mandate and competence of the WTO to enhance the capacity of the multilateral trading system to contribute to a durable solution to the problem of external indebtedness of developing and least-developed countries, and to strengthen the coherence of international trade and financial policies, with a view to safeguarding the multilateral trading system from the effects of financial and monetary instability. The General Council shall report to the Fifth Session of the Ministerial Conference on progress in the examination.

Trade and Transfer of Technology

37. We agree to an examination, in a Working Group under the auspices of the General Council, of the relationship between trade and transfer of technology, and of any possible recommendations on steps that might be taken within the mandate of the WTO to increase flows of technology to developing countries. The General Council shall report to the Fifth Session of the Ministerial Conference on progress in the examination.

Technical Cooperation and Capacity Building

38. We confirm that technical cooperation and capacity building are core elements of the development dimension of the multilateral trading system, and we welcome and endorse the New Strategy for WTO Technical Cooperation for Capacity Building, Growth and Integration. We instruct the Secretariat, in coordination with other relevant agencies, to support domestic efforts for mainstreaming trade into national plans for economic development and strategies for poverty reduction. The delivery of WTO technical assistance shall be designed to assist developing and least-developed countries and low-income countries in transition to adjust to WTO rules and disciplines, implement obligations and exercise the rights of membership, including drawing on the benefits of an open, rules-based multilateral trading system. Priority shall also be accorded to small, vulnerable, and transition economies, as well as to Members and Observers without representation in Geneva. We reaffirm our support for the valuable work of the International Trade Centre, which should be enhanced.

39. We underscore the urgent necessity for the effective coordinated delivery of technical assistance with bilateral donors, in the OECD Development Assistance Committee and relevant international and regional intergovernmental institutions, within a coherent policy framework and timetable. In the coordinated delivery of technical assistance, we instruct the Director-General to consult with the relevant agencies, bilateral donors and beneficiaries, to identify ways of enhancing and rationalizing the Integrated Framework for Trade-Related Technical Assistance to Least-Developed Countries and the Joint Integrated Technical Assistance Programme (JITAP).

40. We agree that there is a need for technical assistance to benefit from secure and predictable funding. We therefore instruct the Committee on Budget, Finance and Administration to develop a plan for adoption by the General Council in December 2001 that will ensure long-term funding for WTO technical assistance at an overall level no lower than that of the current year and commensurate with the activities outlined above.

41. We have established firm commitments on technical cooperation and capacity building in various paragraphs in this Ministerial Declaration. We reaffirm these specific commitments contained in paragraphs 16, 21, 24, 26, 27, 33, 38-40, 42 and 43, and also reaffirm the understanding in paragraph 2 on the important role of sustainably financed technical assistance and capacity-building programmes. We instruct the Director-General to report to the Fifth Session of the Ministerial Conference, with an interim report to the General Council in December 2002 on the implementation and adequacy of these commitments in the identified paragraphs.

Least-Developed Countries

42. We acknowledge the seriousness of the concerns expressed by the least-developed countries (LDCs) in the Zanzibar Declaration adopted by their Ministers in July 2001. We recognize that the integration of the LDCs into the multilateral trading system requires meaningful market access, support for the diversification of their production and export base, and trade-related technical assistance and capacity building. We agree that the meaningful integration of LDCs into the trading system and the global economy will involve efforts by all WTO Members. We commit ourselves to the objective of duty-free, quota-free market access for products originating from LDCs. In this regard, we welcome the significant market access improvements by WTO Members in advance of the Third UN Conference on LDCs (LDC-III), in Brussels, May 2001. We further commit ourselves to consider additional measures for progressive improvements in market access for LDCs. Accession of LDCs remains a priority for the Membership. We agree to work to facilitate and accelerate negotiations with acceding LDCs. We instruct the Secretariat to reflect the priority we attach to LDCs' accessions in the annual plans for technical assistance. We reaffirm the commitments we undertook at LDC-III, and agree that the WTO should take into account, in designing its work programme for LDCs, the trade-related

elements of the Brussels Declaration and Programme of Action, consistent with the WTO's mandate, adopted at LDC-III. We instruct the Sub-Committee for Least-Developed Countries to design such a work programme and to report on the agreed work programme to the General Council at its first meeting in 2002.

43. We endorse the Integrated Framework for Trade-Related Technical Assistance to Least-Developed Countries (IF) as a viable model for LDCs' trade development. We urge development partners to significantly increase contributions to the IF Trust Fund and WTO extra-budgetary trust funds in favour of LDCs. We urge the core agencies, in coordination with development partners, to explore the enhancement of the IF with a view to addressing the supply-side constraints of LDCs and the extension of the model to all LDCs, following the review of the IF and the appraisal of the ongoing Pilot Scheme in selected LDCs. We request the Director-General, following coordination with heads of the other agencies, to provide an interim report to the General Council in December 2002 and a full report to the Fifth Session of the Ministerial Conference on all issues affecting LDCs.

Special and Differential Treatment

44. We reaffirm that provisions for special and differential treatment are an integral part of the WTO Agreements. We note the concerns expressed regarding their operation in addressing specific constraints faced by developing countries, particularly least-developed countries. In that connection, we also note that some Members have proposed a Framework Agreement on Special and Differential Treatment (WT/GC/W/442). We therefore agree that all special and differential treatment provisions shall be reviewed with a view to strengthening them and making them more precise, effective and operational. In this connection, we endorse the work programme on special and differential treatment set out in the Decision on Implementation-Related Issues and Concerns.

ORGANIZATION AND MANAGEMENT OF THE WORK PROGRAMME

45. The negotiations to be pursued under the terms of this Declaration shall be concluded not later than 1 January 2005. The Fifth Session of the Ministerial Conference will take stock of progress in the negotiations, provide any necessary political guidance, and take decisions as necessary. When the results of the negotiations in all areas have been established, a Special Session of the Ministerial Conference will be held to take decisions regarding the adoption and implementation of those results.

46. The overall conduct of the negotiations shall be supervised by a Trade Negotiations Committee under the authority of the General Council. The Trade Negotiations Committee shall hold its first meeting not later than 31 January 2002. It shall establish appropriate negotiating mechanisms as required and supervise the progress of the negotiations.

47. With the exception of the improvements and clarifications of the Dispute Settlement Understanding, the conduct, conclusion and entry into force of the outcome of the negotiations shall be treated as parts of a single undertaking. However, agreements reached at an early stage may be implemented on a provisional or a definitive basis. Early agreements shall be taken into account in assessing the overall balance of the negotiations.

48. Negotiations shall be open to:

(i) all Members of the WTO; and

(ii) States and separate customs territories currently in the process of accession and those that inform Members, at a regular meeting of the General Council, of their intention to negotiate the terms of their membership and for whom an accession working party is established.

Decisions on the outcomes of the negotiations shall be taken only by WTO Members.

49. The negotiations shall be conducted in a transparent manner among participants, in order to facilitate the effective participation of all. They shall be conducted with a view to ensuring benefits to all participants and to achieving an overall balance in the outcome of the negotiations.

50. The negotiations and the other aspects of the Work Programme shall take fully into account the principle of special and differential treatment for developing and least-developed countries embodied in: Part IV of the GATT 1994; the Decision of 28 November 1979 on Differential and More Favourable Treatment, Reciprocity and Fuller Participation of Developing Countries; the Uruguay Round Decision on Measures in Favour of Least-Developed Countries; and all other relevant WTO provisions.

51. The Committee on Trade and Development and the Committee on Trade and Environment shall, within their respective mandates, each act as a forum to identify and debate developmental and environmental aspects of the negotiations, in order to help achieve the objective of having sustainable development appropriately reflected.

52. Those elements of the Work Programme which do not involve negotiations are also accorded a high priority. They shall be pursued under the overall supervision of the General Council, which shall report on progress to the Fifth Session of the Ministerial Conference.

Annex 6
2002 Plan of Implementation of the World Summit on Sustainable Development (excerpts)

31. To achieve sustainable fisheries, the following actions are required at all levels:

(a) Maintain or restore stocks to levels that can produce the maximum sustainable yield with the aim of achieving these goals for depleted stocks on an urgent basis and where possible not later than 2015;

(b) Ratify or accede to and effectively implement the relevant United Nations and, where appropriate, associated regional fisheries agreements or arrangements, noting in particular the Agreement for the Implementation of the Provisions of the United Nations Convention on the Law of the Sea of 10 December 1982 relating to the Conservation and Management of Straddling Fish Stocks and Highly Migratory Fish Stocks and the 1993 Agreement to Promote Compliance with International Conservation and Management Measures by Fishing Vessels on the High Seas;

(c) Implement the 1995 Code of Conduct for Responsible Fisheries, taking note of the special requirements of developing countries as noted in its article 5, and the relevant international plans of action and technical guidelines of the Food and Agriculture Organization of the United Nations;

(d) Urgently develop and implement national and, where appropriate, regional plans of action, to put into effect the international plans of action of the Food and Agriculture Organization of the United Nations, in particular the International Plan of Action for the Management of Fishing Capacity by 2005 and the International Plan of Action to Prevent, Deter and Eliminate Illegal, Unreported and Unregulated Fishing by 2004. Establish effective monitoring, reporting and enforcement, and control of fishing vessels, including by flag States, to further the International Plan of Action to Prevent, Deter and Eliminate Illegal, Unreported and Unregulated Fishing;

(e) Encourage relevant regional fisheries management organizations and arrangements to give due consideration to the rights, duties and interests of coastal States and the special requirements of developing States when addressing the issue of the allocation of share of fishery resources for straddling stocks and highly migratory fish stocks, mindful of the provisions of the United Nations Convention on the Law of the Sea and the Agreement for the Implementation of the Provisions of the United Nations Convention on the Law of the Sea of 10 December 1982 relating to the Conservation and Management of Straddling Fish Stocks and Highly Migratory Fish Stocks, on the high seas and within exclusive economic zones;
(f) Eliminate subsidies that contribute to illegal, unreported and unregulated fishing and to over-capacity, while completing the efforts undertaken at the World Trade Organization to clarify and improve its disciplines on fisheries subsidies, taking into account the importance of this sector to developing countries;
(g) Strengthen donor coordination and partnerships between international financial institutions, bilateral agencies and other relevant stakeholders to enable developing countries, in particular the least developed countries and small island developing States and countries with economies in transition, to develop their national, regional and subregional capacities for infrastructure and integrated management and the sustainable use of fisheries;
(h) Support the sustainable development of aquaculture, including small-scale aquaculture, given its growing importance for food security and economic development.

Annex 7
2005 WTO Hong Kong Ministerial Declaration (excerpts)

World Trade Organization

WT/MIN(05)/DEC/1
22 December 2005
(05-6248)

MINISTERIAL CONFERENCE
Sixth Session
Hong Kong, 13–18 December 2005

DOHA WORK PROGRAMME
MINISTERIAL DECLARATION

Adopted on 18 December 2005

Annex D
Rules

I. Anti-Dumping and Subsidies and Countervailing Measures including Fisheries Subsidies

We:

1. *acknowledge* that the achievement of substantial results on all aspects of the Rules mandate, in the form of amendments to the Anti-Dumping (AD) and

Subsidies and Countervailing Measures (SCM) Agreements, is important to the development of the rules-based multilateral trading system and to the overall balance of results in the DDA;

2. *aim* to achieve in the negotiations on Rules further improvements, in particular, to the transparency, predictability and clarity of the relevant disciplines, to the benefit of all Members, including in particular developing and least-developed Members. In this respect, the development dimension of the negotiations must be addressed as an integral part of any outcome;

3. *call on* Participants, in considering possible clarifications and improvements in the area of anti-dumping, to take into account, *inter alia*, (a) the need to avoid the unwarranted use of anti-dumping measures, while preserving the basic concepts, principles and effectiveness of the instrument and its objectives where such measures are warranted; and (b) the desirability of limiting the costs and complexity of proceedings for interested parties and the investigating authorities alike, while strengthening the due process, transparency and predictability of such proceedings and measures;

4. *consider* that negotiations on anti-dumping should, as appropriate, clarify and improve the rules regarding, *inter alia*, (a) determinations of dumping, injury and causation, and the application of measures; (b) procedures governing the initiation, conduct and completion of antidumping investigations, including with a view to strengthening due process and enhancing transparency; and (c) the level, scope and duration of measures, including duty assessment, interim and new shipper reviews, sunset, and anti-circumvention proceedings;

5. *recognize* that negotiations on anti-dumping have intensified and deepened, that Participants are showing a high level of constructive engagement, and that the process of rigorous discussion of the issues based on specific textual proposals for amendment to the AD Agreement has been productive and is a necessary step in achieving the substantial results to which Ministers are committed;

6. *note* that, in the negotiations on anti-dumping, the Negotiating Group on Rules has been discussing in detail proposals on such issues as determinations of injury/causation, the lesser duty rule, public interest, transparency and due process, interim reviews, sunset, duty assessment, circumvention, the use of facts available, limited examination and all others rates, dispute settlement, the definition of dumped imports, affiliated parties, product under consideration, and the initiation and completion of investigations, and that this process of discussing proposals before the Group or yet to be submitted will continue after Hong Kong;

7. *note*, in respect of subsidies and countervailing measures, that while proposals for amendments to the SCM Agreement have been submitted on a number of issues, including the definition of a subsidy, specificity, prohibited subsidies, serious prejudice, export credits and guarantees, and the allocation of benefit, there is a need to deepen the analysis on the basis of specific textual proposals in order to ensure a balanced outcome in all areas of the Group's mandate;

8. *note* the desirability of applying to both anti-dumping and countervailing measures any clarifications and improvements which are relevant and appropriate to both instruments;

9. *recall* our commitment at Doha to enhancing the mutual supportiveness of trade and environment, *note* that there is broad agreement that the Group should strengthen disciplines on subsidies in the fisheries sector, including through the prohibition of certain forms of fisheries subsidies that contribute to overcapacity and over-fishing, and *call on* Participants promptly to undertake further detailed work to, *inter alia*, establish the nature and extent of those disciplines, including transparency and enforceability. Appropriate and effective special and differential treatment for developing and least-developed Members should be an integral part of the fisheries subsidies negotiations, taking into account the importance of this sector to development priorities, poverty reduction, and livelihood and food security concerns;

10. *direct* the Group to intensify and accelerate the negotiating process in all areas of its mandate, on the basis of detailed textual proposals before the Group or yet to be submitted, and complete the process of analysing proposals by Participants on the AD and SCM Agreements as soon as possible;

11. *mandate* the Chairman to prepare, early enough to assure a timely outcome within the context of the 2006 end date for the Doha Development Agenda and taking account of progress in other areas of the negotiations, consolidated texts of the AD and SCM Agreements that shall be the basis for the final stage of the negotiations.

II. Regional Trade Agreements

1. We welcome the progress in negotiations to clarify and improve the WTO's disciplines and procedures on regional trade agreements (RTAs). Such agreements, which can foster trade liberalization and promote development, have become an important element in the trade policies of virtually all Members. Transparency of RTAs is thus of systemic interest as are disciplines that ensure the complementarity of RTAs with the WTO.

2. We commend the progress in defining the elements of a transparency mechanism for RTAs, aimed, in particular, at improving existing WTO procedures for gathering factual information on RTAs, without prejudice to the rights and obligations of Members. We instruct the Negotiating Group on Rules to intensify its efforts to resolve outstanding issues, with a view to a provisional decision on RTA transparency by 30 April 2006.

3. We also note with appreciation the work of the Negotiating Group on Rules on WTO's disciplines governing RTAs, including *inter alia* on the "substantially all the trade" requirement, the length of RTA transition periods and RTA developmental aspects. We instruct the Group to intensify negotiations, based on text proposals as soon as possible after the Sixth Ministerial Conference, so as to arrive at appropriate outcomes by end 2006.

Annex 8
WTO Chair's Text on Fisheries Subsidies (TN/RL/W/213, 30 November 2007, Annex VIII, pp.87–93)

WORLD TRADE ORGANIZATION

TN/RL/W/213
30 November 2007
(07-5291)

Negotiating Group on Rules

DRAFT CONSOLIDATED CHAIR TEXTS OF THE AD AND SCM AGREEMENTS

ANNEX VIII
FISHERIES SUBSIDIES

Article I
Prohibition of Certain Fisheries Subsidies

I.1 Except as provided for in Articles II and III, or in the exceptional case of natural disaster relief,[1] the following subsidies within the meaning of paragraph 1 of Article 1, to the extent they are specific within the meaning of paragraph 2 of Article 1, shall be prohibited:

(a) Subsidies the benefits of which are conferred on the acquisition, construction, repair, renewal, renovation, modernization, or any other modification of fishing vessels[2] or service vessels[3], including subsidies to boat building or shipbuilding facilities for these purposes.
(b) Subsidies the benefits of which are conferred on transfer of fishing or service vessels to third countries, including through the creation of joint enterprises with third country partners.
(c) Subsidies the benefits of which are conferred on operating costs of fishing or service vessels (including licence fees or similar charges, fuel, ice, bait, personnel, social charges, insurance, gear, and at-sea support); or of landing, handling or in- or nearport processing activities for products of marine wild capture fishing; or subsidies to cover operating losses of such vessels or activities.
(d) Subsidies in respect of, or in the form of, port infrastructure or other physical port facilities exclusively or predominantly for activities related to marine wild capture fishing (for example, fish landing facilities, fish storage facilities, and in- or near-port fish processing facilities).
(e) Income support for natural or legal persons engaged in marine wild capture fishing.
(f) Price support for products of marine wild capture fishing.
(g) Subsidies arising from the further transfer, by a payer Member government, of access rights that it has acquired from another Member government to fisheries within the jurisdiction of such other Member.[4]
(h) Subsidies the benefits of which are conferred on any vessel engaged in illegal, unreported or unregulated fishing.[5]

I.2 In addition to the prohibitions listed in paragraph 1, any subsidy referred to in paragraphs 1 and 2 of Article 1 the benefits of which are conferred on any fishing vessel or fishing activity affecting fish stocks that are in an unequivocally overfished condition shall be prohibited.

Article II
General Exceptions

Notwithstanding the provisions of Article I, and subject to the provision of Article V:

(a) For the purposes of Article I.1(a), subsidies exclusively for improving fishing or service vessel and crew safety shall not be prohibited, provided that:
  (1) such subsidies do not involve new vessel construction or vessel acquisition;
  (2) such subsidies do not give rise to any increase in marine wild capture fishing capacity of any fishing or service vessel, on the basis of gross tonnage, volume of fish hold, engine power, or on any other basis, and do not have the effect of maintaining in operation any such vessel that otherwise would be withdrawn; and

(3) the improvements are undertaken to comply with safety standards.

(b) For the purposes of Articles I.1(a) and I.1(c) the following subsidies shall not be prohibited:
subsidies exclusively for: (1) the adoption of gear for selective fishing techniques; (2) the adoption of other techniques aimed at reducing the environmental impact of marine wild capture fishing; (3) compliance with fisheries management regimes aimed at sustainable use and conservation (e.g., devices for Vessel Monitoring Systems); provided that the subsidies do not give rise to any increase in the marine wild capture fishing capacity of any fishing or service vessel, on the basis of gross tonnage, volume of fish hold, engine power, or on any other basis, and do not have the effect of maintaining in operation any such vessel that otherwise would be withdrawn.

(c) For the purposes of Article I.1(c), subsidies to cover personnel costs shall not be interpreted as including:
(1) subsidies exclusively for re-education, retraining or redeployment of fishworkers[6] into occupations unrelated to marine wild capture fishing or directly associated activities; and
(2) subsidies exclusively for early retirement or permanent cessation of employment of fishworkers as a result of government policies to reduce marine wild capture fishing capacity or effort.

(d) Nothing in Article I shall prevent subsidies for vessel decommissioning or capacity reduction programmes, provided that:
(1) the vessels subject to such programmes are scrapped or otherwise permanently and effectively prevented from being used for fishing anywhere in the world;
(2) the fish harvesting rights associated with such vessels, whether they are permits, licences, fish quotas or any other form of harvesting rights, are permanently revoked and may not be reassigned;
(3) the owners of such vessels, and the holders of such fish harvesting rights, are required to relinquish any claim associated with such vessels and harvesting rights that could qualify such owners and holders for any present or future harvesting rights in such fisheries; and
(4) the fisheries management system in place includes management control measures and enforcement mechanisms designed to prevent overfishing in the targeted fishery. Such fishery-specific measures may include limited entry systems, catch quotas, limits on fishing effort or allocation of exclusive quotas to vessels, individuals and/or groups, such as individual transferable quotas.

(e) Nothing in Article I shall prevent governments from making user-specific allocations to individuals and groups under limited access privileges and other exclusive quota programmes.

Article III
Special and Differential Treatment of Developing Country Members

III.1 The prohibition of Article 3.1(c) and Article I shall not apply to least-developed country ("LDC") Members.
III.2 For developing country Members other than LDC Members:

(a) Subsidies referred to in Article I.1 shall not be prohibited where they relate exclusively to marine wild capture fishing performed on an inshore basis (i.e., within the territorial waters of the Member) with non-mechanized net-retrieval, provided that (1) the activities are carried out on their own behalf by fishworkers, on an individual basis which may include family members, or organized in associations; (2) the catch is consumed principally by the fishworkers and their families and the activities do not go beyond a small profit trade; and (3) there is no major employer-employee relationship in the activities carried out. Fisheries management measures aimed at ensuring sustainability, such as the measures referred to in Article V, should be implemented in respect of the fisheries in question, adapted as necessary to the particular situation, including by making use of indigenous fisheries management institutions and measures.

(b) In addition, subject to the provisions of Article V:

(1) Subsidies referred to in Articles I.1(d), I.1(e) and I.1(f) shall not be prohibited.

(2) Subsidies referred to in Articles I.1(a) and I.1(c) shall not be prohibited provided that they are used exclusively for marine wild capture fishing employing decked vessels not greater than 10 meters or 34 feet in length overall, or undecked vessels of any length.

(3) For fishing and service vessels of such Members other than the vessels referred to in paragraph (b)(2), subsidies referred to in Article I.1(a) shall not be prohibited provided that (i) the vessels are used exclusively for marine wild capture fishing activities of such Members in respect of particular, identified target stocks within their Exclusive Economic Zones ("EEZ"); (ii) those stocks have been subject to prior scientific status assessment conducted in accordance with relevant international standards, aimed at ensuring that the resulting capacity does not exceed a sustainable level; and (iii) that assessment has been subject to peer review in the relevant body of the United Nations Food and Agriculture Organization ("FAO").[7]

III.3 Subsidies referred to in Article I.1(g) shall not be prohibited where the fishery in question is within the EEZ of a developing country Member, provided that the agreement pursuant to which the rights have been acquired is made public, and contains provisions designed to prevent overfishing in the area covered by the agreement based on internationally-recognized best practices for fisheries management and conservation as reflected in the relevant provisions of international instruments aimed at ensuring the sustainable use and conservation of marine species, such as, inter alia, the Agreement for the Implementation of the Provisions of the United Nations Convention on the Law of the Sea of 10 December 1982 Relating to the Conservation and Management of Straddling Fish Stocks and Highly Migratory Fish Stocks ("Fish Stocks Agreement"), the Code of Conduct on Responsible Fisheries of the Food and Agriculture Organization ("Code of Conduct"), the Agreement to Promote Compliance with International Conservation and Management Measures by Fishing Vessels on the High Seas ("Compliance Agreement"), and technical guidelines and plans of action (including criteria and precautionary reference points) for the implementation of these instruments, or other related or successor instruments. These provisions shall include requirements and support for science-based stock assessment before fishing is undertaken pursuant to the agreement and for regular assessments thereafter, for management and control measures, for vessel registries, for reporting of effort, catches and discards to the national authorities of the host Member and to relevant international organizations, and for such other measures as may be appropriate.

III.4 Members shall give due regard to the needs of developing country Members in complying with the requirements of this Annex, including the conditions and criteria set forth in this Article and in Article V, and shall establish mechanisms for, and facilitate, the provision of technical assistance in this regard, bilaterally and/or through the appropriate international organizations.

Endnotes

1 Subsidies referred to in this provision shall not be prohibited when limited to the relief of a particular natural disaster, provided that the subsidies are directly related to the effects of that disaster, are limited to the affected geographic area, are time-limited, and in the case of reconstruction subsidies, only restore the affected area, the affected fishery, and/or the affected fleet to its pre-disaster state, up to a sustainable level of fishing capacity as established through a science-based assessment of the post-disaster status of the fishery. Any such subsidies are subject to the provisions of Article VI.

2 For the purposes of this Agreement, the term "fishing vessels" refers to vessels used for marine wild capture fishing and/or on-board processing of the products thereof.

3 For the purposes of this Agreement, the term "service vessels" refers to vessels used to tranship the products of marine wild capture fishing from fishing vessels to on-shore facilities; and vessels used for at-sea refuelling, provisioning and other servicing of fishing vessels.

4 Government-to-government payments for access to marine fisheries shall not be deemed to be subsidies within the meaning of this Agreement.

5 The terms "illegal fishing", "unreported fishing" and "unregulated fishing" shall have the same meaning as in paragraph 3 of the International Plan of Action to Prevent, Deter and Eliminate Illegal Unreported and Unregulated Fishing of the United Nations Food and Agricultural Organization.

6 For the purpose of this Agreement, the term "fishworker" shall refer to an individual employed in marine wild capture fishing and/or directly associated activities.

7 If the Member in question is not a member of the FAO, the peer review shall take place in another recognized and competent international organization.

Annex 9
WTO Fisheries Roadmap (TN/RL/W/236, 19 December 2008, Annex VIII, pp.85–94)

WORLD TRADE ORGANIZATION

TN/RL/W/236
19 December 2008
(08-6255)

Negotiating Group on Rules

NEW DRAFT CONSOLIDATED CHAIR TEXTS OF THE AD AND SCM AGREEMENTS

ANNEX VIII
FISHERIES SUBSIDIES

Note: The text on fisheries subsidies will be revised following discussion of the issues identified in the annexed fisheries subsidies roadmap.

FISHERIES SUBSIDIES – ROADMAP FOR DISCUSSIONS

THE NEGOTIATING MANDATE

1. The Ministerial mandate from Hong Kong directs this Negotiating Group to '<u>strengthen disciplines</u> on subsidies in the fisheries sector, including through the <u>prohibition of certain forms of fisheries subsidies that contribute to overcapacity and over-fishing</u>' and, as an integral part of the negotiations, to establish '<u>appropriate and effective special and differential treatment</u> for developing and least-developed Members'.

GENERAL CONSIDERATIONS

2. The issue of fisheries subsidies continues to be the subject of a vigorous debate. My sense is that all participants recognize the global crisis of overcapacity and overfishing, with its consequent negative economic and environmental effects, and are committed to ensuring that the disciplines ultimately developed, whatever their form, must be *effective* in fulfilling the negotiating mandate from Ministers. That said, since I tabled my first draft text in document TN/RL/W/213, participants' views have continued to differ widely.

3. In my view, the existing differences are mainly due to varied perceptions among participants as to the exact scope and meaning of the mandate. Ministers in Hong Kong identified as the central focus of these negotiations the strengthening of disciplines, in particular through a prohibition, on subsidies that contribute to overcapacity or overfishing. Thus it is clear that crafting such new disciplines is at the core of the negotiations. During the negotiations, much of the debate in practice has been about other considerations that should modulate the impact of any new disciplines. In this regard, while the Group had many useful discussions on the basis of my first draft text, of both possible disciplines and other considerations, that text did not move us closer to a common understanding as to subsidies that should be prohibited and exempted from disciplines. I thus fully recognize that I will need to revise that text in view of the continued differences of view. The discussions to date have not, however, generated the necessary elements that would provide me with the basis for a revision that could lead to greater convergence.

4. As we resume our work, therefore, I believe that in the first instance in our discussions we should take a step back and reflect on the fundamental issues raised by the mandate. In particular, in my view the Group should work to identify those subsidies that contribute to overcapacity or overfishing, with a view to determining which of these should and should not be prohibited, while considering at the same time how to effectively address the needs and particularities of developing Members.

5. In this context, wherever certain participants may consider that certain subsidies should not be prohibited, the Group would need to consider the reasons advanced: for example, because such subsidies contribute only minimally to overcapacity or overfishing; because the effects of such subsidies could be adequately controlled by fisheries management or other means; because the small scale of certain subsidized operations would limit or eliminate the potential contribution to overcapacity or overfishing of the subsidies; or because of their importance to development priorities. In respect of any such proposed exemptions, the Group would need to consider how the integrity of the mandate would be ensured, such that any such subsidies would not in practice contribute to overcapacity or overfishing.

6. A basic issue that would need to be taken up is how the existence of overcapacity or overfishing can be established as objectively and precisely as possible.

Given that many questions of judgement would be involved in such determinations, one important question is the appropriateness of leaving it to each Member to judge its own situation, or if not the Member, who else could or should make such judgements.

7. A further fundamental issue is how to ensure adequate implementation, monitoring and surveillance. As is the case for all WTO rules, fisheries subsidies disciplines would need to include provisions for such mechanisms. Given the potential, as recognized in the mandate, for fisheries subsidies to contribute to overcapacity and overfishing, coupled with the mobile and undomesticated nature of the resources, a principal question would be how adequate surveillance of the effectiveness of the operation of the rules could be ensured, and whether the rules therefore should include enhanced surveillance mechanisms.

8. I believe that the best approach for the Negotiating Group to take in seeking a common understanding in respect of these basic conceptual issues is to discuss these issues on the basis of a number of more detailed questions. In this regard, to facilitate this process, in the next sections of this paper I raise a series of such questions, all of which are aimed at clarifying participants' positions on different aspects of the mandate.

9. I call upon all participants to re-engage in, and deepen, the debate of the basic concepts that must be addressed, and especially to come to grips with how the various elements of a new discipline will work together in a coherent way to effectively fulfil the mandate that we have received from the Ministers.

PROHIBITION

10. On the basis of this mandate, a core task for the Negotiating Group is to seek a common understanding as to the subsidies that "contribute to overcapacity and overfishing". As participants are aware, my first draft text took a "bottom-up" approach, listing particular specific subsidies that would be prohibited.[1] This list was drawn up based on submissions by and discussions among participants identifying those subsidies that the submitters considered contribute to overcapacity and/or overfishing. It is clear from the discussions that participants' views still differ widely regarding which subsidies should be prohibited.

11. I therefore would seek first a detailed discussion by the Group on which particular subsidies they believe should be prohibited (including but not limited to those referred to in my first draft text) in the light of the Ministerial mandate, addressing *inter alia* the following points:

(a) For each subsidy that in your view should be prohibited, how does it contribute to overcapacity or overfishing?

(b) If you believe that a given subsidy should not be prohibited, why? If this is because you do not consider that it contributes to overcapacity or overfishing, what are the specific reasons for this view?

(c) Where you consider that a given subsidy that has been proposed for prohibition should not be included in a list of prohibited subsidies, because in your view it does not contribute to overcapacity or overfishing, how would you address the question of interchangeability of subsidies? Specifically, how could non-prohibited subsidies that reduce costs to the fisheries sector be prevented from contributing to overcapacity or overfishing?

(d) What should the scope of any prohibitions be, i.e., how far upstream or downstream should they reach, and should the focus be marine wild capture fisheries?

(e) Should any or all proposed prohibitions be conditional on anything? If so, on what?

(f) What role should there be for a provision such as the prohibition of non-listed subsidies in respect of "unequivocally overfished" fisheries?

(g) If such a provision is viewed as unnecessary,

(i) How otherwise could it be ensured that a bottom-up list of prohibited subsidies would be effective in fulfilling the mandate?

(ii) What would be the justification for permitting subsidies in respect of overexploited fisheries?

GENERAL EXCEPTIONS

12. In my first draft text, the list of subsidies that would be prohibited is modulated by general exceptions, mainly for subsidies identified by participants as helping to reduce overcapacity and overfishing, and to improve environmental conditions in marine wild capture fisheries. In the light of the mandate from Ministers, i.e., to prevent them from undermining the effectiveness of the prohibition, these proposed exceptions as envisaged would be conditioned on the establishment and operation of fisheries management systems and measures.

13. Participants are asked to consider the following:

(a) What measures are appropriate for inclusion in a list of general exceptions?

(i) For each such measure, is the rationale that it would be unlikely to contribute to, or would counteract, overcapacity or overfishing? If so, why?

(ii) For any measure proposed for a general exception in spite of possibly contributing to overcapacity or overfishing, what would be the rationale for excepting it from prohibition?

(b) Where you consider that a given measure that has been proposed for a general exception should instead be prohibited, what are the reasons? Do you consider that the measure would contribute to overcapacity or overfishing, and if so, how specifically?

(c) Should general exceptions be conditional in principle on having a fisheries management system in place (leaving for the section below the details concerning fisheries management):

(i) Should each general exception be conditional on fisheries management?

- Are there additional or alternative conditions that should apply?
- Are there general exceptions that in your view should be unconditional? Which ones?
- How could it be ensured that any unconditional exceptions would not contribute to overcapacity or overfishing?

(d) If most WTO Members are committed to and already have implemented or are in the process of implementing fisheries management systems, what would be the difficulties, in practical terms, of agreeing to having fisheries management in place as a condition for making use of general exceptions?

SPECIAL AND DIFFERENTIAL TREATMENT

14. In the light of the mandate from Ministers, my first draft text contains special and differential treatment provisions whereby developing Members could provide certain capacity- and effort-enhancing subsidies that otherwise would be subject to prohibition. During the negotiations, developing Members have consistently acknowledged that subsidies by any country can contribute to the global problems of overcapacity and overfishing, and thus have indicated that they are not seeking a "blank check" allowing unlimited and unconditional subsidization. The difficulty in the negotiations on S&D provisions thus far is how to find a balance between flexibility for developing Members to provide subsidies to develop their fisheries sector on the one hand, and ensuring that such subsidies do not contribute to overcapacity and overfishing, on the other hand.

15. Participants are asked to reflect in detail on the following questions:

(a) Do participants support an essentially full carve-out from the disciplines for LDCs?

(b) For developing Members other than LDCs, is it appropriate and consistent with the mandate that S&D exceptions be broadest and subject to the fewest conditions for subsidies to the smallest-scale, closest to shore, and least commercial fishing operations, with exceptions becoming progressively narrower and subject to more conditions as the subsidized operations become larger-scale, further from shore, and more commercial?

(i) If so, how could such different types, scales and/or geographic areas of operations be defined and differentiated?

(ii) How would the resulting categories relate to the mandate to discipline subsidies that contribute to overcapacity or overfishing?

(iii) What types of conditionalities would apply, to which categories?

(c) Are there other bases on which fisheries operations of developing Members could be categorized for the purpose of S&D exceptions, which would provide the necessary flexibility to developing Members without contributing to overcapacity or overfishing?

(d) If no other dividing lines among types, scales and/or geographic areas of operations can be identified, would all non-LDC developing Members receive the same S&D treatment in respect of their fisheries subsidies to all types, scales and geographic areas of fisheries operations?

(i) In such a situation, what should the exceptions be, and to what conditions should they be subject?
(ii) If all non-LDC developing Members were fully exempted from eventual prohibitions on, for example, subsidies for vessel construction/modification, and operating costs, how could it be ensured that such subsidies would not contribute to overcapacity or overfishing?
(iii) On what other basis could S&D treatment be structured?
(e) Should some or all exemptions for developing Members be conditional on fisheries management?
(i) If so, which exemptions should be subject to such conditionality?
(ii) What sort of fisheries management conditionalities should these be and how could their effectiveness at preventing overcapacity and overfishing be ensured? Would self-certification that management was effective be sufficient?
(f) If S&D exceptions were not conditioned on fisheries management, what other conditions should there be, if any, and how would those operate to prevent the subsidies from contributing to overcapacity or overfishing?
(g) What is the appropriate role for technical assistance for developing Members to implement new disciplines?
(i) How can effective technical assistance for the implementation of management conditionalities be ensured while not indirectly making more resources available to subsidize?
(ii) How can developing Members' needs and donor Member's capabilities be reconciled in a way that contributes most efficiently to fulfilling the mandate?

GENERAL DISCIPLINE/ACTIONABILITY

16. Under the existing provisions of the SCM Agreement, specific subsidies benefiting the fisheries sector are actionable, and thus can be the subject of multilateral adverse effects challenges (e.g. serious prejudice) and countervailing measures if the applicable conditions are present. These provisions are not, however, designed to clearly identify any contribution that such subsidies may make to overcapacity or overfishing. With a view to creating predictability and certainty concerning how to identify and address such effects, my first draft text contains certain guidelines on the basis of which a Member could challenge another Member's subsidies if those subsidies caused depletion of or harm to, or creation of overcapacity in respect of, stocks whose range extended into the challenging Member's EEZ, or particular stocks in which that Member had identifiable fishing interests.

(a) Should new fisheries subsidies rules define any fisheries-specific negative effects from subsidies that could be challenged by other Members?
(b) If so, how should such effects be defined?
(i) To what extent, if at all, should implementation by the subsidizing Member of sound fisheries management be considered to be relevant evidence for a determination of whether its subsidies have caused such effects? Why?

(ii) What additional or alternative evidence would be relevant? Why?

(c) If no parameters are established in respect of relevant evidence, what would be alternative approaches to ensure predictability and certainty in the application of general subsidy disciplines?

FISHERIES MANAGEMENT CONDITIONALITIES

17. The discussions in the Negotiating Group indicate that participants generally believe that exceptions – both general and S&D – should not be unconditional, given their potential to undercut the effectiveness of the disciplines on subsidies that contribute to overcapacity and overfishing.

18. The discussions also have indicated a widely-shared view that the principal conditionalities should pertain to fisheries management, a central component of which would be stock assessments, in part because of the difficulties of directly measuring the effect of particular subsidies on particular wild capture fisheries due to the mobile and undomesticated nature of the resource.

19. There are, however, differing views as to how much detail WTO fisheries subsidies rules should contain on fisheries management, in respect of both the substantive basis of such conditionalities and the appropriate fora and mechanisms for monitoring and enforcing their implementation. There has been considerable debate on the approach taken to these issues in my first draft text.

20. I would ask participants to reflect on the following questions:

(a) Are there other conditionalities that should be applicable to exceptions (general and S&D), either in addition to or instead of fisheries management conditionalities?

(b) How important is it for the effective operation of the disciplines that all Members' fisheries management systems and measures adhere to a common standard, and how prescriptive should that standard be?

(c) If a common standard is not necessary or not acceptable, how could the effectiveness of different Members' systems in controlling overcapacity and overfishing be monitored and enforced?

– What would prevent one Member with ineffective management from overfishing stocks that were safeguarded/replenished by another Member's effective management measures?

(d) Given that an international consensus already exists in respect of a substantial number of international fisheries management instruments, and those instruments themselves take account of the capacity constraints of developing countries, does it make sense to draw inspiration from those instruments for the substantive content of management conditionalities for using exceptions from the prohibition?

(i) If not, why not?

(ii) If so, how could the typically non-binding nature and relatively general and flexible wording of those instruments be reconciled with a binding prohibition of subsidies that contribute to overcapacity or overfishing, and binding conditionalities concerning fisheries management where exceptions are used?

(e) What role should stock assessments play in any management conditionalities?

(i) If stock assessments are considered unnecessary, why, and how could overfishing and overcapacity be monitored in the absence of stock assessments?

(ii) If stock assessments are considered to be a necessary element, how could the rules take into account Members' different capabilities while ensuring that the assessments are as reliable and robust as possible?

(iii) To what extent if at all should the results of stock assessments form part of any conditionalities?

(f) Is it logical to require a stock assessment before a capacity-enhancing subsidy is provided?

(i) If not, why not?

(ii) If so, what practical problems would need to be resolved?

(iii) What timing and review mechanism for such stock assessments could best reconcile a Member's need to implement a given subsidy on the one hand, and other Members' need for multilateral surveillance/transparency in respect of the fisheries resources that would be affected by it on the other hand?

(g) Given the existing role of the FAO in discussing FAO members' substantive implementation of various international fisheries instruments, what specific problems/concerns would there be if that role were enhanced in respect of stock assessments and/or fisheries management systems?

– Would the problems be the same if the review at the FAO were similar to reviews of notifications by WTO bodies (i.e., multilateral review among members, for transparency, rather than a mechanism for approval, whether by a panel of experts or a multilateral body)?

(h) Why, if at all, would the WTO be better-positioned to perform such reviews of fisheries-related information?

– How could the necessary expertise be built into any WTO-based review of fisheries management, without the WTO becoming a fisheries management organization?

TRANSPARENCY

21. Most participants appear to believe that enhanced transparency in respect of fisheries subsidies should be an important outcome of the negotiations, as a necessary element in ensuring that the disciplines operate effectively. In this regard, my first draft text includes specific transparency provisions that would apply where exceptions are used; in particular, information related to subsidies for which exceptions were invoked would be notified to, and reviewed by, the WTO, and there would be consequences for non-notification. Concerning transparency in respect of fisheries management, given the technical questions that would arise in review of Members' fisheries management systems, as outlined in

the previous section, my first draft text envisages a multilateral mechanism that would make use of outside fisheries expertise and institutions.

22. I would ask participants to reflect on the following issues:

(a) Assuming that notifications would be required in respect of subsidies for which exceptions from the prohibition were invoked, is it logical for such notifications to include information addressing the conformity of the subsidies with the specific criteria and characteristics of the invoked exceptions?

(b) Would a requirement to notify such subsidies after rather than before they were implemented be effective in ensuring that the pertinent criteria were respected? Are there effective alternatives?

(c) What mechanisms could be built into the rules to create incentives to notify?

(d) If a presumption that non-notified subsidies are prohibited is either not workable or not acceptable, what other approaches would be effective in encouraging the submission of complete and timely notifications, so as to provide meaningful information to Members?

DISPUTE SETTLEMENT

23. On dispute settlement, depending on the specific disciplines agreed, disputes clearly could arise which would focus on technical questions related to the fisheries sector rather than on purely legal issues. My first draft text suggests a mechanism that could be used to bring fisheries expertise into the dispute settlement process as needed.

(a) Is it necessary for WTO fisheries subsidies rules to provide for a specific mechanism by which panels can obtain scientific and other technical expertise related to fisheries issues that may arise in disputes?

(i) If so, why would the provisions in Article 13 of the DSU not be sufficient?

(ii) Should resort to any such mechanism be mandatory for all panels dealing with disputes under the fisheries subsidies rules, or should such resort be at the discretion of each panel?

(b) If such a specific mechanism is considered unnecessary, how could necessary expertise be assured in disputes involving technical fisheries issues?

(c) How could it be ensured that panels would make use of available mechanisms, in particular the provisions in Article 13 of the DSU, when necessary?

IMPLEMENTATION

24. Given that any mechanisms ultimately agreed as part of new disciplines on fisheries subsidies will imply a certain degree of institutional infrastructure, expertise and resources, it is clear that implementation of the disciplines would pose certain difficulties, especially for developing Members. My first draft text

envisages certain flexibilities in this respect, such as the possibility for regional bodies to fulfil the roles otherwise envisaged for national mechanisms, at least for developing Members, as well as provisions on technical assistance.

25. Participants are asked to consider the following issues in respect of the implementation of new disciplines:

(a) What concrete mechanisms could be included in fisheries subsidies disciplines to facilitate their implementation to the maximum possible extent?

(b) Would such mechanisms be limited to implementation of fisheries management, or would they address other aspects of the disciplines as well?

(c) To the extent that disciplines foresee implementation of management or other obligations on a regional basis, how could it be ensured that the obligations were fulfilled in respect of and by each Member individually?

TRANSITION RULES

26. It is clear that whatever the new disciplines, Members would need a certain amount of time to bring their existing measures – subsidies and potentially fisheries management systems – into conformity. My first draft text parallels the approach in the SCM Agreement, envisaging a longer transition period for developing than developed Members.

27. I would ask Members to reflect on the following points:

(a) What would be the specific purpose of a transition period in the context of new disciplines on fisheries subsidies?

(b) Should a transition period for bringing otherwise prohibited subsidies into conformity with new rules be available only in respect of such subsidies that have been notified?

(c) To what extent should the length of a transition period be linked to the strength of the disciplines (i.e., the degree of the new obligations imposed on Members)?

(d) How should a transition rule for developing Members take into account the nature and extent of special and differential treatment?

(e) Should a transition period be linked to implementation of fisheries management systems on which exceptions would be conditioned, or should it only relate to bringing otherwise prohibited subsidies into conformity with the new disciplines?

(f) Should transition periods vary based on the particular type of subsidy involved? What would be the implications of such differentiation on overcapacity and overfishing?

(g) Should a transition rule require Members to report during the transition period on their progress in implementing new obligations?

Endnotes

1 On a fundamental point related to this list of measures, I wish to be clear that my first draft text would apply only to subsidies that are specific on any of the bases provided for in Article 2 of the Agreement (i.e., the list does not include subsidies that would not meet the definition of specificity in Article 2, but is not limited to subsidies that are specific on a sectoral basis to fisheries).

Annex 10
WTO Agreement on Subsidies and Countervailing Measures (ASCM) (excerpts)

PART I: GENERAL PROVISIONS

Article 1
Definition of a Subsidy

1.1 For the purpose of this Agreement, a subsidy shall be deemed to exist if:

(a)(1) there is a financial contribution by a government or any public body within the territory of a Member (referred to in this Agreement as "government"), i.e. where:

(i) a government practice involves a direct transfer of funds (e.g. grants, loans, and equity infusion), potential direct transfers of funds or liabilities (e.g. loan guarantees);

(ii) government revenue that is otherwise due is foregone or not collected (e.g. fiscal incentives such as tax credits);[1]

(iii) a government provides goods or services other than general infrastructure, or purchases goods;

(iv) a government makes payments to a funding mechanism, or entrusts or directs a private body to carry out one or more of the type of functions illustrated in (i) to (iii) above which would normally be vested in the government and the practice, in no real sense, differs from practices normally followed by governments;

or

(a)(2) there is any form of income or price support in the sense of Article XVI of GATT 1994;

and

(b) a benefit is thereby conferred.

1.2 A subsidy as defined in paragraph 1 shall be subject to the provisions of Part II or shall be subject to the provisions of Part III or V only if such a subsidy is specific in accordance with the provisions of Article 2.

Article 2
Specificity

2.1 In order to determine whether a subsidy, as defined in paragraph 1 of Article 1, is specific to an enterprise or industry or group of enterprises or industries (referred to in this Agreement as "certain enterprises") within the jurisdiction of the granting authority, the following principles shall apply:

(a) Where the granting authority, or the legislation pursuant to which the granting authority operates, explicitly limits access to a subsidy to certain enterprises, such subsidy shall be specific.

(b) Where the granting authority, or the legislation pursuant to which the granting authority operates, establishes objective criteria or conditions[2] governing the eligibility for, and the amount of, a subsidy, specificity shall not exist, provided that the eligibility is automatic and that such criteria and conditions are strictly adhered to. The criteria or conditions must be clearly spelled out in law, regulation, or other official document, so as to be capable of verification.

(c) If, notwithstanding any appearance of non-specificity resulting from the application of the principles laid down in subparagraphs (a) and (b), there are reasons to believe that the subsidy may in fact be specific, other factors may be considered. Such factors are: use of a subsidy programme by a limited number of certain enterprises, predominant use by certain enterprises, the granting of disproportionately large amounts of subsidy to certain enterprises, and the manner in which discretion has been exercised by the granting authority in the decision to grant a subsidy.[3] In applying this subparagraph, account shall be taken of the extent of diversification of economic activities within the jurisdiction of the granting authority, as well as of the length of time during which the subsidy programme has been in operation.

2.2 A subsidy which is limited to certain enterprises located within a designated geographical region within the jurisdiction of the granting authority shall be specific. It is understood that the setting or change of generally applicable tax rates by all levels of government entitled to do so shall not be deemed to be a specific subsidy for the purposes of this Agreement.

2.3 Any subsidy falling under the provisions of Article 3 shall be deemed to be specific.

2.4 Any determination of specificity under the provisions of this Article shall be clearly substantiated on the basis of positive evidence.

PART II: PROHIBITED SUBSIDIES

Article 3
Prohibition

3.1 Except as provided in the Agreement on Agriculture, the following subsidies, within the meaning of Article 1, shall be prohibited:

(a) subsidies contingent, in law or in fact,[4] whether solely or as one of several other conditions, upon export performance, including those illustrated in Annex I;[5]

(b) subsidies contingent, whether solely or as one of several other conditions, upon the use of domestic over imported goods.

3.2 A Member shall neither grant nor maintain subsidies referred to in paragraph 1.

Article 4
Remedies

4.1 Whenever a Member has reason to believe that a prohibited subsidy is being granted or maintained by another Member, such Member may request consultations with such other Member.

4.2 A request for consultations under paragraph 1 shall include a statement of available evidence with regard to the existence and nature of the subsidy in question.

4.3 Upon request for consultations under paragraph 1, the Member believed to be granting or maintaining the subsidy in question shall enter into such consultations as quickly as possible. The purpose of the consultations shall be to clarify the facts of the situation and to arrive at a mutually agreed solution.

4.4 If no mutually agreed solution has been reached within 30 days[6] of the request for consultations, any Member party to such consultations may refer the matter to the Dispute Settlement Body ("DSB") for the immediate establishment of a panel, unless the DSB decides by consensus not to establish a panel.

4.5 Upon its establishment, the panel may request the assistance of the Permanent Group of Experts[7] (referred to in this Agreement as the "PGE") with regard to whether the measure in question is a prohibited subsidy. If so requested, the PGE shall immediately review the evidence with regard to the existence and nature of the measure in question and shall provide an opportunity for the Member applying or maintaining the measure to demonstrate that the measure in question is not a prohibited subsidy. The PGE shall report its conclusions to the panel within a time limit determined by the panel. The PGE's conclusions on the issue of whether or not the measure in question is a prohibited subsidy shall be accepted by the panel without modification.

4.6 The panel shall submit its final report to the parties to the dispute. The report shall be circulated to all Members within 90 days of the date of the composition and the establishment of the panel's terms of reference.

4.7 If the measure in question is found to be a prohibited subsidy, the panel shall recommend that the subsidizing Member withdraw the subsidy without delay. In this regard, the panel shall specify in its recommendation the time-period within which the measure must be withdrawn.

4.8 Within 30 days of the issuance of the panel's report to all Members, the report shall be adopted by the DSB unless one of the parties to the dispute formally notifies the DSB of its decision to appeal or the DSB decides by consensus not to adopt the report.

4.9 Where a panel report is appealed, the Appellate Body shall issue its decision within 30 days from the date when the party to the dispute formally notifies its intention to appeal. When the Appellate Body considers that it cannot provide its report within 30 days, it shall inform the DSB in writing of the reasons for the delay together with an estimate of the period within which it will submit its report. In no case shall the proceedings exceed 60 days. The appellate report shall be adopted by the DSB and unconditionally accepted by the parties to the dispute unless the DSB decides by consensus not to adopt the appellate report within 20 days following its issuance to the Members.[8]

4.10 In the event the recommendation of the DSB is not followed within the time period specified by the panel, which shall commence from the date of adoption of the panel's report or the Appellate Body's report, the DSB shall grant authorization to the complaining Member to take appropriate[9] countermeasures, unless the DSB decides by consensus to reject the request.

4.11 In the event a party to the dispute requests arbitration under paragraph 6 of Article 22 of the Dispute Settlement Understanding ("DSU"), the arbitrator shall determine whether the countermeasures are appropriate.[10]

4.12 For purposes of disputes conducted pursuant to this Article, except for time periods specifically prescribed in this Article, time periods applicable under the DSU for the conduct of such disputes shall be half the time prescribed therein.

PART III: ACTIONABLE SUBSIDIES

Article 5
Adverse Effects

No Member should cause, through the use of any subsidy referred to in paragraphs 1 and 2 of Article 1, adverse effects to the interests of other Members, i.e.:

(a) injury to the domestic industry of another Member;[11]

(b) nullification or impairment of benefits accruing directly or indirectly to other Members under GATT 1994 in particular the benefits of concessions bound under Article II of GATT 1994;[12]

(c) serious prejudice to the interests of another Member.[13]

This Article does not apply to subsidies maintained on agricultural products as provided in Article 13 of the Agreement on Agriculture.

Article 6
Serious Prejudice

6.1 Serious prejudice in the sense of paragraph (c) of Article 5 shall be deemed to exist in the case of:

(a) the total ad valorem subsidization[14] of a product exceeding 5 per cent;[15]

(b) subsidies to cover operating losses sustained by an industry;

(c) subsidies to cover operating losses sustained by an enterprise, other than one time measures which are non recurrent and cannot be repeated for that enterprise and which are given merely to provide time for the development of long term solutions and to avoid acute social problems;

(d) direct forgiveness of debt, i.e. forgiveness of government held debt, and grants to cover debt repayment.[16]

6.2 Notwithstanding the provisions of paragraph 1, serious prejudice shall not be found if the subsidizing Member demonstrates that the subsidy in question has not resulted in any of the effects enumerated in paragraph 3.

6.3 Serious prejudice in the sense of paragraph (c) of Article 5 may arise in any case where one or several of the following apply:

(a) the effect of the subsidy is to displace or impede the imports of a like product of another Member into the market of the subsidizing Member;

(b) the effect of the subsidy is to displace or impede the exports of a like product of another Member from a third country market;

(c) the effect of the subsidy is a significant price undercutting by the subsidized product as compared with the price of a like product of another Member in the same market or significant price suppression, price depression or lost sales in the same market;

(d) the effect of the subsidy is an increase in the world market share of the subsidizing Member in a particular subsidized primary product or commodity[17] as compared to the average share it had during the previous period of three years and this increase follows a consistent trend over a period when subsidies have been granted.

6.4 For the purpose of paragraph 3(b), the displacement or impeding of exports shall include any case in which, subject to the provisions of paragraph 7, it has been demonstrated that there has been a change in relative shares of the market to the disadvantage of the non subsidized like product (over an appropriately representative period sufficient to demonstrate clear trends in the development of the market for the product concerned, which, in normal circumstances, shall be at least one year). "Change in relative shares of the market" shall include any of the following situations: *(a)* there is an increase in the market share of the subsidized product; *(b)* the market share of the subsidized product remains constant in circumstances in which, in the absence of the subsidy, it would have declined; *(c)* the market share of the subsidized product declines, but at a slower rate than would have been the case in the absence of the subsidy.

6.5 For the purpose of paragraph 3(c), price undercutting shall include any case in which such price undercutting has been demonstrated through a comparison of prices of the subsidized product with prices of a non subsidized like product supplied to the same market. The comparison shall be made at the same level of trade and at comparable times, due account being taken of any other factor affecting price comparability. However, if such a direct comparison is not possible, the existence of price undercutting may be demonstrated on the basis of export unit values.

6.6 Each Member in the market of which serious prejudice is alleged to have arisen shall, subject to the provisions of paragraph 3 of Annex V, make available to the parties to a dispute arising under Article 7, and to the panel established pursuant to paragraph 4 of Article 7, all relevant information that can be obtained as to the changes in market shares of the parties to the dispute as well as concerning prices of the products involved.

6.7 Displacement or impediment resulting in serious prejudice shall not arise under paragraph 3 where any of the following circumstances exist[18] during the relevant period:

(a) prohibition or restriction on exports of the like product from the complaining Member or on imports from the complaining Member into the third country market concerned;

(b) decision by an importing government operating a monopoly of trade or state trading in the product concerned to shift, for non commercial reasons, imports from the complaining Member to another country or countries;

(c) natural disasters, strikes, transport disruptions or other *force majeure* substantially affecting production, qualities, quantities or prices of the product available for export from the complaining Member;

(d) existence of arrangements limiting exports from the complaining Member;

(e) voluntary decrease in the availability for export of the product concerned from the complaining Member (including, *inter alia*, a situation where firms in the complaining Member have been autonomously reallocating exports of this product to new markets);

(f) failure to conform to standards and other regulatory requirements in the importing country.

6.8 In the absence of circumstances referred to in paragraph 7, the existence of serious prejudice should be determined on the basis of the information submitted to or obtained by the panel, including information submitted in accordance with the provisions of Annex V.

6.9 This Article does not apply to subsidies maintained on agricultural products as provided in Article 13 of the Agreement on Agriculture.

Article 7
Remedies

7.1 Except as provided in Article 13 of the Agreement on Agriculture, whenever a Member has reason to believe that any subsidy referred to in Article 1, granted or maintained by another Member, results in injury to its domestic industry, nullification or impairment or serious prejudice, such Member may request consultations with such other Member.

7.2 A request for consultations under paragraph 1 shall include a statement of available evidence with regard to *(a)* the existence and nature of the subsidy in question, and *(b)* the injury caused to the domestic industry, or the nullification or impairment, or serious prejudice[19] caused to the interests of the Member requesting consultations.

7.3 Upon request for consultations under paragraph 1, the Member believed to be granting or maintaining the subsidy practice in question shall enter into such consultations as quickly as possible. The purpose of the consultations shall be to clarify the facts of the situation and to arrive at a mutually agreed solution.

7.4 If consultations do not result in a mutually agreed solution within 60 days,[20] any Member party to such consultations may refer the matter to the DSB for the establishment of a panel, unless the DSB decides by consensus not to establish a panel. The composition of the panel and its terms of reference shall be established within 15 days from the date when it is established.

7.5 The panel shall review the matter and shall submit its final report to the parties to the dispute. The report shall be circulated to all Members within 120 days of the date of the composition and establishment of the panel's terms of reference.

7.6 Within 30 days of the issuance of the panel's report to all Members, the report shall be adopted by the DSB[21] unless one of the parties to the dispute formally notifies the DSB of its decision to appeal or the DSB decides by consensus not to adopt the report.

7.7 Where a panel report is appealed, the Appellate Body shall issue its decision within 60 days from the date when the party to the dispute formally notifies its intention to appeal. When the Appellate Body considers that it cannot provide its report within 60 days, it shall inform the DSB in writing of the reasons for the delay together with an estimate of the period within which it will submit its report. In no case shall the proceedings exceed 90 days. The appellate report shall be adopted by the DSB and unconditionally accepted by the parties to the dispute unless the DSB decides by consensus not to adopt the appellate report within 20 days following its issuance to the Members.[22]

7.8 Where a panel report or an Appellate Body report is adopted in which it is determined that any subsidy has resulted in adverse effects to the interests of another Member within the meaning of Article 5, the Member granting or maintaining such subsidy shall take appropriate steps to remove the adverse effects or shall withdraw the subsidy.

7.9 In the event the Member has not taken appropriate steps to remove the adverse effects of the subsidy or withdraw the subsidy within six months from the date when the DSB adopts the panel report or the Appellate Body report, and in the absence of agreement on compensation, the DSB shall grant authorization to the complaining Member to take countermeasures, commensurate with the degree and nature of the adverse effects determined to exist, unless the DSB decides by consensus to reject the request.

7.10 In the event that a party to the dispute requests arbitration under paragraph 6 of Article 22 of the DSU, the arbitrator shall determine whether the countermeasures are commensurate with the degree and nature of the adverse effects determined to exist.

PART IV: NONACTIONABLE SUBSIDIES

Article 8
Identification of NonActionable Subsidies

8.1 The following subsidies shall be considered as nonactionable:[23]

(a) subsidies which are not specific within the meaning of Article 2;

(b) subsidies which are specific within the meaning of Article 2 but which meet all of the conditions provided for in paragraphs 2(a), 2(b) or 2(c) below.

8.2 Notwithstanding the provisions of Parts III and V, the following subsidies shall be nonactionable:

(a) assistance for research activities conducted by firms or by higher education or research establishments on a contract basis with firms if:[24,25,26]
the assistance covers[27] not more than 75 per cent of the costs of industrial research[28] or 50 per cent of the costs of precompetitive development activity;[29,30] and provided that such assistance is limited exclusively to:
(i) costs of personnel (researchers, technicians and other supporting staff employed exclusively in the research activity);
(ii) costs of instruments, equipment, land and buildings used exclusively and permanently (except when disposed of on a commercial basis) for the research activity;
(iii) costs of consultancy and equivalent services used exclusively for the research activity, including bought in research, technical knowledge, patents, etc.;
(iv) additional overhead costs incurred directly as a result of the research activity;
(v) other running costs (such as those of materials, supplies and the like), incurred directly as a result of the research activity.

(b) assistance to disadvantaged regions within the territory of a Member given pursuant to a general framework of regional development[31] and non specific (within the meaning of Article 2) within eligible regions provided that:
(i) each disadvantaged region must be a clearly designated contiguous geographical area with a definable economic and administrative identity;
(ii) the region is considered as disadvantaged on the basis of neutral and objective criteria,[32] indicating that the region's difficulties arise out of more than temporary circumstances; such criteria must be clearly spelled out in law, regulation, or other official document, so as to be capable of verification;
(iii) the criteria shall include a measurement of economic development which shall be based on at least one of the following factors:
– one of either income per capita or household income per capita, or GDP per capita, which must not be above 85 per cent of the average for the territory concerned;

– unemployment rate, which must be at least 110 per cent of the average for the territory concerned;

as measured over a three year period; such measurement, however, may be a composite one and may include other factors.

(c) assistance to promote adaptation of existing facilities[33] to new environmental requirements imposed by law and/or regulations which result in greater constraints and financial burden on firms, provided that the assistance:

(i) is a onetime nonrecurring measure; and

(ii) is limited to 20 per cent of the cost of adaptation; and

(iii) does not cover the cost of replacing and operating the assisted investment, which must be fully borne by firms; and

(iv) is directly linked to and proportionate to a firm's planned reduction of nuisances and pollution, and does not cover any manufacturing cost savings which may be achieved; and

(v) is available to all firms which can adopt the new equipment and/or production processes.

8.3 A subsidy programme for which the provisions of paragraph 2 are invoked shall be notified in advance of its implementation to the Committee in accordance with the provisions of Part VII. Any such notification shall be sufficiently precise to enable other Members to evaluate the consistency of the programme with the conditions and criteria provided for in the relevant provisions of paragraph 2. Members shall also provide the Committee with yearly updates of such notifications, in particular by supplying information on global expenditure for each programme, and on any modification of the programme. Other Members shall have the right to request information about individual cases of subsidization under a notified programme.[34]

8.4 Upon request of a Member, the Secretariat shall review a notification made pursuant to paragraph 3 and, where necessary, may require additional information from the subsidizing Member concerning the notified programme under review. The Secretariat shall report its findings to the Committee. The Committee shall, upon request, promptly review the findings of the Secretariat (or, if a review by the Secretariat has not been requested, the notification itself), with a view to determining whether the conditions and criteria laid down in paragraph 2 have not been met. The procedure provided for in this paragraph shall be completed at the latest at the first regular meeting of the Committee following the notification of a subsidy programme, provided that at least two months have elapsed between such notification and the regular meeting of the Committee. The review procedure described in this paragraph shall also apply, upon request, to substantial modifications of a programme notified in the yearly updates referred to in paragraph 3.

8.5 Upon the request of a Member, the determination by the Committee referred to in paragraph 4, or a failure by the Committee to make such a determination, as well as the violation, in individual cases, of the conditions set out in a notified programme, shall be submitted to binding arbitration. The arbitration body shall present its conclusions to the Members within 120 days from the date when the matter was referred to the arbitration body. Except as otherwise provided in this paragraph, the DSU shall apply to arbitrations conducted under this paragraph.

Article 9
Consultations and Authorized Remedies

9.1 If, in the course of implementation of a programme referred to in paragraph 2 of Article 8, notwithstanding the fact that the programme is consistent with the criteria laid down in that paragraph, a Member has reasons to believe that this programme has resulted in serious adverse effects to the domestic industry of that Member, such as to cause damage which would be difficult to repair, such Member may request consultations with the Member granting or maintaining the subsidy.

9.2 Upon request for consultations under paragraph 1, the Member granting or maintaining the subsidy programme in question shall enter into such consultations as quickly as possible. The purpose of the consultations shall be to clarify the facts of the situation and to arrive at a mutually acceptable solution.

9.3 If no mutually acceptable solution has been reached in consultations under paragraph 2 within 60 days of the request for such consultations, the requesting Member may refer the matter to the Committee.

9.4 Where a matter is referred to the Committee, the Committee shall immediately review the facts involved and the evidence of the effects referred to in paragraph 1. If the Committee determines that such effects exist, it may recommend to the subsidizing Member to modify this programme in such a way as to remove these effects. The Committee shall present its conclusions within 120 days from the date when the matter is referred to it under paragraph 3. In the event the recommendation is not followed within six months, the Committee shall authorize the requesting Member to take appropriate countermeasures commensurate with the nature and degree of the effects determined to exist.

PART V: COUNTERVAILING MEASURES

Article 15
Determination of Injury[35]

15.1 A determination of injury for purposes of Article VI of GATT 1994 shall be based on positive evidence and involve an objective examination of both *(a)* the volume of the subsidized imports and the effect of the subsidized imports on prices in the domestic market for like products[36] and *(b)* the consequent impact of these imports on the domestic producers of such products.

15.2 With regard to the volume of the subsidized imports, the investigating authorities shall consider whether there has been a significant increase in subsidized imports, either in absolute terms or relative to production or consumption in the importing Member. With regard to the effect of the subsidized imports on prices, the investigating authorities shall consider whether there has been a significant price undercutting by the subsidized imports as compared with the price of a like product of the importing Member, or whether the effect of such imports is otherwise to depress prices to a significant degree or to prevent price increases, which otherwise would have occurred, to a significant degree. No one or several of these factors can necessarily give decisive guidance.

15.3 Where imports of a product from more than one country are simultaneously subject to countervailing duty investigations, the investigating authorities may cumulatively assess the effects of such imports only if they determine that *(a)* the amount of subsidization established in relation to the imports from each country is more than *de minimis* as defined in paragraph 9 of Article 11 and the volume of imports from each country is not negligible and *(b)* a cumulative assessment of the effects of the imports is appropriate in light of the conditions of competition between the imported products and the conditions of competition between the imported products and the like domestic product.

15.4 The examination of the impact of the subsidized imports on the domestic industry shall include an evaluation of all relevant economic factors and indices having a bearing on the state of the industry, including actual and potential decline in output, sales, market share, profits, productivity, return on investments, or utilization of capacity; factors affecting domestic prices; actual and potential negative effects on cash flow, inventories, employment, wages, growth, ability to raise capital or investments and, in the case of agriculture, whether there has been an increased burden on government support programmes. This list is not exhaustive, nor can one or several of these factors necessarily give decisive guidance.

15.5 It must be demonstrated that the subsidized imports are, through the effects[37] of subsidies, causing injury within the meaning of this Agreement. The demonstration of a causal relationship between the subsidized imports and the injury to the domestic industry shall be based on an examination of all relevant evidence before the authorities. The authorities shall also examine any known factors other than the subsidized imports which at the same time are injuring the domestic industry, and the injuries caused by these other factors must not be attributed to the subsidized imports. Factors which may be relevant in this respect include, *inter alia*, the volumes and prices of non subsidized imports of the product in question, contraction in demand or changes in the patterns of consumption, trade restrictive practices of and competition between the foreign and domestic producers, developments in technology and the export performance and productivity of the domestic industry.

15.6 The effect of the subsidized imports shall be assessed in relation to the domestic production of the like product when available data permit the separate identification of that production on the basis of such criteria as the production process, producers' sales and profits. If such separate identification of that production is not possible, the effects of the subsidized imports shall be assessed by the examination of the production of the narrowest group or range of products, which includes the like product, for which the necessary information can be provided.

15.7 A determination of a threat of material injury shall be based on facts and not merely on allegation, conjecture or remote possibility. The change in circumstances which would create a situation in which the subsidy would cause injury must be clearly foreseen and imminent. In making a determination regarding the existence of a threat of material injury, the investigating authorities should consider, *inter alia*, such factors as:

(i) nature of the subsidy or subsidies in question and the trade effects likely to arise therefrom;

(ii) a significant rate of increase of subsidized imports into the domestic market indicating the likelihood of substantially increased importation;

(iii) sufficient freely disposable, or an imminent, substantial increase in, capacity of the exporter indicating the likelihood of substantially increased subsidized exports to the importing Member's market, taking into account the availability of other export markets to absorb any additional exports;

(iv) whether imports are entering at prices that will have a significant depressing or suppressing effect on domestic prices, and would likely increase demand for further imports; and

(v) inventories of the product being investigated.

No one of these factors by itself can necessarily give decisive guidance but the totality of the factors considered must lead to the conclusion that further subsidized exports are imminent and that, unless protective action is taken, material injury would occur.

15.8 With respect to cases where injury is threatened by subsidized imports, the application of countervailing measures shall be considered and decided with special care.

PART VI: INSTITUTIONS

Article 24
Committee on Subsidies and Countervailing Measures and Subsidiary Bodies

24.1 There is hereby established a Committee on Subsidies and Countervailing Measures composed of representatives from each of the Members. The Committee shall elect its own Chairman and shall meet not less than twice a year and otherwise as envisaged by relevant provisions of this Agreement at the request of any Member. The Committee shall carry out responsibilities as assigned to it under this Agreement or by the Members and it shall afford Members the opportunity of consulting on any matter relating to the operation of the Agreement or the furtherance of its objectives. The WTO Secretariat shall act as the secretariat to the Committee.

24.2 The Committee may set up subsidiary bodies as appropriate.

24.3 The Committee shall establish a Permanent Group of Experts composed of five independent persons, highly qualified in the fields of subsidies and trade relations. The experts will be elected by the Committee and one of them will be replaced every year. The PGE may be requested to assist a panel, as provided for in paragraph 5 of Article 4. The Committee may also seek an advisory opinion on the existence and nature of any subsidy.

24.4 The PGE may be consulted by any Member and may give advisory opinions on the nature of any subsidy proposed to be introduced or currently maintained by that Member. Such advisory opinions will be confidential and may not be invoked in proceedings under Article 7.

24.5 In carrying out their functions, the Committee and any subsidiary bodies may consult with and seek information from any source they deem appropriate. However, before the Committee or a subsidiary body seeks such information from a source within the jurisdiction of a Member, it shall inform the Member involved.

PART VII: NOTIFICATION AND SURVEILLANCE

Article 25
Notifications

25.1 Members agree that, without prejudice to the provisions of paragraph 1 of Article XVI of GATT 1994, their notifications of subsidies shall be submitted not later than 30 June of each year and shall conform to the provisions of paragraphs 2 through 6.

25.2 Members shall notify any subsidy as defined in paragraph 1 of Article 1, which is specific within the meaning of Article 2, granted or maintained within their territories.

25.3 The content of notifications should be sufficiently specific to enable other Members to evaluate the trade effects and to understand the operation of notified subsidy programmes. In this connection, and without prejudice to the contents and form of the questionnaire on subsidies,[38] Members shall ensure that their notifications contain the following information:

(i) form of a subsidy (i.e. grant, loan, tax concession, etc.);
(ii) subsidy per unit or, in cases where this is not possible, the total amount or the annual amount budgeted for that subsidy (indicating, if possible, the average subsidy per unit in the previous year);
(iii) policy objective and/or purpose of a subsidy;
(iv) duration of a subsidy and/or any other time limits attached to it;
(v) statistical data permitting an assessment of the trade effects of a subsidy.

25.4 Where specific points in paragraph 3 have not been addressed in a notification, an explanation shall be provided in the notification itself.

25.5 If subsidies are granted to specific products or sectors, the notifications should be organized by product or sector.

25.6 Members which consider that there are no measures in their territories requiring notification under paragraph 1 of Article XVI of GATT 1994 and this Agreement shall so inform the Secretariat in writing.

25.7 Members recognize that notification of a measure does not prejudge either its legal status under GATT 1994 and this Agreement, the effects under this Agreement, or the nature of the measure itself.

25.8 Any Member may, at any time, make a written request for information on the nature and extent of any subsidy granted or maintained by another Member (including any subsidy referred to in Part IV), or for an explanation of the reasons for which a specific measure has been considered as not subject to the requirement of notification.

25.9 Members so requested shall provide such information as quickly as possible and in a comprehensive manner, and shall be ready, upon request, to provide additional information to the requesting Member. In particular, they shall provide sufficient details to enable the other Member to assess their compliance with the terms of this Agreement. Any Member which considers that such information has not been provided may bring the matter to the attention of the Committee.

25.10 Any Member which considers that any measure of another Member having the effects of a subsidy has not been notified in accordance with the provisions of paragraph 1 of Article XVI of GATT 1994 and this Article may bring the matter to the attention of such other Member. If the alleged subsidy is not thereafter notified promptly, such Member may itself bring the alleged subsidy in question to the notice of the Committee.

25.11 Members shall report without delay to the Committee all preliminary or final actions taken with respect to countervailing duties. Such reports shall be available in the Secretariat for inspection by other Members. Members shall also submit, on a semi annual basis, reports on any countervailing duty actions taken within the preceding six months. The semiannual reports shall be submitted on an agreed standard form.

25.12 Each Member shall notify the Committee *(a)* which of its authorities are competent to initiate and conduct investigations referred to in Article 11 and *(b)* its domestic procedures governing the initiation and conduct of such investigations.

Article 26
Surveillance

26.1 The Committee shall examine new and full notifications submitted under paragraph 1 of Article XVI of GATT 1994 and paragraph 1 of Article 25 of this Agreement at special sessions held every third year. Notifications submitted in the intervening years (updating notifications) shall be examined at each regular meeting of the Committee.

26.2 The Committee shall examine reports submitted under paragraph 11 of Article 25 at each regular meeting of the Committee.

PART VIII: DEVELOPING COUNTRY MEMBERS

Article 27
Special and Differential Treatment of Developing Country Members

27.1 Members recognize that subsidies may play an important role in economic development programmes of developing country Members.

27.2 The prohibition of paragraph 1(a) of Article 3 shall not apply to:

(a) developing country Members referred to in Annex VII.

(b) other developing country Members for a period of eight years from the date of entry into force of the WTO Agreement, subject to compliance with the provisions in paragraph 4.

27.3 The prohibition of paragraph 1(b) of Article 3 shall not apply to developing country Members for a period of five years, and shall not apply to least developed country Members for a period of eight years, from the date of entry into force of the WTO Agreement.

27.4 Any developing country Member referred to in paragraph 2(b) shall phase out its export subsidies within the eight year period, preferably in a progressive manner. However, a developing country Member shall not increase the level of its export subsidies,[39] and shall eliminate them within a period shorter than that provided for in this paragraph when the use of such export subsidies is inconsistent with its development needs. If a developing country Member deems it necessary to apply such subsidies beyond the 8 year period, it shall not later than one year before the expiry of this period enter into consultation with the Committee, which will determine whether an extension of this period is justified, after examining all the relevant economic, financial and development needs of the developing country Member in question. If the Committee determines that the extension is justified, the developing country Member concerned shall hold annual consultations with the Committee to determine the necessity of maintaining the subsidies. If no such determination is made by the Committee, the developing country Member shall phase out the remaining export subsidies within two years from the end of the last authorized period.

27.5 A developing country Member which has reached export competitiveness in any given product shall phase out its export subsidies for such product(s) over a period of two years. However, for a developing country Member which is referred to in Annex VII and which has reached export competitiveness in one or more products, export subsidies on such products shall be gradually phased out over a period of eight years.

27.6 Export competitiveness in a product exists if a developing country Member's exports of that product have reached a share of at least 3.25 per cent in world trade of that product for two consecutive calendar years. Export competitiveness shall exist either *(a)* on the basis of notification by the developing country Member having reached export competitiveness, or *(b)* on the basis of a computation undertaken by the Secretariat at the request of any Member. For the purpose of this paragraph, a product is defined as a section heading of the Harmonized System Nomenclature. The Committee shall review the operation of this provision five years from the date of the entry into force of the WTO Agreement.

27.7 The provisions of Article 4 shall not apply to a developing country Member in the case of export subsidies which are in conformity with the provisions of paragraphs 2 through 5. The relevant provisions in such a case shall be those of Article 7.

27.8 There shall be no presumption in terms of paragraph 1 of Article 6 that a subsidy granted by a developing country Member results in serious prejudice, as defined in this Agreement. Such serious prejudice, where applicable under the terms of paragraph 9, shall be demonstrated by positive evidence, in accordance with the provisions of paragraphs 3 through 8 of Article 6.

27.9 Regarding actionable subsidies granted or maintained by a developing country Member other than those referred to in paragraph 1 of Article 6, action may not be authorized or taken under Article 7 unless nullification or impairment of tariff concessions or other obligations under GATT 1994 is found to exist as a result of such a subsidy, in such a way as to displace or impede imports of a like product of another Member into the market of the subsidizing developing country Member or unless injury to a domestic industry in the market of an importing Member occurs.

27.10 Any countervailing duty investigation of a product originating in a developing country Member shall be terminated as soon as the authorities concerned determine that:

(a) the overall level of subsidies granted upon the product in question does not exceed 2 per cent of its value calculated on a per unit basis; or

(b) the volume of the subsidized imports represents less than 4 per cent of the total imports of the like product in the importing Member, unless imports from developing country Members whose individual shares of total imports represent less than 4 per cent collectively account for more than 9 per cent of the total imports of the like product in the importing Member.

27.11 For those developing country Members within the scope of paragraph 2(b) which have eliminated export subsidies prior to the expiry of the period of eight years from the date of entry into force of the WTO Agreement, and for those developing country Members referred to in Annex VII, the number in paragraph 10(a) shall be 3 per cent rather than 2 per cent. This provision shall apply from the date that the elimination of export subsidies is notified to the Committee, and for so long as export subsidies are not granted by the notifying developing country Member. This provision shall expire eight years from the date of entry into force of the WTO Agreement.

27.12 The provisions of paragraphs 10 and 11 shall govern any determination of *de minimis* under paragraph 3 of Article 15.

27.13 The provisions of Part III shall not apply to direct forgiveness of debts, subsidies to cover social costs, in whatever form, including relinquishment of government revenue and other transfer of liabilities when such subsidies are granted within and directly linked to a privatization programme of a developing country Member, provided that both such programme and the subsidies involved are granted for a limited period and notified to the Committee and that the programme results in eventual privatization of the enterprise concerned.

27.14 The Committee shall, upon request by an interested Member, undertake a review of a specific export subsidy practice of a developing country Member to examine whether the practice is in conformity with its development needs.

27.15 The Committee shall, upon request by an interested developing country Member, undertake a review of a specific countervailing measure to examine whether it is consistent with the provisions of paragraphs 10 and 11 as applicable to the developing country Member in question.

PART XI: FINAL PROVISIONS

Article 31
Provisional Application

The provisions of paragraph 1 of Article 6 and the provisions of Article 8 and Article 9 shall apply for a period of five years, beginning with the date of entry into force of the WTO Agreement. Not later than 180 days before the end of this period, the Committee shall review the operation of those provisions, with a view to determining whether to extend their application, either as presently drafted or in a modified form, for a further period.

ANNEX I
ILLUSTRATIVE LIST OF EXPORT SUBSIDIES

(a) The provision by governments of direct subsidies to a firm or an industry contingent upon export performance.

(b) Currency retention schemes or any similar practices which involve a bonus on exports.

(c) Internal transport and freight charges on export shipments, provided or mandated by governments, on terms more favourable than for domestic shipments.

(d) The provision by governments or their agencies either directly or indirectly through government mandated schemes, of imported or domestic products or services for use in the production of exported goods, on terms or conditions more favourable than for provision of like or directly competitive products or services for use in the production of goods for domestic consumption, if (in the case of products) such terms or conditions are more favourable than those commercially available[40] on world markets to their exporters.

(e) The full or partial exemption remission, or deferral specifically related to exports, of direct taxes[41] or social welfare charges paid or payable by industrial or commercial enterprises.[42]

(f) The allowance of special deductions directly related to exports or export performance, over and above those granted in respect to production for domestic consumption, in the calculation of the base on which direct taxes are charged.

(g) The exemption or remission, in respect of the production and distribution of exported products, of indirect taxes[41] in excess of those levied in respect of the production and distribution of like products when sold for domestic consumption.

(h) The exemption, remission or deferral of prior stage cumulative indirect taxes on goods or services used in the production of exported products in excess of the exemption, remission or deferral of like prior stage cumulative indirect taxes on goods or services used in the production of like products when sold for domestic consumption; provided, however, that prior stage cumulative indirect taxes may be exempted, remitted or deferred on exported products even when not exempted, remitted or deferred on like products when sold for domestic consumption, if the prior stage cumulative indirect taxes are levied on inputs that are consumed in the production of the exported product (making normal allowance for waste).[43] This item shall be interpreted in accordance with the guidelines on consumption of inputs in the production process contained in Annex II.

(i) The remission or drawback of import charges in excess of those levied on imported inputs that are consumed in the production of the exported product (making normal allowance for waste); provided, however, that in particular cases a firm may use a quantity of home market inputs equal to, and having the same quality and characteristics as, the imported inputs as a substitute for them in order to benefit from this provision if the import and the corresponding export operations both occur within a reasonable time period, not to exceed two years. This item shall be interpreted in accordance with the guidelines on consumption of inputs in the production process contained in Annex II and the guidelines in the determination of substitution drawback systems as export subsidies contained in Annex III.

(j) The provision by governments (or special institutions controlled by governments) of export credit guarantee or insurance programmes, of insurance or guarantee programmes against increases in the cost of exported products or of exchange risk programmes, at premium rates which are inadequate to cover the long term operating costs and losses of the programmes.

(k) The grant by governments (or special institutions controlled by and/or acting under the authority of governments) of export credits at rates below those which they actually have to pay for the funds so employed (or would have to pay if they borrowed on international capital markets in order to obtain funds of the same maturity and other credit terms and denominated in the same currency as the export credit), or the payment by them of all or part of the costs incurred by exporters or financial institutions in obtaining credits, in so far as they are used to secure a material advantage in the field of export credit terms.

Provided, however, that if a Member is a party to an international undertaking on official export credits to which at least twelve original Members to this Agreement are parties as of 1 January 1979 (or a successor undertaking which has been adopted by those original Members), or if in practice a Member applies the interest rates provisions of the relevant undertaking, an export credit practice which is in conformity with those provisions shall not be considered an export subsidy prohibited by this Agreement.

(l) Any other charge on the public account constituting an export subsidy in the sense of Article XVI of GATT 1994.

Endnotes

1 In accordance with the provisions of Article XVI of GATT 1994 (Note to Article XVI) and the provisions of Annexes I through III of this Agreement, the exemption of an exported product from duties or taxes borne by the like product when destined for domestic consumption, or the remission of such duties or taxes in amounts not in excess of those which have accrued, shall not be deemed to be a subsidy.

2 Objective criteria or conditions, as used herein, mean criteria or conditions which are neutral, which do not favour certain enterprises over others, and which are economic in nature and horizontal in application, such as number of employees or size of enterprise.

3 In this regard, in particular, information on the frequency with which applications for a subsidy are refused or approved and the reasons for such decisions shall be considered.

4 This standard is met when the facts demonstrate that the granting of a subsidy, without having been made legally contingent upon export performance, is in fact tied to actual or anticipated exportation or export earnings. The mere fact that a subsidy is granted to enterprises which export shall not for that reason alone be considered to be an export subsidy within the meaning of this provision.

5 Measures referred to in Annex I as not constituting export subsidies shall not be prohibited under this or any other provision of this Agreement.

6 Any timeperiods mentioned in this Article may be extended by mutual agreement.

7 As established in Article 24.

8 If a meeting of the DSB is not scheduled during this period, such a meeting shall be held for this purpose.

9 This expression is not meant to allow countermeasures that are disproportionate in light of the fact that the subsidies dealt with under these provisions are prohibited.

10 This expression is not meant to allow countermeasures that are disproportionate in light of the fact that the subsidies dealt with under these provisions are prohibited.

11 The term "injury to the domestic industry" is used here in the same sense as it is used in Part V.

12 The term "nullification or impairment" is used in this Agreement in the same sense as

it is used in the relevant provisions of GATT 1994, and the existence of such nullification or impairment shall be established in accordance with the practice of application of these provisions.

13 The term "serious prejudice to the interests of another Member" is used in this Agreement in the same sense as it is used in paragraph 1 of Article XVI of GATT 1994, and includes threat of serious prejudice.

14 The total ad valorem subsidization shall be calculated in accordance with the provisions of Annex IV.

15 Since it is anticipated that civil aircraft will be subject to specific multilateral rules, the threshold in this subparagraph does not apply to civil aircraft.

16 Members recognize that where royalty based financing for a civil aircraft programme is not being fully repaid due to the level of actual sales falling below the level of forecast sales, this does not in itself constitute serious prejudice for the purposes of this subparagraph.

17 Unless other multilaterally agreed specific rules apply to the trade in the product or commodity in question.

18 The fact that certain circumstances are referred to in this paragraph does not, in itself, confer upon them any legal status in terms of either GATT 1994 or this Agreement. These circumstances must not be isolated, sporadic or otherwise insignificant.

19 In the event that the request relates to a subsidy deemed to result in serious prejudice in terms of paragraph 1 of Article 6, the available evidence of serious prejudice may be limited to the available evidence as to whether the conditions of paragraph 1 of Article 6 have been met or not.

20 Any time periods mentioned in this Article may be extended by mutual agreement.

21 If a meeting of the DSB is not scheduled during this period, such a meeting shall be held for this purpose.

22 If a meeting of the DSB is not scheduled during this period, such a meeting shall be held for this purpose.

23 It is recognized that government assistance for various purposes is widely provided by Members and that the mere fact that such assistance may not qualify for nonactionable treatment under the provisions of this Article does not in itself restrict the ability of Members to provide such assistance.

24 Since it is anticipated that civil aircraft will be subject to specific multilateral rules, the provisions of this subparagraph do not apply to that product.

25 Not later than 18 months after the date of entry into force of the WTO Agreement, the Committee on Subsidies and Countervailing Measures provided for in Article 24 (referred to in this Agreement as "the Committee") shall review the operation of the provisions of subparagraph 2(a) with a view to making all necessary modifications to improve the operation of these provisions. In its consideration of possible modifications, the Committee shall carefully review the definitions of the categories set forth in this subparagraph in the light of the experience of Members in the operation of research programmes and the work in other relevant international institutions.

26 The provisions of this Agreement do not apply to fundamental research activities independently conducted by higher education or research establishments. The term "fundamental research" means an enlargement of general scientific and technical knowledge not linked to industrial or commercial objectives.

27 The allowable levels of non actionable assistance referred to in this subparagraph shall be established by reference to the total eligible costs incurred over the duration

of an individual project.

28 The term "industrial research" means planned search or critical investigation aimed at discovery of new knowledge, with the objective that such knowledge may be useful in developing new products, processes or services, or in bringing about a significant improvement to existing products, processes or services.

29 The term "pre-competitive development activity" means the translation of industrial research findings into a plan, blueprint or design for new, modified or improved products, processes or services whether intended for sale or use, including the creation of a first prototype which would not be capable of commercial use. It may further include the conceptual formulation and design of products, processes or services alternatives and initial demonstration or pilot projects, provided that these same projects cannot be converted or used for industrial application or commercial exploitation. It does not include routine or periodic alterations to existing products, production lines, manufacturing processes, services, and other on-going operations even though those alterations may represent improvements.

30 In the case of programmes which span industrial research and pre-competitive development activity, the allowable level of non-actionable assistance shall not exceed the simple average of the allowable levels of non-actionable assistance applicable to the above two categories, calculated on the basis of all eligible costs as set forth in items (i) to (v) of this subparagraph.

31 A "general framework of regional development" means that regional subsidy programmes are part of an internally consistent and generally applicable regional development policy and that regional development subsidies are not granted in isolated geographical points having no, or virtually no, influence on the development of a region.

32 "Neutral and objective criteria" means criteria which do not favour certain regions beyond what is appropriate for the elimination or reduction of regional disparities within the framework of the regional development policy. In this regard, regional subsidy programmes shall include ceilings on the amount of assistance which can be granted to each subsidized project. Such ceilings must be differentiated according to the different levels of development of assisted regions and must be expressed in terms of investment costs or cost of job creation. Within such ceilings, the distribution of assistance shall be sufficiently broad and even to avoid the predominant use of a subsidy by, or the granting of disproportionately large amounts of subsidy to, certain enterprises as provided for in Article 2.

33 The term "existing facilities" means facilities which have been in operation for at least two years at the time when new environmental requirements are imposed.

34 It is recognized that nothing in this notification provision requires the provision of confidential information, including confidential business information.

35 Under this Agreement the term "injury" shall, unless otherwise specified, be taken to mean material injury to a domestic industry, threat of material injury to a domestic industry or material retardation of the establishment of such an industry and shall be interpreted in accordance with the provisions of this Article.

36 Throughout this Agreement the term "like product" ("produit similaire") shall be interpreted to mean a product which is identical, i.e. alike in all respects to the product under consideration, or in the absence of such a product, another product which, although not alike in all respects, has characteristics closely resembling those of the product under consideration.

37 As set forth in paragraphs 2 and 4.

38 The Committee shall establish a Working Party to review the contents and form of the questionnaire as contained in BISD 9S/193–194.

39 For a developing country Member not granting export subsidies as of the date of entry into force of the WTO Agreement, this paragraph shall apply on the basis of the level of export subsidies granted in 1986.

40 The term "commercially available" means that the choice between domestic and imported products is unrestricted and depends only on commercial considerations.

41 For the purpose of this Agreement:
The term "direct taxes" shall mean taxes on wages, profits, interests, rents, royalties, and all other forms of income, and taxes on the ownership of real property;
The term "import charges" shall mean tariffs, duties, and other fiscal charges not elsewhere enumerated in this note that are levied on imports;
The term "indirect taxes" shall mean sales, excise, turnover, value added, franchise, stamp, transfer, inventory and equipment taxes, border taxes and all taxes other than direct taxes and import charges;
"Prior stage" indirect taxes are those levied on goods or services used directly or indirectly in making the product;
"Cumulative" indirect taxes are multi staged taxes levied where there is no mechanism for subsequent crediting of the tax if the goods or services subject to tax at one stage of production are used in a succeeding stage of production;
"Remission" of taxes includes the refund or rebate of taxes;
"Remission or drawback" includes the full or partial exemption or deferral of import charges.

42 The Members recognize that deferral need not amount to an export subsidy where, for example, appropriate interest charges are collected. The Members reaffirm the principle that prices for goods in transactions between exporting enterprises and foreign buyers under their or under the same control should for tax purposes be the prices which would be charged between independent enterprises acting at arm's length. Any Member may draw the attention of another Member to administrative or other practices which may contravene this principle and which result in a significant saving of direct taxes in export transactions. In such circumstances the Members shall normally attempt to resolve their differences using the facilities of existing bilateral tax treaties or other specific international mechanisms, without prejudice to the rights and obligations of Members under GATT 1994, including the right of consultation created in the preceding sentence.

43 Paragraph (h) does not apply to value-added tax systems and border-tax adjustment in lieu thereof; the problem of the excessive remission of value-added taxes is exclusively covered by paragraph (g).

Annex 11
Developing Country Categories and S&D Benefits under the ASCM*

A characteristic of S&D provisions in the SCM Agreement is that they effectively provide for different S&D benefits to developing countries on the basis of specific categories of developing countries.

These developing country "categories" and their corresponding S&D "package" are as follows:

1. Least-developed countries designated as such by the United Nations; and
2. Developing country WTO Members whose GNP per capita is less than US$1,000 per annum.

Category	*S&D Benefits under Article 27 SCM*
Least-developed countries designated as such by the United Nations	Allowed to provide subsidies otherwise prohibited under Article 3:1(a) of the SCM Agreement,[1] and are allowed a longer (eight years) transition period to phase-out subsidies prohibited under Article 31:1(b) from the date of the entry into force of the WTO Agreement (other developing countries have only a five-year transition period with respect to subsidies prohibited under Article 31:1(b));[2]
Developing country WTO Members whose GNP per capita is less than US$1,000 per annum[3]	Allowed to provide subsidies otherwise prohibited under Article 3:1(a) of the SCM Agreement for as long as they remain below the US$1,000 GNP per capita per annum level based on the most recent data from the World Bank.[4] Once they reach the US$1,000 GNP per capita per annum level, or when they attain export competitiveness[5] in a given product, they will then be required to progressively phase-out their subsidies prohibited under Article 3:1(a) over an eight-year period from the time they reach the US$1,000 threshold or, with respect to a specific export subsidy that helped them attain export competitiveness in a given product, from the time that they attained such export competitiveness in such a given product.

* This annex reproduces annex 3 of UNEP's publication *Reflecting Sustainable Development and Special and Differential Treatment for Developing Countries in the Context of New WTO Fisheries Subsidies Rules. An Issue and Options Paper.*

Endnotes

1 Paragraph 10.5 of the 2001 Doha Implementation Decision reaffirmed that LDCs "are exempt from the prohibition on export subsidies set forth in Article 3.1(a) ... and thus have flexibility to finance their exporters, consistent with their development needs."
2 By virtue of Paragraph 10.5 of the Ministerial Decision on Implementation Issues and Concerns, the phase-out period starts running from the first year an LDC under this paragraph reaches export competitiveness in an industry: "... *It is understood that the eight-year period in Article 27.5 within which a least-developed country member must phase out its export subsidies in respect of a product in which it is export-competitive begins from the date export competitiveness exists within the meaning of Article 27.6.*"
3 At the time of the conclusion of the Uruguay Round's conclusion, these were Bolivia, Cameroon, Congo, Côte d'Ivoire, Dominican Republic, Egypt, Ghana, Guatemala, Guyana, India, Indonesia, Kenya, Morocco, Nicaragua, Nigeria, Pakistan, Philippines, Senegal, Sri Lanka and Zimbabwe. See SCM Agreement, Annex VII.
4 Inclusion in the list in Annex VII(b) of the SCM Agreement was clarified by paragraph 10.1 of the 2001 Doha Ministerial Decision on Implementation-Related Issues and Concerns. This paragraph essentially states that the WTO Members listed in Annex VII(b) will remain in that list until their GNP per capital reaches US$1,000 in constant 1990 dollars for three consecutive years based on the most recent data from the World Bank. The methodology for calculating constant 1990 dollars is contained in Appendix 2 of G/SCM/38, 26 October 2001, In addition, Paragraph 10.4 of the Doha Implementation Decision clarified that WTO Members listed in Annex VII(b) who may have reached or exceeded the US$1,000 threshold for inclusion will be re-included in the list should its GNP per capital fall back below US$1,000.00.
5 "Export competitiveness" in a given product exists, under Art. 27.6 of the SCM Agreement, "if a developing country Member's exports of that product have reached a share of at least 3.25 per cent in world trade of that product for two consecutive calendar years ... either (a) on the basis of notification by the developing country Member having reached export competitiveness, or (b) on the basis of a computation undertaken by the Secretariat at the request of any Member."

Annex 12
Artisanal Fishing: A Few Examples of Existing Usages and Definitions*

DEFINITIONS OF KEY WORDS AND TERMS

This appendix gives a short sampling of current usages and definitions of the term 'artisanal' in the fisheries context. These are not meant to be representative or exhaustive, but merely illustrative. [Author's emphasis throughout]

FAO, 1993[1]

Artisanal or Small-scale Fisheries (SSF): generally a labour-intensive fishing subsector whose operators use simple and practical technology, work in decentralized coastal areas, experience fluctuating production and low incomes, live in isolated areas usually under difficult conditions and occupy a relatively low social status in many countries. It is composed of private sector entrepreneurs operating at different organisational levels from single person operations, through informal micro-enterprises to formal sector business. It represents a mix of several entrepreneurs in the fish capture, processing and marketing areas and also in ancillary industries such as boat building, engine supply and repairs, ice plants, net manufacturing, fuel and fuelwood supplies and money lending; each contributing especially to food self-sufficiency and the creation of numerous jobs for both women and men.

FAO, 2005[2]

Definition: 'Artisanal Fisheries' – Traditional fisheries involving fishing households (as opposed to commercial companies), using relatively small amounts of

* This annex reproduces annex B of UNEP's publication *Artisanal Fishing: Promoting Poverty Reduction and Community Development Through New WTO Rules on Fisheries Subsidies. An Issue and Options Paper.*

capital and energy, relatively small fishing vessels (if any), making short fishing trips, close to shore, mainly for local consumption. In practice, definition varies between countries, e.g. from gleaning or a one-man canoe in poor developing countries, to more than 20 m. trawlers, seiners, or long-liners in developed ones. Artisanal fisheries can be subsistence or commercial fisheries, providing for local consumption or export. Sometimes referred to as small-scale fisheries.

Asian Development Bank, Policy on Fisheries[3]

From the glossary:

Artisanal fishery – Small-scale fishery generally limited to nearshore waters and inland water bodies, and employing labor-intensive fishing technologies.

From the text:

Artisanal (**or small-scale**) **fisheries** are generally limited to nearshore waters and inland water bodies, and employ labor-intensive fishing technologies. Artisanal fishing operations are typically family-based, using small craft (usually smaller than 12 meters [m] long) and fishing gear such as beach seine and gill nets, hook and line, and traps. In the Region, artisanal fisheries are estimated to contribute at least 50 percent of total fisheries production. The artisanal subsector is strategically significant to Bank operations, as it supports extensive rural employment in the DMCs, particularly in countries where fisheries have become the employer of last resort.

United Nations Code of Conduct for Responsible Fisheries

6.18 Recognizing the important contributions of artisanal and small-scale fisheries to employment, income and food security, States should appropriately protect the rights of fishers and fishworkers, particularly those engaged in **subsistence, small-scale and artisanal** fisheries, to a secure and just livelihood, as well as preferential access, where appropriate, to traditional fishing grounds and resources in the waters under their national jurisdiction.

International Development Research Center, 2001[4]

Various terminologies are used to label the range of fisheries (Table 1.1). The terms differ in the details of definition but not in substance. It is useful, however, to distinguish the largescale (commercial/industrial) from the small-scale (commercial, artisanal, subsistence) ends of the spectrum. Strictly speaking, all fisheries are commercial. Even the smallest artisanal fishery sells what is surplus to household needs. Today there are very few fisheries in which none of the catch is sold, and these are usually termed subsistence fisheries.

Traditional, artisanal, and subsistence fisheries are also in the category of small-scale fisheries, exploiting many of the stocks harvested by commercial fisheries. In addition, they exploit a great variety of very small stocks distributed over numerous management units (Figure 1.1). Some of these fisheries are mechanized but most use traditional fishing gear, such as small nets, traps, lines, spears, and hand-collection methods. Of all the fisheries, biodiversity of the catch is highest in these. For that reason, and because low gear used is unselective, these harvests include a greater variety of species than do those of the larger commercial fisheries. Traditional, artisanal, and subsistence fisheries tend to target the following groups of species:

- fish and invertebrates of coral reefs, typically with traps, spears, lines, and by hand;
- fish and invertebrates of coastal lagoons and estuaries, typically using nets;
- stream and river fisheries, typically using nets;
- aquarium species in all habitats, using nets and noxious substances.

Chilean fisheries regulation, c. 1996 (described)[5]

If they want to continue as artisanal fishermen, according to the Fishery Law, their boats must be less than 18 m in length and no more than 50 gross registered tonnes (GRT). Such a boat can no longer be managed with only the help of relatives.

Artisanal Fishing in France[6]

Artisanal fishing refers to the use of boats under 25 metres in length. A subdivision specifies that boats under 12 metres represent 74% of the whole fleet, and those comprised between 12 and 25 metres correspond to 23% (Ofimer, 2000).

Endnotes

1 FAO, 1993, p.v.
2 FAO, 2005 (online glossary).
3 Asian Development Bank, 1997, pp.iv and 6.
4 Berkes et al, 2001, Ch. 1.4.
5 This text is taken from *Samudra*, Issue No. 16, November 1996, p.18 (available at www.icsf.net/jsp/publication/samudra/pdf/english/issue_16/ALL.pdf), which in turn is presenting a translation by Brian O'Riordan of an article from *Chile Pesquero*. *Samudra* presents the article without detailed attribution and the original is no longer available online (see www.chilepesquero.cl/).
6 Gouin 2000, fn. 7.

Annex 13
Summary of Proposed Sustainability Criteria*

* This annex reproduces annex 1 of UNEP–WWF's publication *Sustainability Criteria for Fisheries Subsidies. Options for the WTO and Beyond.*

Stock- and Capacity-related Criteria

	Examples of Possible Best Practices	*Minimum Recommended Conditions*	*Minimum International Requirements*
Stock-related Criteria	Biomass is known quantitatively with high levels of confidence on the basis of assessments that include scientific surveys; Biomass is significantly above formal science-based precautionary threshold reference points below which additional limits on fishing would be imposed; and Biomass is stable or rising.	Science-based assessments reveal that the stock is "underexploited" (per FAO definition); and Bio-economic data for at least [three] previous years does not reveal any "red flag" trends, such as those listed on p. 13.	Science-based assessments reveal that the stock is "underexploited" (per FAO definition). (Absence of "red flag" bio-economic trends could also be incorporated into WTO rules)
Capacity-related Criteria	Capacity is known quantitatively with high levels of confidence on the basis of assessments that include direct scientific observations of fleet characteristics, fishing practices, and stock conditions; Capacity is far below full capacity (e.g., <50 per cent); and Capacity has been flat or declining (or, in new or very underdeveloped fisheries, has been growing very slowly (e.g., 5 per cent per year) such that the 50 per cent threshold would not be breached during the economic life of the subsidies).	(Same as "best practices", with some possible relaxation of intensity of scientific data collecting)	(Same as "best practices" except that quantitative science-based assessments could rely on crude fleet inventories and indirect bio-economic indicators (see p.18) rather than dedicated scientific surveys) (for a much weaker alternative, see discussion on p.19)

Management-related Criteria

	Examples of Possible Best Practices	*Minimum Recommended Conditions*	*Minimum International Requirements*
Assessment	Scientific survey stock assessments to be conducted in all subsidized fisheries, supplementing fisheries-dependent assessments carried out on an ongoing basis; Direct measurement and calculation of capacity through scientific surveys of capacity to be carried out in every subsidized fishery.	Stock assessments to be supplemented wherever possible by scientific surveys and by investigations into ecosystem or coastal zone considerations, including changes in trophic levels of catches; Capacity assessments to be supplemented by scientific surveys and direct capacity assessment techniques.	Science-based stock assessments based on catch data or catch and effort data for at least three years prior to subsidization, and annually during the life of the subsidies; Science-based capacity assessments (may be based on crude inventories & indirect indicators) resulting in quantitative estimates of total fleet capacity and trends in capacity for at least three years prior to subsidization, and annually during life of the subsidies.
Controls	In addition to "recommended" and "required" controls, legally binding reference points to include "threshold" reference points triggering restrictions on subsidies.	In addition to "required" controls, management plans to include ecosystem-based management and, where appropriate, coordination with integrated coastal zone management plans.	Formal management plan in place, including a capacity management plan consistent with the FAO International Plan of Action for the Management of Fishing Capacity; Legally binding precautionary target and limit reference points for both stocks and capacity based on science-based assessments, taking MSY equilibrium as the outer limit of acceptable limit for stock biomass; Pre-determined regulatory responses to be taken in the event target reference points are breached.
Enforcement	In addition to "recommended" and "required" conditions, enforcement procedures to include independent public review of enforcement actions and effectiveness thereof.	In addition to "required" conditions, enforcement procedures to include a public record of enforcement actions.	Procedures sufficient to permit reasonably effective action against illegal fishing activities in the target fishery, and to prevent significant patterns of illegal fishing therein; Mandatory withdrawal and repayment of subsidies to any vessel once engaged in (non de minimis) illegal fishing activities.

	Examples of Possible Best Practices	*Minimum Recommended Conditions*	*Minimum International Requirements*
MCS Administrative Infrastructure	In addition to all "required" and "recommended" conditions: Vessel registry information to include "additional" information sought by the HSVAR database; Full on-board observer coverage of all vessels active in target fisheries.	In addition to all "required" conditions: Vessel registry information to include all "optional" information sought by the HSVAR database; Mandatory reporting of catches (including all discards), to be verified by at least partial observer coverage in target fisheries.	Mandatory registration of all active vessels, providing "mandatory" information required by the HSVAR database, and provision of all requisite information to any applicable international registry system; Mandatory licensing of all vessels/fishers, detailing authorization to fish and license information kept in a public license registry; Mandatory reporting of catches or landings (and effort, where applicable).
Rapid Evaluation (per Appendix 2)	(Used in context of overall detailed evaluation).	Structured use of these or similar benchmarks as ongoing evaluation of fishery.	Rapid evaluation questions could provide additional information during rules implementation.

Annex 14
Benchmarks for Rapid Overall Evaluations*

Appendix 2
Benchmarks for Rapid Overall Evaluations

This "condensed questionnaire" to establish the state of management and exploitation of marine resources was prepared by John Caddy on commission to UNEP and WWF for purposes of this paper. It should be viewed as illustrative, and as a basis for further development.[1]

Characteristics of the fishery for resource A over the last decade	*Yes (Green)*	*Maybe/partially (Yellow)*	*No (Red)*
OUTPUTS			
1) Landings are still above 50 per cent of the average for the best 3 years landings on record (FAO Statistics)?			
2) Landings have not continued to decline significantly over the last 5 years?			
3) Catch rates have not declined significantly over the last 5 years (by standard vessel category)?			
4) The fleet capacity utilizing the resource has not grown by more than 10–15 per cent since the second of the best 3 years landings on record?			
5) Prices for the product on the domestic market of the coastal state have not grown by more than 25 per cent over the last 5 years?			
6) Biological data are collected in port, OR in-port interviews are carried out, OR copies of catch log books are completed and collected by port officials?			
7) The capture of protected species is actively discouraged?			
8) The diversity of resources/habitats is being actively maintained and protected?			

* This annex reproduces annex 2 of UNEP–WWF's publication *Sustainability Criteria for Fisheries Subsidies. Options for the WTO and Beyond.*

9) Illegal or unreported fishing is being kept under strict control by active at-sea surveillance?

Characteristics of the fishery for resource A over the last decade	*Yes (Green)*	*Maybe/ partially (Yellow)*	*No (Red)*
INPUTS			
10) Research vessel surveys are carried out at regular intervals?			
11) There is a limited license system in operation that covers all vessels fishing the resource, and an up-to-date registry of active fishing vessels and their characteristics is maintained?			
12) There is a formal system of licence transfers on vessel replacement that ensures that fleet capacity is not increasing?			
13) There is a system of at-sea surveillance of the fleet operation or on-board observers?			
14) Biologists are employed to evaluate the fishery with at least Masters in Science education?			
15) A management plan exists for the fishery?			
16) Closed areas or MPAs are in effect? Such areas within the stock range are still unfished or form refugia, nursery areas or spawning areas?			
17) For shared, straddling and highly migratory stocks, there are fisheries agreements or negotiations in course with other users of the resource?			
18) The government fisheries agency meets regularly with local community or fishing industry representatives?			
19) Sports fishing or diving activities are revenue earners on the fishing grounds for the resource?			
20) If there are foreign access agreements, do these specify avoidance of national fishing areas/resources, and are these provisions policed?			
21) Are the provisions of ecosystem management/biodiversity conservation applied?			
22) Is there an integrated coastal area management plan in effect, protecting coastal resources from pollution/unwise developments?			

Endnotes

1 Although exact criteria are not proposed, if scored impartially by persons familiar with the fishery, traffic light criterion (for example, Caddy and Surette 2005) should return a high percentage of 'Green' responses if the fishery is properly managed. A proportion of 'Yellow' responses can be accepted, but any 'Red' responses should be seen as the basis for urgent improvements.

Annex 15 Model WTO Language on Sustainability Criteria*

[reprinted from *Sustainability Criteria for Fisheries Subsidies: Options for the WTO and Beyond* (UNEP 2007), Appendix 3]

The following model language assumes that new WTO fisheries subsidies disciplines will include a prohibition that, at a minimum, covers all directly capacity- or effort-enhancing subsidies. This language illustrates one possible approach to minimum fisheries-related criteria for permitting the use of otherwise prohibited subsidies. This language is not comprehensive, and is meant to illustrate how some of the criteria discussed above could be integrated into WTO terms. Some of the options for criteria discussed above are not reflected here (e.g., the use of indirect criteria such as those set forth in Appendix 2 to inform judgments about stock health or fleet capacity), but are not therefore meant to be given lower priority. This model language also does not cover new institutional mechanisms that may be advisable to facilitate the involvement of external fisheries experts and authorities in the implementation of the proposed rules.

ARTICLE [X]
[preconditions based on minimum fisheries-related criteria]

1. The subsidies referred to in paragraph __ [establishing the right to S&DT or other exemptions from a broad prohibition] shall be maintained or granted only on condition that:

   a. All fish stocks affected by subsidized fishing operations are subject to regular science-based stock assessments, and assessments[1] conducted for at least [three] years prior to the use of subsidies provide a reasonable basis to conclude that the target stocks are underexploited.[2]

---

* This annex reproduces annex 3 of UNEP–WWF's publication *Sustainability Criteria for Fisheries Subsidies. Options for the WTO and Beyond.*

b. All fleets affecting the target stocks are subject to regular science-based capacity assessments,[3] and assessments conducted for at least [three] years prior to the use of subsidies provide a reasonable basis to conclude that total capacity in the fishery is [substantially less than] [less than [50 per cent] of] full capacity.[4]
c. Every fishery to be affected by subsidized fishing operations is subject to an effective fisheries management system consistent with the UN Code of Conduct for Responsible Fisheries, in accordance with paragraph 2 of this article.

2. Except in the case of artisanal fisheries, a fisheries management system shall be presumed to fulfill the requirements of paragraph 1.c if [its objectives are consistent with the objectives set forth in article 6 of the UN Code of Conduct for Responsible Fisheries, and]:
a. it is designed and implemented on the basis of regular science-based assessments of target fish stocks and fishing fleets;
b. it includes a formal fishery management plan, including a capacity management plan consistent with the FAO International Plan of Action for the Management of Fishing Capacity;
c. it includes legally binding regulations setting precautionary limits on fishing capacity, fishing effort, and/or fishing production on the basis of target reference points designed to maintain populations of harvested species at levels no lower than necessary to produce the maximum sustainable yield, taking into account ecosystem interactions, where applicable;
d. it includes regulations requiring the mandatory withdrawal and repayment of subsidies received by any vessel or enterprise found to have engaged in illegal fishing activities;[5]
e. it requires all vessels active in the fishery to be registered in a national vessel registry (and/or, where applicable, international vessel registries);[6]
f. it requires all vessels active in the fishery to hold a valid fishing license, with such licenses maintained in a public registry;
g. it includes a system for the regular collection of data on catches and/or landings, and all vessels active in the fishery are required to participate in that system;
h. it includes mechanisms for monitoring and enforcement of its regulations, and reasonable monitoring and enforcement activities are regularly undertaken;[7] and
i. it prohibits all vessels and enterprises benefiting from subsidies to seek or receive licenses to fish in, or otherwise to be transferred to, any fishery not meeting all conditions set forth in this article.

3. In the case of an artisanal fishery,[8] the requirements of paragraph 1.c shall be deemed to be fulfilled if:
a. the status of its fish stocks and its fleets are subject to regular investigation by qualified fisheries experts, including through informal, non-quantitative methods, and there is a reasonable basis to conclude that

the target stocks are underexploited and the fleets are at substantially less than full capacity;

b. there is in place a formal plan for the management and development of the fishery which includes the identification of precautionary target reference points for both stocks and fleets designed to maintain populations of harvested species at levels no lower than necessary to produce the maximum sustainable yield, taking into account ecosystem interactions, where applicable;

c. there are in place, or there are being developed, community-based mechanisms for the control of fishing activities within the fishery;

d. subsidized increases in fishing capacity or effort are carefully monitored and recorded; and

e. other government expenditures in proportion to the subsidies affecting the fishery are achieving continuous improvements in the management of the fishery, including through the increased availability of data relevant to the quantitative assessment of its stocks and fleets.

Endnotes

1 Except in the case of artisanal fisheries, "science-based stock assessments" shall be quantitative assessments based, at a minimum, on regularly collected and publicly documented data about catches or landings. Members are encouraged to use the best available scientific methods for stock assessments, including full biological surveys of target stocks, where appropriate.

2 "Underexploited" shall mean that the biomass of the target stocks is significantly below "maximum sustainable yield" equilibrium levels, such as in a new or developing fishery, and that there is a reasonable basis to conclude that the stock has significant potential for expanded production. A finding or report by the FAO regarding the status of fishery, such as those routinely contained in the FAO's biannual State of World Fisheries and Aquaculture, shall be presumed valid in the absence of clear and convincing evidence to the contrary.

3 Except in the case of artisanal fisheries, "science-based capacity assessments" shall, to the maximum reasonable extent, be quantitative assessments based on regularly collected and publicly documented data. Members are encouraged to use the best available scientific methods for capacity assessments, including direct surveys of fleets and stocks, where appropriate.

4 "Substantially less than full capacity" shall mean that the total capacity of all active fleets is significantly below the level necessary to achieve catch levels consistent with long-term "maximum sustainable yield" when fleets are fully employed, and that there is a reasonable basis to conclude that the stock has significant potential for expanded production.

5 For purposes of this paragraph, illegal fishing activities shall not include de minimis violations of applicable laws or regulations.

6 To qualify as a "national vessel registry" for purposes of this article, the registry must be open to the public and include all types of information considered "mandatory"

under the High Seas Vessel Authorization Records (HSVAR) database maintained by the FAO.

7 The existence of laws and administrative mechanisms for the monitoring and enforcement of fisheries regulations shall raise a rebuttable presumption that reasonable enforcement activities are being undertaken. A finding that reasonable enforcement activities are not being undertaken shall require clear and convincing proof of (i) a consistent pattern of illegal activity having biologically [or commercially] significant impacts on the fishery or its surrounding ecosystem or on markets for its products, and (ii) an unreasonable inattention to enforcement activities on the part of public authorities, taking account of economic circumstances, including the economic value of the fishery.

8 "Artisanal fisheries" are fisheries in developing countries consisting of a large number of small, owner-operated vessels using low-tech fishing gear (such as manual net retrieval) in nearby inshore fisheries, whose products are destined for consumption by the fishers' own households or for sale in highly localized markets, and whose poverty, geographic location, traditional social organization, diffuse patterns of fishing and landing, and disconnection from centralized markets make them particularly difficult to manage through data-intensive, command and control techniques.

Annex 16
Resource Material: Contents of the CD-ROM

The documents contained in the CD-ROM were reproduced with the permission of the listed organizations or other publishers/journals. Further reproduction of the listed works shall require advance written permission from the listed organizations or other publishers/journals.

UNEP

UNEP Events

UNEP, WWF, CPPS. 2009. *Regional Symposium on "Sustainability Criteria for Fisheries Subsidies: The Latin American Context", Regional Symposium on "Sustainability Criteria for Fisheries Subsidies: The Latin American Context"*, Guayaquil, Ecuador, 29-30 July 2009.

UNEP-WWF. 2009. *"The WTO Fisheries Subsidies Negotiations: Update and Introductory Briefing for New Delegates"*, Geneva, 1 April 2009.

UNEP-WWF, in collaboration with ICTSD and OCEANA. 2008. *Technical and Informal Workshop on WTO Disciplines on Fisheries Subsidies: Elements of the Chair's Draft, Geneva, Switzerland, 29 January 2008.*

UNEP-WWF. 2007. *UNEP-WWF Symposium on Disciplining Fisheries Subsidies: Incorporating Sustainability at the WTO & Beyond, Geneva, Switzerland, 1-2 March 2007.*

UNEP, ICTSD, WWF. 2006. *Joint UNEP-ICTSD-WWF Workshop on Development and Sustainability in the WTO Fishery Subsidies Negotiations: Issues and Alternatives, Geneva, Switzerland, 11 May 2006.*

UNEP-WWF. 2005. UNEP-WWF High-Level Event and Panel Discussion at the Sixth WTO Ministerial Conference - Fisheries Subsidies Disciplines in the WTO: Challenges and Opportunities, Hong Kong, China, 14-15 December 2005: *Time to Draw in the Net on Fishing Subsidies*. Klaus Toepfer and James Leape.

UNEP. 2005. *Roundtable on Promoting Development and Sustainability in Fisheries Subsidies Disciplines, Geneva, 30 June 2005.*

UNEP. 2004. *UNEP Workshop on Fisheries Subsidies and Sustainable Fisheries Management, Geneva, 26-27 April 2004.*

UNEP. 2003. *UNEP Informal Expert Consultations on Fisheries Subsidies, Geneva, 16th July 2003.*

UNEP, in consultation with the WTO, FAO and OECD Secretariats. *Workshop on the Impacts of Trade-Related Policies on Fisheries and Measures Required for their Sustainable Management Geneva, Switzerland, 15 March 2002.*

UNEP. 2001. *UNEP Fisheries Subsidies Workshop, Geneva, 12 February* 2001.

UNEP. 1999. *Working Group Meeting on Trade and Environment, Geneva, 21 October, 1999.*

Analytical Studies

UNEP. 2009. *Certification and Sustainable Fisheries.* Graeme Macfadyen, Tim Huntington.

UNEP. 2008. *Fisheries Subsidies: A Critical Issue for Trade and Sustainable Development at the WTO: An Introductory Guide.*

UNEP. 2008. *Towards Sustainable Fisheries Access Agreements – Issues and Options at the World Trade Organization.* Marcos A. Orellana.

UNEP-WWF. 2007. *Sustainability Criteria for Fisheries Subsidies. Options for the WTO and Beyond.* David K. Schorr and John F. Caddy.

UNEP. 2005. *Artisanal Fishing: Promoting Poverty Reduction and Community Development Through New WTO Rules on Fisheries Subsidies. An Issue and Options Paper.* David K. Schorr

UNEP. 2005. *Reflecting Sustainable Development and Special and Differential Treatment for Developing Countries in the Context of New WTO Fisheries Subsidies Rules. An Issue and Options Paper.* Vincente Paolo B. Yu III and Darlan Fonseca-Marti.

UNEP. 2004. *Analyzing the Resource Impact of Fisheries Subsidies: A Matrix Approach.* Gareth Porter.

UNEP. 2004. *Incorporating Resource Impact into Fisheries Subsidies Disciplines: Issues and Options. A discussion Paper.* Gareth Porter.

UNEP. 2004. *A UNEP Update on Fisheries Subsidies and Sustainable Fisheries Management.*

UNEP. 2003. *Fisheries Subsidies and Overfishing: Towards a Structured Discussion,* Fisheries and the Environment. Gareth Porter.

Country Studies

UNEP-CPPS. 2009. *The Impact of Fisheries Subsidies on the Ecuadorian Tuna's Sustainability and Trade.* Iván Prieto Bowen. (in Spanish) (Executive Summary in English)

UNEP, VIFEP, WWF. 2009. *Fisheries Subsidies, Supply Chain and Certification in Vietnam - Summary Report.* Tuong Phi Lai, Pham Ngoc Tuan, Nguyen Thi Dieu Thuy, Duong Long Tri, Pham Thi Hong Van.

UNEP. 2004. *Fisheries Subsidies and Marine Resources Management: Lessons learned from Studies in Argentina and Senegal.* Fisheries and the Environment. Maria Onestini, Karim Dahou.

UNEP. 2004. *Policy Implementation and Fisheries Resource Management: Lessons from Senegal.* Fisheries and the Environment. Senegal Fisheries Ministry and ENDA.

UNEP. 2004. *Fisheries Subsidies and Marine Resource Management: Lessons from Bangladesh.* Fisheries and the Environment. Fahmida A Khatun, Mustafizur Rahman, Debapriya Bhattacharya.

UNEP. 2002. *Evaluation de l'impact de la libéralisation du commerce: Une étude de cas sur le secteur des pêches de la République Islamique de Mauritanie*. Pêche et Environnement. Pierre Failler, Mika Diop, Cheikh Abdallahi O/ Inejih, Mamoudou Aliou Dia, Abou Daim Dia.

UNEP. 2002. *Integrated Assessment of Trade Liberalization and Trade-Related Policies A Country Study on the Argentina Fisheries Sector*. New York and Geneva.

UNEP. 2002. *Integrated Assessment of Trade Liberalization and Trade-Related Policies: A Country Study on the Fisheries Sector in Senegal*. New York and Geneva.

Other

UNEP film "Caught Out...The way forward for fisheries subsidies"

OTHER INTER-GOVERNMENTAL ORGANISATIONS

APEC

APEC. 2000. *Study Into the Nature and Extent of Subsidies in the Fisheries Sector of APEC Members Economies. "Reproduced with the permission of the APEC Secretariat, Singapore."[Study Into the Nature and Extent of Subsidies in the Fisheries Sector of APEC Members Economies]" ISBN 1-896633-10-2 . APEC#00-FS- 01.1*

FAO

Publications

FAO. 2009. *The State of World Fisheries and Aquaculture 2008*. Food and Agriculture Organization of the United Nations. Rome.

FAO. 2004. *Guide for identifying, assessing and reporting on subsidies in the fisheries sector*. FAO Fisheries Technical Paper 438. Food and Agriculture Organization of the United Nations. Rome.

FAO. 2003. *Introducing Fisheries Subsidies*. FAO Fisheries Technical Paper 437. Food and Agriculture Organization of the United Nations. Rome.

FAO. 2001. *International Plan of Action To Prevent, Deter and Eliminate Illegal, Unreported and Unregulated Fishing*. Food and Agriculture Organization of the United Nations. Rome.

FAO. 1999. *International Plan of Action for the Management of Fishing Capacity*. Food and Agriculture Organization of the United Nations. Rome.

FAO. 1995. *Code of Conduct for Responsible Fisheries*. Food and Agriculture Organization of the United Nations. Rome.

FAO.1993. *Marine Fisheries and the Law of the Sea: A Decade of Change*. Special Chapter (revised) of *The state of food and agriculture 1992*. FAO Fisheries Circular No.853. Food and Agriculture Organization of the United Nations. Rome.

Meetings

Extracts of Reports of COFI and COFI Sub-Committee on Trade Meetings Related to Fisheries Subsidies (2009-1999). Food and Agriculture Organization of the United Nations.

FAO. 2008. *Status and Important Recent Events Concerning International Trade in Fishery Products. Committee on Fisheries, Sub-Committee on Trade, Eleventh*

Session, Bremen, Germany, 2-6 June 2008. Food and Agriculture Organization of the United Nations.

FAO. 2004. *Report of the Technical Consultation on the Use of Subsidies in the Fisheries Sector, Rome, 30 June – 2 July 2004*. FAO Fisheries Report No. 752. Food and Agriculture Organization of the United Nations. Rome.

FAO. 2004. *TC SUB /2004/Inf.4 - A global Project for the Study of Impacts of Fisheries Subsidies*. Food and Agriculture Organization of the United Nations.

FAO. 2004. *TC SUB /2004/Inf.3 - A summary of recent work on subsidies in the fishing sector*. Food and Agriculture Organization of the United Nations.

FAO. 2004. *TC SUB/2004/2 - Working Document - A Global Technical Initiative on Fisheries Subsidies*. Food and Agriculture Organization of the United Nations.

FAO. 2003. *Report of the Third Ad Hoc Meeting of Intergovernmental Organizations on Work Programmes Related to Subsidies in Fisheries, Rome, 23-25 July 2003*. FAO Fisheries Report No. 719. Food and Agriculture Organization of the United Nations.

FAO. 2003. *Report of the Expert Consultation on Identifying, Assessing and Reporting on Subsidies in the Fishing Industry, Rome, 3-6 December 2002*. FAO Fisheries Report No. 698. Food and Agriculture Organization of the United Nations.

FAO. 2002. *Report of the Second Ad Hoc Meeting of Intergovernmental Organizations on Work Programmes Related to Subsidies in Fisheries, Rome, 4-5 July 2002*. FAO Fisheries Report No. 688. Food and Agriculture Organization of the United Nations.

FAO. 2001. *Report of the Ad Hoc Meeting of Intergovernmental Organizations on Work Programmes Related to Subsidies in Fisheries, Rome, 21-22 May 2001*. FAO Fisheries Report No. 649. Food and Agriculture Organization of the United Nations.

FAO. 2001. *Papers presented at the Expert Consultation on Economic Incentives and Responsible Fisheries. Rome, 28 November – 1 December 2000*. FAO Fisheries Report No. 638, Supplement. Food and Agriculture Organization of the United Nations.

FAO. 2000. *Report of the Expert Consultation on Economic Incentives and Responsible Fisheries. 28 November – 1 December 2000*. FAO Fisheries Report No. 638. Food and Agriculture Organization of the United Nations. Rome.

OECD

OECD. 2006. *Subsidy Reform and Sustainable Development: Economic, Environmental and Social Aspects*.

OECD. 2006. *Financial Support to Fisheries: Implications for Sustainable Development*.

OECD. 2005. *Environmentally Harmful Subsidies: Challenges for Reform*.

OECD. 2005. *Subsidies: A Way Towards Sustainable Fisheries?* Policy Brief December 2005.

FURTHER LINKS FOR DOWNLOAD OR FOR PURCHASE

Sutinen, J. G. 2008. "*Major Challenges for Fishery Policy Reform: A Political Economy Perspective*", *OECD Food, Agriculture and Fisheries Working Papers*, No. 8, OECD Publishing. www.oecd.org/dataoecd/37/3/41305791.pdf

OECD. 2007. *Structural Change in Fisheries: Dealing with the Human Dimension*. www.browse.oecdbookshop.org/oecd/pdfs/browseit/5307051E.PDF (OECD Browse_it editions)

OECD. 2007. *Subsidy Reform and Sustainable Development: Political Economy Aspects. Chapter 5: Fisheries*.www.oecd.org/document/26/0,3343,en_2649_37425_38516570_1_1_1_1,00.html

OECD. 2006. *Review of Fisheries in OECD countries, Country Statistics 2002-2004.* Paris.www.oecd.org/document/15/0,3343,en_2649_33901_31686479_1_1_1_1,00.html

OECD. 2005. *Environmentally Harmful Subsidies: Challenges for Reform.*
www.browse.oecdbookshop.org/oecd/pdfs/browseit/5105081E.PDF (OECD Browse_it editions)

OECD. 2005. *Subsidies: A Way Towards Sustainable Fisheries?* Policy Brief December 2005. http://www.oecd.org/dataoecd/63/54/35802686.pdf

OECD. 2003. Liberalising Fisheries Markets: Scope and Effects. Paris.
www.browse.oecdbookshop.org/oecd/pdfs/browseit/5303021E.PDF (OECD_Browse it edition)

OECD. 2000. Transition to Responsible Fisheries: Economic and Policy Implications. Paris. www.browse.oecdbookshop.org/oecd/pdfs/browseit/5300021E.PDF (OECD_Browse it edition)

Pacific Islands Forum Fisheries Agency

Pacific Islands Forum Fisheries Agency (FFA). 2007. Liam Campling, Elizabeth Havice and Vina Ram-Bidesi. *Pacific Island Countries, the Global Tuna Industry and the International Trade Regime - A Guidebook.*

Pacific Islands Forum Fisheries Agency (FFA). 2007. Elizabeth Havice. *The State of Play of Access Agreements With Distant Water Fishing Partners: Implications and Options for Pacific Island Countries.*

World Bank

The International Bank for Reconstruction and Development/ TheWorld Bank, FAO. 2008. *The Sunken Billions: The Economic Justification for Fisheries Reform.*

The International Bank for Reconstruction and Development/ The World Bank. 2004. Cathy A. Roheim: *Trade Liberalization in Fish Products: Impacts on Sustainability of International Markets and Fish Resources.*

The International Bank for Reconstruction and Development/ The World Bank. 1998. Matteo Milazzo: *Subsidies in World Fisheries: A Reexamination.* World Bank Technical Paper No.406.

WTO

Publications

WTO. 2006. *World Trade Report - Exploring the links between subsidies, trade and the WTO.*

Negotiations

Compilation of country submissions to WTO

NON-GOVERNMENTAL ORGANISATIONS

CIEL

CIEL. 1996. "*Natural Resource Subsidies, Trade and Environment: The Cases of Forest and Fisheries*", Center for International Environmental Law (CIEL), Washington, DC. Porter, G..

Greenpeace

Greenpeace. 2006. *Deadly Subsidies – How government funds are killing oceans and forests and why the CBD rather than the WTO should stop this perverse use of public money*. Juergen Knirsch, Daniel Mittler, Martin Kaiser, Karen Sack, Christoph Thies, Larry Edwards.

ICSF

ICSF. 2006. *Untangling Subsidies, Supporting Fisheries: The WTO Fisheries Subsidies Debate and Developing-country Priorities*, ICSF Occasional Paper. Prepared for ICTSD, John Kurien.

ICTSD

Campling, L. 2008. *Fisheries Aspects of ACP-EU Interim Economic Partnership Agreements: Trade and Sustainable Development Implications.* ICTSD Series on Fisheries, Trade and Sustainable Development. Issue Paper No. 6, International Centre for Trade and Sustainable Development, Geneva, Switzerland.

Mbithi Mwikya, S. 2006. *Fisheries Access Agreements: Trade and Development Issues,* ICTSD Natural Resources, International Trade and Sustainable Development Series Issue Paper No.2, International Centre for Trade and Sustainable Development, Geneva, Switzerland.

Roheim, C. A. and Sutinen, J. 2006. *Trade and Marketplace Measures to Promote Sustainable Fishing Practices*, ICTSD Natural Resources, International Trade and Sustainable Development Series Issue Paper No. 3, International Centre for Trade and Sustainable Development and the High Seas Task Force, Geneva, Switzerland.

Ahmed, M. 2006. *Market Access and Trade Liberalisation in Fisheries*, ICTSD Natural Resources, International Trade and Sustainable Development Series Issue Paper No. 4, International Centre for Trade and Sustainable Development, Geneva, Switzerland.

ICTSD. 2006. *Fisheries, International Trade and Sustainable Development: Policy Discussion Paper.* ICTSD Natural Resources, International Trade and Sustainable Development Series. International Centre for Trade and Sustainable Development, Geneva, Switzerland.

IISD

International Institute for Sustainable Developments (IISD) – Global Subsidies Initiative (GSI). 2007. *A new template for notifying subsidies to the WTO*. Ronald Steenblik and Juan Simón.

IUCN/SRCF

Mfodwo K., *"Negotiating equitable fisheries access agreements. A capacity-building and reference manual for developing coastal states"*, IUCN/SRFC, Dakar, January 2008.

MRAG

MRAG/DFID. 2009. *Fisheries and Subsidies*. Policy Brief 9.

MRAG/DFID. 2009. *Fisheries and the WTO Negotiations*. Insert to Policy Brief 9.

MRAG/DFID. 2007. *Fisheries and Access Agreements*. Policy Brief 6.

The Nature Conservancy, Coral Triangle Center

The Nature Conservancy – Coral Triangle Center. 2008. *"Selected Indonesian Fisheries Subsidies: Quantitative and Qualitative Assessment of Policy Coherence and Effectiveness."* A. Ghofar, D.K.Schorr, A. Halim.

Oceana

Oceana and ICTSD. 2009. *Healthier Oceans, Healthier Economies.* Info Note Number 7.

Oceana. 2007. *Towards Sustainable Fishing.*

Royal Commission on Environmental Pollution

RCEP. 2004. *Turning the Tide. Addressing the Impact of Fisheries on the Marine Environment.* Twenty-fifth report.

UBC

Sharp, R. and Sumaila U.R. 2009. *Quantification of U.S. Marine Fisheries Subsidies.* North American Journal of Fisheries Management, 29, 18-32.

Sumaila, U.R., Khan, A., Watson, R., Munro, G., Zeller, D., Baron, N., Pauly, D. 2007. *The World Trade Organization and global fisheries sustainability.* Fisheries Research, 88, pp.1-4.

Abdallah, P.R. and Sumaila, U.R. 2007. *A historical account of Brazilian policy on fisheries subsidies. Marine Policy 31, 444-450.*

Sumaila, U.R. and Keith H.. 2006. *Regulating fisheries subsidies - A role for RFMOs.* Bridges Monthly, No. 2, March-April 2006. ICTSD, Geneva. pp. 21-22.

Sumaila, U.R., Pauly, D. (eds.). 2006. *Catching more bait: A bottom-up re-estimation of global fisheries subsidies.* Fisheries Centre Research Reports 14(6), 114 pp. Fisheries Centre, the University of British Columbia, Vancouver, Canada.

FURTHER LINKS FOR PURCHASE

Clark, C.M., Munro, G., and Sumaila, U.R. 2007. *Buyback, subsidies, the time consistency problem and the ITQ alternative.* Land Economics, 83(1), pp. 50-58.

Munro, G. and Sumaila, U.R. 2002. *The impact of subsidies upon fisheries management and sustainability: the case of the North Atlantic.* Fish and Fisheries, 3, 233-250.

WWF

WWF. 2008. *Small Boats, Big Problems.*

WWF. 2007. *Fisheries Subsidies: WWF Statement on the Chairman's Draft.*

WWF. 2007. *WWF Statement on Recent Submissions Regarding Fisheries Subsidies To the WTO Negotiating Group on Rules.*

WWF. 2007. *S&DT for Developing Country Fisheries Subsidies: Technical Refinements + Replies to Chair's Questions.* 6 June 07, Geneva, Switzerland.

WWF. 2006. *The Best of Texts, the Worst of Texts – Will the first draft of new WTO rules on fisheries subsidies be strong? Or will governments just settle for the "weakest common denominator"?*

WWF. 2005. *What's the Catch? New WTO rules that support healthy fisheries and sustainable trade are within reach... but will governments really come home with their nets full?*

WWF. 2005. *Fishing Subsidies: Issues for ACP Countries.*

WWF. 2004. *Healthy Fisheries, Sustainable Trade: Crafting New Rules on Fishing Subsidies in the World Trade Organization.* A WWF Position Paper and Technical Resource.

WWF. 2002. *Turning the Tide On Fishing Subsidies – Can the World Trade Organization Play a Positive Role?*

WWF. 2001. *Hard Facts, Hidden Problems: A Review of Current Data on Fishing Subsidies.* A WWF Technical Paper.

WWF. 2000. *Fishing in the Dark: A Symposium on Access to Environmental Information and Government Accountability in Fishing Subsidy Programmes, 28-29 November 2000, Brussels,* Belgium, Symposium Proceedings.

WWF. 2000. Fishing in the Dark. An Issue Brief.

WWF.1999. *Underwriting Overfishing. Issue Summary No.1 9/99.*

WWF. 1998. *Towards Rational Disciplines on Subsidies to the Fishery Sector: A Call for New International Rules and Mechanisms.* A WWF discussion paper.

Bibliography

Acheampong, A. (1997) *Coherence Between EU Fisheries Agreements and EU Development Cooperation: The Case of West Africa*, ECDPM Working Paper 52, European Centre for Development Policy Management, Maastricht, The Netherlands

Abdallah, P.R. and Sumaila, U.R. (2006) 'An historical account of Brazilian public policy on fisheries subsidies', in Sumaila, U.R. and Pauly, D. (eds), *Catching More Bait: A Bottom-up Re-estimation of Global Fisheries Subsidies*, Fisheries Centre Research Reports 14(6), pp.67–76, Fisheries Centre, University of British Columbia, Vancouver, Canada

ADB (1997) 'Policy on Fisheries', available at www.adb.org/Documents/Policies/Fisheries/default.asp?p=policies, last accessed 10/03/2010

ADE-PWC-EPU (2002) *Evaluation of the Relationship between Country Programmes and Fisheries Agreements*, Final Report, prepared for European Commission, available at www.ec.europa.eu/europeaid/how/evaluation/evaluation_reports/reports/sector/951637_final_en.pdf, last accessed 10/03/2010

Ahmed, I. and Lipton, M. (1997) *Impact of Structural Adjustment on Sustainable Rural Livelihoods: A Review of the Literature*, IDS Working Paper 62, University of Sussex, Institute of Development Studies

Ahmed, M. (2006) *Market Access and Trade Liberalisation in Fisheries*, ICTSD Natural Resources, International Trade and Sustainable Development Series Issue Paper No. 4, International Centre for Trade and Sustainable Development, Geneva, Switzerland

Andrew, N.L., Béné, C., Hall, S.J., Allison, E.H., Heck, S. and Ratner, B.D. (2007) 'Diagnosis and management of small-scale fisheries in developing countries', in *Fish and Fisheries*, Vol. 8, No. 3, September 2007, pp.227–240, Wiley-Blackwell

Anonymous (2001) 'Fiji fisheries in bad shape: overfishing could close the industry', *Pacific Magazine* (June 2003)

APA (1999) 'Preliminary Assessment of the Pollock Conservation Cooperative', Seattle and Anchorage, APA

APEC (2000) 'Study into the nature and extent of subsidies in the fisheries sector of APEC member economies', Asia-Pacific Economic Co-operation, Singapore

Apostle, R., Barret, G., Holm, P., Jentoft, S., Manzany, L., McCay, B. and Mikalsen, K. (1998) *Community, State, and Market on the North Atlantic Rim*, Toronto, University of Toronto Press

Aqorau, T. and Bergin, A. (1997) 'Ocean Governance in the Western Pacific Purse Seine Fishery – the Palau Arrangement', *Marine Policy* 21/2: pp.173–186

Aqorau, T. and Bergin, A. (1998) 'The UN Fish Stocks Agreement – A New Era for International Cooperation to Conserve Tuna in the Central Western Pacific', *Ocean Development and International Law*, 29, pp.21–42

Arnason, R. (1998) 'Fisheries and Economic losses', in Hatcher, Aaron and Robinson, Kate (eds) *Overcapacity, Overcapitalisation and Subsidies in European Fisheries*, proceedings of the first workshop held in Portsmouth, UI, 28–30 October, Portsmouth, UK, CEMARE

Asche, F., Bjorndal, T. and Gordon, D.V. (1998) 'Quota Regulation, Rent and Value of Licenses in the Norwegian Pelagic Fisheries', in Hatcher, Aaron and Robinson, Kate (eds) *Overcapacity, Overcapitalisation and Subsidies in European Fisheries*, proceedings of the first workshop held in Portsmouth, UI, 28–30 October, Portsmouth, UK, CEMARE

Ashley, C. (2000) *Applying Livelihood Approaches to Natural Resource Management Initiatives: Experiences in Namibia and Kenya*, working paper 134, London, UK, ODI

Ashley, C. and Carney, D. (1999) *Sustainable Livelihoods: Lessons from Early Experience*, London, UK: Department for International Development (Dfid)

Auditor General of Canada (1997) 'Fisheries and Oceans Canada: Rationalization and Renewal: Atlantic Groundfish', chapter 15 of the 1997 October Report of the Auditor General of Canada, available at www.oag-bvg.gc.ca/internet/English/parl_oag_199710_15_e_8096.html, last accessed 10/03/2010

Australian Trade Commission (2003) 'Building and Construction to Indonesia', available at www.austrade.gov.au/home, last accessed 11/03/2010

Banks, R. (1999) 'Subsidising EU Fleets: Capacity Reduction or Capital Subsidisation', in Hatcher, Aaron and Robinson, Kate (eds) *Overcapacity, Overcapitalisation and Subsidies in European Fisheries*, proceedings of the first workshop held in Portsmouth, UI, 28–30 October, Portsmouth, UK: CEMARE

Bartels et al., *Policy Coherence for Development and the Effects of EU Fisheries Policies on Development in West Africa*. Draft Report, submitted to the European Parliament, 2007.

Beddington, J.R., Agnew, D.H. and Clark, C.W. (2007) 'Current Problems in the Management of Marine Fisheries', in *Science Magazine*, vol. 316, pp.1713 et seq

Binet, T. and Failler, P. (2009) 'Évolution des migrations de pêcheurs artisans en Afrique de l'Ouest entre 1988 et 2008, État des lieux et évolution récente des migrations de pêcheurs artisans dans les pays de la CSRP', IUCN, Juillet

Blasé, F.W. (1982) 'Coastal Village Development in Four Fishing Communities of Adirampattinam, Tamil Nadu, India', FAO Bay of Bengal Programme, available at www.fao.org/documents/show_cdr.asp?url_file=/docrep/007/ad959e/ad959e00.htm, last accessed 11/03/2010

Bonfil, R., Munro, G., Sumaila, U.R., Valtysson, H., Wright, M., Pitcher, T., Preikshot, D., Haggan, N. and Pauly, D. (1998) 'Impacts of Distant Water Fleets: An Ecological, Economic and Social Assessment', in *The Footprint of Distant Water Fleets on World Fisheries*, Godalming, UK: WWF International, WWF's Endangered Seas Campaign

Brack, D. (ed.) (1998) *Trade and Environment: Conflict or Compatibility?*, Trade and Environment Series – Royal Institute for International Affairs, Earthscan, London, UK

Bradsher, K. (2005) 'Collective Stance at WTO: Activists ally with nations on fishing aid', *International Herald Tribune*, 15 December 2005 (Hong Kong edition)

Brock, K. (1999) *Implementing a Sustainable Livelihoods Framework for Policy-directed Research: Reflections from Practice in Mali*, working paper 90, Brighton, UK: IDS

Caddy, J.F. (1997) 'Establishing A Consultative Mechanism or Arrangement for Managing Shared Stocks Within the Jurisdiction of Contiguous States", in D. Hancock (ed.) *Taking Stock: Defining and Managing Shared Resources*, Australian Society for Fish Biology and Aquatic Resource Management Association of

Australasia Joint Workshop Proceedings, Darwin, NT, 15-16 June 1997, Sydney, Australian Society for Fish Biology: pp. 81–123.

Caddy, J.F. (1996) 'A Checklist for Fisheries Resource Management Issues Seen from the Perspective of the FAO Code of Conduct for Responsible Fisheries', *FAO Fisheries Circular*, No. 917, Rome

Caddy, J.F. (1994) *The Age Structure of Fishing Fleets and its Relevance for Reconstructing Past Fishery Trends and Forecasting*, paper presented at Lowell Wakefield Fishery Symposium on 'Management strategies for exploited fish populations', Anchorage, Alaska, 21–24 October 1992

Caddy, J.F. and Mahon, R. (1995) 'Reference Points for Fisheries Management', *FAO Fisheries Technical Paper*, No. 347, Rome

Caddy, J.F. and Surette, T. (2005) 'In retrospect the assumption of sustainability for Atlantic fisheries has proved an illusion', *Reviews in Fish Biology and Fisheries* 15, pp. 313–337

Cameron, H. 'The Evolution of the Trade and Environment Debate at the WTO', in Najam, A., Halle, M. and Meléndez-Ortiz, R. (eds.) (2007) *Trade and the Environment: A Resource Book*, IISD, ICTSD, The Ring, Geneva. Winnipeg, London

Cancun Declaration (1992) 'Declaration of the International Conference on Responsible Fishing', Cancun, Mexico, 6–8 May

Catarci, C. (2004) 'The world tuna industry – an analysis of imports and prices, and of their combined impact on catches and tuna fishing capacity', in 'Second Meeting of the Technical Advisory Committee of the FAO Project Management of Tuna Fishing Capacity: Conservation and Socio-economics', 15–18 March 2004, Madrid, FAO Fisheries Proceedings No.2, Food and Agriculture Organization of the United Nations, Rome.

CCAMLR Conservation Measure No. 10-05 (2006) (available on CCAMLR website, www.ccamlr.org, last accessed 11/03/2010).

CEC (2008) Council Regulation (EC) No. 1005/2008 of 29 September 2008 establishing a community system to prevent, deter and eliminate illegal, unreported and unregulated fishing, OJ L 286 of 29.10.2008, available at www.eur-lex.europa.eu/LexUriServ/LexUriServ.do?uri=OJ:L:2008:286:0001:0032:EN:PDF, last accessed 18/03/2010

CFFA (2005) 'EU fisheries subsidies: Significance for developing countries, Coalition for Fair Fisheries Arrangement', Brussels, available at www.cape-cffa.org/pub_CFP/CFFA%20subsidies.doc, last accessed 10/03/2010

Chambers, R. and Conway, G. (1992) *Sustainable Rural Livelihoods: Practical Concepts for the 21st Century*, IDS Discussion Paper No. 296, Brighton, IDS

Chauveau, J.P. (1988) 'Note sur l'histoire de la motorization dans la pêche artisanale maritime senegalaise, préconditions et rancons d'un success precoce', in *Economie de la mécanisation en région chaude*, International Cooperation Centre in Agronomic Research for Development CIRAD-MESRU

Chang, S.W. (2003) 'WTO Disciplines on Fisheries Subsidies: A Historic Step Towards Sustainability?', *Journal of International Economic Law*, Oxford University Press, Vol. 6, No. 4, pp.879–921 (available at www.jiel.oxfordjournals.org/cgi/reprint/6/4/879.pdf, last accessed 10/03/2010)

Christy, F.T. (1997) *The Development and Management of Marine Fisheries in Latin America and the Caribbean (IADB)*, Inter-American Development Bank, Washington D.C., USA, available at www.iadb.org/sds/ENV/publication/publication_205_92_e.htm, last accessed 10/03/2010

Chuang, C. and Zhang, X. (1999) 'Review of Vessel Buyback Schemes and Experience in Chinese Taipei', in Riepen, M. (ed.) *The Impact of Government Financial Transfers on Fisheries Management, Resource Sustainability and International Trade*, Report of the Proceedings of the PECC Workshop held 17–19 August 1998, Manila, Philippines', Singapore, Pacific Economic Co-operation Council

Clark, I. (1993) 'Individual Transferable Quotas: the New Zealand Experience', *Marine Policy*, September, pp.340–343

Clark, C.W. and Munro, G.R. (1994) 'Renewable Resources as Natural Capital: The Fishery', in Jansson, A., Hammer, M., Folke, C. and Constanza, R. (eds) *Investing in Natural Capital*, Island Press, Washington, D.C.

Clark, C.W., Munro, G. and Sumaila, U.R. (2005) 'Subsidies, buybacks, and sustainable fisheries', *Journal of Environmental Economics and Management*, Vol. 50, pp.47–58

Clark, C.W., Munro, G.R. and Sumaila, U.R. (2006) 'Buyback subsidies, the time consistency problem and the ITQ alternative', Fisheries Centre Working Paper #2006–08, The University of British Columbia, Vancouver, B.C., Canada [In press: *Land Economics*, (vol. 83, No. 1, pp. 50–58, 2007].

Clark, L. (2006) 'Perspectives on Fisheries Access Agreements: Developing Country View', prepared for OECD Workshop on Policy Coherence for Development in Fisheries, COM/AGR/DCD/PCDF(2006)2

Clover, C. (2006) 'EU split over return of fishing subsidies' in *The Telegraph* (22 May), available at www.telegraph.co.uk/news/worldnews/europe/1519089/EU-split-over-return-of-fishing-subsidies.html, last accessed 10/03/2010

CNP (2008) Informe técnico. Análisis de impacto en la industria atunera y pesquera nacional ante potenciales cambios en el sistema de intermediación laboral y tercerización de servicios especializados, de Prieto, Velasco y Anastacio. Ecuador

Coan, Jr., A. L., Williams, P., Staish, K. and Yamasaki, G. (2000) 'The 1999 U.S. Central-Western Pacific Tropical Tuna Purse Seine Fishery', Administrative Report LJM-00-10, document prepared for the annual meeting of the South Pacific Regional Tuna Treaty, 3–10 March 2000, Niue

Coglan, L., Pascoe, S. and Mardle, S. (2000) 'Physical vs. Harvest Based Measures of Capacity: The Case of the UK Vessel Capacity Unit System', in *Microbehavior and Macroresults: Proceedings of the Tenth Biennial Conference of the International Institute of Fisheries Economics and Trade Presentations*, Corvallis, Oregon, IIFET, Oregon State University

Commission of the European Communities (1991) *Report 1991 from the Commission to the Council and the European Parliament on the Common Fisheries Policy*, Brussels, Commission of the European Communities

Corten, A. (1996) 'The Widening Gap between Fisheries Biology and Fisheries Management in the European Union', *Fisheries Research*, Vol. 27, pp.1–15

Court of Auditors (1994) 'Special Report No. 33/93 Concerning the Implementation of the Measures for the Restructuring, Modernization and Adaptation of the Capacities of Fishing Fleets in the Community Together with the Commission`s Replies', *Official Journal of the European Communities*, 4 January

CPPS (2004) Informe de Subsidios Pesqueros (also available in www.cpps-int.org/asambleas/iiiasamblea/primerasesion/Informe%20Subsidios%20Pesqueros%20OMC.pdf, last accessed 17/03/2010

CRODT (2006) 'Etude de base de la pêche pour une gestion intégrée des ressources marines et côtières sénégalaises: caractérisation des zones marines et côtières, identification des sites prioritaires, stratégies d'intervention, manuel de procédure', Rapport final, projet GIRMac

Crutsinger, M. (1999) 'Five nations call for end to harmful fishing subsidies', Associated Press wire story, March 13

Dahou, K. and Dème, M. (2002) 'Accords de pêche UE-Sénégal et commerce international: Respects des règlements internationaux, gestion durable des ressources et sécurité alimentaire, Enda Tiers Monde, Centre de Recherches Océanographiques de Dakar-Thiaroye (CRODT), Senegal : Dakar

Dallmeyer (1989) 'Traditional fishermen being caught in Australia's net', *Hobart Mercury*, 5 January

Danielsen, J.F. (2004) Further Examination of Economic Aspects Relating to the Transition to Sustainable Fisheries: A Case Study of Norway, OECD Committee for Fisheries, AGR/FI(2004)5/PART2, OECD, Paris

Danish Directorate of Fisheries (2001) *Yearbook of Fishery Statistics 2001*, Copenhagen, Ministry of Food, Agriculture and Fisheries

Daures, F. and Guyader, O. (2000) *Economic Analysis of the Impact of Buyback Programs and the Role of Financial Incentives Schemes: Application to a Limited Entry French Fishery*, paper for XIIth EAFE Annual Conference, Esbjerg, Denmark, 13–14 April 2000

Delgado, C.L., Wada, N., Rosegrant, M.W., Meijer, S. and Ahmed, M. (2003) *Fish to 2020: Supply and Demand in Changing Global Markets*, International Food Policy Research Institute and WorldFish Center, USA and Malaysia

Dème, M. (2007) 'Etude des coûts et revenus des principales unités de pêche artisanale sénégalaises (forthcoming, UNIVAL/ISRA)

Dème, M. (2009) 'Pêche maritime et pression environnementale: un risque pour le premier poste d'exportation' (forthcoming)

De Wilde, J.W. (1999) 'Effects of Subsidies on Distant Water and Coastal Fisheries of the Netherlands' in Hatcher, Aaron and Robinson, Kate (eds) *Overcapacity, Overcapitalisation and Subsidies in European Fisheries*, proceedings of the first workshop held in Portsmouth, UI, 28–30 October, Portsmouth, UK, CEMARE

De Wilde, J.W. and van Beek, F. (1996) 'North Sea Fishery for Flatfish', in Salz, P. (ed.), *Bio-Economic Evaluation of Multi-Species and Multi-Annual Fisheries Management Measures*,The Hague, Agricultural Economics Research Institute

DOPM (2007) 'Résultats des pêches maritimes 1990 à 2006'. Dakar, Sénégal

Eggert, H. (2001) 'Technical Efficiency in the Swedish Trawl Fishery for Norway Lobster', *Working Papers in Economics*, No. 53. Goteborg, Sweden, Department of Economics, Goteborg University

European Commission (n.d.) 'Bilateral fisheries partnership agreements between the EC and third countries', available at www.ec.europa.eu/fisheries/cfp/external_relations/bilateral_agreements_en.htm, last accessed 11/03/2010

European Commission, *Countervailing Measures on DRAM Chips*, Panel Report

European Court of Auditors (1992) 'Common Policy on Fisheries and the Sea', Chapter 4, *Official Journal of the European Communities*, 15 December, pp.107–115

Failler, P. (2001) 'The impact of the EU fishing agreements on the African fish market supply', Dfid Responsible Project/CEMARE WR No. 1, Portsmouth, UK, CEMARE

Failler P. et al. (2002), La recherche halieutique et le développement durable des ressources naturelles marines en Afrique de l'Ouest, quels enjeux?, Initiative de recherche halieutique ACP/UE, Rapport Recherche Halieutique ACP/UE, n° 11, EUR20188, 145 p.

Failler, P. and Kane, A. (2003) 'Sustainable livelihood approach and improvement of the living conditions of fishing communities: relevance, applicability and applications', in Neiland, A. and Béné, C. (eds.) *In: Poverty and Small-scale Fisheries in West Africa*, Kluwer Academic Publishers/ FAO, Rome (Italy), Fisheries Department, 2004, pp.121–149

Failler, P., Kane, A., Cissé, A.T., Dème, M., Samb, M., Wally, M. and Sagna, P. (2001) 'Impact of policies, institutions and processes on the livelihoods of fishing communities in Senegal', summary report, Cotonou, Bénin: Dfid/FAO

Failler, P., Ndiaye, P.G. and Bakanova, D. (2004) 'Politiques commerciales et durabilité des secteurs halieutiques en Afrique de l'Ouest, Compte rendu de l'atelier Conakry, Guinée', 1–2 October, Dakar, Programme Pêche, Commerce et Environnement en Afrique de l'Ouest (PCEAO), ENDA-Diapol

FAO (1983) 'Approaches to the Regulation of Fishing Effort', Fisheries Technical Paper No. 243, United Nations Food and Agriculture Organization, Rome, Beddington, J.R. and Rettig, R.B.

FAO (1982) 'Coastal Village Development in Four Fishing Communities of Adirampattinam, Tamil Nadu, India', FAO Bay of Bengal Programme, available at www.fao.org/docrep/007/ad959e/ad959e00.htm, last accessed 11/03/2010, Blase, F.W.

FAO (1993a) 'Marine fisheries and the law of the sea: A decade of change', Special Chapter (revised) of *The State of Food and Agriculture 1992*, FAO Fisheries Circular No. 853, United Nations Food and Agriculture Organization, Rome

FAO (1993b) 'Papers Presented at the FAO/Japan Consultation on the Development of Community-Based Coastal Fishery Management Systems for Asia and the Pacific, Kobe, Japan', 8–12 June, 1992, FAO Fisheries Report No. 474, Supplement Vol. 1, United Nations Food and Agriculture Organization, Rome

FAO (1993c) *Ten Years of Integrated Development of Artisanal Fisheries in West Africa*, Technical Report No. 50, United Nations Food and Agriculture Organization, Rome, Satia, B.P.

FAO (1993d) 'The State of Food and Agriculture 1992', United Nations Food and Agriculture Organization, Rome

FAO (1995a) *Code of Conduct for Responsible Fisheries*, FAO Fisheries Technical Paper No. 350(1), United Nations Food and Agriculture Organization, Rome, available at www.fao.org/DOCREP/005/v9878e/v9878e00.htm#1, last accessed 11/03/2010

FAO (1995b) *Agreement to Promote Compliance with International Conservation and Management Measures by Fishing Vessels on the High Seas*, United Nations Food and Agriculture Organization, Rome

FAO (1997) *Fisheries Management, FAO Technical Guidelines for Responsible Fisheries*, No. 4, United Nations Food and Agriculture Organization, Rome

FAO (1998) *Introduction to Tropical Fish Stock Assessment, Part 1 (Manual)*, FAO Fisheries Technical Paper No. 306, United Nations Food and Agriculture Organization, Rome Sparre, P. and Venema, S.C.

FAO (1999a) *Indicators for Sustainable Development of Marine Capture Fisheries*, FAO Technical Guidelines for Responsible Fisheries No. 8, United Nations Food and Agriculture Organization, Rome

FAO (1999b) *Guidelines for the Routine Collection of Capture Fishery Data*, FAO Fisheries Technical Paper No. 382, United Nations Food and Agriculture Organization, Rome

FAO (1999c) *The International Plan of Action for the Management of Fishing Capacity*, Food and Agricultural Organization of the United Nations, Rome

FAO (1999d) *Measuring Capacity and Capacity Utilization in Fisheries*, in D. Greboval (ed.) *Managing Fishing Capacity: Selected Papers on Underlying Concepts and Issues*, United Nations Food and Agriculture Organization, Rome, Kirkley, J. and Squires, D.

FAO (1999e) *Measuring Capacity and Capacity Utilization in Fisheries: The Case of the Danish Gill-net Fleet*, prepared for the FAO Technical Consultation on the Measurement of Fishing Capacity, Mexico, Vestergaard, N., Squires D. and Kirkley, J.

FAO (2000a) *Report of the Technical Consultation on the Measurement of Fishing Capacity*, FAO Fisheries Report No. 615, United Nations Food and Agriculture Organization, Rome

FAO (2000b) *Report of the Expert Consultation on Economic Incentives and Responsible Fisheries*, Rome, 28 November–1 December 2000, FAO Fisheries Report No. 638, United Nations Food and Agriculture Organization, Rome

FAO (2001a) *International Plan of Action to Prevent, Deter and Eliminate Illegal, Unreported and Unregulated Fishing*, United Nations Food and Agriculture Organization, Rome, available at www.fao.org/DOCREP/003/y1224e/y1224e00.HTM, last accessed 11/03/2010

FAO (2001b) *Inventory of Artisanal Fishery Communities in the Western-Central Mediterranean*, United Nations Food and Agriculture Organization, Rome, available at www.fao.org/docrep/009/a0824e/a0824e00.htm, last accessed 11/03/2010, Coppola, S.R.

FAO (2001c) *Report of the Ad Hoc Meeting of Intergovernmental Organizations on Work Programmes Related to Subsidies in Fisheries*, Fisheries Report No. 649, United Nations Food and Agriculture Organization, Rome

FAO (2001d) *Understanding the Cultures of Fishing Communities: A Key to Fisheries Management and Food Security*, FAO Fisheries Technical Paper No. 401, United Nations Food and Agriculture Organization, Rome, McGoodwin, J.R.

FAO (2001e) *Subsidies and the Fisheries Sector: Facilitating the International Discussion*, in papers presented at the expert consultation on economic incentives and responsible fisheries, United Nations Food and Agriculture Organization, Rome, 28 November–December 2000

FAO (2002a) *Promoting the Contribution of the Sustainable Livelihoods Approach and the Code of Conduct for Responsible Fisheries in Poverty Alleviation*, FAO Fisheries Report No. 678, United Nations Food and Agriculture Organization, Rome, available at www.fao.org/DOCREP/005/Y3910E/Y3910E00.HTM, last accessed 11/03/2010

FAO (2002b) *Fishing Vessels Operating under Open Registers and the Exercise of Flag State Responsibilities – Information and Options*, FAO Fisheries Circular No. 980, United Nations Food and Agriculture Organization, Rome

FAO (2002c) *A Fishery Manager's Guidebook. Management Measures and Their Application*, FAO Fisheries Technical Paper No. 424, United Nations Food and Agriculture Organization, Rome, Cochrane, K.L. (ed.)

FAO (2002d) *Recent Trends in Monitoring, Control and Surveillance Systems for Capture Fisheries*, FAO Fisheries Technical Paper. No. 415, United Nations Food and Agriculture Organization, Rome, Flewwelling, P., Cullinan, C., Balton, D., Sautter, R.P. and Reynolds, J.E.

FAO (2003b) *Report on the Expert Consultation on Identifying, Assessing and Reporting on Subsidies in the Fishing Industry*, FAO Fisheries Report No. 698, United Nations Food and Agriculture Organization, Rome

FAO (2003c) 'Small-scale Fisheries Perspectives on an Ecosystem-based Approach to Fisheries Management', in Sinclair, M. and Valdimarsson, G. (eds.) *Responsible Fisheries in the Marine Ecosystem*, Mathew, S.

FAO (2003d) *The WTO Doha Round and Fisheries: What's at Stake*, FAO Fact Sheet for WTO Ministerial Conference in Cancun, United Nations Food and Agriculture Organization, Rome, Lem, A.

FAO (2003e) *Introducing Fisheries Subsidies*, FAO Fisheries Technical Paper No. 437, United Nations Food and Agriculture Organization, Rome, Schrank, W.

FAO (2003f) *Yearbooks of Fishery Statistics: Summary Tables*, United Nations Food and Agriculture Organization, Rome

FAO (2004a) *Report of the Technical Consultation on the Use of Subsidies in the Fisheries Sector*, Rome, 30 June–2 July 2004, FAO Fisheries Report No. 752, United Nations Food and Agriculture Organization, Rome, available at www.fao.org/docrep/007/y5689e/y5689e00.HTM, last accessed 11/03/2010

FAO (2004b) Ward, J.M., Kirkley, J.E., Metzner, R. and Pascoe, S., *Measuring and Assessing Capacity in Fisheries*, FAO Fisheries Technical Paper No. 433/1, Rome

FAO (2004c) *Contributions of Fisheries and Aquaculture in the Asia-Pacific Region*, United Nations Food and Agriculture Organization, Rome, Sugiyama, S., Staples, D. and Funge-Smith, S.

FAO (2004d) *Overview of Fish Production, Utilization, Consumption and Trade*, (based on 2002 data) United Nations Food and Agriculture Organization, Rome, Vannuccini, S.

FAO (2004e) *Guide for Identifying, Assessing and Reporting on Subsidies in the Fisheries Sector*, FAO Fisheries Technical Paper No. 438, United Nations Food and Agriculture Organization, Rome, Westlund, L.

FAO (2004f) *Fish Trade Issues in WTO and ACP-EU Negotiations, Globefish*, United Nations Food and Agriculture Organization, Rome

FAO (2005a) *FAO Fisheries Department Glossary*, available at www.fao.org/fi/glossary/default.asp, last accessed 11/03/2010

FAO (2005b) *Review of the State of World Marine Fishery Resources*, FAO Fisheries Technical Paper No. 457, United Nations Food and Agriculture Organization, Rome

FAO (2005c) *Guidelines for the Ecolabelling of Fish and Fishery Products from Marine Capture Fisheries*, United Nations Food and Agriculture Organization, Rome

FAO (2005d) *Report of the FAO/WorldFish Center Workshop on Interdisciplinary Approaches to the Assessment of Small-Scale Fisheries*, Rome, 20–22 September 2005, FAO Fisheries Report No. 787, United Nations Food and Agriculture Organization, Rome

FAO (2005e), Emerson, W., 'Trends in Fisheries Trade and the Role of Developing Countries', FAO Presentation at the ICTSD Trade and Development Symposium in Hong Kong, 15 December 2005

FAO (2005f), Halls, A.S., Arthur, R.I., Bartley, D., Felsing, M., Grainger, R., Hartmann, W., Lamberts, D., Purvis, J., Sultana, P., Thompson, P. and Walmsley, S., *Guidelines for Designing Data Collection and Sharing Systems for Co-managed Fisheries, Part 2: Technical Guidelines*, FAO Fisheries Technical Paper No. 494/2, Rome

FAO (2006a) 'FAO-CITES agreement promotes sustainable fish trade: Collaborative relationship formalized in MoU', United Nations Food and Agriculture Organization, Rome, newsroom release dated 3 October 2006, available at www.fao.org/newsroom/en/news/2006/1000410/index.html, last accessed 11/03/2010

FAO (2006b), Hoggarth, D.D., Abeyasekera, S., Arthur, R.I., Beddington, J.R., Burn, R.W., Halls, A.S., Kirkwood, G.P., McAllister, M., Medley, P., Mees, C.C., Parkes, G.B., Pilling, G.M., Wakeford, R.C. and Welcomme, R.L., *Stock Assessment for Fishery Management – A Framework Guide to the Stock Assessment Tools of the Fisheries Management Science Programme (FMSP)*, FAO Fisheries Technical Paper No. 487, Rome

FAO (2006c) *The State of World Highly Migratory, Straddling and Other High Seas Fishery Resources and Associated Species*, United Nations Food and Agriculture Organization, Rome

FAO (2007a) *Report of the Twenty-Seventh Session of the Committee on Fisheries*, Rome, 5–9 March 2007, FAO Fisheries Report No. 830, United Nations Food and Agriculture Organization, Rome

FAO (2007b) *Report on the Development of a Comprehensive Record of Fishing Vessels*, COFI/2007/Inf.12, United Nations Food and Agriculture Organization, Rome

FAO (2007c) *Report of the Twenty-Second Session of the Coordinating Working Party on Fishery Statistics*, Rome, 27 February–2 March 2007, United Nations Food and Agriculture Organization, Rome

FAO (n.d.) 'Regional Fishery Bodies (RFB) Fact Sheets', available at www.fao.org/fishery/rfb/search/en, last accessed 11/03/2010

FAO (n.d.) 'FISHSTAT PLUS', available at www.fao.org/fishery/statistics/software/fishstat, last accessed 11/03/2010

FAO COFI (2008) 'Status and Important Recent Events Concerning International Trade in Fishery Products', COFI Sub-Committee on Fisheries, Eleventh Session, Bremen, Germany, 2–6 June 2008, United Nations Food and Agriculture Organization, Rome

FAO SOFIA (1992) *The State of World Fisheries and Aquaculture Report*, United Nations Food and Agriculture Organization Rome

FAO SOFIA (2002) *The State of World Fisheries and Aquaculture Report*, United Nations Food and Agriculture Organization, Rome

FAO SOFIA (2004) *The State of World Fisheries and Aquaculture Report*, United Nations Food and Agriculture Organization, Rome

FAO SOFIA (2006) *The State of World Fisheries and Aquaculture Report*, United Nations Food and Agriculture Organization, Rome

FAO SOFIA (2008) *The State of World Fisheries and Aquaculture Report*, United Nations Food and Agriculture Organization Rome

Ferris, J.S. and Plourde, C.G. (1982) 'Labour mobility, seasonal unemployment insurance and the Newfoundland inshore fishery', *Canadian Journal of Economics*, Vol. 15, No. 3, pp.426–441

Fisheries and Oceans Canada (DFO) (2001) 'Backgrounder: Selective Fishing', BG-PR-01-003E, 24 January, available at www.dfo-mpo.gc.ca/Library/252358.pdf, last accessed 19/03/2010

Fishery Committee for the Eastern Central Atlantic (1992) 'Twelfth Session of the Fishery Committee for the Eastern Central Atlantic Accra, Ghana, 27 April–1 May 1992', CECAF/ XII/92/3, United Nations Food and Agriculture Organization, Rome

Fiskeridirektoratet (Norwegian Directorate of Fisheries) (2009) *Economic and Biological Key Figures from Norwegian Fisheries*, Bergen

Flaaten, O., Heen, K. and Salvanes, K.G. (1995) 'The Invisible Resource Rent in Limited Entry and Quota Managed Fisheries: The Case of Norwegian Purse Seine Fisheries', *Marine Resources Economics*, Vol. 10, pp.341–356

Flaaten, O. and Wallis P. (2000) 'Government Financial Transfers to Fishing Industries in OECD Countries', available at www.oecd.org/dataoecd/2/32/1917911.pdf, last accessed 11/03/2010

Frost, H., Lanters, R., Smit, J. and Sparre, P. (1996) *An Appraisal of the Effects of the Decommissioning Scheme in the Case of Denmark and the Netherlands*, DIFER Report Series 16/96, South Jutland, Denmark, South Jutland University Press

Fundacion Patagonia Natural (2005) 'Fisheries in Argentine Patagonia', available at www.costapatagonica.org.ar/proyecto_pesca.htm, last accessed 11/03/2010

Garrod, B. and Whitmarsh, D. (1991) 'Overcapitalization and Structural Adjustment: Recent Developments in UK Fisheries', *Marine Policy*, Vol. 15, July

Gascuel D., Laurans, M., Sidibe, M. and Barry, M.D. (2004) 'Diagnostic comparatif de l'état des stocks & évolutions d'abondance des ressources démersales dans les pays de la C.S.R.P.', in Chavance, P., Ba., M., Gascuel, D., Vakily, M., Pauly, D. (eds) *Pêcheries maritimes, écosystèmes & sociétés en Afrique de l'Ouest: Un demi-siècle de changement* [Marine Fisheries, Ecosystems and Societies in West Africa: Half a Century of Change], actes du symposium international, Dakar (Sénégal), 24–28 June 2002, Bruxelles, Office des publications officielles des Communautés européennes, pp.208–222

Gates, J., Holland, D. and Gudmundsson, E. (1997a) 'Theory and Practice of Fishing Vessel Buy-back Program', in *Subsidies and Depletion of World Fisheries: Case Studies*, Burns, S. (ed.), Washington, D.C., WWF's Endangered Seas Campaign

Gates, J., Holland, D. and Gudmundsson, E. (1997b) 'Theory and Practice of Fishing Vessel Buy-back Program: Appendix', unpublished paper for the Endangered Seas Campaign, World Wildlife Fund

Gauvin, J. R., Ward, J. M. and Burgess, E.E. (1994) 'Description and Evaluation of the Wreckfish Fishery under Individual Transferable Quotas', *Marine Resource Economics*, Vol. 9, pp.99–118

Geen, G. and Nayar, M. (1988) 'Individual Transferable Quotas in the Southern Bluefin Tuna Fishery: An Economic Appraisal', *Marine Resource Economics*, Vol. 5, pp.365–387

Giguelay, T. and Piot-Lepetit, I. (2000) *Decommissioning Schemes in the French Fishing Industry: An Evaluation of the Performance of a Public Policy*, paper for the International Atlantic Economic Conference, Charleston, Va., 5–18 October 2000

GIRMaC (2007) 'Revue des dépenses publiques et l'analyse de la rentabilité, économique du secteur de la pêche', Final Report, November 2007

Gorez, B. (2005) 'EU-ACP Fisheries Agreements, Coalition of Fair Fisheries Agreements (CFFA)', Policy Study, Brussels

Gouin, S. (2000) 'Seafood Products Enhancement: the Case of Artisanal Fishing in France', paper presented at the Institute of Fisheries and Trade, University of Corvallis (Oregon 2000) (www.oregonstate.edu/dept/IIFET/2000/papers/gouin.pdf, last accessed 18/03/2010)

Grafton, R.Q. (1996) 'Individual Transferable Quotas: Theory and Practice', *Reviews in Fish Biology and Fisheries*, Vol. 6, pp.109–112

Grafton, R.Q., Squires, D. and Kirkley, J. E. (1996) 'Private Property Rights and Crises in World Fisheries: Turning the Tide?', *Contemporary Economic Policy*, Vol. 14, pp.90–99

Guyader, O., Daures, F. and Fifas, S. (2000) 'A Bio-economic Analysis of Buyback Programs: Application to a Limited Entry Scallop French Fishery', paper presented at the Biennial Conference of the International Institute of Fisheries Economics and Trade, Oregon State University, Corvallis, Oregon, 10–15 July 2000

Hannesson, R. (1991) *Economic Support of the Fishing Industry: Effects on Efficiency and Trade*, AGR/FI/EG(91)1, Paris, OECD Directorate for Food, Agriculture and Fisheries, Committee on Fisheries

Hannesson, R. (1992) 'Fishery Management in Norway', in Loayza, E.A. (ed.) *Managing Fishery Resources*, World Bank Discussion Paper No. 217, Washington, D.C., World Bank

Hannesson, R. (1996) *Fisheries Mismanagement: the Case of Atlantic Cod*, Oxford, Fishing News Books

Hannesson, R. (2001) *Effects of Liberalizing Trade in Fish, Fishing Services and Investment in Fishing Vessels*, OECD Papers Offprint No. 8, Paris OECD

Hatcher, A. and Pascoe, S. (1998) *Charging in the UK Fishing Industry: A Report to the Ministry of Agriculture, Fisheries and Food*, Report 49, Portsmouth, UK, (CEMARE), University of Portsmouth

Hayes, J., Geen, G. and Wilks, L. (1986) *Beneficiaries of Fisheries Management*, discussion paper 86.1, Canberra, Bureau of Agricultural Economics, Australian Government Publishing Service

HELCOM (n.d.) 'Summary of the four main segments of the HELCOM Baltic Sea Action Plan, detailing goals, objectives, and actions', available at www.helcom.fi/BSAP/ActionPlan/en_GB/SegmentSummary/#maritime, last accessed 11/03/2010

Hersoug, B. (2005) *Closing the Commons: Norwegian Fisheries from Open Access to Private Property*, Eburon, Delft, The Netherlands

Hesselmark, O. (2003) 'Why fish quota levies should be increased', *Namibia Economist*, 28 February

Holland, D. and Sutinen, J. (1998) 'Draft Guidelines on Fishing Capacity', unpublished background paper for FAO Technical Workshop on Capacity Management, La Jolla, California, 14–18 April 1998

Holm P., Raanes, S.A. and Hershog, B. (1996) 'Political Attributes of Right-Based Management Systems: The Case of Individual Vessel Quotas in the Norwegian Coastal Cod Fishery', in Symes, D. (ed.) *Property Rights and Regulatory Systems in Fisheries*, Oxford, Fishing News Books

Hunt, C. (1997) 'Management of the South Pacific Tuna Fishery', *Marine Policy*, Vol. 21/2, pp.155–171

IATTC (2008) 'Document (79-07) RFMOs Performance Revision', 79 Meeting – La Jolla, California, US, 6–7 November 2008

Ibsen T. (2000) 'Iceland's Proposal about Fisheries Subsidies', in Nordquist, M.H. and Moore, J.N. (eds) *Current Fisheries Issues and the Food and Agriculture Organization of the United Nations*, Klewer Law International, The Netherlands

ICSF (2006) 'Untangling Subsidies, Supporting Fisheries: The WTO Fisheries Subsidies Debate and Developing Country Priorities', ICSF Occasional Paper, prepared for ICTSD, John Kurien, Chennai, India

ICTSD (1997) Osakwe, C., 'Implementing MEAs: Trade and Positive Measures', *Bridges Monthly Review*, Vol. 1, No. 4, p.7, International Centre for Trade and Sustainable Development, Geneva

ICTSD (2000) Dommen, C., *Fish Scales: Fisheries, International Trade and Sustainable Development: A Reference Guide to Legal Frameworks*, draft manuscript, International Centre for Trade and Sustainable Development, Geneva

ICTSD (2001) Cosbey, A.,'The WTO and PPMs: Time to drop a taboo' *Bridges Monthly Review*, No.1–3, January–April 2001, International Centre for Trade and Sustainable Development, Geneva

ICTSD (2006a) *Fisheries, International Trade and Sustainable Development*: Policy Discussion Paper, ICTSD Natural Resources, International Trade and Sustainable Development Series. International Centre for Trade and Sustainable Development, Geneva, Switzerland.

ICTSD (2006b) Asche, F. and Khatun, F. (2006) *Aquaculture: Issues and Opportunities for Sustainable Production and Trade*, ICTSD Natural Resources, International Trade and Sustainable Development Series Issue Paper No. 5, International Centre for Trade and Sustainable Development, Geneva, Switzerland.

ICTSD (2006c) Sumaila, U.R. and Keith, H. Regulating Fisheries Subsidies – A Role for RFMOs?', *Bridges Monthly Review*, No. 2, pp.21–22, International Centre for Trade and Sustainable Development, Geneva.

ICTSD (2006d) Mbithi Mwikya, S. (2006) *Fisheries Access Agreements: Trade and Development Issues*, ICTSD Natural Resources, International Trade and Sustainable Development Series Issue Paper No. 2, International Centre for Trade and Sustainable Development, Geneva, Switzerland.

ICTSD (2006e) Roheim, C. A. and Sutinen, J. (2006) *Trade and Marketplace Measures to Promote Sustainable Fishing Practices*, ICTSD Natural Resources, International Trade and Sustainable Development Series Issue Paper No. 3, International Centre for Trade and Sustainable Development and the High Seas Task Force, Geneva, Switzerland.

IEEP (2003) *Combating Subsidies, Developing Precaution: Institutional Interplay and Responsible Fisheries*, Project Deliverable No. D 9, Final Draft, The Fridtjof Nansen Institute and Institute for European Environmental Policy, Stokke, O.S. and Coffey, S.

IISD (1998) *Perverse Subsidies: Tax $$ Undercutting Our Economies and Environments Alike*, International Institute for Sustainable Development, Winnipeg, Canada

IISD (2007) 'Summary of the Sixth Informal Consultations of States Parties to the UNFSA: 23–24 April 2007', in *Earth Negotiations Bulletin*, Vol. 7, No. 62, April, International Institute for Sustainable Development, available at www.iisd.ca/download/pdf/enb0762e.pdf, last accessed 11/03/2010

IMF (2001) 'IMF approves Third annual PRGF Loan for Senegal', Press Release No. 01/5, Washington

Israel, D.C. and Roque, R.M.G.R. (2000) *Analysis of Fishing Ports in the Philippines*, Legaspi Village, Philippines, Philippine Institute for Development Studies

ITA (1986) 'Etude et identification des caractéristiques techniques et socio-économiques de l'amélioration du secteur de traitement artisanal du poisson au Sénégal', Phase I du Projet ITA – ALTERSIAL de développement de la technologie au Sénégal

IUCN (2003) De Fontaubert, C. and Lutchman, I. *Achieving Sustainable Fisheries: Implementing the New International Legal Regime*, Gland, Switzerland

Japan MAFF (1994) 'Draft Ninth Long-Term Improvement Plan for Fishing Ports Announced', MAFF Update, No. 68, March 4, available at www.maff.go.jp/mud/68.html, last accessed 11/03/2010

Jensen, F. and Vestergaard, N. (2000) 'Moral Hazard Problems in Fisheries Regulation: The Case of Illegal Landings', *Resource and Energy Economics*, Volume 24, Number 4, 1 November 2002 , pp. 281–-299(19)

Jentoft, S. and Mikalsen, K.H. (1987) 'Government Subsidies in Norwegian Fisheries: Regional Development or Political Favouritism?', *Marine Policy*, July, pp.,217–228

Johnston, N. (1996) 'Economics of fisheries access agreements: perspectives on the EU-Senegal Case', discussion paper 96–02, International Institute for Environment and Development, London

Jorgensen, H. and Jensen, C. (1999) 'Overcapacity, Subsidies and Local Stability', in Hatcher, Aaron and Robinson, Kate (eds) *Overcapacity, Overcapitalisation and Subsidies in European Fisheries*, proceedings of the first workshop held in Portsmouth, UI, 28–30 October, Portsmouth, UK, CEMARE, pp.239–252

Kaczynski, V.M. and Fluharty, D.L. (2002) 'European policies in West Africa: who benefits from fisheries agreements?', *Marine Policy*, Vol. 26, pp.75–93

Karagiannakos, A. (1996) 'Total Allowable Catch and Quota Management in the European Union', *Marine Policy*, Vol. 20/3, pp.335–372

Kebe, M. and Dème, M. (1996) 'Filière pêche artisanale: rentabilité, exportations et consommation locale. Atelier de restitution « Impact de la dévaluation sur les revenus et la sécurité alimentaire au Sénégal», Novotel-Dakar, 23 février 1996, ISRA-PASE/ Institut du Sahel-PRISAS, DT 96–02

Khan, A.S., Sumaila, U.R., Watson, R., Munro, G. and Pauly, D. (2006) 'The Nature and Magnitude of Global Non-Fuel Fisheries Subsidies', in Sumaila, U.R. and Pauly, D. (eds) *Catching More Bait: A Bottom-up Re-estimation of Global Fisheries Subsidies*, Fisheries Centre Research Reports, Vol. 14, No. 6, pp.5–37. Fisheries Centre, the University of British Columbia, Vancouver

Kitts, A., Thunberg, E. and Sheppard, G. (1999) 'The Northeast Groundfish Fishery Buyout Program', in Clark, S. H. (ed.) *Status of Fishery Resources of the Northeastern US for 1998*, Woods Hole, Mass., Resource Evaluation and Assessment Division, Northeast Fisheries Science Center

Kurien, J. (2004) *Responsible Fish Trade and Food Security: Report of the Study on the Impact of International Trade in Fishery Products on Food Security*, United Nations Food and Agriculture Organization and the Norwegian Ministry of Foreign Affairs, Rome

Lindebo, E. (1999) *Fishing Capacity and EU Fleet Adjustment*, SJFI Working Paper No. 19, Frederiksberg, Danish Institute of Agricultural and Fisheries Economics

Lindebo, E. (2000) *Capacity Development of the EU and Danish Fishing Fleets*, Frederiksberg, Danish Institute of Agricultural and Fisheries Economics

Macy, W.K. and Brodziak, J.K.T. (2001) 'Seasonal maturity and size at age of Loligo pealeii in waters of southern New England', *ICES Journal of Marine Science*, Vol. 58, p.4

Mathew, S. (2003) 'Fishing for Subsidies', *Samudra*, Vol. 36, November 2003, Chennai

Mayo, R. and O'Brien, L. (2000) *Atlantic Cod: In Status of Fisheries Resources off Northeastern United States*, Woods Hole, Mass., Resource Evaluation and Assessment Division, Northeast Fisheries Science Center, available at www.nefsc.noaa.gov/ sos/spsyn/pg/cod/, last accessed 11/03/2010

McCleod, R. (1996) *Market Access Issues for the New Zealand Seafood Trade*, Wellington, N.Z., New Zealand Fishing Industry Board

MEM (2004) Ajustement des capacités de pêche de la pêche industrielle au Sénégal, Dakar, Sénégal, 67 p.

MEMTM (2006) Programme d'ajustement des capacités de pêche au Sénégal, Dakar, Sénégal, 112 p.

McGoodwin, J.R. (1990) *Crisis in the World's Fisheries*, Stanford, CA, Stanford University Press

Michaud, P. (2003) 'Experience from the bilateral fisheries access agreement, impact on the economy and implications for Seychelles of the outcome of the WTO medication on the case of tuna between the EU and Thailand and the Philippines', Seminar on ACP-EU fisheries relations: towards a greater sustanability ACP, Brussels, Belgium; 7–9 April 2003, Victoria: Seychelles Fishing Authority

Mikalsen, K.H. and Jentoft, S. (2003) 'Limits to Participation? On the History, Structure and Reform of Norwegian Fisheries Management', *Marine Policy*, Vol. 27, pp.397–407

Milazzo, M. J. (1997) *Reexamining Subsidies in World Fisheries*, Office of Sustainable Fisheries, National Marine Fisheries Service, National Oceanic and Atmospheric Administration, US Department of Commerce, Washington D.C.

Miller, J.W. (2007) 'Global Fishing Trade Depletes African Waters', *Wall Street Journal*, 18 July

Morón J. (2006) *Tuna Fishing Capacity: Perspective of Purse Seine Fishing Industry on Factors Affecting It and Its Management*, available at www.iattc.org/PDFFiles2/P15-Tuna-Fishing-Capacity.pdf, last accessed 17/03/2010.

MSC (2002) *Principles and Criteria for Sustainable Fishing*, Marine Stewardship Council, London

Munro, G.R. (1999) 'The Economics of Overcapitalization and Fishery Resource Management: A Review', in Hatcher, Aaron and Robinson, Kate (eds) *Overcapacity, Overcapitalisation and Subsidies in European Fisheries*, proceedings of the first workshop held in Portsmouth, UI, 28–30 October, Portsmouth, UK, CEMARE, pp.7–23

Munro, G. and Sumaila, U.R. (1999) 'Subsidies and their potential impact on the management ecosystems of the North Atlantic', in: Pitcher T., Sumaila, U.R., Pauly, D. (eds.) (2002) *Fisheries Impacts on North Atlantic Ecosystems: Evaluations and Policy Explorations* University of British Columbia Fisheries Centre Research Reports 9(5). 10–27

Munro, G. and Sumaila, U.R. (2002) 'The impact of subsidies upon fisheries management and sustainability: The case of the North Atlantic', *Fish and Fisheries*, Vol 3

Myrstad, B. (1996) *The Norwegian Fishing Industry*

Nautilus Consultants (1997) *The Economic Evaluation of the Fishing Vessels (Decommissioning) Schemes*, report on behalf of the United Kingdom Fisheries Department, Edinburgh, Nautilus Consultants

Nielsen, J.R. (1992) 'Structural Problems in the Danish Fishing Industry: Institutional and Socio-economic Factors as Barriers to Adjustment', *Marine Policy*, September

Nielsen, J.R. and Joker, L. (1995) *Fisheries Management and Enforcement in Danish Perspective: Paper prepared for the Fifth Annual Conference of the International Society for the Study of Common Property*, Bodo, Norway, 24–28 May 1995

NOAA (1999) *Federal Fisheries Investment Taskforce: Report to US Congress. Part 1: Background and Conceptual Basis*, Baltimore, M.D.

Nordstrom, H., and Vaughan, S. (1999) 'Trade and Environment', *Special Studies* 4, Geneva, World Trade Organization

NRI (2006), Kleih, U., Greenhalgh, P., Marter, D., *Sustainability Impact Assessments of Proposed WTO Negotiations*, final report of the Fisheries Sector Study, Natural Resources Institute, University of Greenwich, UK with Nigel Peacock, NAP Fisheries, UK. In association with: Impact Assessment Research Centre, Institute for Development Policy and Management University of Manchester

OECD (1993a) *Economic Assistance to the Fishing Industry: Observations and Findings*, Organization for Economic Cooperation and Development, Paris

OECD (1993b) *Committee for Fisheries. Economic Assistance to the Fishing Industry: Observations and Finding*, AGR/FI(93)11/Rev. 1, Organization for Economic Cooperation and Development, Paris

OECD (1996) *Integrating Environment and Economy: Progress in the 1990s*, Organization for Economic Cooperation and Development, Paris

OECD (1997a) *Toward Sustainable Fisheries: Economic Aspects of the Management of Living Resources*, Organization for Economic Cooperation and Development, Paris

OECD (1997b) *Eco-Labelling: Actual Effects of Selected Programmes*, Organization for Economic Cooperation and Development, Paris

OECD (2000a) *Transition to Responsible Fisheries, Government Financial Transfers and Resource Sustainability: Case Studies*, AGR/FI(2000)10/FINAL, Organization for Economic Cooperation and Development, Paris

OECD (2000b) *Transition to Responsible Fisheries: Economic and Policy Implications*, Organization for Economic Cooperation and Development, Paris

OECD (2000c) *The Impact on Fisheries Resource Sustainability of Government Financial Transfers*, Organization for Economic Co-operation and Development, Paris

OECD (2000d) *Briefing on the OECD Study on Government Financial Transfers and Resource Sustainability: Further Work on Fisheries Trade, Resource Sustainability, and Government Financial Transfer*, presented at the WTO Committee on Trade and Environment, November 2000, Organization for Economic Cooperation and Development, Paris

OECD (2001) *Review of Fisheries in OECD Countries: Policies and Summary Statistics*, Organization for Economic Co-operation and Development, Paris

OECD (2003) *Liberalising Fisheries Markets: Scope and Effects*, Organization for Economic Cooperation and Development, Paris

OECD (2004) *Policy Coherence for Development in Fisheries*, AGR/FI(2004)3/REV2, Organization for Economic Co-operation and Development, Paris

OECD (2005a) *Environmentally Harmful Subsidies: Challenges for Reform*, Organization for Economic Cooperation and Development, Paris

OECD (2005b) *Subsidies: A Way Towards a Sustainable Fisheries?*, Policy Brief, Organization for Economic Cooperation and Development, Paris

OECD (2006) *Financial Support to the Fisheries Sector. Organization for Economic Cooperation and Development*, Organization for Economic Cooperation and Development, Paris

OECD (2008) *Review of Fisheries in OECD Countries: Policies and Summary Statistics*, Organization for Economic Cooperation and Development, Paris

OECD (2009) *Reducing Fishing Capacity: Guidelines for the Design and Implementation of Decommissioning Schemes*, Organization for Economic Cooperation and Development, Paris

OECD (2010) *Fisheries Policy Reform: National Experiences*, OECD, Paris (forthcoming)

Ostrom, E. (1990) *Governing the Commons: The Evolution of Institutions for Collective Action*, Cambridge University Press

Pascoe, S. and Coglan, L. (2002) 'Contribution of Unmeasurable Factors to the Efficiency of Fishing Vessels', *American Journal of Agricultural Economics*, Vol. 84/3, pp.585–597

Pascoe, S., Tingley, D. and Mardle, S. (2002) *Appraisal of Alternative Policy Instruments to Regulate Fishing Capacity*, Final Report, Economic Research Project funded by the UK Department for Environment, Food and Rural Affairs, University of Portsmouth, available at www.statistics.defra.gov.uk/esg/reports/capman/finalrep.pdf, last accessed 11/03/2010

Petersen, L. (2001) 'Governance of the South Pacific Tuna Fishery', paper presented to the 30th Annual Conference of Economists, Perth, Australia, 23–26 September 2001

Pieters, J. (2002) 'What Makes a Subsidy Environmentally Harmful: Developing a Checklist Based on the Conditionality of Subsidies', paper prepared for the OECD Workshop on Environmentally Harmful Subsidies, 7–8 November 2002, Paris

Poole, E. (2000) 'Income Subsidies and Incentives to Overfish: Microbehavior and Macroresults: Proceedings of the Tenth Biennial Conference of the International Institute of Fisheries Economics and Trade Presentations', Corvallis, Oregon, IIFET, Oregon State University

Reijnders, L. (1990) 'Subsidies and the Environment', in Gerritse, R. (ed.) *Producer Subsidies*, London and New York, Pinter

Rome Consensus on World Fisheries (1995) Adopted by the FAO Ministerial Conference on Fisheries, Rome, 14–15 March 1995

Runolfsson, B. and Arnason, R. (2001) 'Changes in Fleet Capacity and Ownership of Harvesting Rights in the Icelandic Fisheries', in Shotton, R. (ed.) *Case Studies on the Effect of Transferable Property Rights on Fleet Capacity and Concentration of Quota Ownership*, FAO Fisheries Technical Paper No. 412, Rome, FAO

Salz, P. (1991) *The European Atlantic Fisheries*, The Hague, Agricultural Economics Research Institute

Schrank, W.E. (1997) 'The Newfoundland Fishery: Past, Present and Future, in *Subsidies and Depletion of World Fisheries: Case Studies*, Godalming, UK, WWF's Endangered Seas Campaign

Schrank, W.E. (1998) 'The Failure of Canadian Seasonal Fisherman's Unemployment Insurance Reform During the 1960s and 1970s', *Marine Policy*, Vol. 22, No. 1, pp.67–81

Schrank, W.E. (2000) *Subsidies for Fisheries: A Review of Concepts*, in papers presented at the FAO Expert Consultation on Economic Incentives and Responsible Fisheries, FAO Fisheries Report No. 638, Supplement. Rome

SFLP Dakar Declaration (2001) 'Declaration: Statement from the Workshop on Problems and Prospects for Developing Artisanal Fish Trade in West Africa', Centre Social Derklé, Dakar, Sénégal, 30 May to 1st June 2001, Sustainable Fisheries Livelihoods Programme, available at www.icsf.net/icsf2006/jspFiles/wif/english/pdfs/West%20Africa%202001.pdf, last accessed 11/03/2010

Silbert, J. (1999) 'Status of South Pacific Tuna Stocks', *Pelagic Fisheries Research Program Newsletter*, University of Hawaii at Manoa, Honolulu 4/4, October–December 1999

Smith, C. and Hanna, S.S. (1990) 'Measuring Fleet Capacity and Capacity Utilization', *Canadian Journal of Fisheries and Aquatic Science*, Vol. 47, p.2086

Sporrong, N., Coffey, C. and Bevins, K. (2002) *Fisheries Agreements With Third Countries – Is the EU Moving Towards Sustainable Development?*, Institute for European Environmental Policy for WWF, London

Squires, D. and Kirkley, J. (1991) 'Production Quota in Multi-product Pacific Fisheries', *Journal of Environmental Economics and Management*, Vol. 21/2, pp.109–126

Steenblik, R. (2009), Background paper prepared for Panel 5: Subsidies and Pricing, of the Conference Global Challenges at the Intersection of Trade, Energy and the Environment 22nd and 23 October 2009, organised by the Centre for Trade and Economic Integration (CTEI) at the Graduate Institute of International and Development Studies, Geneva, in collaboration with the World Trade Organization.

Stilwell J., Samba, A., Laloë, F. and Failler, P. (2009) 'Sustainable Development Consequences of European Union Participation in Senegal's Maritime Fishery', *Marine Policy* Volume 34 (2010), pages 616–623

Stone, C. (1997) 'Too many fishing boats, too few fish: Can trade laws trim subsidies and restore the balance in world fisheries?', *Ecology Law Quarterly*, Vol. 24, pp.505–544

Sub-Regional Fisheries Commission and United Nations Development Program (1997) 'Sub-Regional Fisheries Roundtable Conference', Praia, Cape Verde, SRFC and UNDP

Sumaila, U.R. (2003) 'A fish called Subsidy', *Science and the Environment*, Vol. 12, No. 12, available at www.downtoearth.org.in/fullprint.asp (last accessed 25 August 2006)

Sumaila, U.R. and Pauly, D. (2006) 'Catching more bait: a bottom-up re-estimation of global fisheries subsidies', *Fisheries Centre Research Reports*, Vol. 14, No. 6, Fisheries Centre, University of British Columbia, Vancouver

Sumaila, U., Teh, L., Watson, R., Tyedmers, P. and Pauly, D. (2006) 'Fuel subsidies to fisheries globally: Magnitude and impacts on resource sustainability', in Sumaila, U.R. and Pauly, D. (eds) 'Catching more bait: A bottom-up re-estimation of global fisheries subsidies', Fisheries Centre Research Reports, Vol. 14, No.6, pp.38–48. Fisheries Centre, University of British Columbia, Vancouver

Sutinen, J., Rieser, A. and Gauvin, J.R. (1990) 'Measuring and Explaining Noncompliance in Federally Managed Fisheries', *Ocean Development and International Law*, Vol. 21, pp.335–372

Thomson, A.M. (2000) 'Sustainable livelihoods approaches at the policy level', paper prepared for FAO e-conference and forum on operationalising participatory ways of applying a sustainable livelihoods approach, Rome: FAO

Thorpe, A., Reid, C., van Anrooy, R. and Brugere, C. (2004) 'Integrating fisheries into the national development plans of Small, Island Developing States (SIDS): Ten Years on from Barbados', *Natural Resources Forum*, vol. 29(1): pages 51–69.'

Thunberg, E. (2000) 'Latent Fishing Effort and Vessel Ownership Transfer in the Northeast US Groundfish Fishery', paper presented at the Biennial Conference of the International Institute of Fisheries Economics and Trade, Oregon State University, Corvallis, Oregon

Touray, I. (1996) 'Gender Issues in the Fisheries Sector and Effective Participation', paper for the 'Workshop on Gender Roles and Issues in Artisanal Fisheries in West Africa', Gender Trainer in the Management Development Institute, Lomé, 11–13 December 1996

Townsend, R.E. (1985) 'On Capital Stuffing in Regulated Fisheries', *Land Economics*, Vol. 61/2, pp.195–197

Townsend, R.E. (1992) 'A Fractional Licensing Program for Fisheries', *Land Economics*, Vol. 68, pp.185–190

Trondsen, T. (1999) 'Market Orientation in Fisheries Management: Catching the Market Value by Seasonal Quota Auctions (SQA)', paper presented to Management Institutions and Governance Systems in European Fisheries, Vigo, Spain, 28–30 October 1999

UNDP (1998) *Investing for Sustainable Livelihoods. United Nations Development Programme*, New York, Navie, L., Gilman, J. and Singh, N.

UNDP (2000) *A Better Life...Thanks to Nature; Combat Poverty While Improving the Environment: Practical Recommendations*, Poverty Environment Initiative', United Nations Development Programme, New York

UNDP (2001) *Overcoming Human Poverty*, UNDP Poverty Report 2000, New York

UNEP (1998) *Fisheries Subsidies and Overfishing: Towards a structured Discussion*, United Nations Environment Programme, Geneva, Porter, G.

UNEP (2000) 'Study of the costs and incomes of small-scale fishing units', in Dahou, K. and Dème, M. (eds) *Environmental and Socio-economic Impact of Trade Liberalisation and Policies for Sustainable Management of Natural Resources: A Case Study on Senegalese's Fisheries Sector*, United Nations Environment Programme, Geneva. Dème, M.

UNEP (2001a) *Ecosystem Based Management of Fisheries – Opportunities and Challenges for Coordination between Marine Regional Fisheries Bodies and Regional Seas Conventions*, UNEP Regional Seas Reports and Studies No. 175, United Nations

Environment Programme, Geneva, available at www.unep.org/regionalseas/Publications/Reports/RSRS/pdfs/rsrs175.pdf, last accessed 11/03/2010

UNEP (2001b) *Economic Reforms, Trade Liberalization and the Environment: A Synthesis of UNEP Country Projects*, United Nations Environment Programme, Geneva

UNEP (2002a) *Integrated Assessment of Trade Liberalization and Trade Related Policies. A Country Study on Fisheries Sector in Senegal*, UNEP/ETB/2002/10, United Nations Environment Programme, Geneva

UNEP (2002b) *Fisheries Subsidies and Overfishing: Toward a Structured Discussion*, paper prepared for the United Nations Environment Programme Fisheries Workshop, United Nations Environment Programme, Geneva, Porter, G.

UNEP (2004a) *Analyzing the Resource Impact of Fisheries Subsidies: A Matrix Approach United Nations Environment Programme*, United Nations Environment Programme, Geneva, Porter, G.

UNEP (2004b) *Incorporating Resource impact into Fisheries Subsidies Disciplines: Issues and Options*, United Nations Environment Programme, Geneva, Porter, G.

UNEP (2004c) *Fisheries Subsidies and Marine Resource Management: Lessons Learned from Studies in Argentina and Senegal*, United Nations Environment Programme, Geneva, Maria Onestini, Karim Dahoo

UNEP (2004d) *Policy Implementation and Fisheries Resource Management: Lessons from Senegal Fisheries and the Environment*, United Nations Environment Programme, Geneva, Senegal Fisheries Ministry and ENDA

UNEP (2004e) 'Workshop on Fisheries Subsidies and Sustainable Fisheries Management Documentation', United Nations Environment Programme, Geneva, available at www.unep.ch/etb/events/FishMeeting2004.php, last accessed 11/03/2010

UNEP (2004f) 'Fisheries Subsidies and Sustainable Fisheries Management', *UNEP/ETB Bulletin*, June 2004, United Nations Environment Programme, Geneva, available at www.unep.ch/etb/publications/etbBriefs/bulletinPeche.pdf, last accessed 11/03/2010

UNEP (2004g) 'Experiences with Subsidies and Fisheries Management: The Case of EU-ACP Fisheries Access Agreements', prepared for UNEP Workshop on Fisheries Subsidies and Sustainable Fisheries Management, 26–27 April 2004, United Nations Environment Programme, Geneva

UNEP (2005a) *Reflecting Sustainable Development and Special and Differential Treatment for Developing Countries in the Context of New WTO Fisheries Subsidies Rules: Some Issues and Options*, United Nations Environment Programme, Geneva, Yu V. P. and Fonseca, D.

UNEP (2005b) *Artisanal Fishing: Promoting Poverty Reduction and Community Development through New WTO Rules on Fisheries Subsidies*, Economics and Trade Branch, An option and issue paper, United Nations Environment Programme, Geneva, Schorr, D.

UNEP (2005c) 'Chair's Summary of UNEP Roundtable: Promoting Development and Sustainability in Fishery Subsidies Disciplines', 30 June 2005, United Nations Environment Programme, Geneva, available at www.unep.ch/etb/events/Events2005/midTermReview/unepChairsSummary.pdf, last accessed 11/03/2010

UNEP (2006a) *Evaluation de l'Impact de la Libéralisation du Commerce : Une étude de cas sur le secteur des pêches de la République Islamique de Mauritanie*, United Nations Environment Programme, Geneva, Failler, P.

UNEP (2006b) 'Chair's Summary of UNEP-ICTSD-WWF Workshop on Development and Sustainability in the WTO Fisheries Subsidies Negotiations: Issues and Alternatives', 11 May 2006, United Nations Environment Programme, Geneva, available at www.unep.ch/etb/events/pdf/Chair%27s%20summary_11May.pdf, last accessed 11/03/2010

UNEP (2008a) *Towards Sustainable Fisheries Access Agreements: Issues and Options at the World Trade Organization*, United Nations Environment Programme, Geneva, Orellana, M.A.

UNEP (2008b) *Fisheries Subsidies: A Critical Issue for Trade and Sustainable Development at the WTO: An Introductory Guide*, United Nations Environment Programme, Geneva

UNEP (2009) *Certification and Sustainable Fisheries*, United Nations Environment Programme, Geneva, Macfadyen, G. and Huntington, T.

UNEP (n.d) 'The Regional Seas Programme – About', available at www.unep.org/regionalseas/about/default.asp, last accessed 11/03/2010

UNEP-CPPS (2009) *The Impact of Fisheries Subsidies on the Ecuadorian Tuna's Sustainability and Trade*, Bowen, I.P. (in Spanish), Guayaquil, Ecuador

UNEP-WWF (1997) Porter, G. 'The Euro-African Fishing Agreements: Subsidizing Overfishing in African Waters', Background Paper for UNEP/WWF Workshop, in Burns S. (ed.) *Subsidies and Depletion of World Fisheries: Case Studies*, World Wildlife Fund, Washington, DC. 1997, pp.7–33.

UNEP-WWF (2005) 'Time to draw in the net on fishing subsidies', Opinion Editorial, Leape, J. and Toepfer, K., available at www.unep.ch/etb/events/Events2005/pdf/WWF-UNEP_opinion_editorial.pdf, last accessed 12/03/2010

UNEP–WWF (2007a) 'Symposium on Disciplining Fisheries Subsidies: Incorporating Sustainability at the WTO & Beyond documentation', available at www.unep.ch/etb/events/2007fish_symposium.php, last accessed 11/03/2010

UNEP-WWF (2007b) *Sustainability Criteria for Fisheries Subsidies. Options for the WTO and Beyond*, United Nations Publication, Geneva, Schorr, D.K. and Caddy, J.F.

United Nations (1992a) 'United Nations General Assembly Resolution S/19–2', UN Doc. No. A/RES/S-19/2, 28 June 1992, available at www.un.org/documents/ga/res/spec/aress19-2.htm, last accessed 11/03/2010

United Nations (1992b) 'Rio Declaration on Environment and Development, Annex 1' in *Report of the United Nations Conference on Environment and Development*, Rio de Janeiro, 3–14 June 1992, available at www.un.org/documents/ga/conf151/aconf15126-1annex1.htm, last accessed 11/03/2010

United Nations (2002) 'World summit on sustainable development plan of implementation', in *Report of the World Summit on Sustainable Development*, Johannesburg, South Africa, 26 August–4 September 2002, available at www.un.org/esa/sustdev/documents/WSSD_POI_PD/English/POIChapter4.htm

United Nations (n.d.) *Atlas of the Oceans* (online atlas), available at www.oceansatlas.org/index.jsp, last accessed 11/03/2010

United Nations (n.d.) *United Nations Convention of the Law of the Sea (UNCLOS)*, United Nations, New York, available at www.un.org/Depts/los/convention_agreements/convention_overview_convention.htm, last accessed 11/03/2010

United Nations Division for Ocean Affairs and the Law of the Sea (n.d.) *United Nations Convention on the Law of the Sea of 10 December 1982*, overview and full text, available at www.un.org/Depts/los/convention_agreements/convention_overview_convention.htm, last accessed 11/03/2010

US Energy Information Administration (n.d.) 'Petroleum Navigator', (online price history) available at www.tonto.eia.doe.gov/dnav/pet/hist/wtotworldw.htm (last accessed 20 August 2009)

US General Accounting Office (GAO) (2000) 'Entry of Fishermen Limits Benefits of Buyback Programs: Report to House Committee on Resources', GAO/RCED–00–120, available at www.gao.gov/archive/2000/rc00120.pdf, last accessed 11/03/2010

White House (1999) 'The Clinton Administration Agenda For The Seattle WTO' press release of 24 November, available at www.clinton4.nara.gov/WH/New/html/19991124.html, last accessed 11/03/2010

van Bogaert, O. (2003) 'Senegalese fishermen mourn loss of species' *Science in Africa* (online magazine), June–July 2003, available at www.scienceinafrica.co.za/2003/june/fish.htm, last accessed 11/03/2010

von Moltke, A. (2007) *Fisheries Subsidies, in Trade and Environment – A Resource Book*, ICTSD, IISD, The Ring, Geneva, Winnipeg, London

von Moltke, A. (2009) 'WTO Negotiations on Fisheries Subsidies: A Critical Issue for Commonwealth Countries', in Bourne, R. and Collins, M. (eds.) (2009) *From Hook to Plate: The State of Marine Fisheries. A Commonwealth Perspective*, London: Commonwealth Foundation.

Walden, J.B. and Kirkley, J. (2000) 'Measuring Capacity in the New England Otter Trawl Fleet', paper presented at the Biennial Conference of the International Institute of Fisheries Economics and Trade, Oregon State University, Corvallis, Oregon, 10–15 July 2000

Watson, R., Kitchingman, A., Gelchu, A. and Pauly, D. (2004) 'Mapping Global Fisheries: Sharpening Our Focus in Fish and Fisheries', in Watson, R. and Pauly, D. 'Systematic distortions in world fisheries catch trends', *Nature*, Vol. 414, pp.534–536

Wang, S.D. and Tang, V.H. (1996) 'The Surf Claim ITQ Management: An Evaluation', in *Our Living Oceans: Economic Status of US Fisheries 1996*, Silver Spring, Md., National Marine Fisheries Service

World Bank (1998) Milazzo, M., *Subsidies in World Fisheries – A Reexamination*, World Bank Technical Paper No. 406, April 1998, The World Bank. Washington, D.C.

World Bank (2003) *Madagascar Diagnostic Trade Integration Study*, The World Bank, Washington, D.C.

World Bank (2004a) Roheim, C. *Trade Liberalisation in Fish Products: Impacts on Sustainability of International Markets and Fish Resources*, Washington, D.C.

World Bank (2004b) *Saving Fish and Fisheries: Towards Sustainable and Equitable Governance of Global Fishing Sector*, World Bank Report No. 29090-GLB, Washington, D.C.

World Bank/FAO (2008) *The Sunken Billions: The Economic Justification for Fisheries Subsidies Reform*, The World Bank, Washington, D.C.

Williams, F. (1999) 'Nations Call for Fishing Subsidy Ban', *Financial Times*, 2 August 1999

Williams, F. (2004) 'Japan changes tack on fish subsidies', *Financial Times*, 9 June 2004

WTO (1994a) 'Trade and Environment, Decision of 14 April 1994', Meeting at Ministerial Level, Palais des Congrès, Marrakesh (Morocco), 12–15 April 1994, MTN.TNC/45(MIN), 6 May 1994, Annex II, World Trade Organization, Geneva

WTO (1994b) 'Uruguay Round Agreement on Subsidies and Countervailing Measures', Articles 1–32, World Trade Organization, Geneva

WTO (1997a) 'The Fisheries Sector – Submission by New Zealand',WT/CTE/W/52, 21 May 1997, World Trade Organization, Geneva

WTO (1997b) 'Environmental and trade benefits of removing subsidies in the fisheries sector – submission by the United States', WT/CTE/W/51, 19 May 1997, World Trade Organization, Geneva

WTO (1998a) 'Comments by the European Community on the Document of the Secretariat of the Committee on Trade and Environment (WT/CTE/W/80) on Subsidies and Aids Granted in the Fishing Industry', WT/CTE/W/99, 6 November 1998, World Trade Organization, Geneva

WTO (1998b) 'Appellate Body, United States – Import Prohibition of Certain Shrimp and Shrimp Products (AB–1998–4)', WT/DS58/AB/R, 12 October 1998, World Trade Organization, Geneva

WTO (1999a) 'Statement by H.E. Mr. Alami Tazi, Minister of Commerce, Industry and Handicrafts, Morocco' (opening plenary of Seattle Ministerial), WT/MIN(99)/ST/29, 1 December 1999, World Trade Organization, Geneva

WTO (1999b) 'Preparations for the 1999 Ministerial Conference – Fisheries Subsidies – Communication from Australia, Iceland, New Zealand, Norway, Peru, Philippines and United States', WT/GC/W/303, 6 August 1999, World Trade Organization, Geneva

WTO (1999c) 'Preparations for the 1999 Ministerial Conference – Elimination of Trade Distorting and Environmentally Damaging Subsidies in the Fisheries[...]Communication from New Zealand', WT/GC/W/292, 5 August 1999, World Trade Organization, Geneva

WTO (1999d) 'Preparations for the 1999 Ministerial Conference – Fisheries Subsidies – Communication from Iceland', WT/GC/W/229, 6 July 1999, World Trade Organization, Geneva

WTO (1999e) 'Preparations for the 1999 Ministerial Conference – Negotiations on Forestry and Fishery Products – Communication from Japan', WT/GC/W/221, 28 June 1999, World Trade Organization, Geneva

WTO (2000) 'Environmentally-Harmful and Trade-Distorting Subsidies in Fisheries – Communication from the United States', 4 July 2000, WT/CTE/W/154, World Trade Organization, Geneva

WTO (2001a) 'Cuba, Dominican Republic, Honduras, India, Indonesia, Kenya, Malaysia, Pakistan, Sri Lanka, Tanzania, Uganda, and Zimbabwe – Proposal for a Framework Agreement on Special and Differential Treatment', WT/GC/W/442, 19 September 2001, World Trade Organization, Geneva

WTO (2001b) 'Doha Conference. Doha Ministerial Declaration', WT/MIN (01)/DEC/1, 20 November 2001, World Trade Organization, Geneva, available at www.wto.org/english/thewto_e/minist_e/min01_e/mindecl_e.htm, last accessed 11/03/2010

WTO (2001c) 'Ministerial Conference – 2001 Ministerial Decision on Implementation-Related Issues and Concerns', WT/MIN(01)/17, 20 November 2001, World Trade Organization, Geneva

WTO (2001d) 'Secretariat 2001 – Implementation of Special and Differential Treatment Provisions in WTO Agreement and Decisions', WT/COMTD/W/77/Rev.1, 21 September 2001, World Trade Organization, Geneva

WTO (2001e) 'Secretariat 2001 – Compilation of Outstanding Implementation Issues Raised by Members', Job(01)/152/Rev.1, 27 October 2001, World Trade Organization, Geneva

WTO (2002a) 'Japan. Negotiating Group on Rules – Japan's Basic Position on the Fisheries Subsidies Issue', 2 July 2002, TN/RL/W/11, World Trade Organization, Geneva

WTO (2002b) 'The Doha mandate to address fisheries subsidies: Issues – submission from Australia, Chile, Ecuador, Iceland, New Zealand, Peru, Philippines, and United States', TN/RL/W/3, 24 April, 2002, World Trade Organization, Geneva

WTO (2002c) Summary Report of the Meeting Held on 6 and 8 May 2002 – Note by the Secretariat, TN/RL/M/2, 11 June 2002, World Trade Organization, Geneva
WTO (2003a) 'Possible Approaches to Improved Disciplines on Fisheries Subsidies – Communication from the United States', 19 March, TN/RL/W/77, World Trade Organization, Geneva
WTO (2003b) *Fisheries Subsidies*, TN/RL/W/136, 14 July 2003, World Trade Organization, Geneva
WTO (2003c) 'Submission of the European Communities to the Negotiating Group on Rules – Fisheries Subsidies', TN/RL/W/82, 23 April 2003, World Trade Organization, Geneva
WTO (2003d) Summary Report of the Meeting Held on 18 – 19 June 2003 – Note by the Secretariat, TN/RL/M/10, 17 July 2003, World Trade Organization, Geneva
WTO (2003e) Possible Approaches to Improved Disciplines on Fisheries Subsidies – Communication from Chile, TN/RL/W/115, 10 June 2003, World Trade Organization, Geneva
WTO (2003f) Summary Report of the Meeting Held on 19-21 March 2003 – Note by the Secretariat, TN/RL/M/7, 11 April 2003, World Trade Organization, Geneva
WTO (2003g) Summary Report of the Meeting Held on 21 – 22 July 2003 – Note by the Secretariat, TN/RL/M/11, 8 September 2003, World Trade Organization, Geneva
WTO (2004a) 'Fisheries Subsidies – Communication from Argentina, Chile, Ecuador, New Zealand, Philippines, Peru', TN/RL/W/166, 2 November 2004, World Trade Organization, Geneva
WTO (2004b) 'Proposal on Fisheries Subsidies – Paper by Japan', TN/RL/W/164, 27 September 2004, World Trade Organization, Geneva
WTO (2004c) 'Fisheries Subsidies: UNEP Workshop on Fisheries Subsidies and Sustainable Fisheries Management – Communication from New Zealand', TN/RL/W/161, 8 June 2004, World Trade Organization, Geneva
WTO (2004d) 'Fisheries Subsidies: Proposed Structure of the Discussion – Communication from Japan', TN/RL/W/159, 7 June 2004, World Trade Organization, Geneva
WTO (2004e) 'Fisheries Subsidies: Overcapacity and Over-Exploitation – Communication from New Zealand', TN/RL/W/154, 26 April 2004, World Trade Organization, Geneva
WTO (2004f) Questions and Comments from Korea on New Zealand's Communication on Fisheries Subsidies (TN/RL/W/154), TN/RL/W/160, 8 June 2004, World Trade Organization, Geneva
WTO (2005a) 'Antigua and Barbuda; Barbados; Dominican Republic; Fiji; Grenada; Guyana; Jamaica; Papua New Guinea; St. Kitts and Nevis; St. Lucia; Solomon Islands; and Trinidad and Tobago – Architecture on Fisheries Subsidies Disciplines', TN/RL/GEN/57, 7 July 2005 (revisions dated 4 August (Rev.1) and 13 September 2005 (Rev.2)) World Trade Organization, Geneva
WTO (2005b) 'Brazil, Chile, Colombia, Ecuador, Iceland, New Zealand, Pakistan, Peru and the United States. Negotiating Group on Rules – Fisheries', TN/RL/W/196, World Trade Organization, Geneva
WTO (2005c) 'Brazil – Contribution to the discussion on the framework for disciplines on fisheries subsidies', TN/RL/GEN/56, 4 July 2005, World Trade Organization, Geneva
WTO (2005d) 'Brazil – Contribution to the Discussion on the Framework for Disciplines on Fisheries Subsidies', TN/RL/W/176, 21 March 2005, World Trade Organization, Geneva

WTO (2005e) 'Hong Kong Ministerial Declaration', WT/MIN(05)/DEC, World Trade Organization, Geneva
WTO (2005f) 'Japan, the Republic of Korea; and the Separate Customs Territory of Taiwan, Penghu, Kinmen and Matsu. Negotiating Group on Rules – Contribution to the Discussion on the Framework for the Disciplines on the Fisheries Subsidies', TN/RL/W/172, World Trade Organization, Geneva
WTO (2005g) 'United States of America. Negotiating Group on Rules – Fisheries Subsidies: Programmes for Decommissioning of Vessels and Licence Retirement', TN/RL/GEN/41, World Trade Organization, Geneva
WTO (2005h) 'Further Contribution to the Discussion on the Framework for Disciplines on Fisheries Subsidies – Paper from Brazil', TN/RL/GEN/79, 16 November 2005, World Trade Organization, Geneva
WTO (2005i) 'Contribution to the Discussion on the Framework for Disciplines on Fisheries Subsidies – Paper from Brazil', TN/RL/W/176, 31 March 2005, World Trade Organization, Geneva
WTO (2005j) Fisheries Subsidies to Management Services – Paper from New Zealand, TN/RL/GEN/36, 23 March 2005, World Trade Organization, Geneva
WTO (2005k) Definitions Related to Artisanal, Small-Scale and Subsistence Fishing – Note by the Secretariat, TN/RL/W/197, 24 November 2005, World Trade Organization, Geneva
WTO (2006a) 'Possible Disciplines on Fisheries Subsidies – Paper from Brazil – Revision', TN/RL/GEN/79/Rev.3, 2 June 2006, World Trade Organization, Geneva
WTO (2006b) 'D.G. Lamy: time out needed to review options and positions', WTO press release of 24 July 2006, World Trade Organization, Geneva, available at www.wto.org/english/news_e/news06_e/tnc_dg_stat_24july06_e.htm, last accessed 11/03/2010
WTO (2006c) 'Fisheries Subsidies – Exhaustive List of Non-Prohibited Fisheries Subsidies – Paper from New Zealand', TN/RL/GEN/141, 6 June 2006, World Trade Organization, Geneva
WTO (2006d) 'Fisheries Subsidies: Special and Differential Treatment – Paper from Argentina', TN/RL/GEN/138, 1 June 2006, (revision 26 January 2007) World Trade Organization, Geneva
WTO (2006e) 'Fisheries Subsidies – Submission of the European Communities', TN/RL/GEN/134, 24 April 2006, World Trade Organization, Geneva
WTO (2006f) 'Fisheries Subsidies – Communication from the United States', TN/RL/GEN/127, 24 April 2006, World Trade Organization, Geneva
WTO (2006g) 'Fisheries Subsidies – Framework for Disciplines – Communication from Japan; the Republic of Korea; and the Separate Customs Territory of Chinese Taipei, Penghu, Kinmen and Matsu', TN/RL/GEN/114, 21 April 2006, World Trade Organization, Geneva
WTO (2006h) 'Further Contribution to the Discussion on the Framework for Disciplines on Fisheries Subsidies – Paper from Brazil – Revision, TN/RL/GEN/79/Rev.2, 21 April 2006, World Trade Organization, Geneva
WTO (2006i) 'Small Scale, Artisanal Fisheries – Submission by India', TN/RL/W/203, 6 March 2006, World Trade Organization, Geneva
WTO (2006j) 'Fisheries Subsidies – Framework for Disciplines – Paper from New Zealand', TN/RL/GEN/100, 3 March 2006, World Trade Organization, Geneva
WTO (2006k) 'Further Contribution to the Discussion on the Framework for Disciplines on Fisheries Subsidies – Paper from Brazil – Revision', TN/RL/GEN/79/Rev.1, 21 February 2006, World Trade Organization, Geneva

WTO (2007a) 'Draft Consolidated Chair Texts of the AD and SCM Agreements', TN/RL/W/213, 30 November 2007, World Trade Organization, Geneva
WTO (2007b) 'Special and Differential Treatment – Proposal from Argentina and Brazil', TN/RL/GEN/151, 17 September 2007 (revision 26 November 2007) World Trade Organization, Geneva
WTO (2007c) 'Fisheries Subsidies: Fisheries Adverse Effects and S&D Treatment – Paper from Brazil', TN/RL/W/212, 29 June 2007, World Trade Organization, Geneva
WTO (2007d) 'S&DT in the Fisheries Subsidies Negotiations: Views of the Small, Vulnerable Economies (SVEs) – Communication from Antigua and [...]Solomon Islands – Revision', TN/RL/W/210/Rev.2, 22 June 2007, World Trade Organization, Geneva
WTO (2007e) 'Fisheries Subsidies: Continuation of Work on Special and Differential Treatment – Paper from Argentina', TN/RL/W/211, 19 June 2007, World Trade Organization, Geneva
WTO (2007f) 'S&DT in the Fisheries Subsidies Negotiations: Views of the Small, Vulnerable Economies (SVEs) – Communication from Barbados, Cu[...]New Guinea and Solomon Islands', TN/RL/W/210, 6 June 2007 (revisions dated 18 June and 22 June) World Trade Organization, Geneva
WTO (2007g) 'Access Fees in Fisheries Subsidies Negotiations – Communication from the ACP Group', TN/RL/W/209, 5 June 2007, World Trade Organization, Geneva
WTO (2007h) 'Fisheries Subsidies: Framework for Disciplines – Communication from Japan; the Republic of Korea and the Separate Customs Terri[...]u, Kinmen and Matsu – Revision', TN/RL/GEN/114/Rev.2, 5 June 2007, World Trade Organization, Geneva
WTO (2007i) 'Fisheries Subsidies: Proposed New Disciplines – Proposal from the United States', TN/RL/GEN/145, 22 March 2007, World Trade Organization, Geneva
WTO (2007j) 'Possible Disciplines on Fisheries Subsidies – Paper from Brazil – Revision', TN/RL/GEN/79/Rev.4, 13 March 2007, World Trade Organization, Geneva
WTO (2007 k) Fisheries Subsidies: Proposed New Disciplines – Proposal from the Republic of Indonesia, TN/RL/GEN/150, 22 July 2007, World Trade Organization, Geneva
WTO (2007 l) Fisheries Subsidies – Proposal by Norway, TN/RL/GEN/144, 26 January 2007, World Trade Organization, Geneva
WTO (2007 m) Fisheries Subsidies: Special and Differential Treatment – Paper from Argentina – Revision, TN/RL/GEN/138/Rev.1, 26 January 2007, World Trade Organization, Geneva
WTO (2008a) 'Need for Effective Special and Differential Treatment for Developing Country Members in the Proposed Fisheries Subsidies Text – [...]Indonesia and China – Revision', TN/RL/GEN/155/Rev.1, 19 May 2008, World Trade Organization, Geneva
WTO (2008b) 'Fisheries Subsidies – De Minimis Exemption – Communication from Canada', TN/RL/GEN/156, 2 May 2008, World Trade Organization, Geneva
WTO (2008c) 'Drafting Proposal on Issues Relating to Article V (Fisheries Management) of the Fisheries Subsidies Annex to the SCM Agreement [...]13 – Communication from Norway', TN/RL/W/231, 24 April 2008, World Trade Organization, Geneva
WTO (2008d) 'New Draft Consolidated Chair Texts of the AD and SCM Agreements', TN/RL/W/236, 19 December 2008, Annex VIII: Fisheries Subsidies – Roadmap for Discussion, World Trade Organization, Geneva
WTO (2009a) 'Fisheries Subsidies – Communication from Argentina, Australia, Chile, Colombia, the United States, New Zealand, Norway, Iceland, Peru and Pakistan',

TN/RL/W/243, 7 October 2009, World Trade Organization, Geneva

WTO (2009b) 'Fisheries Subsidies – Communication from Brazil, China, Ecuador and Mexico', TN/RL/W/241, 28 September 2009 and Rev.1 (adding Venezuela) of 16 October 2009, World Trade Organization, Geneva

WTO (n.d.) 'Agreement on Subsidies and Countervailing Measures (ASCM)', World Trade Organization, Geneva available at www.wto.org/English/docs_e/legal_e/24-scm_01_e.htm, last accessed 11/03/2010

WTO (n.d.) 'Appellate Body Report, US – Countervailing Duty Investigation on DRAMs', World Trade Organization, Geneva

WTO (n.d.) 'Appellate Body Report, US – Softwood Lumber IV', World Trade Organization' Geneva

WTO (1999) Appellate Body Report, Canada-Aircraft, Canada – Measures Affecting the Export of Civilian Aircraft – AB-1999-2 – Report of the Appellate Body, WT/DS70/AB/R

WTO (2003) General Council Chairman – Proposal on an Approach for Special and Differential Treatment: Agreement-Specific S&D Proposals (Divided into Three Categories), Job3404, 5 May 2003, pp. 30–-32

WWF (1997) 'Euro-African Fishing Agreements: Subsidising Overfishing in African Waters', in Porter, G., *Subsidies and Depletion of World Fisheries: Case Studies*, Washington, D.C., WWF Endangered Seas Campaign

WWF (1998a) *The Footprint of Distant Water Fleets on World Fisheries*, WWF International, Godalming (United Kingdom)

WWF (1998b) *Estimating overcapacity in the global fishing fleets*, WWF, Washington, D.C.

WWF (1998c) 'Towards rational disciplines on subsidies to the fishery sector: A call for new international rules and mechanisms', initially printed in Porter, G., Schorr, D.K. *The Footprint of Distant Water Fleets on World Fisheries*, WWF, Godalming (United Kingdom)

WWF (2001) *Hard Facts, Hidden Problems: A Review of Current Data on Fishing Subsidies*, WWF Technical Paper, Gland. Virdin, J.

WWF (2003) *Fisheries Subsidies and the Marine Environment, the Spanish Case*, WWF, Washington, D.C.

WWF (2004) *Healthy Fisheries, Sustainable Trade: Crafting New Rules on Fishing Subsidies in the World Trade Organization*, WWF Position Paper and Technical Resource, June 2004, p.5 Schorr, D.K.

WWF (2004) *Fishing Subsidies: Issues for ACP Countries*, available at www.assets.panda.org/downloads/euacpschorrpresentation.doc

WWF (2006a) *Fisheries Subsidies: Will the EU Turn its Back on the 2002 Reforms?*, WWF, Washington, D.C.

WWF (2006b) *The Best of Texts, The Worst of Texts*, WWF, Washington, D.C., available at www.panda.org/about_our_earth/blue_planet/publications/?72220/The-Best-of-Texts-The-Worst-of-Texts, last accessed 11/03/2010

WWF (2007a) 'Fisheries Subsidies: WWF Statement on the Chairman's Draft', 12 December 2007, Geneva

WWF (2007b) *Evaluation des impacts socio-économiques et environnementaux de la pêche sénégalaise*, WWF technical document

Index

access agreements 235–64
Agreement on Subsidies and Countervailing Measures 241, 252–62
environment 235–6
'financial contribution' interpretation 241–51
foreign financial flows 85–7
global fisheries governance 6
issue/scope/terminology 238–9
Law of the Sea 239
Special and Differential Treatment 191
subsidy clarification 252, 254–7
UN Environment Programme 237–8
WTO history/law/submissions 152, 160–1, 240–52
ACP *see* African, Caribbean and Pacific countries
'actionable' subsidies 146
Agreement on Subsidies and Countervailing Measures 369–72
artisanal fisheries 217
Special and Differential Treatment 206–7, 209–10
Subsidies and Countervailing Measures 358–9
AD *see* anti-dumping
African, Caribbean and Pacific (ACP) countries 72, 83, 165, 250
Agreement on the Conservation and Management of Straddling and Highly Migratory Stocks 321–2
Agreement on Subsidies and Countervailing Measures (ASCM) 172–3, 353–63
access agreements 238, 240–4, 245, 248–9, 252–62
actionability 358–9
actionable subsidies 369–72
adverse effects 369
artisanal fisheries 220–1
developing countries 380–3, 389–90
dispute settlements 361
export subsidies 383–5
final provisions 383
general provisions 365–6
Hong Kong Ministerial Declaration 343–5
implementation 361–2
'inaction' 252, 254
injury determination 376–8
institutions 378
management regimes 359–62
negotiating mandates 353
nonactionable subsidies 372–5
notification 379–80
prejudice 369–70
prohibition 355–6, 367
remedies 367–8, 371–2, 375
Special and Differential Treatment 196, 197–200, 357–8, 389–90
specificity 366
subsidies definition 12–3
sustainability criteria 272
transition rules 362
transparency 252, 261–2
WTO history 140–1
Agreement to Promote Compliance with International Conservation and Management Measures by Fishing Vessels on the High Seas 321
anti-dumping (AD) 165, 168, 172, 194, 343–5
APEC *see* Asia-Pacific Economic Cooperation

Appellate Body 187, 241–51, 254, 256, 368, 372
aquaculture 3, 15, 61, 68, 74, 114, 146, 166, 188, 321, 342
artisanal fisheries 9, 21, 74, 81, 86, 87, 88, 106, 110, 111, 149, 183, 184, 185, 202, 213–34, 283, 284, 391–3, 402
  definitional debate 222–5, 391–3
  objectives/practices 225–8
  practical implications 229–33
  reform challenges 17
  Senegalese fishing trends 73
  Special and Differential Treatment 188, 191–3, 196, 220–2
  sustainability criteria 268, 292–5
  WTO history 150, 152, 165
ASCM *see* Agreement on Subsidies and Countervailing Measures
Asia-Pacific Economic Cooperation (APEC) 13–5, 20, 134, 141–2, 163
Asian Development Bank 229, 392
Atlantic fisheries 54–5

'bad subsidies' 14, 15
below market loans 78–9
benchmark criteria 273–5, 288, 399–400
'benefits', interpretation 246–51
Bering Sea Pollock Fishery 41–2
'best practices' sustainability criteria 267, 274–5, 279–83, 288–91
biomass dynamic modeling 283–5
'blank cheque' 155, 157, 169, 171, 201, 357–8
bloated fleets 10
  see also overcapacity
'bottom-up' approaches 146–53, 166
  environment 152–3
  politics 149–51
  Special and Differential Treatment 199
  substantive issues clarification 147–9
'broad ban' approaches 156–8
'buy-in'/'buy-out' policies 45, 138
buyback programmes 41–4, 148, 158, 195

capacity 275–81, 396
  analytical framework 26, 27, 29
  assessment 277–9
  Doha Ministerial Declaration 336–7
  global fisheries governance 8, 9
  impact analyses 32
  'overfishing' 276–7
  reduction programmes 88–9
  Special and Differential Treatment 194–6, 210–1
  subsidized transfers 88
  WTO history 150
  see also overcapacity
'capital costs' 11–2, 24–5, 46–50
'capital stuffing' 38
carte blanche *see* blank cheque
'catch control' fisheries
  capital cost subsidies 47–9
  decommissioning subsidies/license retirement 42–5
  foreign country water access 36–7
  infrastructure subsidies 33–4
  price support subsidies 57
  subsidies to income 54–6
  tuna sustainability 100–1
  variable cost subsidies 51
catch and effort controls 26, 27
catch per unit of effort (CPUE) 69–70
catfishes 69
CBD *see* Convention on Biological Diversity
CCAMLR *see* Commission for the Conservation of Antarctic Marine Living Resources
ceiling limits 158–9, 171–2, 205–6, 209
CFA Franc devaluation 83–4
CFP *see* Common Fisheries Policy
Chair's Texts 347–51
  anti-dumping 353–63
  developing countries 350–1
  general exceptions 348–9
  prohibition 347–8
  Special and Differential Treatment 350–1
  Subsidies and Countervailing Measures 353–63
closed access fisheries 119, 120, 122
coastal areas
  access agreements 260–1
  Norwegian reform processes 123
  Senegalese fisheries 69, 73, 74
  Special and Differential Treatment 191
cod fisheries 118–9
Code of Conduct for Responsible Fishing 322

access agreements 239–40
artisanal fishing 392
global fisheries governance 7–9
reform challenges 17
sustainability criteria 269–70, 277, 282–3, 287, 291–2
WTO history 133
command and control management regimes 293–4
commercial fisheries 52, 188
commercially isolated fisheries 219
Commission for the Conservation of Antarctic Marine Living Resources (CCAMLR) 288, 323
Committee on Trade and Environment (CTE) 135, 137, 141, 222
Common Fisheries Policy (CFP) 145–6, 156–7
community development 226, 229–30
community-based management 27, 28
competition 103–4, 105, 332–3
conservation and management 201–2
Convention on Biological Diversity (CBD) 321
Cotonou Agreement 83
CPUE *see* catch per unit of effort
credit systems 78–9
CTE *see* Committee on Trade and Environment
culture preservation 227

Dakar Industrial Free Zones 84–5
Dakar Senegalese fishing central markets 80
data-intensive management regimes 293–4
de minimis approaches 158–9, 171–2, 205–6, 209
debt 336, 369, 382
decommissioning schemes 40–5, 46
impact analyses 38–46
Norwegian reform processes 123
Special and Differential Treatment 195–6
subsidies categorization 21, 24
demersal species 69, 73
depleted fisheries 270, 273
developing countries
Chair Texts 350–1
export earnings 189–92
food security 189
livelihoods 188–9
reform challenges 16
Special and Differential Treatment 187–200, 203–10
subsidies problems 15
trade 4
WTO history 144–6, 149–54, 157–60, 167–74
diesel *see* fuel subsidies
'differentiation' 158–9
direct access agreements 255
'direct capacity assessment' 278–9
direct subsidies
access agreements 243–4
below market loans 78–9
export subsidies 81–5
foreign financial flows 85–7
free tax systems 76
fuel subsidies 77–8
infrastructure 79–80
micro-credit systems 78–9
Senegalese fisheries 75–87
small-scale fishing 80–1
trade liberalization 82–5
Dispute Settlement 198, 334, 361, 367, 368, 372
Distant Water Fishing Nations (DWFNs) 191, 237–8, 241–2, 246, 248–51, 259–61
distant water fleets (DWF) 215
Doha Ministerial Declaration (DMD) 138–43, 327–39
agriculture 330
capacity building 336–7
competition policy 332–3
debt 336
dispute settlement understanding 334
electronic commerce 335
environment 334–5
finance 336
government procurement 333
implementation-related concerns/issues 329–30
intellectual property rights 331–2
investment 332
least-developed countries 337–8
management regimes 338–9
market access 331
non-agricultural products 331
services 330–1

small economies 336
special and differential treatment 194, 195, 338
technical cooperation 336–7
technology transfer 336
trade 332–6
Trade-Related Aspects of Intellectual Property Rights 331–2
transparency 333
WTO rules 334
'dolphin-safe' tuna 137
duty-free exporting enterprises 85
DWF *see* distant water fleets
DWFNs *see* Distant Water Fishing Nations
'dynamite' fishing 215

Earth Summit 135
Eastern Pacific Ocean (EPO) 96, 97, 99–101
eco-labels 111
Economic Partnership Agreements (EPAs) 72
Ecuador xxvi, 65, 66, 128, 309
environmental impacts 100–1
foreign subsidies 103–4, 105
main fisheries subsidies 96–7
management systems 105–6, 107–9, 110
reform processes 110–1
Regional Fishery Management Organizations 110
trade 101–3
tuna sustainability 95–112
WTO history 173, 174
EEZs *see* 'exclusive economic zones'
EFF *see* European Fisheries Fund
'effective management' fisheries
capital cost subsidies 47
decommissioning subsidies 40–2
fisheries infrastructure subsidies 33
foreign country water access 36
license retirement 40–2
price support subsidies 56–7
subsidies to income 53–4
variable cost subsidies 51
engine subsidies 49–51
EPAs *see* Economic Partnership Agreements
EPO *see* Eastern Pacific Ocean
EU *see* European Union
European Fisheries Fund (EFF) 145–6, 156–7
European Union (EU)
African fishing agreements 37
Doha convention/Seattle Round 139
fishing capacity 88
foreign country water access 37
foreign financial flows 85–7
Senegalese export trends 72
WTO history 144, 145, 147–8, 150, 156–7
excess capacity 26, 27, 29, 32
see also overcapacity
'exclusive economic zones' (EEZs)
access agreements 237–42, 245–6, 249–51, 255, 261–2
Chair Texts 350
Ecuadorian tuna sustainability 96–7
global fisheries governance 6–7
sustainability criteria 277
WTO history 150, 157
exploited fisheries 270, 273
export subsidies 218–9
Agreement on Subsidies and Countervailing Measures 383–5
Ecuadorian tuna exports 103–4, 105
Senegalese fisheries 70–2, 81–5
Special and Differential Treatment 189–92
WTO history 133
extensive marine boundary states 191

FAO *see* Food and Agriculture Organization
'financial contribution', interpretation 241–51
Financial Instrument for Fisheries Guidance (FIFG) 145–6, 156–7
'Finnmark model' 122
Fishery Resources Monitoring System (FIRMS) 287
fishing gear-based free tax systems 76
'fleet adjustment' programmes 48–9
Food and Agriculture Organization (FAO) 11, 321, 322
access agreements 239–40
artisanal fishing 391–2
Chair Texts 350
global fisheries governance 7–9
Special and Differential Treatment 188

subsidies categorization/definition 12, 13, 20
sustainability criteria 270, 273–6, 282–3, 287
trade dynamics/scale 4
WTO history 132–41, 144–5, 159–64, 167, 170
see also Code of Conduct for Responsible Fishing
food security 68, 189, 226
foreign subsidies 85–7
access agreements 152
Ecuadorian tuna exports 103–5
water access 12, 23, 36–8
foreign-oriented industrial fleets 218–9
Franc (CFA) devaluation 83–4
free tax systems 76
free zone enterprises 84–5
'Friends of Fish' coalition 144, 146–56, 165–8
fuel subsidies
Ecuadorian tuna sustainability 97
problems 15
Senegalese fisheries 77–8
variable costs 50–2
WTO history 157
fully exploited fisheries 270, 273

G8 Declaration 2
GATT *see* General Agreement on Tariffs and Trade
General Agreement on Tariffs and Trade (GATT) 137, 142, 187, 208, 245, 331, 333, 334, 339, 365, 369, 376, 379–80, 382, 385
gender relations 188
global fisheries governance 5–10
globalization 112, 131, 136–43, 154, 312, 314, 329
'good subsidies' 14, 15, 24
'goods/services', interpretation 12–3, 244–6
'Green Box' 200–2, 205–7, 209–10, 314
green economy vi, xii, xxiii, 1–3, 4, 145, 314
Group of Latin America and Caribbean Countries (GRULAC) 174
grouper fish 69–70

haddock fisheries 123
herring fisheries 123
High Seas Vessel Authorization Records (HSVAR) 287
high-sea pelagic resources 69
Hong Kong
artisanal fisheries 215
Ministerial Declaration 343–5
Special and Differential Treatment 195
WTO history 152–5
HSVAR *see* High Seas Vessel Authorization Records

IATTC *see* Inter American Tropical Tuna Commission
ICTSD *see* International Centre for Trade and Sustainable Development
Illegal, Unregulated, and Unreported (IUU) fishing 8, 195–6, 216
impact analyses 32–58, 65–130
capital cost subsidies 46–50
decommissioning subsidies/license retirement 38–46
Ecuadorian tuna sustainability 95–112
foreign country water access 36–8
infrastructure subsidies 32–6
Norwegian reforms 113–30
price support subsidies 56–8
Senegalese fisheries 67–94
subsidies to income 52–6
variable cost subsidies 50–2
'implicit subsidies' 35
income subsidies 25–6, 52–6
income support programmes 12–3
indirect capacity assessments 278–9
indirect stock-criteria indicators 273–5
indirect subsidies 12–3
access agreements 243–4, 255
below market loans 78–9
exports 81–5
foreign financial flows 85–7
free tax systems 76
fuel subsidies 77–8
infrastructure 79–80
micro-credit systems 78–9
Senegalese fisheries 75–87
small-scale fishing 80–1
trade liberalization 82–5
Individual Transferable Quota (ITQ) systems 26–8, 31, 41, 119–20
Individual Vessel Quotas (IVQs) 120–1
industrial fishing fleets 218–9

Industrial Free Zones (IFZs) 84–5
infrastructure subsidies 33–4
  impact analyses 32–6
  landing/marketing/processing 79–80
  resource impacts 21–2
  sustainability criteria 286–8, 294
'input stuffing' 40–1
'input-based' capacity-related criteria 275–6, 278
input-based evaluation benchmarks 400
inshore small-scale fisheries 52
insurance programmes 25, 53–5
intellectual property rights 331–2
Inter American Tropical Tuna Commission (IATTC) 106, 110
International Centre for Trade and Sustainable Development (ICTSD) 141–2, 160–1, 237
International Convention for the Regulation of Whaling 321
International Development Research Center 392–3
International Plan of Action (IPOA) 7–9, 11, 322
  sustainability criteria 277–8, 287
  WTO history 134, 136
International Union for Conservation of Nature (IUCN) 141–2
ITQ *see* Individual Transferable Quota systems
IUU *see* Illegal, Unregulated, and Unreported fishing
IVQs *see* Individual Vessel Quotas

Jakarta Mandate of the Convention on Biological Diversity 321
Johannesburg World Summit 143

Kaolack Senegalese fishing central markets 80

Lamy, P. 164, 172
landing procedures 79–80
Law of the Sea (LOS) 239
  see also UN Convention on the Law of the Sea
'laying up' subsidies (LUs) 53–6
least-developed countries (LDCs)
  anti-dumping 357–8
  Chair Texts 350
  Doha Ministerial Declaration 337–8
  Special and Differential Treatment 191
  Subsidies and Countervailing Measures 357–8
lesser African threadfin 69–70
liberalization
  CFA Franc devaluation 83–4
  Cotonou Agreement 83
  free zone enterprises 84–5
  Lomé Convention 83
  Senegalese fisheries 82–5
  WTO history 134, 136–8
license retirement 21, 24, 38–46
LIFDCs *see* Low-Income Food-Deficit Countries
lifestyle preservation 227
'limit reference points' (LRPs) 271
livelihoods 188–9
Lomé Convention 83
longline fisheries 8, 11
LOS *see* Law of the Sea
Low-Income Food-Deficit Countries (LIFDCs) 3, 190
LRPs *see* 'limit reference points'
lump-sum income transfers 55
LUs *see* 'laying up' subsidies

mâchoiron 69
management procedures 323
  artisanal fisheries 226–7
  assessment 283–5
  community-based 27, 28
  conservation 201–2
  controls 285
  data-intensive 293–4
  Doha Ministerial Declaration 338–9
  Ecuadorian tuna sustainability 105–10
  enforcement/surveillance 286
  evaluation 288
  global fisheries governance 9–10
  infrastructure subsidies 34–6
  international fisheries 289
  monitoring, control and surveillance 286–8, 294
  Norwegian reform processes 119–25
  responsible management 282–3
  Senegalese trends 73–5
  'simple reference to international norms' approaches 291–2
  Special and Differential Treatment 195–6
  subsidies categorization 21, 22–3

Subsidies and Countervailing Measures 359–62
sustainability criteria 266, 281–94, 397–8
Marine Stewardship Council (MSC) 281–2
market-based approaches
access agreements 201–2, 248–51, 331
infrastructure 79–80
Norwegian reform processes 119–25
Senegalese fisheries 78–9
matrix approaches
analytical subsidy frameworks 29–32
capital cost subsidies 46–50
decommissioning subsidies/license retirement 38–46
foreign country water access 36–8
infrastructure subsidies 32–6
price support subsidies 56–8
subsidies to income 52–6
tuna sustainability 100–2
variable cost subsidies 50–2
WTO history 148
Maximum Economic Yields (MEYs)
analytical subsidy frameworks 29
impact analyses 47
price support subsidies 57
sustainability criteria 271
Maximum Sustainable Yields (MSYs)
analytical subsidy frameworks 29
global fisheries governance 6
Senegalese fisheries 69, 70
sustainability criteria 269–71, 273–5, 292–3
MCS *see* monitoring, control and surveillance infrastructure
MEYs *see* Maximum Economic Yields
micro-credit systems 78–9
'minimum international requirements' 267–8, 274, 280–1, 284–6, 289–92
'minimum recommended conditions' 267–8, 274–5, 280, 288–91
Ministerial Declaration 327–39, 343–5
agriculture 330
capacity building 336–7
competition policy 332–3
debt 336
dispute settlement understanding 334
electronic commerce 335
environment 334–5
finance 336
government procurement 333
implementation-related concerns/issues 329–30
intellectual property rights 331–2
investment 332
least-developed countries 337–8
management regimes 338–9
market access 331
non-agricultural products 331
regional trade agreements 345
services 330–1
small economies 336
Special and Differential Treatment 195, 338
technical cooperation 336–7
technology transfer 336
trade 332–6
Trade-Related Aspects of Intellectual Property Rights 331–2
transparency 333
WTO rules 334
'ministerial' meetings 155–6, 161–2, 171
monitoring, control and surveillance (MCS) infrastructure 286–8, 294
Mottled grouper fish 69
MSC *see* Mutual of Saving and Credit
MSC *see* Marine Stewardship Council
MSYs *see* Maximum Sustainable Yields
'multilateral environmental agreements' (MEAs) 140
Mutual of Saving and Credit 79

NAFTA *see* North American Free Trade Agreement
NCFA *see* Norwegian Coastal Fishers' Association
nearshore artisanal fleets 218
'negative list' approaches 147–8
Negotiating Group on Market Access (NGMA) 194
Negotiating Group on Rules (NGR) 143–4, 161–8, 172–3, 194
New England Fisheries Management Council 45
NFA *see* Norwegian Fishermen's Association
NGMA *see* Negotiating Group on Market Access
NGR *see* Negotiating Group on Rules

'no blank cheque' approach 169
'non-actionable' subsidies 145–6, 206–7, 372–5
non-agricultural products 331
non-tariff barriers (NTBs) 194
North American Free Trade Agreement (NAFTA) 136–7
Northeast Multi-species Groundfish Fishery 44, 45
Northwest Groundfish Fishery 45
Norway xxvi, 35, 65, 66, 139, 150, 309
  market-based approaches 119–25
  price support subsidies 57
  subsidy fall/rise 115–9
  subsidy reforms 113–30
  WTO history 171, 173
Norwegian Coastal Fishers' Association (NCFA) 115
Norwegian Fishermen's Association (NFA) 115, 121
Norwegian Purse Seine decommissioning scheme 41
NTBs *see* non-tariff barriers

Oceana (non-governmental organization) 172
OECD *see* Organization for Economic Co-operation and Development
Official Development Assistance (ODA) 87–9
offshore fisheries 218–9
'open access' fisheries
  capital cost subsidies 49–50
  decommissioning subsidies/license retirement 46
  fisheries infrastructure subsidies 34
  fisheries sector subsidies 10
  foreign country water access 37–8
  Norwegian reform processes 117, 119, 123
  price support subsidies 58
  subsidies to income 56
  variable cost subsidies 51–2
'optimal' exploitation levels 269–71
Organization for Economic Co-operation and Development (OECD)
  analytical subsidy frameworks 27, 28, 29–32
  capital cost subsidies 46–50
  decommissioning subsidies 38–46
  foreign country water access 36–8
  infrastructure subsidies 32–6
  license retirement 38–46
  price support subsidies 56–8
  subsidies categorization/definition 13, 20
  subsidies problems 14, 15
  subsidies to income 52–6
  variable cost subsidies 50–2
  WTO history 133, 141–2, 163
  see also matrix approaches
outboard motors 76
'output-based' capacity-related criteria 275–6, 278
output-based evaluation benchmarks 399–400
overcapacity
  artisanal fisheries 217–8
  fisheries sector subsidies 11
  impact analyses 42–3, 45
  Norwegian reform processes 117, 123
  reduction proposals 88–9
  Subsidies and Countervailing Measures 27, 29
  sustainability criteria 279, 281
  variable cost subsidies 52
  WTO history 151
overcapitalization 26, 27, 29, 41–2, 49
overexploitation
  artisanal fisheries 229
  foreign financial flows 87
  impact analyses 45, 54
  price support subsidies 58
overfishing
  artisanal fisheries 217–8
  capacity-related criteria 276–7
  fisheries sector subsidies 10–1
  Subsidies and Countervailing Measures 357–8
pelagic species 73
Permanent Group of Experts (PGE) 367, 378
Plan of Implementation of the World Summit 341–2
'policy space' 152, 169, 185, 187, 197, 199, 201, 210–2, 220, 228, 232, 266
'positive list' approach 147–8, 166, 204–5
'positive subsidies' 166–7
'possibly harmful' subsidies 30
poverty alleviation/reduction 68–9, 226, 229–30
predictive power criteria 267

'prevailing market price' 257
price support subsidies 12–4, 26, 56–8
'prior authorization' regimes 203–4
processing subsidies 15, 79–80
'production distortions' 218, 230, 231
prohibited subsidies 146
access agreements 257–61
Agreement on Subsidies and Countervailing Measures 367
Chair Texts 347–8
Special and Differential Treatment 195–6, 204, 206–7, 209–10
Subsidies and Countervailing Measures 355–6
property rights 26–8, 30–1, 41, 119–20, 331–2
'provides', interpretation 246
Purse Seine decommissioning scheme 41

qualitative assessments 208
qualitative sustainability criteria indicators 272–3
quantitative assessments 208, 278–81, 283–5
quantitative limits 205
quantitative sustainability criteria indicators 272–3
Quota Exchange Systems 121, 123

Ramsar Convention on Wetlands of International Importance Especially as Waterfowl Habitat 321
rapid overall evaluation benchmarks 399–400
recovering fisheries 270, 273
'red flag' benchmarks 274–5, 281
reform processes 16–7, 65–130
Ecuadorian tuna sustainability 110–1
market-based approaches 119–25
Norway 113–30
Senegalese fisheries 67–94
subsidy fall/rise 116–9
'regional fisheries bodies' (RFBs) 9–10, 16, 323–4
Regional Fishery Management Organizations (RFMOs) 323
artisanal fisheries 223
Ecuadorian tuna sustainability 106, 110
global fisheries governance 6–7, 9, 10
reform challenges 16
sustainability criteria 269, 281, 287–8
WTO history 140
Regional Seas Conventions (RSCs) 9–10
regional seas programmes and conventions 325–6
regional trade agreements (RTAs) 345
regional treaties 9–10
research activities 22–3, 34–6
resource impacts 19–63
access to water payments 23
analytical subsidy frameworks 26–32
capital costs 24–5, 46–50
'decommissioning' of vessels 21, 24
impact analyses 32–58
income subsidies 25–6
infrastructure subsidies 21–2, 32–6
license retirement 21, 24
'management services' 21, 22–3
Norwegian reform processes 118–9
price support subsidies 26
Special and Differential Treatment 195–6
subsidies categorization 20–6
variable costs 25
RFBs *see* 'regional fisheries bodies'
RFMOs *see* Regional Fishery Management Organizations
'Roadmap for Discussions' 172–3, 353–9
rouget 69
RSCs *see* Regional Seas Conventions
RTAs *see* regional trade agreements
Rubberlip grunt 69–70
rural areas 68–9, 189

saithe fisheries 123
sanitary and phytosanitary measures (SPS) 72
science-based assessments 280–1, 283–5
SCM *see* Agreement on Subsidies and Countervailing Measures
scrapping schemes 118
sea catfishes 69
sea-bird bycatches 8, 11
seabream 69
Seattle Round 137–43
Senegal xxvi, 2, 5, 37, 51, 65, 66, 128, 142, 159, 190, 216, 309
access agreements 235–6
below market loans 78–9
capital cost subsidies 49–50
direct/indirect subsidies 75–87
export subsidies 70–2, 82–5

export trends 70–2
foreign financial flows 85–7
free tax systems 76
fuel subsidies 77–8
impact analyses 67–94
infrastructure 79–80
management policy trends 73–5
micro-credit systems 78–9
Official Development Assistance 87–9
reform processes 67–94
small-scale fishing 73, 80–1
socio-economic fishing importance 68–9
state of fisheries 69–70
trade liberalization 82–5
variable cost subsidies 52
WTO history 144
shark management 8, 11
shrimp 69–70, 74
SIDS *see* Small Island Developing Countries
'simple blanket bans' 257
'simple reference to international norms' approaches 291–2
Small Island Developing Countries (SIDS)
access agreements 254, 258–61
artisanal fisheries 217
'green economy' 3
Special and Differential Treatment 191
WTO history 160–1
small island states 149
small pelagic species 69, 73
small vulnerable economies (SVEs) 149, 160–1
small-scale fisheries (SSF)
artisanal fisheries 217, 219, 223, 391–3
below market loans 78–9
capital cost subsidies 49–52
micro-credit systems 78–9
reform challenges 17
Senegalese fisheries 80–1
Special and Differential Treatment 188, 196
sustainability criteria 268
trends 73
variable cost subsidies 52
WTO history 150
social development 226, 229–30
social insurance programmes 53
'soft law' international standards 290–1
Softwood Lumber case 243–6, 249, 256–7
Special and Differential Treatment (S&DT) 149–53, 158–61, 165, 169–70, 173–4
Agreement on Subsidies and Countervailing Measures 389–90
artisanal fisheries 220–2
beneficiaries 210
capacity 194–6, 210–1
Chair Texts 350–1
developing countries 188–200, 203–10, 380–3, 389–90
Doha Ministerial Declaration 338
export earnings 189–92
food security 189
key issues 187–212
prospective importance 192–3
Subsidies and Countervailing Measures 196, 197–200, 357–8
technical assistance 210–1
trade 194–6
transitioning out subsidies 211
SPS *see* sanitary and phytosanitary measures
SQSs *see* Structural Quota Systems
SSF *see* small-scale fisheries
'standing timber' 244–5
static quantitative benchmarks 273–5
stock-related criteria 268–75, 396
Structural Quota Systems (SQSs) 121, 123–4
Subsidies and Countervailing Measures (SCM) *see* Agreement on Subsidies and Countervailing Measures
subsidies to income 25–6, 52–6
subsistence fisheries 392–3
subsistence livelihoods 189
sustainability criteria 265–306, 395–404
artisanal fisheries 268, 292–5
'best practices' 267, 274–5, 279–83, 288–91
capacity-related criteria 275–81
'good indicator' choice 267–8
management processes 266, 281–92
'minimum international requirements' 267–8, 274, 280–1, 284–6, 289–92
'minimum recommended conditions' 267–8, 274–5, 280, 288–91
stock-related criteria 268–75
technical discussion 268–92

WTO model language 401–4
SVEs *see* small vulnerable economies

TACs *see* Total Allowable Catches
'target reference points' (TRPs) 271
technical barriers to trade (TBTs) 194
technical and financial assistance 201–2
technology
Doha Ministerial Declaration 336
Norwegian reform processes 117–8
variable cost subsidies 52
territorial use rights 28
'threshold reference points' (ThRPs) 271
'top-down' approach 147–53, 166
environment 152–3
politics 149–51
Special and Differential Treatment 199
substantive issues clarification 147–9
Total Allowable Catches (TACs) 27, 28, 31, 35, 51, 59, 115, 120
trade
artisanal fisheries 217–8, 230, 231
CFA Franc devaluation 83–4
Cotonou Agreement 83
Doha Ministerial Declaration 332–6
dynamics/scale 3–5
Ecuador 95–112
free zone enterprises 84–5
globalization 136–8
liberalization 82–5
Lomé Convention 83
NAFTA 136–7
Senegalese fisheries 82–5
Special and Differential Treatment 194–6
tuna sustainability 95–112
WTO history 152–3
see also Committee on Trade and Environment; General Agreement on Tariffs and Trade; International Centre for Trade and Sustainable Development; UN Conference on Trade and Sustainable Development
Trade-Related Aspects of Intellectual Property Rights (TRIPS) 286, 331–2
'traditional' fisheries 215–6, 392–3
'traffic light' approach 146
transparency 261–2, 333, 360–1
TRIPS *see* Trade-Related Aspects of Intellectual Property Rights
tropical reef 'dynamite' fishing 215
TRPs *see* 'target reference points'
tuna sustainability
Ecuador 95–112
environmental impacts 100–1
foreign subsidies 103–4, 105
main fisheries subsidies 96–7
management systems 105–10
reform processes 110–1
Regional Fishery Management Organizations 110
trade 101–3
'tuna-dolphin' dispute 137

UBC *see* University of British Columbia
'ugly subsidies' 14, 15
UN Commission on Sustainable Development (UNCSD) 138
UN Compliance Agreement 287
UN Conference on Trade and Sustainable Development (UNCTAD) 159–61
UN Convention on the Law of the Sea (UNCLOS) 321
access agreements 239
global fisheries governance 5–7
reform challenges 16, 17
sustainability criteria 269–70, 277, 283, 287
WTO history 150
UN Environment Programme (UNEP) v–vi, xi–xii, xvii, xxv–xxvii, 10, 13–14, 20, 30, 50–2, 65, 66, 73, 86, 96, 100, 106–9, 132, 135, 136, 142, 144, 145, 147, 148, 152, 154, 159, 160, 161, 163, 164, 165, 172, 174, 183–86, 193, 207, 216, 237, 265, 307, 310, 311, 313, 314, 318, 319, 328, 389, 391, 395, 399, 401
access agreements 237–8
administered programmes 325
analytical subsidy frameworks 29–30
artisanal fisheries 216
evaluation benchmarks 399–400
global fisheries governance 9–10
subsidies categorization/definition 13–4, 20
sustainability criteria 395–404
WTO history 134–6, 141–8, 152–5, 159–65, 172–4
see also matrix approaches
UNCLOS *see* UN Convention on the Law of the Sea

UNCSD *see* UN Commission on Sustainable Development
under-reporting of subsidies 14, 15
underexploited fisheries 196, 270, 273
unemployed fishermen 54–5
Unemployment Insurance (UI) programmes 54–5
Unit Quota Systems (UQSs) 121, 122
University of British Columbia (UBC) 13–5, 99–101, 163
UQSs *see* Unit Quota Systems

'variable costs' 12, 25, 50–2
vessel quota regimes 120–1
voluntary international standards 290–1
voluntary sectoral liberalization 134

water access 23, 36–8
Western and Central Pacific Ocean (WCPO) 99
workshops 144, 147, 151, 158–9, 174, 237
World Bank 4, 14, 163
World Summit on Sustainable Development 143, 266, 341–2
World Wildlife Fund (WWF)
  access agreements 237–8
  early analysis/preliminary international action 134–6
  rapid overall evaluation benchmarks 399–400
  subsidies definition 13
  sustainability criteria 395–404
  WTO history 141–4, 147, 154, 160–5, 172–4